Fungi Bio-Prospects in Sustainable Agriculture, Environment and Nano-Technology

Volume 1: Fungal Diversity of Sustainable Agriculture

Fungi Bio-Prospects in Sustainable Agriculture, Environment and Nano-Technology

Volume 1: Fungal Diversity of Sustainable Agriculture

Edited by

Vijay Kumar Sharma
Medical Microbiology Laboratory,
Medical School of Kunming University of Science and Technology,
Yunnan Province, P.R. China

Maulin P. Shah
Environmental Microbiology Lab, Bharuch, Gujarat, India

Shobhika Parmar
Medical Microbiology Laboratory,
Medical School of Kunming University of Science and Technology,
Yunnan Province, P.R. China

Ajay Kumar
Agricultural Research Organization - Volcani Centre,
Rishon LeZion, Israel

ELSEVIER

ACADEMIC PRESS
An imprint of Elsevier

Academic Press is an imprint of Elsevier
125 London Wall, London EC2Y 5AS, United Kingdom
525 B Street, Suite 1650, San Diego, CA 92101, United States
50 Hampshire Street, 5th Floor, Cambridge, MA 02139, United States
The Boulevard, Langford Lane, Kidlington, Oxford OX5 1GB, United Kingdom

British Library Cataloguing-in-Publication Data
A catalogue record for this book is available from the British Library

Library of Congress Cataloging-in-Publication Data
A catalog record for this book is available from the Library of Congress

ISBN: 978-0-12-821394-0

For Information on all Academic Press publications
visit our website at https://www.elsevier.com/books-and-journals

Publisher: Andre Gerhard Wolff
Acquisitions Editor: Linda Versteeg-buschman
Editorial Project Manager: Mona Zahir
Production Project Manager: Swapna Srinivasan
Cover Designer: Greg Harris

Typeset by MPS Limited, Chennai, India

Contents

List of Contributors

B. Aiyamperumal CAS in Marine Biology, Faculty of Marine Sciences, Annamalai University, Chidambaram, India

Sabeena Aliyarukunju Microbiology Division, KSCSTE -Jawaharlal Nehru Tropical Botanic Garden and Research Institute, Thiruvananthapuram, India

N. Amaresan C.G. Bhakta Institute of Biotechnology, Uka Tarsadia University, Surat, India

P. Anantharaman CAS in Marine Biology, Faculty of Marine Sciences, Annamalai University, Chidambaram, India

Moh. Sajid Ansari Department of Botany, Aligarh Muslim University, Aligarh, India

Randhir Kumar Bharti University School of Environmental Management, Guru Gobind Singh Indraprastha University, New Delhi, India

Mahesh Borde Department of Botany, Savitribai Phule Pune University, Pune, India

Shivani Chandra Amity Institute of Biotechnology, Amity University, Noida, India

Nitin Chauhan Department of Microbiology, Shaheed Rajguru College of Applied Sciences for Women, University of Delhi, New Delhi, India

Premchand Subhash Chigadannavar Department of Biotechnology, Ramaiah Institute of Technology, Bengaluru, India

Mahendra Chinthala Department of Chemical Engineering, Ramaiah Institute of Technology, Bengaluru, India

Priyadarshini Dey Department of Biotechnology, Ramaiah Institute of Technology, Bengaluru, India

Madhankumar Dhakshinamoorthy Endophytic Fungal Metabolite Research Laboratory, Department of Biotechnology, Bannari Amman Institute of Technology, Sathyamangalam, Erode, India

Raunak Dhankar Department of Biological Sciences, School of Basic and Applied Sciences, GD Goenka University, Gurugram, India

Mamta Dhiman Department of Biosciences, Manipal University Jaipur, Jaipur, India

Khirood Doley Department of Botany, Savitribai Phule Pune University, Pune, India

P. Ghosh Birla Institute of Scientific Research, Statue Circle, Jaipur, India

Reeta Goel Department of Microbiology, College of Basic Sciences & Humanities, GBPUAT, Pantnagar, India

Deepak Gola Department of Biotechnology, Noida Institute of Engineering and Technology, Noida, India

Raj Kumar Gothwal Birla Institute of Scientific Research, Statue Circle, Jaipur, India; Department of BioEngineering, Birla Institute of Technology, Mesra, Jaipur, India

Lavanya Gowda Department of Biotechnology, Ramaiah Institute of Technology, Bengaluru, India

Biju Haridas Microbiology Division, KSCSTE -Jawaharlal Nehru Tropical Botanic Garden and Research Institute, Thiruvananthapuram, India

Mohammad Haris Department of Botany, Aligarh Muslim University, Aligarh, India

Touseef Hussain Department of Botany, Aligarh Muslim University, Aligarh, India

J. Immanuel Suresh Department of Microbiology, The American College, Madurai, India

Lata Jain ICAR-National Institute of Biotic Stress Management, Raipur, India

S.R. Joshi Microbiology Laboratory, Department of Biotechnology and Bioinformatics, North-Eastern Hill University, Shillong, India

Sneha Judith Department of Biotechnology, Ramaiah Institute of Technology, Bengaluru, India

Dhriti Kapoor Department of Botany, School of Bioengineering & Biosciences, Lovely Professional University, Phagwara, India

Dhiraj Kapur Department of Botany, Dolphin PG College of Science and Agriculture, Fatehgarh Sahib, India

Arun Karnwal Department of Microbiology, School of Bioengineering & Biosciences, Lovely Professional University, Phagwara, India

Pankaj Kaushal ICAR-National Institute of Biotic Stress Management, Raipur, India

Juhi Kaushik Department of Biotechnology, Ramaiah Institute of Technology, Bengaluru, India

Ashraf Y.Z. Khalifa Biological Sciences Department, College of Science, King Faisal University, Al-Ahsa, Kingdom of Saudi Arabia; Botany and Microbiology Department, Faculty of Science, University of Beni-Suef, Beni-Suef, Egypt

Abrar Ahmad Khan Department of Botany, Aligarh Muslim University, Aligarh, India

Ravindra Nath Kharwar Mycopathology and Microbial Technology Laboratory, CAS in Botany, Institute of Science, Banaras Hindu University, Varanasi, India

Kannan Kilavan Packiam Endophytic Fungal Metabolite Research Laboratory, Department of Biotechnology, Bannari Amman Institute of Technology, Sathyamangalam, Erode, India

Ramya Krishnappa Department of Biotechnology, Ramaiah Institute of Technology, Bengaluru, India

Vinay Kumar ICAR-National Institute of Biotic Stress Management, Raipur, India

Puja Kumari Mycopathology and Microbial Technology Laboratory, CAS in Botany, Institute of Science, Banaras Hindu University, Varanasi, India

Poonam Meena Department of Botany, University of Rajasthan, Jaipur, India

Yash Mishra Department of Biosciences, Manipal University Jaipur, Jaipur, India

Archana Nath Agricultural and Processed Food Products Export Development Authority, Housefed Complex, Guwahati, Assam, India

Sampat Nehra Birla Institute of Scientific Research, Statue Circle, Jaipur, India

A. Noorjahan CAS in Marine Biology, Faculty of Marine Sciences, Annamalai University, Chidambaram, India

Pramod Prasad Regional Station, ICAR-Indian Institute of Wheat and Barley Research, Flowerdale, Shimla, India

Zeba Quadri Department of Biotechnology, Ramaiah Institute of Technology, Bengaluru, India

Arpitha Ramachandraiah Department of Biotechnology, Ramaiah Institute of Technology, Bengaluru, India

Kavita Rambal Department of Botany, M. D. College, Parel, Mumbai, India

Unni Maya Ramesh Department of Biotechnology, Ramaiah Institute of Technology, Bengaluru, India

Balram Sahu Department of Agricultural Microbiology, College of Agriculture, Indira Gandhi Krishi Vishwavidyalaya, Raipur, India

Sardul Singh Sandhu Rani Durgavati University, Jabalpur, India

A. Sankaranarayanan C.G. Bhakta Institute of Biotechnology, Uka Tarsadia University, Surat, India

Adnan Shakeel Department of Botany, Aligarh Muslim University, Aligarh, India

Abhishek Sharma C.G. Bhakta Institute of Biotechnology, Uka Tarsadia University, Surat, India

Indu Sharma Department of Biotechnology, Maharishi Markandeshwar Deemed to be University, Ambala, India

Lakshika Sharma Department of Biosciences, Manipal University Jaipur, Jaipur, India

Madan Mohan Sharma Department of Biosciences, Manipal University Jaipur, Jaipur, India

Arti Singh Mycopathology and Microbial Technology Laboratory, CAS in Botany, Institute of Science, Banaras Hindu University, Varanasi, India

Harpreet Singh Department of Agriculture, Swami Vivekanand Group of Institutes, Banur, India

Sheersha Sivadas Department of Biotechnology, Ramaiah Institute of Technology, Bengaluru, India

Pooran Singh Solanki Birla Institute of Scientific Research, Statue Circle, Jaipur, India; Department of BioEngineering, Birla Institute of Technology, Mesra, Jaipur, India

N.M. Sona Department of Microbiology, The American College, Madurai, India

Ravindra Soni Department of Agricultural Microbiology, College of Agriculture, Indira Gandhi Krishi Vishwavidyalaya, Raipur, India

Shiburaj Sugathan Microbiology Division, KSCSTE -Jawaharlal Nehru Tropical Botanic Garden and Research Institute, Thiruvananthapuram, India

Sushma Department of Botany, Dolphin PG College of Science and Agriculture, Fatehgarh Sahib, India

Deep Chandra Suyal Department of Microbiology, Akal College of Basic Sciences, Eternal University, Sirmour, Himachal Pradesh, India

Nitin Swamy Rani Durgavati University, Jabalpur, India

Shivani Thakur Department of Botany, Dolphin PG College of Science and Agriculture, Fatehgarh Sahib, India

P.C. Trivedi Department of Botany, University of Rajasthan, Jaipur, India

Alok Kumar Varshney Birla Institute of Scientific Research, Statue Circle, Jaipur, India; Department of BioEngineering, Birla Institute of Technology, Mesra, Jaipur, India

Sandeep Venkatesh Department of Biotechnology, Ramaiah Institute of Technology, Bengaluru, India

Rajnish Kumar Verma Department of Agriculture, Swami Vivekanand Group of Institutes, Banur, India

Preface

Fungi are considered to be one of the most diverse life forms on earth that help most of the living entities exist on this planet. The fungi represent huge ecological and taxonomic diversity, varied life-cycle strategies and morphologies. They range from unicellular aquatic chytrids or yeasts to large macrofungi such as well-known mushrooms and toadstools and the largest of fruitbodies, the giant puffball. The exact size of the fungal diversity is still a topic of debate, but according to an estimate, about 1.5 million species of fungi exist out of which currently approx 1.5 lacs species are known. Fungi have attracted significant attention owing to their industrial, agriculture, environmental and medicinal applications. During the last few decades' fungal diversity has attracted the agriculture scientists, microbiologist as well as plant pathologists across the globe, due to advantageous effects linked with enhanced crop yield/productivity and sustainable agriculture. Fungi can serve as the enhancers of crop productivity as well as a suitable option to manage the plant diseases. Further, fungi and mycorrhiza contribute to maintaining crop and soil productivity under environmental stress. Fungi are an important source of industrially important enzymes and have been extensively used over the years. Production of important biomolecules and novel bioactive metabolites has a diverse role in different sectors of agriculture, environment and medicine. Fungal genera surviving in extreme environmental conditions demand analysis of the molecular mechanisms facilitating their existence under harsh conditions. Identification and successful incorporation of desired genes into a suitable host system would be beneficial for particular traits too. The toxigenic fungal groups are important from the perspective of food safety and associated human health hazard. Storage conditions favor the growth of toxigenic fungi reducing global market value. Given the rising application of nanomaterials in different fields of agriculture and environment, fungi are the important environment-friendly alternative to currently practised physico-chemical techniques used for nanomaterials synthesis.

Fungi bio-prospects in sustainable agriculture, environment and nanotechnology is a three-volume series. The book is an attempt to explore the recent advances in fungi from various environments, their diversity and diverse applications to the sustainable development of human life. The volume-I has 19 diverse chapters that offer an up-to-date and detailed account of various aspects of fungal biodiversity including endophytic fungi from diverse habitats, phytopathogenic fungi, arbuscular mycorrhizal fungi and rhizospheric fungi. The chapters cover fungal bioactive metabolites including those from the endophytic origin and their pharmaceutical and agriculture use. Applications of fungi as a bio-fertilizer, a bio-control agent, myco-inoculants, bio-fuel production management of biotic and abiotic stresses were discussed thoroughly. Chapter 1 deals with the influence of survival strategies on the endophytic fungal biodiversity studies. The chapter describes fungal endophytes from subtropics and tropical regions and highlights the importance of fungal endophytes to society. This chapter also gives insights into the various bioactive compounds from fungal endophytes isolated from various host, their uses and modern-era challenges in endophyte research. Chapter 2 presents a detailed account of the phylloplane fungal flora and host plants from the Southern Western Ghats of India. Chapter 3 describes the fungal endophytes from seaweeds and their applications as bio-fertilizer, bio-control agents, phosphate solubilizers and a source of bioactive novel compounds. Chapter 4 explains the fungal endophytes, various modes of their entry and establishment into host plants, diversity, and their contribution to sustainable agriculture. Chapter 5 explains the marine fungal endophytes and their bioactive metabolites. Chapter 6 explains the importance of fungi (endophytic and mycorrhizal) in alleviating different abiotic stresses such as drought, salinity, cold, heat and heavy metal stress; improving plant nutrition and defence. Chapter 7 gives a comprehensive account of the phytopathogenic fungi, the strategies they use to infect the plant and cause disease, and bio-control approaches. Chapter 8 explains the role and applications of myco-inoculants in a sustainable agroecosystem. Chapter 9 discuss Arbuscular Mycorrhizal (AM) fungi, their role and commercial potential in improving agricultural yield and quality. The chapter explains the role of AM fungi for the management of biotic stress such as plant-viral diseases and plant-parasitic nematodes, and abiotic stress such as drought, salinity and metals. Chapter 10 describes the endophytic fungal diversity associated with the medicinal plants, their biological roles, bioactive secondary metabolites, and applications as antiviral, cytotoxic, antidiabetic, antimicrobial and immunosuppressive compounds. Chapter 11 explains the prospect of biofuel production using fungi from waste materials and wastewater. Chapter 12 presents a comprehensive review of the worldwide executed work on applications of fungal endophytes in the area of plant growth promotion and biological control agents. Chapter 13 describes rhizospheric fungi and their interaction with plants, advance tools and techniques to study these fungi, present applications and future possibilities. Chapter 14 describes mycotoxins associated food safety concerns in agricultural crops, their human health hazards, prevention and control in crops. Chapter 15 describes the bio-prospects of endophytic fungi associated with ethnoveterinary plants from different regions of India for the isolation of novel metabolites, their biotechnological applications and future scope. Chapter 16 discusses fungi as an alternative tool in agriculture for sustainable crop production and improvement. Chapter 17 explains the applications of yeast in agricultural for plant growth promotion and biocontrol of phytopathogens. Chapter 18 discuss the different fungi responsible for the phytostimulation and induced systemic resistance in plants. Chapter 19 provides a concise record of the Fusarium diseases associated with fruit crops, their biomanagement strategies, host-pathogen interactions and mode of action of potential biocontrol agents.

This volume is the collaborative work of many people. We appreciate and thank all the experts for readily contributing to various aspects of fungi presented in the book. The views expressed by the authors are their own. We are hopeful that this book will be useful to academics, researchers, postgraduate students, and practitioners in fungal science from agriculture, environment and industries. We express gratitude to Prof. R.N. Kharwar and Prof. Haiyan Li for their constant support and encouragement in bringing out this book.

**Vijay Kumar Sharma, Maulin P. Shah,
Shobhika Parmar and Ajay Kumar**

Chapter 1

Endophytic fungi: A review of survival strategies that influence the biodiversity studies associated with biopotential products

Kannan Kilavan Packiam* **and Madhankumar Dhakshinamoorthy**
Endophytic Fungal Metabolite Research Laboratory, Department of Biotechnology, Bannari Amman Institute of Technology, Sathyamangalam, Erode, India
**Corresponding author.* Email: drkpkannan@gmail.com

Introduction

Endophytic fungi are known to all as symbionts, it able to produce some defence metabolites against plants initially and later for adaptation they secrete challenging bioactive metabolites without disturbing host. Researchers have revealed new microbial species from environment as endophytes and also plant associated metabolites were screened by those endophytic fungi for commercial exploitation. Plant based microbial products derived from endophytic fungi. For the past four decades, we are searching novel drugs from microbial products. Endophyte based metabolites were searched two decades back. Krings et al. (2007) stated that association of endophyte with plant interactions had began early during the evolution of Devonian and cretaceous period. Ascomycetes and Dothideomycetes are common endophytes in all these plants were earlier reported. Not all fungi intrude into their host as endophytes. Some of microbes entered into plants due to survival in ecological changes. Later the ecological balance and ecological fitness leads to expand their territory into host range as well as seasonal variability and other factors of ecosystems role in the environment.

Trees are predominantly present in terrestrial biomes and the endophyte assemblage were highly present as single predominant group to protect the host as well as growth and metabolism of plant and fungus (Duplessis et al.,

Fungi Bio-Prospects in Sustainable Agriculture, Environment and Nano-Technology.
DOI: https://doi.org/10.1016/B978-0-12-821394-0.00001-9

2009; Hacquard et al., 2011; Franco et al., 2017; Raman and Suryanarayanan, 2017). Endophytic fungi are the important predictor for environmental biotic stressors as it is rich in nitrogen source and the plant mediated organisms like insects can access the fungal interactions and utilize the fungus as nitrogen source and also inhibiting the defence response from plant fungal crosstalk (Eberl et al., 2018). The present study focusses on various influencing factors of plant fungal interactions and its survival. It also identifies some of plant based endophytic metabolites and its challenges that reach to societal needs. This study reveals that the critical phase of modern era that survives the challenging unknown diseases into fruitful one. It is hard to differentiate the societal needs without disturbing the ecosystems. Endophytic fungi are key players to overcome these problems into a healthier society in future. Endophytic microbial technology will be boon in all fields of agriculture, medicine, and food industry.

Importance to study the endophytic fungal diversity

Several researches have been focused on to study the endophytic fungal diversity and its importance to this environment. For example, plant tissues in trees can be colonized by endophytic fungi was recently proved by (Baldrian (2017)). Plant–fungal interactions were studied and their properties of mutualistic into antagonistic by Faeth and Hammon (1997) and Stierle and Stierle (2015). Some factors determine the endophytic fungi in treetops is limited due to species specific interact each other to enhance the plant growth at particular environment (Khan et al., 2016). Tree endophytes consists of ascomycetes and some basidiomycetes belong to group of endophytes were reported by Petrini (1986) and Rodriguez et al. (2009). Fungal endophytes studies were highlighted the important contribution of microbial life in plants such as plant thermotolerant in nature, drought resistant and some secondary metabolite produced themselves for defence (Moncrieff et al., 2015; Mishra et al., 2016). Estimation of endophytic fungal diversity is challenged due to global climatic conditions, sampling strategy, methods of isolation temperature, seasonal changes by Suryanarayanan et al. (2011), Massimo et al. (2015) and Singh et al. (2017). Seasonal observations were done to study the relationships with fungi and plant interactions, it leads to increase the productivity of plants in the environment and also increase the plant metabolic activity (Giauque and Hawkes, 2016; Martins et al., 2016). Symbiotic interactions of plant fungal communities can help in their structure, function and its composition to this environment were reported by Loro et al. (2012). Species composition and its frequency are known to vary with different tissues of host plant by Petrini et al. (1992) and Rodrigues 1994). Previous studies have clearly state this, some endophytes are not host specific and endophyte identification is not host specific and its identification should be done through phylogenetic analysis (Liu et al., 2010).

Different environmental factors were influenced the fungal diversity in forest ecosystem (Saikkonen, 2007). Root endophytes played an important role in plant growth (Jumpponen and Trappe, 1998). Endophyte interacted plants are generally often grown faster when compared with other plants due to mutual symbionts and also acts as growth regulators for plants used in the field of agriculture. Researchers have identified the plant-based hormones of auxin, gibberellin, through endophytic metabolites. These endophytes act as keystone species for seasonal changes such as biotic and abiotic stress, it may produce some metabolites which are defence to other host endophytes. It leads to suppress the other host endophyte and cause the other host plants. The discovery of novel metabolites will determine the species richness, population dynamics of both plant and endophytes, and also increase the plant growth and plant health in particular environment predominantly (Yang et al., 1994; Reddy et al., 2016). Investigating the metabolites from endophytes may increase the chance to find the many novel biopotential compounds. There are several advantages to find plant endophyte-based metabolites. We can identify the promising endophytes and stored it easily for further findings. There have been numerous attempts to face the findings of the problems because of microbial advantages. We can use the endophytic fungi with different culture conditions to identify the different compounds which they may be able to provide at that circumstances. Another advantage from microbes is these endophytic fungi have existed for many billion years ago. So, they have their potential to identify the enzyme targets and defence conditions to destruct the illness or diseases. Some endophytes are producing novel anticancer alkaloids like Taxol, Camptothecin and Vincristine, etc. Fungal fermentation provides inexhaustible source of novel metabolites, So, nowadays chemists also turned into endophyte-based metabolites for yield enhancements. Endophytes are host dependent and ability to play a major role in producing metabolites for survival strategies in different environments. The tropical and sub-tropical regions are key player to survive the majority of endophytes due to plant–fungal interactions were highly dense with natural biomes like marine, terrestrial, desert ecosystems can influenced by different factors to suppress this ecosystems (Fig. 1.1A and B).

Fungal endophytes from subtropics and tropical regions

Fungal kingdom is most diversely present in our earth and symbiotically associated with plant nutrition and also disease out-breaker in all terrestrial ecosystems were reported by Dighton. Endophytic fungi is highly present in tropical regions when compared with other biomes by Murali et al. (2007). Few studies have been carried out on plant–fungal interactions in tropical regions. Isolation of endophytic fungi from tropical plants were first reported by Petrini and Dreyfuss (1981). Endophytic fungi could evolve and enhance their ecological adaptation, it leads to ecological fitness to particular host

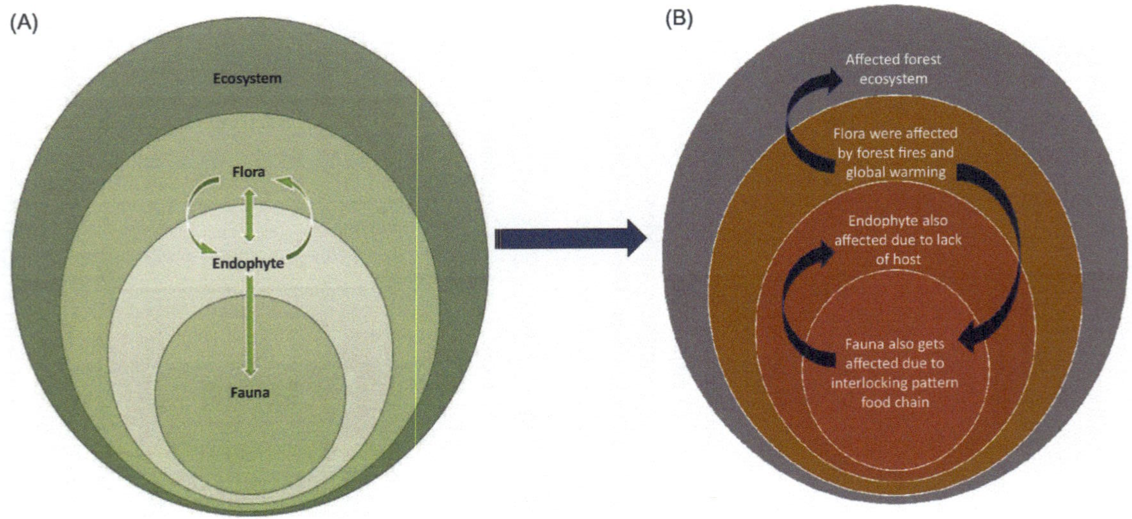

FIGURE 1.1 (A) Forest ecosystem before affected by influencing factors. (B) Forest ecosystems in modern era gets affected.

range (Suryanarayanan et al., 2011). Endophytic fungi can alter the plant signals and plant metabolism by looking at physiological and biochemical aspects of tropical and subtropical regions. The plant defence mechanism is triggered by these endophytes for its survival without affecting from external environments such as herbivorous insects. This will help both plant and fungal from insects. Interactions between phytophagous insects and phyllophane fungi and endophytes would depend upon the host. So, these interactions will support to introduce the novel metabolites to the environment. Diamondback moth larvae, leaves feeding ants and insects also face the challenges to the host–fungal defence mechanism. It leads to find some novel metabolites by endophytes with challenging enzyme targets to suppress the genes of external infectious agents like herbivorous insects and to avoid the leaf-cutting environments in modern society.

There have been many reports of endophytic fungi from tropical regions (Bagchi et al., 2014; Treseder and Lennon, 2015; Suryanarayanan et al., 2016). The available information of endophytes and metabolic activity is limited and able to produce biopotential metabolites in industrial level. The current methods of endophytic metabolites are not suitable for societal needs. The tropical regions have plentiful resource of endophytic fungi from various biomes. Researchers have to find the solutions to enhance the societal needs at cheaper cost. Many researchers have done their excellent findings of antibacterial, antifungal, weedicidal, antimalarial and anticancerous activities from endophytic fungal products (Sachin et al., 2013; Kaushik et al., 2014; Nicoletti and Fiorentino, 2015). But our society demands higher due to technological developments, pollutions, global warming, modified foods due to high demands, usage of pesticides, insecticides, stress environments leads to cause of diseases such as physiological and neurological disorders and cancer also prevalently played a major role in modern societal needs.

Endophytes will be modern microbial technology to fulfill our societal needs

Endophytic fungi may be metabolically arresting the host by secreting some defence chemicals in aggressive environments. This probably means that, the endophytic fungi are predominant to host at some level to adapt into particular environment and later they act as symbionts to each other. Strobel and Daisy (2003) states that in review, plants with new environment and also in unconventional biology, plants may depend upon the endophytes to survive in this environment. So, the endophytes predominantly react and secrets some metabolites or biopotential products to host in particular environment. We also observed some other biopotential compounds which are produced by both plant-fungal interactions in aggressive environment. These endophytes are clearly recognized as main source of novel secondary metabolites by Strobel (2003), Berg and Hallmann (2006), Debbab et al.,

Suryanarayanan et al. (2011), Rosenberg et al. and Martins et al. (2016). Many of biopotential compounds through endophytic fungi of *Pestalotiopsis* sp., *Phomopsis* sp., and *Fusarium* sp., were identified for the past four decades by the researchers have listed in Table 1.1.

The three species were the greatest contributors to this society involved in the biopotential compounds in the tropical regions. They have been acts as plant fitness that benefits the host to native regions and acts as symbionts. This leads to improves the human well-being and also create societal impact to balance the nature with utmost care. The values of research on fungal diversity highlights the importance of influencing factors of seasonality, temperature, plant growth stage and rainfall in those regions (Bateman and Murray, 2001; Aly et al., 2010, 2011; Porras-Alfaro and Bayman, 2011; Loro et al., 2012). The following endophytic metabolite compounds and its structure were collected from Pubchem database (Fig. 1.2).

Significance on understanding the modern era challenges

Natural products of biopotential compounds from endophytic fungi is in high demand due to severity in polluted environment to protect the flora and fauna in earth. In modern era, these endophytes serve the alternate source of novel biopotential compounds (Joseph and Priya, 2011; Xie and Zhou, 2017). Although several compounds showed the good number of pharmacological activities and anticancer activities but they face many challenges in developing the natural compounds which is highly demanded by society. Plant based compounds are very low abundance in nature due to forest fire like Amazon, South America and Australian forest fires and also reported as low yield due to complete destruction of plant environment to get very low number of drugs. The forest fires are not natural phenomena, it affects the healthier society for one's solution into another's problem. Protecting the forest biomes is now challenging one due to climatic conditions and human interference in forest ecosystems. It leads to turned into endophytes based bioactive metabolites. The secondary metabolites from endophytes also face many challenges due to identification of biopotential endophytes and yield enhancement strategies by the researchers. Microbial fermentation is such major advantage to enhance the yield by strain improvement via, genetic engineering tools and also medium parameters and precursors role to increase the productivity under in vitro conditions. Endophytic fungi are the only possibility to enhance this modern era into a fruitful society. So, we critically investigate this endophyte metabolites documentation from undisturbed environment such as forests ecosystem, mangrove and marine habitats. Further studies of optimization for strain improvement, genome shuffling, co-culturing, elicitors addition to enhance the productivity and fungal fermentation strategy (Kusari et al., 2012; Venugopalan et al., 2016; Fangxue Xu et al., 2018; Subban et al., 2019).

TABLE 1.1 Endophytic fungal biopotential products isolated from various host.

Host	Endophytic fungi	Biopotential compounds	References
Apodytes dimidiate	*Fusarium solani*	Camptothecin	Shweta et al. (2010)
Miquelia dentata	*Phomopsis* sp.	Camptothecin	Shweta et al. (2013)
Camptotheca acuminata	*Fusarium solani*	Camptothecin	Ran et al. (2017)
Camptotheca acuminata	*Fusarium nematophilum*	Camptothecin	Su et al. (2014)
Camptotheca acuminata	*Fusarium solani*	Camptothecin	Kusari et al. (2009)
Nothapodytes foetida	*Fusarium oxysporum* NFX06	Camptothecin	Musavi and Balakrishnan (2014)
Taxus cuspidate	*Fusaarium arthrosporioides*	Taxol	Li et al. (2008)
Cardiospermum helicacabum	*Pestalotiopsis pauciseta*	Taxol	Gangadevi et al. (2008)
Taxus mairei	*Fusarium maire*	Taxol	Xu et al. (2006)
Gingko biloba	*Phomopsis* sp. BKH 30	Taxol	Kumaran and Hur (2009)
Terminalia arjuna	*Pestalotiopsis terminaliae*	Taxol	Gangadevi and Muthumary (2009)
Taxus walachiana	*Pestalotiopsis microspora*	Taxol	Strobel et al. (1996)
Rhizophora annamalayan	*Fusarium oxysporum*	Taxol	Elavarasi et al. (2012)
Ervatamia divaricata	*Pestalotiopsis breviseta*	Taxol	
Juniperus recurva	*Fusarium oxysporum*	Podophyllotoxin	Kour et al. (2008)

(Continued)

TABLE 1.1 (Continued)

Host	Endophytic fungi	Biopotential compounds	References
Podophyllum hexandrum	*Fusarium solani*	Podophyllotoxin	Nadeem et al. (2012)
Sinomenium acutum	*Pestalotiopsis palmarum*	Doxorubicin, Sinopestalotiollides	Xiao et al. (2018)
Artemisia japonica	*Pestalotiopsis uvicola*	Kaempferol, Quercetin	Qian et al. (2017)
Camellia sasanqua	*Pestalotiopsis karstenii*	Pestalotin, Hydroxypestalotin	Luo et al. (2012)
Clerodendrum canescens	*Pestalotiopsis adusta*	Uncinatone	Xu et al. (2016)
Salicornia bigelovii Torr.	*Fusarium equiseti*	Cerevisterol, Ergosterol peroxide	Wang et al. (2014)
Pongamia pinnata	*Phomopsis glabrae*	Depsipeptide	Verekar et al. (2014)
Tectona grandis	*Phomopsis* sp.	Phomaxanthona A and B	Isaka et al. (2001)
Camellia sinensis	*Pestalotiopsis fici*	Pestaloficiol L	Liu et al. (2009)
Cinnamomum kanehirae	*Fusarium oxysporum*	Beauvericin	Wang et al. (2011)
Corylus avellana	*Phomopsis amygdale*	Pestalotin	Akay et al. (2014)
Salicornia bigelovii Torr.	*Fusarium equiseti*	Diglucotol	Wang et al. (2014)
Kandelia candel	*Fusarium* sp.	5-O-methyl-2′-methoxy-3′-methylalpiniumisoflavone	Huang et al. (2010)
Rhizophora mucronata	*Pestalotiopsis* sp.	Pestalotiopsone A, B,C,D,E,F	Xu et al. (2009)
Tylophora indica	*Fusarium solani* LCPANF01	Antibacterial compounds	Nomila Merlin et al. (2013)

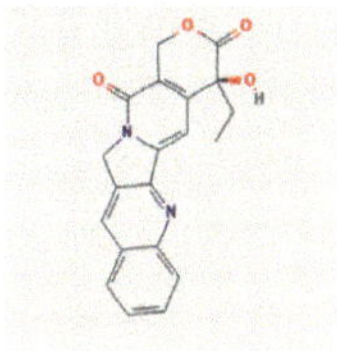

1

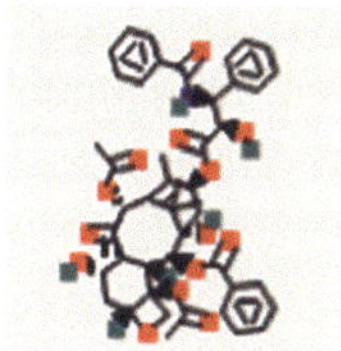

2

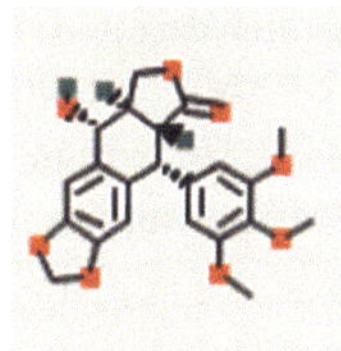

3

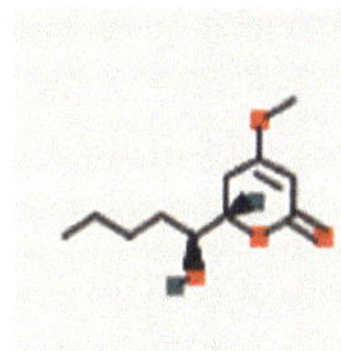

4

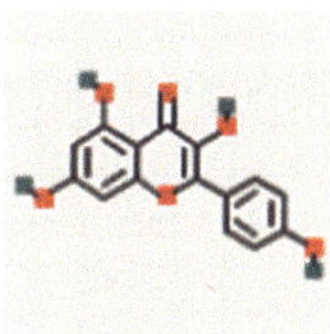

5

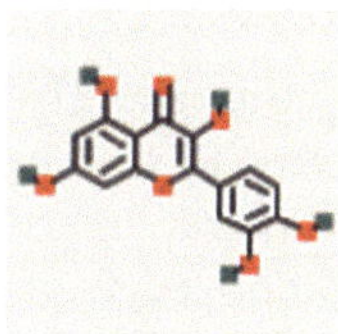

6

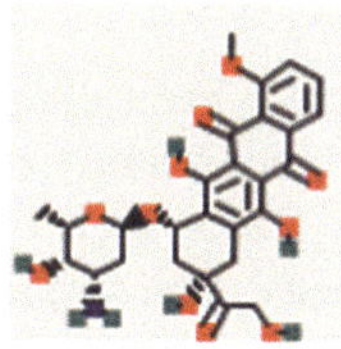

7

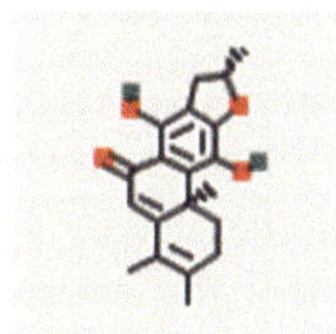

8

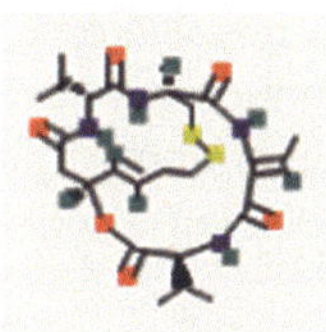

9

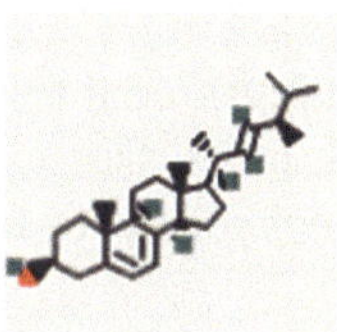

10

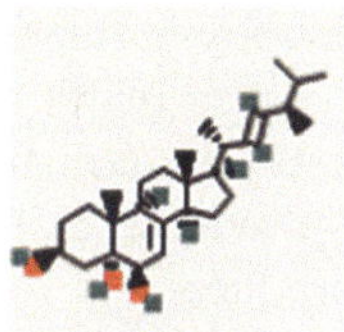

11

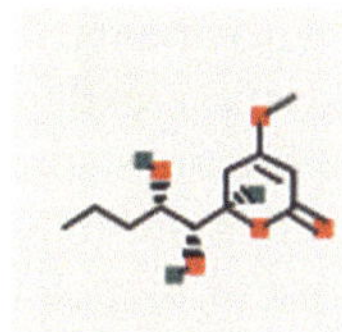

12

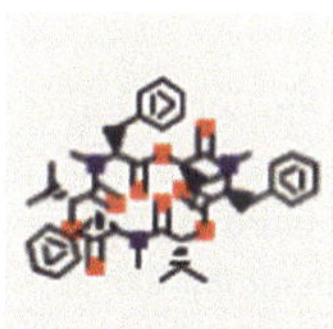

13

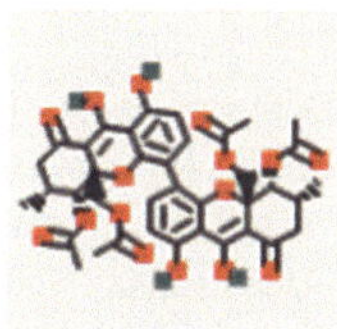

14

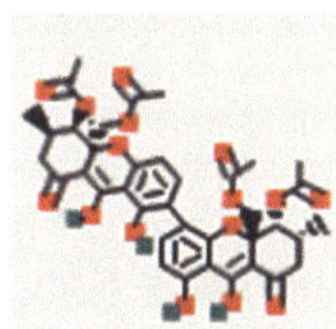

15

FIGURE 1.2 Chemical structure of endophytic fungal metabolites from various host–fungal interactions.
(1) Camptothecin. (2) Taxol. (3) Podophyllotoxin. (4) Pestalotin. (5) Kaempferol. (6) Quercetin. (7) Doxorubicin. (8) Uncinatone. (9) Depsipeptide. (10) Ergosterol peroxide. (11) Cerevisterol. (12) Hydroxypestalotin. (13) Beauvericin. (14) Phomaxanthone A. (15) Phomaxanthona B.

Endophytic fungi are challenged to produce metabolites under mimicked conditions within confined space. This might be due to under expression of particular genes in new environment. So, we have to identify those biosynthetic genes and its expression and enhance the gene clusters with genetic tools and also co-culturing strategies. Another challenge that faced by researchers and societal needs are many reported endophytic metabolites are in data level alone not in products in market. So, the promising endophyte metabolites should be identified and enhance the productivity at cheaper cost to reach the utmost lower level societies to fulfill their needs. Due to modern era and technological advances, we are facing *n*-number of unknown diseases and cancers. For this, we have to find natural metabolites through plant–fungal interactions to reduce the serious side of longer treatment protocols of unknown diseases and will enhance the society into next level of healthy environments.

Conclusion

Endophyte research have been emerging boon and also privileged resource to identify the novel biopotential compounds. Researchers have revealed the challenges in various fields such as agriculture, industry, and medicine to fulfill the societal needs. The unfolding knowledge of traditional medicine were found through endophyte documentation. Biodiversity studies knock off to identify the climatic change, global warming and biotic/abiotic stress environments. The endophytic fungi act as keystone species in forest biomes and these endophytes played major role in nutritional transformation between herbivorous insects, ants and animals with hosts. Our studies examining that endophytic fungi will provide excellent mechanisms to find the novel discoveries of biopotential products.

Acknowledgments

The authors are grateful to the Defence Research Development Organization, ER&IPR, Government of India, New Delhi, India (Project Sanction order No. ERIP/ER/1104597/M/01/1493) for financial support and also thanks to National Biodiversity Authority, Chennai, and Tamil Nadu Forest Department, Sathyamangalam, Tamil Nadu and Bannari Amman Institute of Technology, Sathyamangalam. My special thanks to all my previous researchers and findings for providing fruitful data's to enrich this work valuably.

References

Akay, S., Ekiz, G., Kocabas, F., Hameş Tuna, E.E., Korkmaz, K.S., Bedir, E., 2014. A new 5,6-dihydro-2-pyrone derivative from *Phomopsis amygdali*, an endophytic fungus isolated from hazelnut (*Corylus avellana*). Phytochem. Lett. 7 (Suppl. C), 93–96. Available from: https://doi.org/10.1016/j.phytol.2013.09.012.

Aly, A.H., Debbab, A., Kjer, J., Proksch, P., 2010. Fungal endophytes from higher plants: a prolific source of phytochemicals and other bioactive natural products. Fungal Divers. Available from: https://doi.org/10.1007/s13225-010-0034-4.

Aly, A.H., Debbab, A., Proksch, P., 2011. Fungal endophytes: unique plant inhabitants with great promises. Appl. Microbiol. Biotechnol. 90, 1829–1845. Available from: https://doi.org/10.1007/s00253-011-3270-y.

Bagchi, R., Gallery R.E., Gripenberg S., Gurr, S.J., Narayan, L., Addis, C.E., et al., 2014. Pathogens and insect herbivores drive rainforest plant diversity and composition. 85–88.

Baldrian, P., 2017. Forest microbiome: diversity, complexity and dynamics. FEMS Microbiol. Rev. 41, 109–130. Available from: https://doi.org/10.1093/femsre/fuw040.

Bateman, G.L., Murray, G., 2001. Seasonal variations in populations of *Fusarium* species in wheat-field soil. Appl. Soil Ecol. 18, 117–128. Available from: https://doi.org/10.1016/S0929-1393(01)00158-5.

Berg, G., Hallmann, J., 2006. Control of plant pathogenic fungi with bacterial endophytes. In: Schulz, B., Boyle, C., Sieber, T. (Eds.), Microbial Root Endophytes. Springer-Verlag, Berlin, pp. 53–69.

Duplessis, S., Major, I., Martin, F., Séguin, A., 2009. Poplar and pathogen interactions: insights from populus genome-wide analyses of resistance and defense gene families and gene expression profiling. CRC Crit. Rev. Plant Sci. 28, 309–334. Available from: https://doi.org/10.1080/07352680903241063.

Eberl, F., Hammerbacher, A., Gershenzon, J., Unsicker, S.B., 2018. Leaf rust infection reduces herbivore-induced volatile emission in black poplar and attracts a generalist herbivore. New Phytol. 220, 760–772. Available from: https://doi.org/10.1111/nph.14565.

Elavarasi, A., Rathna, G.S., Kalaiselvam, M., 2012. Taxol producing mangrove endophytic fungi *Fusarium oxysporum* from *Rhizophora annamalayana*. Asian Pac. J. Trop. Biomed. 2, S1081–S1085. Available from: https://doi.org/10.1016/S2221-1691(12)60365-7.

Faeth, S.H., Hammon, K.E., 1997. Fungal endophytes in oak trees: experimental analyses of interactions with leaf miners. Ecology 78, 820–827. Available from: https://doi.org/10.1890/0012-9658(1997)078[0820:Feiote]2.0.Co;2.

Franco, F.P., Moura, D.S., Vivanco, J.M., Silva-Filho, M.C., 2017. Plant–insect–pathogen interactions: a naturally complex ménage à trois. Curr. Opin. Microbiol. Available from: https://doi.org/10.1016/j.mib.2017.04.007.

Kusari, S., Zühlke, S., Spiteller, M., 2009. An endophytic fungus from *Camptotheca acuminata* that produces camptothecin and analogues. J. Nat. Prod. 72, 2–7. Available from: https://doi.org/10.1021/np800455b.

Gangadevi, V., Muthumary, J., 2009. Taxol production by *Pestalotiopsis terminaliae*, an endophytic fungus of *Terminalia arjuna* (arjun tree). Biotechnol. Appl. Biochem. 52, 9–15. Available from: https://doi.org/10.1042/BA20070243.

Gangadevi, V., Murugan, M., Muthumary, J., 2008. Taxol determination from *Pestalotiopsis pauciseta*, a fungal endophyte of a medicinal plant. Chin. J. Biotechnol. 24, 1433–1438. Available from: https://doi.org/10.1016/S1872-2075(08)60065-5.

Giauque, H., Hawkes, C.V., 2016. Historical and current climate drive spatial and temporal patterns in fungal endophyte diversity. Fungal Ecol. 20, 108–114.

Hacquard, S., Petre, B., Frey, P., Hecker, A., Rouhier, N., Duplessis, S., 2011. The poplar–poplar rust interaction: insights from genomics and transcriptomics. J. Pathog. 2011, 1–11. Available from: https://doi.org/10.4061/2011/716041.

Huang, Z., Yang, J., She, Z., Lin, Y., 2010. Isoflavones from the mangrove endophytic fungus *Fusarium* sp. (ZZF41). Nat. Prod. Commun. 5, 1771–1773.

Isaka, M., Jaturapat, A., Rukseree, K., Danwisetkanjana, K., Tanticharoen, M., Thebtaranonth, Y., 2001. Phomoxanthones A and B, novel xanthone dimers from the endophytic fungus *Phomopsis* species. J. Nat. Prod. 64, 1015–1018. Available from: https://doi.org/10.1021/np010006h.

Joseph, B., Priya, R.M., 2011. Bioactive compounds from endophytes and their potential in pharmaceutical effect: a review. Am. J. Biochem. Mol. Biol. 1, 291–309. Available from: https://doi.org/10.3923/ajbmb.2011.291.309.

Jumpponen, A., Trappe, J.M., 1998. Dark septate endophytes: a review of facultative biotrophic root- colonizing fungi. New Phytol. 140, 295–310.

Kaushik, N.K., Murali, T.S., Sahal, D., Suryanarayanan, T.S., 2014. A search for antiplasmodial metabolites among fungal endophytes of terrestrial and marine plants of southern India. 745–757.

Khan, A.L., Al-Harrasi, A., Al-Rawahi, A., Al-Farsi, Z., Al-Mamari, A., Waqas, M., et al., 2016. Endophytic fungi from frankincense tree improves host growth and produces extracellular enzymes and indole acetic acid. PLoS One 11, e0158207. Available from: https://doi.org/10.1371/journal.pone.0158207.

Kour, A., Shawl, A.S., Rehman, S., Sultan, P., Qazi, P.H., Suden, P., et al., 2008. Isolation and identification of an endophytic strain of *Fusarium oxysporum* producing podophyllotoxin from *Juniperus recurva*. World J. Microbiol. Biotechnol. 24, 1115–1121. Available from: https://doi.org/10.1007/s11274-007-9582-5.

Krings, M., Taylor, T.N., Hass, H., Kerp, H., Dotzler, N., Hermsen, E.J., 2007. Fungal endophytes in a 400-million-yr-old land plant: infection pathways, spatial distribution, and host responses. New Phytol. 174, 648–657. Available from: https://doi.org/10.1111/j.1469-8137.2007.02008.x.

Kumaran, R.S., Hur, B.K., 2009. Screening of species of the endophytic fungus *Phomopsis* for the production of the anticancer drug taxol. Biotechnol. Appl. Biochem. 54, 21–30. Available from: https://doi.org/10.1042/BA20080110.

Kusari, S., Hertweck, C., Spiteller, M., 2012. Chemical ecology of endophytic fungi: origins of secondary metabolites. Chem. Biol. Available from: https://doi.org/10.1016/j.chembiol.2012.06.004.

Li, C.-T., Li, Y., Wang, Q.-J., Sung, C.-K., 2008. Taxol production by *Fusarium arthrosporioides* isolated from yew, *Taxus cuspidata*. J. Med. Biochem. 27, 454–458. Available from: https://doi.org/10.2478/v10011-008-0022-3.

Liu, L., Liu, S., Niu, S., Guo, L., Chen, X., Che, Y., 2009. Isoprenylated chromone derivatives from the plant endophytic fungus *Pestalotiopsis fici*. J. Nat. Prod. 72, 1482–1486. Available from: https://doi.org/10.1021/np900308s.

Liu, A.R., Chen, S.C., Wu, S.Y., Xu, T., Guo, L.D., Jeewon, R., et al., 2010. Cultural studies coupled with DNA based sequence analyses and its implication on pigmentation as a phylogenetic marker in Pestalotiopsis taxonomy. Mol. Phylogenet. Evol. 57, 528–535. Available from: https://doi.org/10.1016/j.ympev.2010.07.017.

Loro, M., Valero-Jiménez, C.A., Nozawa, S., Márquez, L.M., 2012. Diversity and composition of fungal endophytes in semi-arid Northwest Venezuela. J. Arid Environ. 85, 46–55.

Luo, D.Q., Zhang, L., Shi, B.Z., Song, X.M., 2012. Two new oxysporone derivatives from the fermentation broth of the endophytic plant fungus *Pestalotiopsis karstenii* isolated from stems of Camellia. Molecules 17, 8554–8560.

Martins, F., Pereira, J.A., Bota, P., Bento, A., Baptista, P., 2016. Fungal endophyte communities in above-and belowground olive tree organs and the effect of season and geographic location on their structures. Fungal Ecol. 20, 193–201.

Massimo, N.C., Nandi Devan, M.M., Arendt, K.R., Wilch, M.H., Riddle, J.M., Furr, S.H., et al., 2015. Fungal endophytes in aboveground tissues of desert plants: infrequent in culture, but highly diverse and distinctive symbionts. Microb. Ecol. 70 (1), 61–76.

Mishra, V.K., Singh, G., Passari, A.K., Yadav, M.K., Gupta, V.K., Singh, B.P., 2016. Distribution and antimicrobial potential of endophytic fungi associated with ethnomedicinal plant *Melastoma malabathricum* L. J. Environ. Biol. 37 (2), 229.

Moncrieff, G.R., Scheiter, S., Slingsby, J.A., Higgins, S.I., 2015. Understanding global change impacts on South African biomes using dynamic vegetation models. South Afr. J. Bot. 101, 16–23.

Murali, T.S., Suryanarayanan, T.S., Venkatesan, G., 2007. Fungal endophyte communities in two tropical forests of southern India: diversity and host affiliation. Mycol. Prog. 6, 191–199.

Musavi, S.F., Balakrishnan, R.M., 2014. A study on the antimicrobial potentials of an endophytic fungus *Fusarium oxysporum* NFX 06. J. Med. Bioeng. 3, 162–166. Available from: https://doi.org/10.12720/jomb.3.3.162-166.

Nadeem, M., Ram, M., Alam, P., Ahmad, M.M., Mohammad, A., Al-Qurainy, F., et al., 2012. *Fusarium solani*, P1, a new endophytic podophyllotoxin producing fungus from roots of *Podophyllum hexandrum*. Afr. J. Microbiol. Res. 6, 2493–2499. Available from: https://doi.org/10.5897/AJMR11.1596.

Nicoletti, R., Fiorentino, A., 2015. Plant bioactive metabolites and drugs produced by endophytic fungi of spermatophyta. 918–970.

Nomila Merlin, J., NimalChristhudas, I.V.S., Praveen Kumar, P., Agastian, P., 2013. Optimization of growth and bioactive metabolite production: *Fusarium solani*. Asian J. Pharm. Clin. Res. 6, 98–103.

Petrini, O., 1986. Taxonomy of endophytic fungi of aerial plant tissues. In: Fokkema, N.J., van den Heuvel, J. (Eds.), Microbiology of the Phyllosphere.

Petrini, O., Dreyfuss, M., 1981. Endophytische Pilze in Epiphytischen Araceae, Bromeliaceae, und Orchidacea. Sydowia 34, 135–148.

Petrini, O., Sieber, T.N., Toti, L., Viret, O., 1992. Ecology, metabolite production, and substrate utilization in endophytic fungi. Nat. Toxins 1, 185–196.

Porras-Alfaro, A., Bayman, P., 2011. Hidden fungi, emergent properties: endophytes and microbiomes. Annu. Rev. Phytopathol. 49, 291–315. Available from: https://doi.org/10.1146/annurev-phyto-080508-081831.

Qian, Y.X., Kang, J.C., Luo, Y.K., He, J., Wang, L., Zhang, X.P., 2017. Secondary metabolites of an endophytic fungus *Pestalotiopsis uvicola*. Chem. Nat. Compd. 53 (4), 756–758.

Raman, A., Suryanarayanan, T.S., 2017. Fungus–plant interaction influences plant-feeding insects. Fungal Ecol. 29, 123–132. Available from: https://doi.org/10.1016/j.funeco.2017.06.004.

Ran, X., Zhang, G., Li, S., Wang, J., 2017. Characterization and antitumor activity of camptothecin from endophytic fungus *Fusarium solani* isolated from *Camptotheca acuminata*. Afr. Health Sci. 17, 566–574. Available from: https://doi.org/10.4314/ahs.v17i2.34.

Reddy, M.S., Murali, T.S., Suryanarayanan, T.S., Govinda Rajulu, M.B., Thirunavukkarasu, N., 2016. *Pestalotiopsis* species occur as generalist endophytes in trees of Western Ghats forests of southern India. Fungal Ecol. 24, 70–75. Available from: https://doi.org/10.1016/j.funeco.2016.09.002.

Rodrigues, K.F., 1994. The foliar fungal endophytes of the Amazonian palm *Euterpe oleracea*. Mycologia 86, 376–385.

Rodriguez, R.J., White Jr, J.F., Arnold, A.E., Redman, R.S., 2009. Fungal endophytes: diversity and functional roles. New Phytol. 182, 314–330. Available from: https://doi.org/10.1111/j.1469-8137.2009.02773.x.

Sachin, N., Manjunatha, B.L., Kumara, P.M., Ravikanth, G., Shweta, S., Suryanarayanan, T.S., et al., 2013. Do endophytic fungi possess pathway genes for plant secondary metabolites? 178–182.

Saikkonen, K., 2007. Forest structure and fungal endophytes. Fungal Biol. Rev. 21, 67–74.

Shweta, S., Zuehlke, S., Ramesha, B.T., Priti, V., Mohana Kumar, P., Ravikanth, G., et al., 2010. Endophytic fungal strains of *Fusarium solani*, from *Apodytes dimidiata* E. Mey. ex Arn (Icacinaceae) produce camptothecin, 10-hydroxycamptothecin and 9-

methoxycamptothecin. Phytochemistry 71, 117–122. Available from: https://doi.org/10.1016/j.phytochem.2009.09.030.

Shweta, S., Gurumurthy, B.R., Ravikanth, G., Ramanan, U.S., Shivanna, M.B., 2013. Endophytic fungi from *Miquelia dentata* Bedd., produce the anti-cancer alkaloid, camptothecine. Phytomedicine 20, 337–342. Available from: https://doi.org/10.1016/j.phymed.2012.11.015.

Singh, D.K., Sharma, V.K., Kumar, J., Mishra, A., Verma, S.K., Sieber, T.N., et al., 2017. Diversity of endophytic mycobiota of tropical tree *Tectonagrandis* Linn. f.: spatiotemporal and tissue type effects. Sci. Rep. 7 (1), 3745.

Su, H., Kang, J., Cao, J., Mo, L., Hyde, K.D., 2014. Medicinal plant endophytes produce analogous bioactive compounds. Chiang Mai J. Sci. 41, 1–13.

Strobel, G.A., 2003. Microb. Infect. 5, 535–544.

Strobel, G., Daisy, B., 2003. Bioprospecting for microbial endophytes and their natural products. Microbiol. Biol. Rev. 67, 491–502.

Strobel, G., Yang, X., Sears, J., Kramer, R., Sidhu, R.S., Hess, W., 1996. Taxol from *Pestalotiopsis microspora*, an endophytic fungus of *Taxus wallachiana*. Microbiology 142, 435–440. Available from: https://doi.org/10.1099/13500872-142-2-435.

Subban, K., Subramani, R., Madambakkam Srinivasan, V.P., Johnpaul, M., Chelliah, J., 2019. Salicylic acid as an effective elicitor for improved taxol production in endophytic fungus *Pestalotiopsis microspora*. PLoS One 14, 1–17. Available from: https://doi.org/10.1371/journal.pone.0212736.

Suryanarayanan, T.S., Govinda Rajulu, M.B., Thirumalai, E., Reddy, M.S., Money, N.P., 2011. Agni's fungi: heat-resistant spores from the Western Ghats, southern India. 833–838.

Suryanarayanan, T.S., Murali, T.S., Thirunavukkarasu, N., GovindaRajulu, M.B., Venkatesan, G., Suryanarayanan, T.S., et al., 2016. Biological control through fungal endophytes: gaps in knowledge hindering success. Curr. Biotechnol. Available from: https://doi.org/10.2174/22115501056661605041303322.

Stierle, A.A., Stierle, D.B., 2015. Bioactive secondary metabolites produced by the fungal endophytes of conifers. Nat. Prod. Commun. Available from: https://doi.org/10.1177/1934578 × 1501001012.

Treseder, K.K., Lennon, J.T., 2015. Fungal traits that drive ecosystem dynamics on land. Microbiol. Mol. Biol. Rev. 243–262.

Venugopalan, A., Potunuru, U.R., Dixit, M., Srivastava, S., 2016. Effect of fermentation parameters, elicitors and precursors on camptothecin production from the endophyte *Fusarium solani*. Bioresour. Technol. 206, 104–111. Available from: https://doi.org/10.1016/j.biortech.2016.01.079.

Verekar, S.A., Mishra, P.D., Sreekumar, E.S., Deshmukh, S.K., Fiebig, H.H., Kelter, G., et al., 2014. Anticancer activity of new depsipeptide compound isolated from an endophytic fungus. J. Antibiot. 67, 697–701.

Wang, Q.X., Li, S.F., Zhao, F., Dai, H.Q., Bao, L., Ding, R., et al., 2011. Chemical constituents from endophytic fungus *Fusarium oxysporum*. Fitoterapia 82, 777–781. Available from: https://doi.org/10.1016/j.fitote.2011.04.002.

Wang, H., Liu, T., Xin, Z., 2014. A new glucitol from an endophytic fungus *Fusarium equiseti* Salicorn 8. Eur. Food Res. Technol. 239, 365–376. Available from: https://doi.org/10.1007/s00217-014-2230-z.

Xiao, J., Hu, J.Y., Sun, H.D., Zhao, X., Zhong, W.T., Duan, D.Z., et al., 2018. Sinopestalotiollides A-D, cytotoxic diphenyl ether derivatives from plant endophytic fungus *Pestalotiopsis palmarum*. Bioorg. Med. Chem. Lett. 28 (3), 515–518.

Xie, S., Zhou, J., 2017. Harnessing plant biodiversity for the discovery of novel anticancer drugs targeting microtubules. Front. Plant Sci. 8, 720. Available from: https://doi.org/10.3389/fpls.2017.00720.

Xu, F., Tao, W., Cheng, L., Guo, L., 2006. Strain improvement and optimization of the media of taxol-producing fungus *Fusarium maire*. Biochem. Eng. J. 31, 67–73. Available from: https://doi.org/10.1016/j.bej.2006.05.024.

Xu, J., Kjer, J., Sendker, J., Wray, V., Guan, H., Edrada, R., et al., 2009. Cytosporones, coumarins, and an alkaloid from the endophytic fungus *Pestalotiopsis* sp. isolated from the Chinese mangrove plant *Rhizophora mucronata*. Bioorg. Med. Chem. 17, 7362–7367. Available from: https://doi.org/10.1016/j.bmc.2009.08.031.

Xu, M.F., Jia, O.Y., Wang, S.J., Zhu, Q., 2016. A new bioactive diterpenoid from Pestalotiopsis adusta, an endophytic fungus from *Clerodendrum canescens*. Nat. Prod. Res. 30 (23), 2642–2647.

Xu, F., Cao, H., Cui, X., Guo, H., Han, C., 2018. Optimization of fermentation condition for echinacoside yield improvement with *penicillium* sp. H1, an endophytic fungus isolated from *Ligustrum lucidum* ait using response surface methodology. Molecules 23, 1–11. Available from: https://doi.org/10.3390/molecules23102586.

Yang, X., Strobel, G., Stierle, A., Hess, W.M., Lee, J., Clardy, J., 1994. A fungal endophyte-tree relationship: *Phoma* sp. in *Taxus wallachiana*. Plant Sci. 102, 1–9. Available from: https://doi.org/10.1016/0168-9452(94)90017-5.

Chapter 2

Evaluation of phylloplane fungal flora and host plants in the Southern Western Ghats

Sabeena Aliyarukunju*, **Biju Haridas and Shiburaj Sugathan**
Microbiology Division, KSCSTE -Jawaharlal Nehru Tropical Botanic Garden and Research Institute, Thiruvananthapuram, India
**Corresponding author.* Email: asabeenarasheed@gmail.com

The diversity and galaxy of fungi and their natural beauty occupy prime place in the biological world and India has been the cradle for such fungi. Only a fraction of total fungal wealth has been subjected to scientific survey and mycologists have to unravel the unexplored and hidden wealth. One third of fungal diversity of the globe exists in India. Out of 2.2–3.8 million fungi, only 8% are characterized until now (Hawksworth and Lucking, 2017). Fungi are not only beautiful but play a significant role in the daily life of human beings besides their utilization in industry, agriculture, medicine, food industry, textiles, bioremediation, natural cycling, as biofertilizers and many other ways. Fungal biotechnology has become an integral part of the human welfare. Microbial sources such as fungi are well recognized to produce a wide variety of chemical structures, several of which are most valuable pharmaceuticals, agrochemicals and industrial products. The world of fungi provides an interesting and almost endless source of biological diversity, which is a rich source for exploitation (Manoharachary et al., 2005).

Fungi are becoming increasingly an important tool in cleaning the environment. The accumulation of pesticides and other chemicals in the environment is destroying many ecosystems and placing many animal and plant species at risk. Good number of fungi found in soil, *e.g. Aspergillus, Fusarium, Rhizopus, Mucor, Penicillium* and *Trichoderma* are used in bioremediation, in which they decompose the organic material and detoxify them.

Fungi differ in their morphology, ecology and life history strategies. The taxonomic work is crucial for understanding the biology in general. Without the work on taxonomy of the organisms in any given area, estimating the

Fungi Bio-Prospects in Sustainable Agriculture, Environment and Nano-Technology.
DOI: https://doi.org/10.1016/B978-0-12-821394-0.00002-0

amount of diversity present and making informed conservation decisions are unrealistic and impossible. As conservation becomes ever more politically important, taxonomic work impacts not only the scientific community, but also the society as a whole.

We know more about large, economically important plants and animals than we do about fungi and bacteria, despite their important ecological roles. Microorganisms have had a tremendous influence on the availability of nutrients in soil through their weathering of rocks and on the composition of the earth's atmosphere, especially the increase in greenhouse gases such as carbon dioxide, nitrous oxide and methane. The microbial diversity is getting degraded in the forest due to forest fragmentation, loss of host plants due to felling of trees, air pollution, use of fungicides, alteration in the microclimates, changes in the frequency of rainfall, droughts and over-exploitation of edible fungi. The forest fragmentation leads to the extinction of symbiotic microbes that are host-specific which have poor dispersal abilities. Each year, hundreds of species are being lost even before they are known to the scientific world, due to their habitat destruction.

The phylloplane (leaf surface) in a forest canopy is regularly exposed to intense sunlight, driving winds, and pouring rain. Fungi inhabiting the leaves are known as phylloplane or foliicolous fungi. They are defined specialized nutritional guilds found on the surface of living plant parts, particularly on leaves; including saprobes, plant parasites, fungal parasites and lichens. Plant diseases have got much importance because of their direct influence on mankind. These fungi influence the fitness of their host plants, either negatively, by acting as pathogens, or positively, by increasing the stress tolerance of the plant, reducing herbivory through the production of secondary metabolites or reducing the infection of plant tissues by pathogens. They are therefore important drivers of the dynamics and diversity of plant populations and communities. Finally, they contribute to nutrient cycling, as early colonizers of leaf litter.

One important part of microbial diversity is as phylloplane fungi, which includes rusts, smuts, powdery mildews, Black mildews, tar spots etc. Most of these are obligate or facultative parasites. Residing on the surface of the leaves and producing special organs and opting special adaptation, they are acting as necrotrophs or biotrophs and their infection may lead to the destruction of the plant. Black mildews are the ectophytic, dematiaceous, black colony forming fungi having ecto or endophytic brown or black mycelium with or without appressoria. The fungi which present appressoria produce haustorium into the epidermal cells of the host plants. The fungi which are devoid of appressoria produce nutritive hyphae into the host cells (Hosagoudar, 2008). Species of these fungi are occurring commonly in the tropical and subtropical regions of the world. They produce colonies, crusts or spots on the leaf surface, and are obligate parasites (biotrophs), that infect mostly leaves and other parts of the plants like stem, petiole and in some cases on fruits too. These fungi occur a month or two after the monsoon.

However, they occur scantily during summer and are abundant during the winter season. None causes enough damage to require control measures, although some species are known to produce dead areas on leaves corresponding in size to the superficial colony of the fungus.

The black colony forming organisms belong to different taxonomic groups, *viz.* Meliolaceous fungi, Schiffnerulaceous fungi, Asterinaceous fungi, Hyphomycetous fungi, etc. Meliolaceous fungi are distinct from others in having brown mycelium with bi-cellular appressoria, setae, perithecial appendages and consistent brown septate ascospores (Hansford, 1961; Hosagoudar, 1996; Hosagoudar and Agarwal, 2008; Hosagoudar, 2008). The fungi with thyriothecium may belong to Asterinales (external mycelium with thyriothecia, which dehisce stellately at the center or vertically all along the length), and Meliolales are having two-celled appressoria and unicellular phialides, *Schiffnerula* are like Asterinales but with unicellular appressoria, having different types of conidial anamorphic states, Hyphomycetes are with the presence of conidia and conidiophores. The potential of phylloplane fungi is not known much, from an application point of view; however, they constitute a fairly big group of micro-fungi. In the light of above, a study was designed to evaluate phylloplane fungi of Southern Western Ghats, that would further help in the conservation and management of biological resources and also increase the agricultural wealth of the nation.

The Indian subcontinent has rich biological diversity in the world owing to its vast geographic area, varied topography and climate; it is classified into 10 biogeographic regions. Among the 34 hotspots of the world, two are mainly found in India extending into neighboring countries - the Western Ghats/Sri Lanka and the eastern Himalaya. The Western Ghats comprises the mountain ranges that runs the ecosystems and are located mainly in the following regions: the tropical wet evergreen forests in Amboli and Radhanagari; the montane evergreen forests in Mahabaleshwar and Bhimashanker; moist deciduous forests in Mulsi and the scrub forest in Mundanthurai. There are two main centers of diversity, the Agashyamalai Hills and the Silent Valley (Hosagoudar, 2013). The complex topography and the heavy rainfall have made certain areas inaccessible and have helped the region to retain its diversity. There are currently seven national parks in the Western Ghats with a total area of 2073 km^2 (equivalent to 1.3% of the region) and 39 wildlife sanctuaries covering an area of about 13,862 km^2 (8.1%). Of about 1.7 million species globally described and recorded in scientific literature, India has about 1,26,200 species. It ranks tenth in the world both in respect of richness of flowering plants (17,000 species) and mammals (372 species). Of India's 49,219 plant species, 1600 endemics (40% of the total number of endemics) are found in a 17,000 km^2 strip of forest along the seaward side of the Western Ghats in Maharashtra, Karnataka, Tamil Nadu, and Kerala. The tropical forests of India especially the Western Ghats is supposed to be rich in microbial diversity. However, in such a rich and the diverse region of Western

Ghats, it is not possible to survey the entire Ghats at a time. Hence, working intensively on a group of microbes by the subject experts by taking smaller region will enable us to bring out the expected result at the end.

The forests in the Kerala part of the Western Ghats support a rich pathogenic fungal flora. The distribution and level of parasitism of various fungi in different forest ecosystems are governed mostly by the climatic and edaphic factors as well as host parasite interrelationships. The tropical climate of Kerala is very congenial for the luxuriant growth of fungi. There are several fungal biodiversity rich places like Neyyar Wildlife Sanctuary, Peppara Wildlife Sanctuary, Agasthyamala Biosphere Reserve, Shendhurny Wildlife Sanctuary, Wayanad Wildlife Sanctuary, Idukki Wildlife Sanctuary, Silent Valley National Park, Periyar Tiger Reserve, forest areas of Palode in Thiruvananthapuram district and Ponthanpuzha reserve in Kottayam district located in the Southern Western Ghats of Kerala.

Study of lower groups like fungi is largely incomplete and has not been taken-up systematically in the southern regions of India. There is a greater need for the documentation of these lower groups of fungi in our forests. Except the few sporadic accounts, a systematic survey of the microbes is yet to be set into. To fill this lacuna, a systematic survey of the phylloplane fungi of the Southern Western Ghats was initiated since 2003. However, none of the works are thorough to state the exact number of fungi from any part of the Western Ghats. One cannot be blamed because the diversity is such a vast that it needs a group or team of experts to study the individual Sanctuary, National park or any protected areas continuously. The forest fragmentation leads to the extinction of symbiotic microbes that are host-specific which have poor dispersal abilities. Since there is a degradation of forests every year, there is a habit and habitat loss resulted in the loss of enormous fungi before they are known to the scientific world, due to their habitat destruction. Hence, for every work, it needs a base to start and the present work starts in this direction in providing the list of fungi published in various floras under Western Ghats which contains mainly Phylloplane fungi of Southern Western Ghats.

Long Term Environmental and Ecological impacts on the multi-purpose River Valley projects of Idukki and Silent Valley were initiated with the assignment of the work on "Foliicolous Fungi of Idukki and Silent Valley Hydroelectric Project Area". Silent Valley National Park is important tropical virgin rain forests of India noted for its biodiversity and species richness. It is a much whispering place, located in Palghat District in Kerala State, believed to be more than fifty million years old (Swaminathan 1999), having an area of 8952 ha, located at an altitude ranging from 750 to 2383 m, receives more than 5000 mm rain fall annually, temperature ranges from 8 to 29 °C. It harbors evergreen forests. Manilal (1988) has given an account of 966 species belonging to 559 genera distributed among 134 families of flowering plants. An account of micro fungi from water and litter has been studied by Subramanian (1986). The systematic study of the foliicolous fungi of this virgin forest is scanty. This work is the consolidated account of the foliicolous fungi known from this area. The Phylloplane fungal collections made in the Silent valley National Park, Palghat, Kerala, since 1985, resulted in recording 139 fungal taxa belonging to 30 fungal genera: *Acremoniula* (1), *Acrodictys* (1), *Amazonia* (3), *Aphanopeltis* (1), *Armatella* (4), *Asteridiella* (12), *Asterina* (32), *Asterostomella* (2), *Asterostomula* (1), *Balladyna* (2), *Didymopsorella* (1), *Diplococcium* (1), *Dysrhynchis* (1), *Echidnodella* (1), *Endophyllum* (1), *Eupelte* (1), *Irenopsis* (1), *Leptosphaerulina* (1), *Meliola* (47), *Meliolina* (1), *Oidium* (1), *Palawaniella* (1), *Phakopsora* (1), *Prataprajella* (1), *Prillieuxina* (1), *Puccini* (1), *Sarcinella* (3), *Schiffnerula* (4), *Spiropes* (2) and *Teratosperma* (1), infected 138 plants belonging to 54 host families (Hosagoudar and Riju, 2013). This area is the type locality for several fungal taxa.

The study of Foliicolous Fungi of Idukki District resulted in recording 337 foliicolous fungal species and infraspecific taxa represent 73 fungal genera infected 289 host plants belonging to 228 genera of flowering plants (Hosagoudar et al., 2011a). All these fungi are grouped under Ascomycetes, Basidiomycetes and Fungi Imperfecti.

Among Ascomycetes, unitunicatae comprises: Erysiphales, Meliolales, Phyllachorales and clavicipitales. Erysiphales represent three genera with 4-species (*Cystotheca*-1, *Oidium*-2 and *Sphaerotheca*-1). Meliolales represent eight genera with 164 species and infraspecific taxa. This order is representing two families, namely, Armatellaceae and Meliolaceae. Armatellaceae represent two genera, namely, *Basavamyces* (1-species) and *Armatella* (9-species). Meliolaceae represents five genera: *Amazonia* (9-species), *Appendiculella* (4-species), *Asteridiella* (16-species), *Irenopsis* (4-species), *Meliola* (119-sepcies) and *Prataprajella* (2-species). Phyllachorales represent four genera, *Ophiodothella* (1-species), *Phyllachora* (34-species), *Rehmidothis* (1-species) and *Sphaerodothis* (1-species) and Clavicipitales represent one genus (1-species). Of the bitunicatae, Asterinales comprises two families, namely Asterinaceae represents the form genus *Asterostomella* (3-species), with the teleomorphs: *Asterina* (40-species), *Ishwaramyces* (1-species) and *Prillieuxina* (1-species); Lembosiaceae represents two genera. *Eupelte* (1-species) and *Lembosia* (3-species). Dothideales represent two families, namely, Dothideaceae with two genera: *Guignardia* (1-species) and *Mycosphaerella* (1-species).

Among Basidiomycetes, Uredinales comprises seventeen genera, namely, *Cerotelium* (1-species), *Cronartium* (1-species), *Dasturella* (1-species), *Didymopsorella* (1-species), *Endophyllum* (1-species), *Hamaspora* (1-species), *Hemileia* (2-species), *Kernkempella* (1-species), *Kuehneola* (1-species), *Kulkarniella* (1-species), *Maravalia* (1-species), *Masseeella* (1-species), *Phakopsora* (2-species), *Puccinia* (16-species), *Ravenelia* (2-species), *Uromyces* (5-species) and *Zaghouania*(1-species). In addition to these, three are the form genera, namely, *Aecidium* (4-species), *Caeoma* (1-species) and *Uredinia* (4-species). Ustilaginales are with four genera: *Cintractia* (2-species), *Sorosporium* (1-species) *Neovossia* (1-species) and *Tillettia* (1-species). Heterobasidiomycetes: Exobasidiales with a genus *Exobasidium* (2-species).

Fungi Imperfecti represent Hyphomycetees and Sphaeropsidales. Former gives an account of twelve genera: *Asperisporium* (1-species), *Cercospora* (5-species), *Cordana* (1-species), *Hadronema* (1-species), *Mycovellosiella* (1-species), *Passalora* (1-species), *Periconiella* (1-species), *Pseudocercospora* (1-species), *Questieriella* (3-species), *Sirosporium* (1-species), *Sporidesmium* (1-species) and *Verrucispora* (1-species). Sphaeropsidales represent: *Ephelis* (1-species), *Phaeoseptoria* (1-species), *Phleospora* (1-species), *Phyllosticta* (3-species) and *Phyllostictina* (1-species).

In addition to these, there are seven fungi which are not identified up to species level and four are hyperparasites or fungicolous fungi.

The present study area forms a type locality for 119 fungi (35%). An intensive study of individual forest may reveal enormous undescribed species.

Kodaikanal, the hill station in Tamil Nadu, located on the Southern ridge of the Palani hills in the Western Ghats. The Study of Foliicolous fungal flora of Kodaikanal region comprises 600 foliicolous fungal collections resulted in recording 101 fungal species belonging to 22 Genera, viz. *Amazonia* (1), *Appendiculella* (3), *Armatella* (4), *Asteridiella* (4), *Asterina* (24), *Balladyna* (2), *Erysiphe* (5), *Irenopsis* (1), *Leveillula* (1), *Meliola* (29), *Meliolina* (2), *Prataprajella* (2), *Phyllachora* (1), *Prillieuxina* (2) and *Uncinula* (3) belong to Ascomycetes; *Gambleola* and *Kernella* to Uredinales, *Asterostomella* (1), *Oidium* (10), *Ovulariopsis* (1), *Questieriella* (2) and *Trochophora* to Fungi Imperfecti (Hosagoudar et al., 2012). The area harboured as type locality for many fungal species and some are endemic to the region

The extensive and systematic field collection trips made to the Palode forest range of Thiruvananthapuram district resulted in an examination of 953 foliicolous fungal collections made from the study area. This study resulted in recording of 240 foliicolous fungal species and infraspecific taxa belonging to 46 fungal genera (Hosagoudar and Archana, 2012). All these fungi are grouped under Ascomycetes, Basidiomycetes and Fungi imperfecti.

Among Ascomycetes, the unitunicatae represent Armatellaceae, Meliolaceae, Englerulaceae, Erysiphaceae, Meliolinaceae and Phyllachoraceae. Armatellaceae represents one genus, *Armatella* (3). Meliolaceae represents five genera: *Amazonia* (3), *Appendiculella* (1), *Asteridiella* (9), *Irenopsis* (5) and *Meliola* (117). Englerulaceae represents two genera *Schiffnerula* (7) and *Questieriella* (1). Erysiphaceae represents two genera: *Microsphaera* (1) and *Oidium* (2). Meliolinaceae represent one genus, *Meliolina* (1) and Phyllachoraceae represent one genus, *Phyllachora* (5). Of the bitunicatae, Asterinaceae represents eight genera: *Asterina* (31), *Asterostomella* (4) *Asterolibertia* (1), *Prillieuxina* (5), *Meliolaster* (1), *Ishwaramyces* (1), *Symphaster* (1) and *Trichasterina* (1). Lembosiaceae represents four genera: *Cirsosia* (1), *Echidnodella* (3), *Lembosia* (7) and *Maheshwaramyces* (1). Magnaporthaceae represents one genus, *Clasterosphaeria* (1). Parmulariaceae represents one genus, *Ferrarisia* (1). Parodiellaceae represents two genera, *Balladynocallia* (1) and *Balladynopsis* (1).

Among Basidiomycetes, Uredinales comprises five genera, namely, *Aecidium* (1), *Coleosporium* (1), *Dasturella* (1), *Puccinia* (2) and *Zaghouania* (1).

Fungi imperfecti represents only Hyphomycetes. Parasites represent seven genera: *Asperisporium* (1), *Ampullifera* (1), *Colemaniella* (1), *Diplococcium* (1), *Tretospora* (1), *Clasterosporium* (1) and *Periconiella* (1). Hyperparasites represent seven genera: *Acremoniula* (1), *Acrodictys* (1), *Domingoella* (1), *Hansfordiellopsis* (1), *Spiropes* (5), *Teratosperma* (1) and *Trichothyrium* (1). In addition to these, there are seven fungi which are not identified up to species level.

Foliicolous fungal flora of Peppara and Neyyar Wildlife Sanctuaries in Kerala is very apt and forms a baseline data for the region and provides a first-hand information on the host-parasite relation, occurrence of the fungal parasites at different altitudes and climates. Peppara and Neyyar are the twin and adjacent Wildlife Sanctuaries, include the hot spot area of Agasthyamala, located in the southern tip and western slope of southern Western Ghats in Thiruvananthapuram district of Kerala State. These sanctuaries lie between 70° 4′ and 77° 17′ east and 8° 7′ and 8° 53′ north, surrounded by Kalakkad and Mundandurai Wildlife Sanctuaries in the east, Palode and Paruthipalli forest range in the north. Both Sanctuaries together have an area of 181 sq. km., with an altitudinal range from 100 to 1864 m, temperature from 16 to 35 °C, annual rain fall about 2800 mm. Agastiar peak (1864 m), Pongalapara (1500 m) and Chemunji (1000 m) are the continuous hill ranges, located towards the eastern side of the study area and steeply descend towards western side. It is drained with Neyyar and Karamanayar perennial rivers by supporting rich riparian vegetation. Since Chemunji is the continuous hill range of Pongalapara, it has been included in the study. The result of the survey work carried out in Peppara and Neyyar Wildlife Sanctuaries, Thiruvananthapuram district of Kerala State during the years from 1996 to 1998. This study resulted in recording 177 taxa distributed among 38 fungal genera (Hosagoudar, 2010).

The study covering the all regions of Peppara and Neyyar Wildlife sanctuaries of Thiruvananthapuram district of Kerala state shows that these regions are a good abode for the Meliolales represented by 7 genera and 160 species (Hosagoudar and Jacob, 2013). Of the 160 species, the genus *Meliola*represents 117 species, the genus *Appendiculella*and *Prataprajella* – 1, *Armatella*– 5, *Asteridiella* – 20, *Amazonia* and *Irenopsis* represents 8 taxa; 97 are endemic (61%) to Southern Western Ghats of Kerala State. The present study area forms the type locality for 25 species and 2 varieties. This reveals that the tropical forests in Kerala are the treasure for these fungi.

The flora of Wayanad are characteristic of the Western Ghats and the plantation crops grown in the cool climate. With the clearing of forests, the diverse and bustling animal life, characteristic of the forests of Western Ghats, has vanished from Wayanad. This district comprises: west coast tropical semi-evergreen forests, southern moist mixed deciduous forest, southern dry mixed deciduous forests and moist bamboo brakes and this rich forest has been protected in the form of Wayanad Wildlife Sanctuary, harboring more than 2000 flowering plant species and it formed the first of its kind for that area. The study gives an account of 580 foliicolous fungal samples from 220 host plants belonging to 70 families of flowering plants from Wayanad District resulted in recording 175 fungal species distributed among 29 fungal genera, belonging to black mildews (Hosagoudar and Sabeena, 2014).

The study, Asterinales of Kerala is the result of an examination of 392 collections (plant parts infected with fungi) belonging to 152 species of flowering plants, distributed among 70 families. This resulted in recording 160 fungal species and infra-specific taxa belonging to 17 genera distributed among two families of the order Asterinales (Hosagoudar et al., 2011b).

The Kodagu (Coorg) district in Karnataka has a mountainous configuration with varied physical features. The main part of the district is composed of mighty ranges of Western Ghats which run parallel to the west coast, rise almost abruptly from the plains in the west and gradually merge into table land of Mysore District. The Western Ghats ranges in this district are more or less crescent shaped and include some of the loftiest peaks, between Himalaya and the Nilgiris. It stretches to about 97 km from Pushpagiri in north-west to Bramhagiri ranges in the south. Kodagu, a tiny district has congenial climate for the luxuriant growth of fungi. This area has high potentiality for the fungal study and other aspects of fungi. Kodagu District exhibits luxuriant vegetation, which harbors, a variety of foliicolous fungi of which Black mildews form the bulk. The tropical climatic conditions prevailing in this hilly district of the Western Ghats, is congenial for the sustenance of a wide range of Black mildews. The vegetation of the study area is hub of fungal activity. Continued explorations could result in recording a number of novel varieties of Black mildews and other foliicolous fungi. Sacred grooves are present in most villages of the Madikeri taluk which are bioreserves and have helped to protect and conserve natural vegetation and are very good source of microbial diversity and floristic composition. All these factors together make the study area as one of the potential locations to understand the rich variety of Black mildews and other phylloplane fungi. The study has resulted in recording 183 black mildews, out of which 24 are new species and 78 are reported for the first time from Karnataka. Meliolales is represented by 3 sp. of *Amazonia*, 3 sp. of *Armatella*, 5 sp. of *Asteridiella* 6 sp. of *Irenopsis* and 26 sp. of *Meliola*; in Asterinaceae 21 sp. of *Asterina*, 1 sp. of *Asterolibertia*, 1 sp. of *Ishwaramyces*, 1 sp. of *Meliolaster* and 1 sp. of *Prillieuxina*; in Lembosiaceae 2 sp. of *Cirsosia*, 1 sp. of *Echidnodella*, 1 sp. of *Eupelte*, and 1 sp. of *Lembosia;* in Anamorphic genera 1 sp. of *Asterostomella*, 1 sp. of *Mahantheshamyces*; in Schiffnerulaceous fungi 2 sp. of *Questieriella*, 1 sp. of *Schiffnerula* have been recorded as first report to Karnataka (Thimmaiah et al., 2013).

The extensive and systematic field collection trips made to the Ponthanpuzha reserve forest of Kottayam district resulted 546 foliicolous fungal collections, resulted in recording of 152 foliicolous fungal species and infraspecific taxa belonging to 22 fungal genera (Robin and Hosagoudar, 2010). All these fungi are grouped under Ascomycetes and Fungi imperfecti.

Among Ascomycetes, unitunicatae represent Armatellaceae, Meliolaceae, Englerulaceae and Phyllachoraceae. Armatellaceae represents one genus, *Armatella* (3). Meliolaceae represents Six genera: *Amazonia* (2), *Appendiculella* (2), *Asteridiella* (8), *Irenopsis* (6) and *Meliola* (86) *Prataprajella* (1). Englerulaceae represent the genus *Schiffnerula* (1) and Phyllachoraceae represents one genus, *Phyllachora* (2). Of the bitunicatae, Asterinaceae represents four genera: *Asterina* (24), *Prillieuxina* (2), *Meliolaster* (1), *Ishwaramyces* (1). Lembosiaceae represents three genera: *Cirsosia* (1), *Lembosia* (2) and *Maheshwaramyces* (1). Parodiellaceae represents two genera, *Balladynocallia* (1) and *Dysrhynchis* (1).

Fungi imperfecti represents only Hyphomycetes. Parasites represent two genera: *Colemaniella* (1) and *Sarcinella* (2). Hyperparasites represents two genera: *Spiropes* (2), *Teratosperma* (1).

The present study area forms the type locality for *Appendiculella elaeocarpicola*, *Meliolaster culicola*, *M. lophopetaligena*, and *M. erumeliensis* and a variety *Meliola gouaniae* Hansf. var. *keralica*. In addition to this, *Meliola garugae* Stev. & Rold. are reported for the first time from India.

The extensive and systematic field collection trips conducted to various places in Thiruvananthapuram district of Kerala State, resulted in recording of 253 foliicolous fungal species and infraspecific taxa belonging to 44 fungal genera (Sabeena et al., 2018). All these fungi are grouped under Ascomycetes, Basidiomycetes and Fungi imperfecti. Among Ascomycetes, the unitunicatae represent Armatellaceae, Meliolaceae, Erysiphaceae, Meliolinaceae and Phyllachoraceae. Armatellaceae represents one genus, *Armatella* (2). Meliolaceae represents five genera: *Amazonia* (2), *Appendiculella* (1), *Asteridiella* (16), *Irenopsis* (9) and *Meliola* (108); Erysiphaceae represents two genera: *Leveillula* (1) and *Oidium* (3). Meliolinaceae represents one genus, *Meliolina* (1) and Phyllachoraceae represents three genus, *Phyllachora* (3), *Ophiodothella* (1) and *Sphaerodothis* (1). Of the bitunicatae, Asterinaceae represents nine genera: *Asterina* (44), *Asterostomella* (4), *Asterostomula* (1) *Asterolibertia* (2), *Prillieuxina* (4), *Meliolaster* (1), *Bheemamyces* (1), *Symphaster* (1), and *Viegasia* (1); Lembosiaceae represents four genera: *Cirsosia* (1), *Echidnodella* (3), *Lembosia* (5) and *Maheshwaramyces* (1). Parmulariaceae represents one genus, *Ferrarisia* (1). Parodiellaceae represents one genera, *Balladynocallia* (1) and Schiffnerulaceae represents three genera *Schiffnerula* (13), Sarcinella (3) and *Questieriella* (1).

Among Basidiomycetes, Uredinales comprises four genera, namely, *Aecidium* (1), *Coleosporium* (1), *Puccinia* (2) and *Zaghouania* (1).

Fungi imperfecti represents only Hyphomycetes. Parasites represent four genera: *Clasterosporium* (2), *Colemaniella* (1), *Diplococcium* (1) and *Periconiella* (1). Hyperparasites represent six genera: *Acremoniula* (1), *Acroconidiellina* (1) *Hansfordiellopsis* (1), *Heteroconium* (1), *Spiropes* (2), and *Teratosperma* (1) (Figs. 2.1–2.3).

FIGURE 2.1 Phylloplane fungal infection on various host plants. (1) ***Meliola gliricidiicola*** Hosag & Agarwal on *Gliricidia sepium*; (2) ***Meliola altissimae*** Hosag. on *Vitex altissima*; (3) ***Meliola anceps*** Sydow & Sydow on *Mussaenda philippica*; (4) ***Meliola jasmini*** Hansf. & Stev. on *Jasminum auriculatum*; (5) ***Meliola pepparaensis*** Hosag. & Abraham on *Tabernaemontana alternifolia L.*; (6) ***Meliola spigiliae*** Hansf. on *Strychnos nux-vomica*.

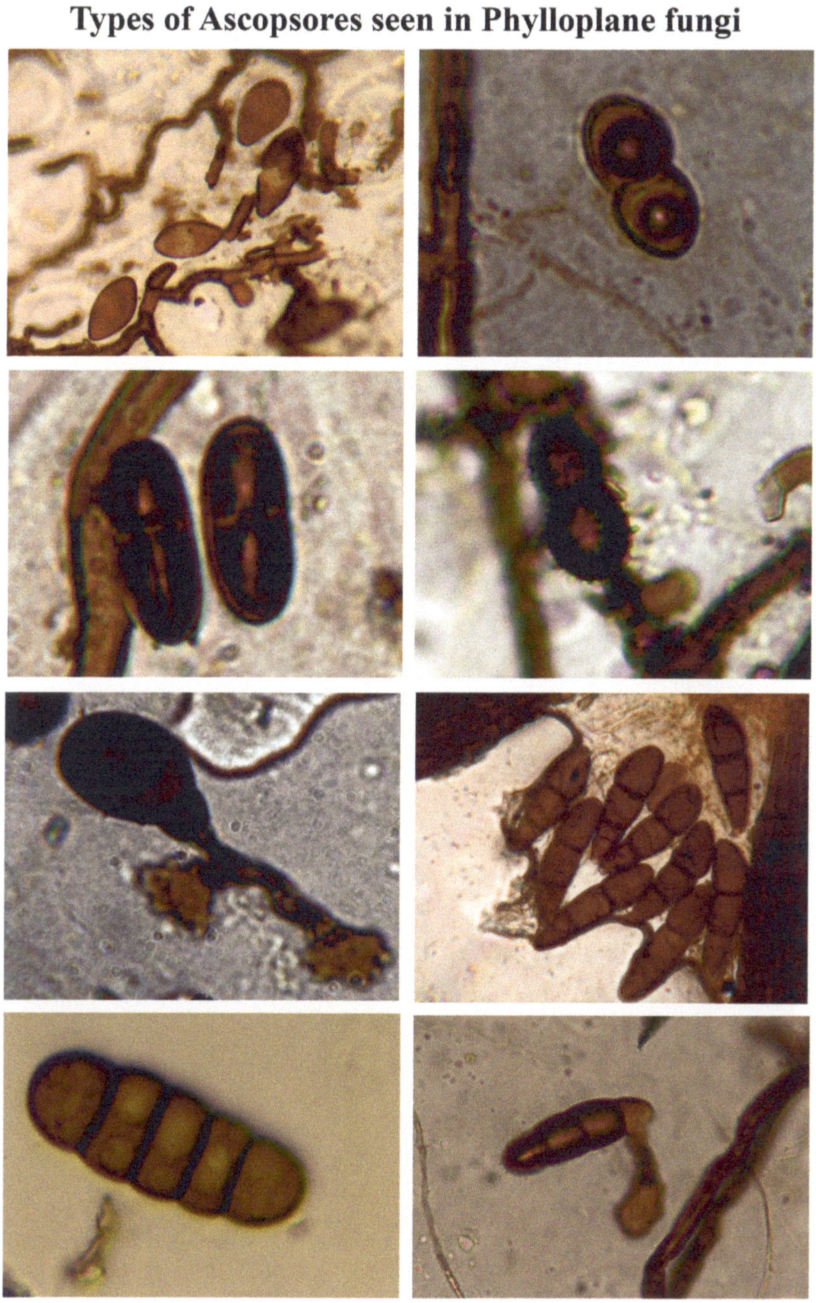

FIGURE 2.2 Types of Ascospores seen in Phylloplane fungi.

Illustrations of different Phylloplane fungi

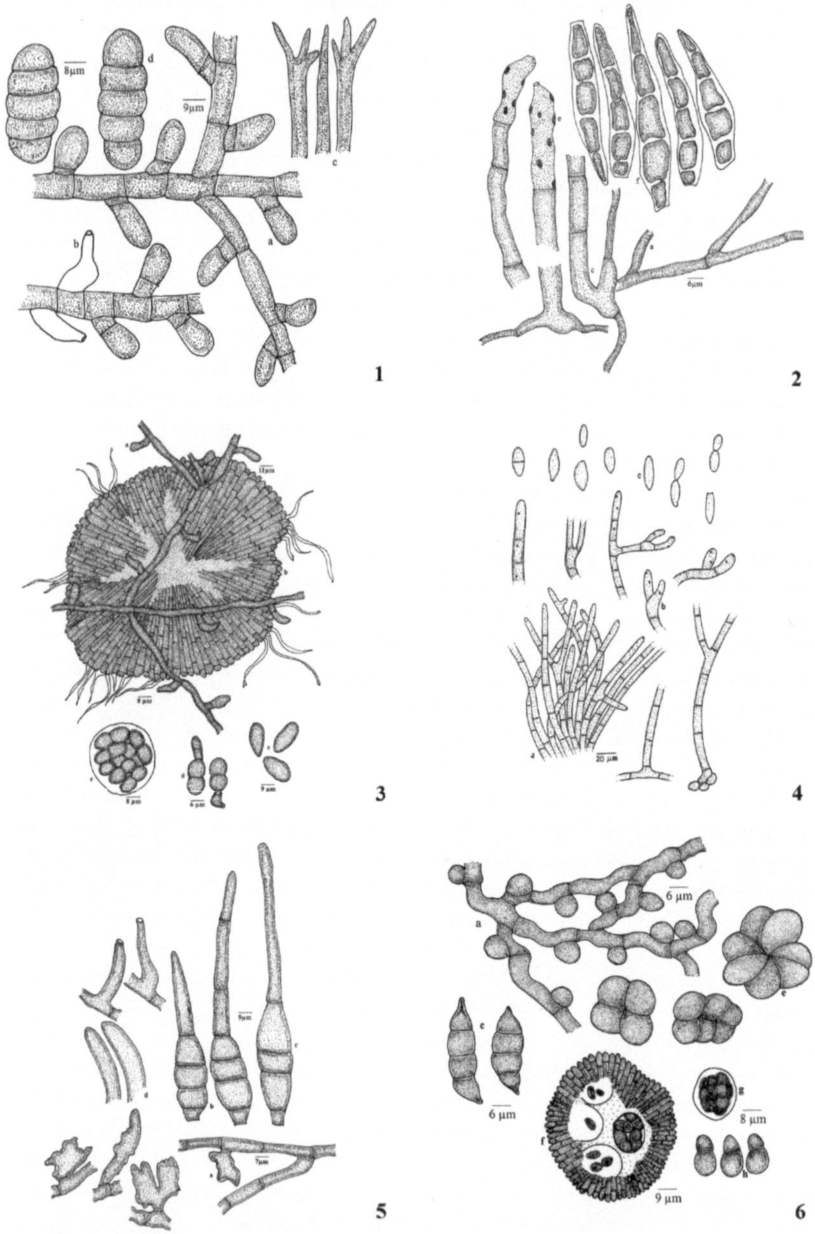

FIGURE 2.3 Illustrations of different phylloplane fungi (1) ***Meliola butleri***, Sydow (Meliolaceae); (2) ***Spiropes capensis*** M. B. Ellis (Hyphomycetes); (3) ***Asterina combreti*** Sydow (Asterinaceae); (4) & (5) ***Diplococcium atrovelutinum*** Brown & Hosag. ***Clasterosporium cyperacearum*** Hosag. (Hyphomycetes); (6) ***Schiffnerula palodensis*** Hosag. & Riju (Schiffnerulaceae).

TABLE 2.1 Phylloplane fungi reported from the Southern Western Ghats.

Fungi	Host plant	Family of the host plant	Locality
Asteridiella diotacanthi	*Diotacanthus* sp	Acanthaceae	Mannavan Shola, Munnar, Kerala
Asteridiella justiciae	*Justicia betonica* var. *betonica*		Bonacaud, Thiruvananthapuram, Kerala
Asteridiella strobilanthicola	*Strobilanthus* sp.		Mannavan Shola, Munnar, Kerala
Asteridiella acanthacearum	Acanthaceae member, *Lepidagathis* sp.		Jodupal, Kodagu, Karnataka
Asterina tertia	*Asystasia* sp. *Justicia adhatoda* *Justicia betonica* *Rhinacanthus nasutus* *Rhinacanthus* sp. *Ruellia prostrata*		Neyyar Wild Life Sanctuary JNTBGRI Campus, Palode, Thiruvananthapuram Dist., Kerala, Kerala
Asterina tertia var. *africana*	Acanthaceae member		Mannavan Shola, Munnar, Kerala
Meliola acanthacearum var. *occidentalis*	*Rungia sisparensis* *Strobilanthes* sp. Acanthaceae member		Kanchiar forest, Nalpathanchu Shola, Munnar Madikari, Kodagu, Karnataka
Meliola nilgirianthi	*Nilgirianthus heyneanus* *Strobilanthes* sp.		Lakshmi Estate, Idukki, Kerala Kuthiraiyar, Kukkal shola forest, Dindugal district, Tamil Nadu.

(*Continued*)

TABLE 2.1 (Continued)

Fungi	Host plant	Family of the host plant	Locality
Meliola strobilanthicola	*Strobilanthus amabilis* *Strobilanthus* sp.,		Mannavan Shola, Munnar, Idukki, Kerala Nalpathanchushola, Munnar, Idukki, Kerala
Puccinia thwaitesii	*Justicia gendarussa*		JNTBGRI campus, Palode, Thiruvananthapuram Dist., Kerala
Asterolibertia hydnocarpi	*Hydnocarpus macrocarpus* *Hydnocarpus pentandrus*	Achariaceae	Peppara wildlife Sanctuary Meenmutty, Thiruvananthapuram Dist., Kerala Kombe. Thiruvananthapuram Dist., Kerala
Asterdiella hydnocarpigena	*Hydnocarpus pentendra*		Madikeri, Karnataka
Asterolibertia mangiferae	*Mangifera indica*	Anacardiaceae	JNTBGRI Campus, Palode, Thiruvananthapuram Dist., Kerala
Asterina nothopegiae	*Nothopegia racemose*		Mandrane, Hoddur, Karnataka Devarakadu, Hoddur, Karnataka
Kuehneola spondiadis	*Spondias pinnata*		Kanchiar forest, Idukki, Kerala
Meliola ardigoosii	*Buchanania cochinchinensis* *Buchanania lanceolata*		Peppara Wildlife Sanctuary, Thiruvananthapuram Dist., Kerala Nanniyode, Thiruvananthapuram Dist., Kerala
Meliola buchananiicola	*Buchanania lanzan*		Calvary Mount, Idukki, Kerala
Meliola geniculate	*Lannea coromandelica*		Peppara Wildlife Sanctuary & JNTBGRI campus, Palode, Thiruvananthapuram Dist., Kerala

Meliola holigarnae	*Holigarna arnottiana* *Holigarna beddomei* *Holigarna* sp.		JNTBGRI Campus, Palode, Thiruvananthapuram Dist., Kerala
Meliola mangiferae	*Mangifera indica*		JNTBGRI Campus, Palode, Thiruvananthapuram Dist., Kerala Manja, Nedumangad, Thiruvananthapuram Dist., Kerala Kallara, Thiruvananthapuram Dist., Kerala Peppara Dam, Peppara Wildlife Sanctuary, Thiruvananthapuram Dist., Kerala
Meliola nothopegiae Hansf.	*Nothopegia racemosa*		Devarakadu, Hoddur, Karnataka
Meliola semecarpi-anacardii	*Semecarpus* sp.		JNTBGRI Campus, Palode, Thiruvananthapuram Dist., Kerala Nedumangad, Thiruvananthapuram Dist., Kerala
Meliola shivarajui	*Holigarna arnottiana* *Semecarpus* sp.		Ponmudi, Thiruvananthapuram Dist., Kerala
Phyllachora thiruvananthapurica	*Semecarpus travancorica*		forest near Bonacaud, Thiruvananthapuram Dist., Kerala
Teratosperma anacardii	*Buchanania barberi*		JNTBGRI campus, Palode, Thiruvananthapuram Dist., Kerala
Meliola ancistrocladi	*Ancistracladus heyneanus*	Ancistrocladaceae	forest along the road from Painavu to Kulamavu, Idukki, Kerala
Meliola angiopteridis	*Angiopteris evecta*	Angiopteridaceae	Bagamandala, Talacauveri, Karnataka
Echidnodella polyalthiae	*Polyalthia longifolia* *Polyalthia* sp.	Annonaceae	Karyavattom campus, Thiruvananthapuram Dist., Kerala
Meliola goniothalamigena	*Goniothalamus cardiopetalus*		Medikari, Hoddur, Karnataka

(*Continued*)

TABLE 2.1 (Continued)

Fungi	Host plant	Family of the host plant	Locality
Phyllachora travancorica	*Polyalthia longifolia*		Museum, Govt. Botanic Garden, Thiruvananthapuram Dist., Kerala Karyavattom, Thiruvananthapuram Dist., Kerala JNTBGRI campus, Palode, Thiruvananthapuram Dist., Kerala
Prillieuxina polyalthiae	*Polyalthia longifolia*		Neyyar Wildlife Sanctuary JNTBGRI campus, Palode, Thiruvananthapuram Dist., Kerala Kallara, Thiruvananthapuram Dist., Kerala Kariyavattom campus, Thiruvananthapuram Dist., Kerala
Asterina toxocarpi	*Toxocarpus* sp.	Apocynaceae	Mannavan Shola, Munnar, Idukki, Kerala
Asterina tylophorae-indicae	*Tylophora indica*		Mannavan shola, Munnar, Idukki, Kerala
Asterina wrightiae	*Wrightia tinctoria*		Peppara Wildlife Sanctuary & JNTBGRI Campus, Palode, Thiruvananthapuram Dist., Kerala
Coleosporium plumeriae	*Plumeria alba* *Plumeria rubra* *Plumeria* sp.		JNTBGRI Campus, Palode, Thiruvananthapuram Dist., Kerala Palayam, Thiruvananthapuram Dist., Kerala
Hemileia wrightiae	*Wrightia tinctoria*		Cheruthoni, Idukki, Kerala
Meliola alstoniae	*Alstonia scholaris*		Lakshmi Estate, Idukki, Kerala
Meliola cariasae var. *indica*	*Carissa carandas*		Forest along the road from Painavu to Kulamavu, Idukki, Kerala

Meliola carissae var. *spinari*	*Carissa carandas*		Kallara, Thiruvananthapuram Dist., Kerala
Meliola ervatamiae	*Tabernaemontana heyneana*		Bhagamandala, Karnataka
Meliola frutescentis	*Ichnocarpus frutescens*		Kombe, Neyyar Wildlife Sanctuary, Thiruvananthapuram Dist., Kerala
Meliola gymnemae	*Gymnema sylvestre* *Gymnema* sp.		JNTBGRI Campus, Karimancode, and Mylamoodu, Palode, Thiruvananthapuram Dist., Kerala Periyakanal shola, Kodaikanal, Tamil Nadu
Meliola gymdesmicola	*Hemidesmus indicus* *Hemidesmus* sp.		JNTBGRI Campus, Palode, Thiruvananthapuram Dist., Kerala
Meliola holarrhenae	*Holarrhena pubescens*		JNTBGRI Campus, Palode, Thiruvananthapuram Dist., Kerala
Meliola ichnocarpi-volubili	*Ichnocarpus frutescense* *Ichnocarpus* sp.		JNTBGRI Campus, Palode, Thiruvananthapuram Dist., Kerala Bonacaud Estate, Thiruvananthapuram Dist., Kerala Palode forest, Palode, Peppara Wildlife Sanctuary
Meliola integripoda	*Carissa carandas*		JNTBGRI Campus, Palode, Thiruvananthapuram Dist., Kerala Eanikara, Karakulam, Thiruvananthapuram Dist., Kerala
Meliola pepparaensis	*Tabernaemontana divaricata* *Tabernaemontana alternifolia* *Tabernaemontana* sp.		Peppara Dam, Peppara Wildlife Sanctuary, Kottoor Reserve Forests, Peppara Wildlife Sanctuary, Thiruvananthapuram Dist., Kerala JNTBGRI Campus, Palode, Thiruvananthapuram Dist., Kerala
Meliola peringamalaensis	*Alstonia scholaris*		Peringamala, Palode, Thiruvananthapuram Dist., Kerala Nanniyode, Thiruvananthapuram Dist., Kerala

(*Continued*)

TABLE 2.1 (Continued)

Fungi	Host plant	Family of the host plant	Locality
Meliola tabernaemontanae var. *wrightiae*	*Wrightia tinctoria*		Chippanchira, Palode, Thiruvananthapuram Dist., Kerala
Meliola toxocarpi	*Toxocarpus* sp.		Nishane motta, Kudagu, Karnataka
Meliola tylophorae	*Tylophora capparidifolia*		Calvary Mount, Idukki, Kerala
Meliola tylophorae-indicae	*Tylophora indica* *Tylophora* sp.		Forest near Kushavoor, Palode, Thiruvananthapuram Dist., Kerala Hoddur, Karnataka
Meliola vallaridis	*Vallaris solanacea*		JNTBGRI campus, Palode & Neyyar Wildlife Sanctuary, Thiruvananthapuram Dist., Kerala
Schiffnerula catharanthi	*Catharanthus roseus*		Karimancode, Thiruvananthapuram Dist., Kerala JNTBGRI Campus, Palode, Thiruvananthapuram Dist., Kerala
Lembosia malabarensis	*Pothos scandens*	Araceae	JNTBGRI campus, Palode, Thiruvananthapuram Dist., Kerala Attayar, Thiruvananthapuram Dist., Kerala
Meliola dichotoma var. *kusanoi*	*Schefflera venulosa*	Araliaceae	Vanachalu, Kodagu
Acroconidiellina arecae	*Areca catechu*	Arecaceae	Mylamood, Thiruvananthapuram Dist., Kerala
Asterina arecacearum	*Calamus* sp.		Munnar, Idukki, Kerala

Asterina aristolochiae	*Aristolochia tagala*	Aristolochiaceae	Peppara Wildlife Sanctuary, Thiruvananthapuram Dist., Kerala
Asterina thotteae	*Thottea ponmudiana* *Thottea siliquosa* *Thottea* sp.		Near Peppara dam, Thiruvananthapuram Dist., Kerala Attayar, Thiruvananthapuram Dist., Kerala JNTBGRI campus, Palode, Thiruvananthapuram Dist., Kerala Ponmudi, Thiruvananthapuram Dist., Kerala Jodupal, Kodagu
Schiffnerula aristolochiae	*Aristolochia tagala*		Devara kadu, Hoddur, Karnataka
Asterina goosii	*Mahonia leschenaultia*	Berberidaceae	Mannavan Shola, Idukki, Kerala
Asterostomella radermacherae	*Radermachera xylocarpa*		Forest between Painavu and Kulamavu, Idukki, Kerala
Gambleola cornuta	*Mahonia leschenaultii*		Then Shola, Kodaikanal, Madurai, Tamil Nadu
Oidium betulacearum	*Alnus nepalensis*	Betulaceae	Bryant Park, Kodaikanal, Tamil Nadu
Meliola crescentiae	*Oroxylum indicum*	Bignoniaceae	near Peppara Dam, Peppara Wildlife Sanctuary, Thiruvananthapuram Dist., Kerala
Oidium hiratae	*Tecomaria capensis*		Bryant Park, Kodaikanal, Tamil Nadu
Phyllachora sikkimensis	*Canarium strictum*	Burseraceae	Manchur, Kodaikanal, Tamil Nadu
Uncinula garugae	*Garuga pinnata*		Kodaikanal, Tamil Nadu
Questieriella sarcococcae	*Sarcococca coriacea*	Buxaceae	Periyakanal Shola, Kodaikanal, Tamil Nadu
Asterina dallasica	*Trema orientalis*	Cannabaceae	Peppara Wildlife Sanctuary, Thiruvananthapuram Dist., Kerala

(*Continued*)

TABLE 2.1 (Continued)

Fungi	Host plant	Family of the host plant	Locality
			Bonacaud, Thiruvananthapuram Dist., Kerala
Asteridiella tremae	*Trema orientalis*		JNTBGRI campus, Palode, Thiruvananthapuram Dist., Kerala
Oidium udaiyanii	*Trema orientalis*		Kombaikadu, Kodaikanal, Tamil Nadu
Schiffnerula hughesii	*Trema orientaslis* *Trema* sp.		JNTBGRI campus, Palode, Thiruvananthapuram Dist. Kerala
Asteridiella capparidigena	Capparaceae member		Nishanemotta, Madikeri, Karnataka
Asterina emciciana	*Capparis sepiaria*	Capparaceae	JNTBGRI campus, Palode, Thiruvananthapuram Dist. Kerala
Meliola goosii	*Viburnum punctatum*	Caprifoliaceae	Mandrane, Hoddur, Karnataka
Meliola aequatoriensis	*Viburnum punctatum*		Samikanal, Kukkal shola forest, Kodaikanal, Tamil Nadu
Cercospora papaya	*Carica papaya*	Caricaceae	Vazhathope, Idukki, Kerala
Bheemamyces capparidis	*Capparis* sp.		JNTBGRI campus, Palode, Thiruvananthapuram Dist. Kerala
Asterina microtropidis	*Microtropis ovalifolia*	Celastraceae	Painavu, Idukki, Kerala
Asterina cassinecola	*Cassine paniculata*		Igguthappa Temple, Kudagu, Karnataka
Schiffnerula celastri	*Celastrus paniculatus*		Tadiandamol hills, Kudagu, Karnataka
Appendiculella calophylli var. *apetali*	*Calophyllum apetalum*	Cluciaceae	JNTBGRI campus, Palode, Thiruvananthapuram Dist., Kerala

Asterina morellae	*Garcinia morella*		Bonacaud, Thiruvananthapuram Dist., Kerala Chemunji, Thiruvananthapuram Dist., Kerala
Asterina clusiacearum	*Garcinia gummigutta*		Mandrane, Hoddur, Karnataka
Cystotheca indica	*Calophyllum apetalum*		Calvary Mount, Idukki, Kerala
Heteroconium solaninum	*Garcinia mangostana*		Kallara, Thiruvananthapuram Dist., Kerala
Lembosia garciniae	*Garcinia gummigutta*		Karnataka, Kodagu, Mandrane, Hoddur
Meliola garciniae	*Garcinia gummigutta*		Hoddur
Amazonia henryi	*Calycopteris floribunda*	Combretaceae	Kombe, Meenmutty, Thiruvananthapuram Dist., Kerala
Asteridiella combreti var. *leonensis*	*Calycopteris floribunda* *Terminalia alata* *Terminalia bellirica* *Terminalia paniculata* *Terminalia* sp.		Palode forest area, JNTBGRI campus, Palode, Neyyar & Peppara Wildlife Sanctuary Thiruvananthapuram Dist., Kerala JNTBGRI campus, Palode, Thiruvananthapuram Dist., Kerala Peppara Wildlife Sanctuary, Thiruvananthapuram Dist., Kerala forest along the road from Painavu to Kulamavu, Idukki, Kerala Meenmutty, Idukki, Kerala Kombe, Thiruvananthapuram Dist., Kerala Peppara Wildlife Sanctuary JNTBGRI campus, Palode, Thiruvananthapuram Dist., Kerala Nanniyode, Thiruvananthapuram Dist., Kerala
Asterina combreti	*Calycopteris floribunda* *Terminalia arjuna*		Peppara Wildlife Sanctuary, Neyyar Wild Life Sanctuary, JNTBGRI campus, Palode, Thiruvananthapuram Dist., Kerala

(*Continued*)

TABLE 2.1 (Continued)

Fungi	Host plant	Family of the host plant	Locality
			Amarakathodu, Thiruvananthapuram Dist., Kerala Nanniyode, Thiruvananthapuram Dist., Kerala
Asterina escharoides	*Quisqualis indica*		Karimancode, Palode, Kallara, Nedumangad and JNTBGRI Campus, Palode, Thiruvananthapuram Dist., Kerala; Chowrira House Garden, Hoddur, Kodagu, Karnataka
Asterostomella terminaliae	*Terminalia* sp.		JNTBGRI Campus, Palode, Thiruvananthapuram Dist., Kerala
Colemaniella ossoorii	*Terminalia bellirica*		Kombe, Meenmutty, Thiruvananthapuram Dist., Kerala Peppara, Thiruvananthapuram Dist., Kerala JNTBGRI Campus, Palode, Thiruvananthapuram Dist., Kerala
Lembosia terminaliae-chebulae	*Terminalia chebula*		Peppara Wildlife Sanctuary, Thiruvananthapuram Dist., Kerala
Sarcinella quesqualidis	*Quisqualis indica*		Peppara Wildlife Sanctuary, Thiruvananthapuram Dist., Kerala
Schiffnerula terminaliae	*Terminalia catappa*		Chippanchira, Palode, Kallara, and JNTBGRI campus, Palode, Thiruvananthapuram Dist., Kerala
Asteridiella cyclopoda	*Vernonia arborea*	Compositae	Forest along the road from Painavu to Kulamavu, Idukki, Kerala
Asteridiella cyclopoda var. *vernoniae*	*Vernonia arborea*		Mannavan Shola, Munnar, Idukki, Kerala
Asterostomella vernoniae	*Vernonia monosis*		Tadiyandamol, Kodagu, Karnataka

Schiffnerula ecliptae	*Eclipta alba*		JNTBGRI Campus, Palode, Thiruvananthapuram Dist., Kerala
Schiffnerula wedeliae	*Wedelia chinensis*		JNTBGRI Campus, Palode, Thiruvananthapuram Dist., Kerala
Ferrarisia ipomoeae	*Ipomoea* sp.	Convolvulaceae	Neyyar dam site, JNTBGRI campus, Palode, Kallara, Thiruvananthapuram Dist., Kerala
Endophyllum kaernbachii	*Merremia vitifolia*		Meenmutty, Idukki, Kerala
Meliola clavulata	*Argyreia populifolia* *Ipomea palmate* *Ipomea* sp.		Vazhathope, Idukki, Kerala Hoddur, Kodagu, Karnataka Hoddur, Kodagu, Karnataka
Meliola erycibe-paniculatae	*Erycibe paniculata*		Chemunji; Forest area near Kusavoor, Palode; Bonacaud, Thiruvananthapuram Dist., Kerala
Meliola malacotricha var. *major*	*Argyreia nervosa*		Kombe, Neyyar Wildlife Sanctuary, Thiruvananthapuram Dist., Kerala
Oidium ipomoeae	*Ipomoea* sp.		Bryant Park, Kodaikanal, Tamil Nadu
Meliola connari	*Connarus sclerocarpus*	Connaraceae	Forest along the road from Painavu to Kulamavu, Idukki, Kerala
Phyllachora connari	*Connarus paniculatus*		Pamba, Pathanamthitta, Kerala
Asterina perpusilla	*Alangium salviifolium*	Cornaceae	JNTBGRI Campus, Palode, Thiruvananthapuram Dist., Kerala
Asteridiella confragosa	*Trichosanthes tricuspidata*		Forest along the road from Painavu to Kulamavu, Idukki, Kerala
Asteridiella triloba	*Mukia maderaspatana* *Mukia* sp.	Cucurbitaceae	Agastiar Hills, Neyyar Wildlife Sanctuary, Thiruvananthapuram Dist., Kerala Mannavan Shola, Munnar, Idukki, Kerala

(*Continued*)

TABLE 2.1 (Continued)

Fungi	Host plant	Family of the host plant	Locality
Asteridiella triloba var. *momordicae*	*Momordica* sp.		Mannavan Shola, Munnar, Idukki, Kerala
Cintractia minor	*Cyperus pangorii*	Cyperaceae	Panamkutty, Idukki, Kerala
Cintractia spicularum	*Rhynchospora corymbosa*		Meenmutty, Idukki, Kerala
Clasterosporium cyperacearum	*Scleria* sp.		Kombe, Thiruvananthapuram Dist., Kerala
Clasterosporium flagellatum	*Scleria* sp.		Forest near Kusavoor, Palode, Thiruvananthapuram Dist., Kerala
Diplococcium atrovelutinum	*Scleria* sp.		Kombe, Meenmutty, Eanikara& Palode forest, Thiruvananthapuram Dist., Kerala
Meliola cyperacearum	*Cyperus* sp.		Kuthiraiyar, Kukkal shola forest, Kodaikanal, Tamil Nadu
Asterostomella daphniphylli	*Daphniphyllum neilgherrense* *Daphniphyllum* sp.	Daphniphyllaceae	Mannavan Shola, Idukki, Kerala Gravel Banks, Munnar, Idukki, Kerala Periyakanal and Kuthiraiyar, Kukkal Shola Forest, Kodaikanal, Tamil Nadu Pooppara and Mannavan Shola, Munnar, Idukki, Kerala
Trochophora simplex	*Daphniphyllum neilgherrense*		Mathikettan Shola, Kodaikanal, Tamil Nadu
Asteridiella dilleniae	*Dillenia pentagyna*	Dilleniaceae	Vithura, Nedumangad, Thiruvananthapuram Dist., Kerala

Asterina acrotremae	*Acrotrema arnottianum*		Chemunji Hills, Thiruvananthapuram Dist., Kerala
Asterostomella dilleniacearum	*Dillenia pentagyna* *Dillenia* sp.		Peppara Wildlife Sanctuary, Meenmutty, Kombe, Thiruvananthapuram Dist., Kerala
Guignardia dioscoreae-bulbiferae	*Dioscorea pentaphylla* *D. bulbifera*	Dioscoreaceae	Forest along the road from Painavu to Kulamavu and Kanchiar Forest, Idukki, Kerala
Echidnodella vateriae	*Vateria indica*	Dipterocarpaceae	Ponmudy and JNTBGRI Campus, Palode, Thiruvananthapuram Dist., Kerala
Aecidium rhytismoideum	*Diospyros malabaricus* *Diospyros* sp.	Ebenaceae	JNTBGRI campus, Palode, Thiruvananthapuram Dist., Kerala
Asterina diospyri	*Diospyros* sp.		Kallar, Thiruvananthapuram Dist., Kerala
Hansfordiellopsis licheniicola	*Diospyros* sp.		JNTBGRI Campus, Palode, Thiruvananthapuram Dist., Kerala
Meliola diospyri	*Diospyros ebenum* *Diospyros malabarica* *Diospyros* sp. *Diospyros sylvatica*		JNTBGRI campus, Palode, Thiruvananthapuram Dist., Kerala Ponmudi, Calvary Mount, Idukki, Kerala
Meliola ebeni	*Diospyros ebenum*		JNTBGRI campus, Palode, Thiruvananthapuram Dist., Kerala
Phyllachora diospyri	*Diospyros melanoxylon*		Forest along Road from Painavu to Kulamavu, Idukki, Kerala
Prillieuxina diospyri	*Diospyros malabarica*		JNTBGRI Campus, Thiruvananthapuram Dist., Kerala
Meliola elaeagni	*Elaeagnus latifolia*	Elaeagnaceae	Devarakadu, Hoddur, Karnataka
Prillieuxina elaeagni	*Elaeagnus kologa*		Mandalpatti, Karnataka, Kodagu, Karnataka
Asterina elaeocarpi var. *ovalis*	*Elaeocarpus tuberculatus*	Elaeocarpaceae	Athirumala and JNTBGRI Campus, Palode, Thiruvananthapuram Dist., Kerala

(*Continued*)

TABLE 2.1 (Continued)

Fungi	Host plant	Family of the host plant	Locality
Asterina gamsii	*Elaeocarpus* sp.		JNTBGRI Campus, Palode, Thiruvananthapuram Dist., Kerala
Asterostomella elaeocarpi-serrati	*Elaeocarpus serratus*		JNTBGRI Campus, Palode, Thiruvananthapuram Dist., Kerala,
Appendiculella gaultheriae	*Gaultheria fragrantissima*	Ericaceae	Mannavan Shola, Munnar, Idukki, Kerala Samikanal, Kukkal Shola Forest, Kodaikanal, Tamil Nadu
Asteridiella gaultheriae	*Gaultheria indica*		Rajamala, Munnar, Idukki, Kerala
Asterina hakgalensis	*Rhododendron arboreum* J sub sp. *nilagiricum Rhododendron* sp.,		Sylvan Valley Hill, Pettumudi, Mannavan Shola, and Rajamala, Munnar, Idukki, Kerala
Asteridiella chowrirae	*Euphorbia pulcherrima*		Chowrira House, Hoddur, Karnataka
Asteridiella crotonis	*Croton zeylanicus*	Euphorbiaceae	Pamba, Pathanamthitta, Kerala
Asteridiella entebbeensis	*Glochidion* sp.		Mannavan Shola, Munnar, Idukki, Kerala
Asteridiella kodavae	*Mallotus* sp.		Hoddur, Medikeri, Kodagu, Karnataka
Asteridiella kombeensis	*Mallotus philippensis*		Kombe and Peppara and Neyyar Wildlife Sanctuaries Thiruvananthapuram Dist., Kerala
Asteridiella macarangicola	*Macaranga peltata*		Meenmutty, Idukki, Kerala
Asteridiella malloti	*Mallotus philippensis*		Lakshmi Estate, Idukki, Kerala

	M. tetracoccus		Pamba, Pathanamthitta, Kerala
Asteridiella sebastianiae	*Sebastiania chamaelea*		JNTBGRI campus, Palode, Thiruvananthapuram Dist., Kerala
Asterina lobulifera	*Glochidion* sp.		Mannavan shola, Munnar, Idukki, Kerala
Asterina ushae	*Glochidion bourdillonii*		Hoddur, Kodagu, Karnataka
Asterostomella baliospermi	*Baliospermum solanifolium*		JNTBGRI campus, Palode, Thiruvananthapuram Dist., Kerala
Cercospora ricinella	*Ricinus communis*		Vazhathope, Idukki, Kerala
Irenopsis crotonis	*Mallotus repandus*		Neyyar Wild life Sanctuary, Thiruvananthapuram Dist., Kerala
Kernkempella breyniae	*Breynia retusa*		Panorama view point, Idukki, Kerala
Leveillula taurica	*Euphorbia heterophylla* *Euphorbia pulcherrima* *Solanum erianthum*		Palinode, Kaniyapuram, JNTBGRI campus, Palode and Kumaranasan Nagar, Ulloor, Thiruvananthapuram Dist., Kerala Perumalmalai, Kodaikanal, Dindigul Anna Dist., Tamil Nadu
Masseeella capparidis	*Securinega virosa*		On the way from Painavu to Kulamavu, Idukki, Kerala
Meliola chandleri	*Excoecaria crenulata* *Excoecaria robusta*		Kanchiar Forest, Idukki, Kerala Mathikettan Shola, Kodaikkanal, Tamil Nadu
Meliola drypeticola	*Drypetes macrophylla*		Forest along the road from Painavu to Kulamavu, Idukki Dist., Kerala
Meliola glochidiicola	*Glochidion ellipticum*		Forest along the road from Painavu to Kulamavu, Calvary Mount, Idukki Dist., Kerala.
Meliola glochidii var. *velutini*	*Glochidion velutinum*		Kanchiar Forest, Idukki Dist., Kerala.

(Continued)

TABLE 2.1 (Continued)

Fungi	Host plant	Family of the host plant	Locality
Meliola ramosii	*Homonoia riparia*		Panamkutty and Pooyamkutty, Idukki Dist., Kerala.
Meliola vivekananthanii	*Antidesma zeylanica*		Chinnar, Munnar, Idukki Dist., Kerala.
Oidium heveae	*Hevea braziliensis*		Near Kattappana, Idukki Dist., Kerala.
Oidium jatrophae	*Jatropha curcas*		Oothu, Kodaikanal, Tamil Nadu
Passalora henningsii	*Manihot esculenta*		Churuly, Idukki Dist., Kerala.
Phakopsora formosana	*Glochidion velutinum*		Savannah along the road from Painavu to Kulamavu, Idukki Dist., Kerala.
Phyllachora shettyi	*Aporusa lindleyana*		Forest along the road from Painavu to Kulamavu, Idukki Dist., Kerala.
Ravenelia emblicae	*Phyllanthus emblica*		Calvary Mount, Idukki Dist., Kerala.
Ravenelia kirganaliae	*Kirganalia reticulata*		Panamkutty, Idukki Dist., Kerala.
Sarcinella bischofiae	*Bischofia javanica*		FMCC Campus, Madikeri, Kodagu, Karnataka
Sarcinella securinegae	*Securinega leucopyrus*		FMCC Campus, Madikeri, Kodagu, Karnataka
Schiffnerula glochidii	*Glochidion* sp.		MPCA, Brahmagiri, Talacauvery, Madikeri,Mandalpatti, Karnataka
Schiffnerula ricini	*Ricinus communis*		Nishane motta, Madikeri, Kodagu, Karnataka Hoddur, Kodagu, Karnataka
Sphaerotheca euphorbiae-hirtae	*Pedilanthus tithymaloides*		K.S.E.B. Quarters, Vazhathope, Idukki Dist., Kerala

Spiropes capensis	*Mallotus philippensis*		Kombe and JNTBGRI campus, Palode, Thiruvananthapuram Dist., Kerala
Verrucispora brideliae	*Bridelia retusa*		Reservoir side of Calvary Mount, Idukki Dist., Kerala
Erysiphe sikkimensis	*Quercus macrocarpa*	Fagaceae	Bryant Park, Kodaikanal, Tamil Nadu
Asteridiella scolopiae	*Scolopia crenata*	Flacourtiaceae	Brahmagiri, Talacauvery, Madikeri, Kodagu, Karnataka
Asteridiella homaligena	*Homalium zeylanicum*		River side, Hoddur, Kodagu, Karnataka
Asterina flacourtiacearum	*Scolopia crenata*		Coorg, Madikeri, Karnataka
Asterina homaligena	*Homalium zeylanica*		Riverside, Hoddur, Kodagu, Karnataka
Asterina talacauveriana	*Scolopia* sp.		Brahmagiri, Talacauvery, Madikeri and Vanachalu, Kodagu, Karnataka
Ishwaramyces flacourtiae	*Flacourtia indica*		Kulamavu, Idukki Dist., Kerala.
Meliola gneti	*Gnetum ula*	Gnetaceae	Adimali Forest Division and forest along the road from Painavu to Kulamavu, Idukki Dist., Kerala. Brahmagiri, Talacauvery and Vanachalu, Kodagu, Karnataka.
Meliola exacigena	*Exacum* sp.	Genteanaceae	Mathikettan Shola, Kodaikanal, Tamil Nadu
Meliola euonymi	*Exacum* sp.		Mathikettan Shola, Kodaikanal, Tamil Nadu
Amazonia gomphandrae	*Gomphandra coriacea*	Icacinaceae	Lakshmi Estate, Idukki Dist., Kerala
Meliola dimidiata	*Apodytes dimidiata* *Nothopodytes* sp., *Mappia foetida*		Forest along the road from Painavu to Kulamavu Murinjapuzha, Erumeli range, Peerumedu and Cheguthan mukku, Adimali, Idukki Dist., Kerala

(*Continued*)

TABLE 2.1 (Continued)

Fungi	Host plant	Family of the host plant	Locality
Asterina gomphandrae	*Gomphandra* sp.		Mannavan Shola, Idukki Dist., Kerala
Meliola chandrasekharanii	*Nothapodytes nimmoniana* *Apodytes dimidiata* *Apodytes* sp., *Nothopodytes nimmoniana* *Mappia foetida* *Nothopodytes nimmoniana*		Forest near Peppara Dam, Peppara Wildlife Sanctuary, Thiruvananthapuram Dist., Kerala Lakshmi Estate, Idukki, Kerala Kannimala Estate, Munnar, Idukki, Kerala Aruvillamchal, Idukki, Kerala Kuttikkanam, Idukki, Kerala Mannavan shola, Munnar, Idukki, Kerala MPCA, Brahmagiri, Talacauvery, Karnataka Nishane motta, and Abbe Falls, Kodagu, Karnataka
Acremoniula sarcinellae	*Tectona grandis*	Lamiaceae	Forest near Peppara Dam, Peppara Wildlife Sanctuary and JNTBGRI campus, Palode, Thiruvananthapuram Dist., Kerala
Asteridiella anastomosans	*Plectranthus deccanicus*		Periyakanal, Kukkal shola forest, Kodaikanal, Tamil Nadu
Asteridiella clerodendricola	*Clerodendrum viscosum*		Forest along the road from Painavu to Kulamavu, Idukki Dist., Kerala
Asteridiella formosensis	*Callicarpa tomentosa* *Callicarpa* sp.		Ponmudi, JNTBGRI campus, Palode, Neyyar Wildlife Sanctuary Thiruvananthapuram Dist., Kerala Forest along the road from Painavu to Kulamavu, Idukki Dist., Kerala Pamba, Pathanamthitta Mallamatti, Kodagu, Karnataka Nishane motta, Kodagu, Karnataka

Asteridiella vivekananthanii	*Clerodendrum viscosum*		Eanikara and JNTBGRI campus, Palode, Thiruvananthapuram Dist., Kerala
Asterina pusilla	*Premna corymbosa* *Premna* sp.		JNTBGRI campus, Palode, Thiruvananthapuram Dist., Kerala
Asterina leucadis	*Leucas* sp.		Kodagu, Karnataka
Erysiphe biocellata	*Salvia plebeia*		Perumalmalai, Kodaikanal, Tamil Nadu
Erysiphe galeopsidis	*Leucas* sp.		Kotagiri, Nilgiris, Tamil Nadu
Erysiphe rabdosiae	*Leucas hirta*		Manchur, Kodaikanal, Tamil Nadu
Meliola altissimae	*Vitex altissima* *Vitex* sp.		Meenmutty, Palode and Kottoor forests, Peppara Wildlife Sanctuary, Thiruvananthapuram Dist., Kerala
Meliola clerodendricola	*Clerodendrum inerme* *Clerodendrum* sp. *Clerodendrum viscosum*		JNTBGRI campus, Karimancode, Palode, Peringamala and Eanikkara, Thiruvananthapuram Dist., Kerala Forest along the road from Painavu to Kulamavu, Idukki
Meliola clerodendricola var. *micromera*	*Gmelina arborea*		Meenmutty, Idukki near Painavu, Idukki
Meliola cookeana	*Clerodendron inerme* *Vitex pinnata*		JNTBGRI campus, Palode, Thiruvananthapuram Dist., Kerala
Meliola cookeana var. *viticis*	*Clerodendrum inerme* *Clerodendrum serratum* *Vitex altissima*		JNTBGRI campus, Palode, and Near Peppara dam, Thiruvananthapuram Dist., Kerala
Meliola hyptidis	*Hyptis suaveolens*		Mylamood, JNTBGRI Campus, forest near Peppara Dam, Kottoor forest, Thiruvananthapuram Dist., Kerala
Meliola pogostemonis	*Pogostemon pubescens*		Forest near Bonacaud, Pandipathu, Peppara Wildlife Sanctuary

(*Continued*)

TABLE 2.1 (Continued)

Fungi	Host plant	Family of the host plant	Locality
	Pogostemon travancoricus *Pogostemon benghalensis*		Pongalappara, Agathyavanam Biological Park, Thiruvananthapuram Dist., Kerala Kuthiraiyar, Kukkal shola forest, Kodaikanal, Tamil Nadu
Ovulariopsis gmelinae-arboreae	*Gmelina arborea*		Kombaikadu, Kodaikanal, Tamil Nadu
Sarcinella pogostemonis	*Pogostemon* sp.		Mandalpatti, Kodagu, Karnataka
Amazonia actinodaphnes	*Actionodaphne malabarica*	Lauraceae	Forest along the road from Painavu to Kulamavu, Idukki, Kerala
Amazonia cinnamomi	*Cinnamomum riparium*		Pooyankutty, Idukki, Kerala
Armatella apolloniadis	*Apollonias arnottii*		Rajamala Hills, Munnar, Idukki, Kerala
Armatella balakrishnanii	*Cinnamomum malabatrum* *Cinnamomum* sp.		Forest along the road from Painavu to Kulamavu, Idukki, Kerala Kombe and Ponmudy, Thiruvananthapuram Dist., Kerala
Armatella caulicola	*Litsea* sp.		Mannavan Shola, Munnar, Idukki, Kerala
Armatella cinnamomicola	*Cinnamomum malabatrum* *C. macrocarpum*		Forest along the road from Painavu to Kulamavu, Idukki, Kerala
Armatella cryptocaryae	*Cryptocarya wightiana* *Litsea glabrata* *Litsea* sp.		Peppara, Thiruvananthapuram Dist., Kerala Kukkal Shola Forest, Kodaikanal Athirumala, Peppara, Thiruvananthapuram Dist., Kerala

	Litsea stocksii *Cryptocarya bourdillonii* *Neolitsea zeylanica* *Litsea* sp.		JNTBGRI campus, Palode, Thiruvananthapuram Dist., Kerala Kuthiraiyar, Kukkal shola forest, Kodaikanal Mathikettan Shola Forest, Kodaikanal Periyakanal, Kukkal shola forest, Kodaikanal Pamba, Pathanamthitta
Armatella indica	*Cinnamomum malabatrum*		Calvary Mount, Idukki, Kerala
Armatella katumotoi	*Cinnamomum malabatrum*		Calvary Mount, Idukki, Kerala
Armatella litseae	*Neolitsea scrobiculata* *Neolitsea* sp. *Neolitsea zeylanica*		Forest along the road from Painavu to Kulamavu, Kundala dam site, Munnar, Idukki, Kerala Kuthiraiyar, Kukkal shola
Armatella phoebecola	*Phoebe lanceolata*		Forest along the road from Painavu to Kulamavu, Idukki, Kerala
Asterina cinnamomicola	*Cinnamomum* sp.		Periyar Tiger Reseve, Idukki, Kerala
Basavamyces litseae	*Litsea* sp.		Mannavan Shola, Munnar, Idukki, Kerala
Caeoma keralense	*Cinnamomum malabatrum*		Calvary Mount, Idukki, Kerala
Exobasidium cinnamomi	*Cinnamomum malabatrum*		Calvary Mount, Idukki, Kerala
Kernella lauricola	*Phoebe* sp.		Then Shola, Kodaikanal, Tamil Nadu
Lembosia perseae	*Persea macrantha*		Rajamala, Munnar, Idukki, Kerala
Mahanteshamyces litseae	*Litsea* sp.		Tadiandamol, Kodagu, Karnataka

(*Continued*)

TABLE 2.1 (Continued)

Fungi	Host plant	Family of the host plant	Locality
Meliola actinodaphnecola	*Actinodaphne wightiana* *Actinodaphne* sp.		Forest near Kombe, Meenmutty, Neyyar Wildlife sanctuary, Peppara, Thiruvananthapuram Dist., Kerala
Meliola beilschmiediae var. *cinnamomicola*	*Cinnamomum malabatrum*		Calvary Mount, Lakshmi Estate, forest along the road from Painavu to Kulamavu, Idukki, Kerala Pamba, Pathanamthitta, Kerala
Meliola cholakadensis	Lauraceae member		Rajamala, Idukki, Kerala
Meliola cinnamomi	*Cinnamomum* sp. *Cinnamomum verum*		Attayar, Neyyar Wildlife Sanctuary JNTBGRI campus, Palode, Thiruvananthapuram Dist., Kerala
Meliola drepanochaeta var. *insignis*	*Litsea insignis* *Litsea* sp. *Litsea stocksi*		Kulamavu, Idukki, Kerala Rajamala, Idukki, Kerala Kulamavu Idukki, Kerala Mannavan shola, Munnar, Idukki, Kerala
Meliola gooseana	*Actinodaphne* sp.		Peppara Wildlife Sanctuary, Thiruvananthapuram Dist., Kerala
Meliola litsea var. *keralensis*	*Litsea stocksii*		Forest along the road from Painavu to Kulamavu Mannavan Shola and Devikulam, Munnar, Idukki, Kerala
Meliola litseae var. *rotundipoda*	*Litsea coriacea* *Litsea* sp.,		Forest near Painavu, Idukki, Kerala Mannavan Shola and Devikulam, Munnar, Idukki, Kerala
Meliola machili	*Persea* sp. *Persea macrantha*		Peppara Wildlife Sanctuary, Thiruvananthapuram, Kerala

Meliola mannavanensis	*Litsea* sp.		Mannavan Shola, Munnar, Idukki, Kerala
Meliola neolitseae	*Neolitsea scrobiculata*		Ponmudi, Thiruvananthapuram Dist., Kerala Lakshmi Estate, Idukki Thondikappu, Adimali, Idukki
Meliola pudukadensis	*Actinodaphne* sp.		Peppara Wildlife Sanctuary, Thiruvananthapuram, Kerala
Meliola tetradeniae	*Neolitsea* sp.		Athirumala, Neyyar Wildlife Sanctuary, Thiruvananthapuram, Kerala
Periconiella perseae-macranthae	*Persea macrantha*		Peppara Wildlife Sanctuary JNTBGRI campus, Palode, Thiruvananthapuram Dist., Kerala
Schiffnerula actinodaphnes	*Actinodaphne* sp.		Palode, JNTBGRI campus, Palode, Thiruvananthapuram Dist., Kerala
Meliola careyae var. *indica*	*Careya arborea*	Lecythidaceae	Eanikara, JNTBGRI campus, Palode, Peppara Wildlife Sanctuary, Thiruvananthapuram Dist., Kerala
Pseudocercospora careyae	*Careya arborea*		Meenmutty, Idukki, Kerala
Irenopsis leeae var. *indica*	*Leea indica*	Leeaceae	Kanchiar Forest, forest along the road from Painavu to Kulamavu, Reservoir side of Calvary Mount, Lakshmi Estate, Idukki, Kerala
Asperisporium pongamiae	*Pongamia pinnata*	Leguminosae	Panamkutty, Idukki, Kerala
Asterina cassiicola	*Cassia fistula*		Karimancode and JNTBGRI campus, Palode, Thiruvananthapuram Dist., Kerala
Asterina cassiigena	*Cassia glauca*		Hoddur, Kodagu, Karnataka

(*Continued*)

TABLE 2.1 (Continued)

Fungi	Host plant	Family of the host plant	Locality
Asterina mezonevronis	*Mezonevron cucullatum*		Karnataka, Kodagu, Karnataka Madikeri, Bharatiya Vidya Bhavan, Kodagu Vidyalaya campus, Karnataka
Asterina saracae	*Saraca asoca*		JNTBGRI Campus, Palode and Peroorkada, Thiruvananthapuram Dist., Kerala
Cercospora canescens	*Vigna catjang*		Katteparatala. Idukki, Kerala
Cercospora erythrinicola	*Erythrina indica*		Vazhathope, Idukki, Kerala
Cercospora nigricans	*Cassia tora*		Moolamattom, Idukki, Kerala
Lembosia humboldtiae	*Humboldtia vahliana* *Humboldtia* sp.		Kulamavu, Idukki, Kerala Neyyar Wildlife Sanctuary, Meenmutty, Kombe, Thiruvananthapuram Dist., Kerala Peppara Wildlife Sanctuary, near Peppara Dam, JNTBGRI Campus, Palode, Thiruvananthapuram Dist., Kerala
Lembosia humboldtiicola	*Humboldtia unijuga*		Athirumala, Peppara Wildlife Sanctuary, Thiruvananthapuram Dist., Kerala
Lembosia humboldtiigena	*Humboldtia vahliana*		JNTBGRI Campus, Palode, Thiruvananthapuram Dist., Kerala
Leptosphaerulina australis	*Crotalaria multiflora*		From Painavu to Kulamavu, Idukki, Kerala

Meliola abrupta	*Derris* sp.		Peppara Dam, Peppara Wildlife Sanctuary Thiruvananthapuram Dist., Kerala
Meliola aethiops	*Cassia fistula*		Eanikara, Karakulam, JNTBGRI campus, Palode, and Peppara Wildlife Sanctuary, Thiruvananthapuram Dist., Kerala
Meliola atylosiae	*Atylosia lineata*		Forest along the road from Painavu to Kulamavu, Idukki, Kerala
Meliola bauhiniicola	*Bauhinia* sp.		JNTBGRI campus, Palode, Thiruvananthapuram Dist., Kerala
Meliola bauhiniicola var. *brevipoda*	*Bauhinia purpurea* *Bauhinia* sp.		JNTBGRI campus, Palode, Thiruvananthapuram Dist., Kerala
Meliola bicornis	*Desmodium latifolium* *Desmodium* sp.		Kombe Meenmutty, Thiruvananthapuram Dist., Kerala Forest area, PalodeThiruvananthapuram, Kerala
Meliola buteae	*Butea monosperma* *Spatholobus parviflorus*		Peppara Wildlife Sanctuary Kombe, Thiruvananthapuram Dist., Kerala JNTBGRI campus, Palode, Thiruvananthapuram Dist., Kerala
Meliola caesalpiniae var. *indica*	*Caesalpinia sappan*		JNTBGRI campus, Palode, Thiruvananthapuram Dist., Kerala
Meliola clitoriae	*Clitoria ternatea*		Puliyanmala, Idukki, Kerala
Meliola desmodii-pulchelli	*Desmodium pulchellum*		Kombe, Neyyar Wildlife Sanctuary JNTBGRI campus, Palode, Thiruvananthapuram Dist., Kerala Nanniyode, Palode, Thiruvananthapuram Dist., Kerala Mylamood, Palode, Thiruvananthapuram Dist., Kerala
Meliola desmodii-triquetri	*Desmodium triquetrum*		Near Peppara Dam, Peppara Wildlife Sanctuary JNTBGRI campus, Palode, Thiruvananthapuram Dist., Kerala Kallara, Thiruvananthapuram Dist., Kerala

(*Continued*)

TABLE 2.1 (Continued)

Fungi	Host plant	Family of the host plant	Locality
Meliola erythrinae	*Erythrina* sp.		JNTBGRI campus, Palode, Thiruvananthapuram Dist., Kerala Nedumangad, Thiruvananthapuram Dist., Kerala
Meliola flemingiicola	*Flemingia* sp.		Chippanchira, Palode, Thiruvananthapuram Dist., Kerala JNTBGRI campus, Palode, Thiruvananthapuram Dist., Kerala In the campus of Mary Matha Engg. College, Neyyattinkara, Thiruvananthapuram Dist., Kerala
Meliola gliricidiicola	*Gliricidia sepium*		Eanikara, Karakulam, Thiruvananthapuram Dist., Kerala JNTBGRI campus, Palode, Thiruvananthapuram Dist., Kerala Karimancode, Palode, Thiruvananthapuram Dist., Kerala Mylamoodu, Palode, Thiruvananthapuram Dist., Kerala
Meliola kingiodendri	*Kingiodendron pinnatum*		Sampaje Ghat, Kodagu, Karnataka
Meliola pongamiae	*Pongamia pinnata*		JNTBGRI Campus, Palode, Thiruvananthapuram Dist., Kerala Kottoor Reserve Forests, Peppara Wildlife Sanctuary, Peringamala, Thiruvananthapuram Dist., Kerala
Meliola pseudarthriae	*Pseudarthria viscida*		Near Peppara Dam, Peppara Wildlife Sanctuary, JNTBGRI Campus, Palode, Thiruvananthapuram Dist., Kerala
Meliola pterocarpi	*Pterocarpus marsupium*		JNTBGRI Campus, Palode, Peppara and Kombe, Thiruvananthapuram Dist., Kerala
Meliola spatholobii	*Spatholobus parviflorus* *Spatholobus* sp.		Forest near Palode, Peppara, Kombe, Meenmutty, Thiruvananthapuram Dist., Kerala

Meliola surattensis	*Cassia glauca*		JNTBGRI Campus, Palode, Thiruvananthapuram Dist., Kerala
Meliola tamarindi	*Tamarindus indica*		Palode forest area; Eanikara, Nedumangad, Thiruvananthapuram Dist., Kerala
Meliola teramnicola	*Teramnus labialis*		Mannavan Shola, Munnar, Idukki, Kerala
Oidium tamarindi	*Tamarindus indica*		Kombe, Thiruvananthapuram Dist., Kerala Near railway station, Kaniyapuram, Thiruvananthapuram Dist., Kerala
Phyllachora pongamiae	*Pongamia pinnata*		Panamkutty, Idukki, Kerala
Prillieuxina humboltiae	*Humboltia* sp.		Sampaje Ghats, Kodagu, Karnataka
Sirosporium gliricidiae	*Gliricidia sepium*		Katteparatala, Idukki, Kerala
Teratosperma anacardii	*Dalbergia sissoides*		JNTBGRI campus, Palode, Thiruvananthapuram Dist., Kerala
Meliola toreniae	*Torenia travancorica*	Linderniaceae	Neyyar Wildlife Sanctuary Pongalappara, Neyyar Wildlife Sanctuary Athirumala, Peppara Wildlife Sanctuary, Thiruvananthapuram Dist., Kerala
Meliola cannonii	*Strychnos nux-vomica*	Loganiaceae	Eanikara, Karakulam, Thiruvananthapuram Dist., Kerala
Meliola gamsii	*Strychnos nux-vomica*		Vellalloor, Kallambalam, Thiruvananthapuram Dist., Kerala Eanikara, Thiruvananthapuram Dist., Kerala
Meliola kodaguensis	*Strychnos colubrina*		Madikeri, Kodagu, Karnataka
Meliola spigeliae	*Strychnos nux-vomica*		JNTBGRI campus, Palode, Thiruvananthapuram Dist., Kerala
Meliola strychnacearum	*Strychnos* sp.		Athirumala forest, Neyyar Wildlife Sanctuary Nettukalthery, Neyyar Wildlife Sanctuary
Questieriella strychni	*Strychnos nuxvomica* *Strychnos* sp.		Eanikara, Karakulam, Thiruvananthapuram Dist., Kerala

(*Continued*)

TABLE 2.1 (Continued)

Fungi	Host plant	Family of the host plant	Locality
Asterina deightonii	*Helixanthera* sp. *Loranthus falcatus* *Loranthus* sp. *Dendrophthoe* sp. *Dendrophthoe falcate* *Taxillus* sp.	Loranthaceae	Peppara Wildlife Sanctuary, Bonacaud, Thiruvananthapuram Dist., Kerala Gravel Banks, Munnar, Chinnar Wildlife Sanctuary, Idukki, Kerala Near Mannavan shola, Munnar, Idukki, Kerala
Meliola loranthacearum	*Dendrophthoe falcata* *Loranthus* sp.		Peppara Wildlife Sanctuary, Thiruvananthapuram Dist., Kerala Chippanchira, Palode, Thiruvananthapuram Dist., Kerala
Meliola prataprajii	*Loranthus falcatus* *Loranthus* sp.		Peppara Wildlife Sanctuary JNTBGRI Campus, Palode, Thiruvananthapuram Dist., Kerala Eanikara, Thiruvananthapuram Dist., Kerala
Asterina lawsoniae	*Lawsonia inermis*	Lythraceae	Kallara, Thiruvananthapuram Dist., Kerala
Ophiodothella lagerstromiae	*Lagerstroemia microcarpa*		Peppara, Attayar, Thiruvananthapuram Dist., Kerala
Asteridiella micheliae	*Michelia champaka*	Magnoliaceae	JNTBGRI campus, Palode, Thiruvananthapuram Dist., Kerala
Asterostomella micheliae	*Michelia nilagirica* *Michelia champaka*		Mannavan Shola, Munnar, Idukki, Kerala Hoddur, Kodagu, Karnataka
Meliola cariappae	*Michelia champaca*		Sampaje Ghat, Karnataka
Asterina diplocarpa	*Sida cordata*	Malvaceae	Kushavoor Forest area, Thiruvananthapuram Dist., Kerala

Asterina helicteridis	*Helicteres isora*		Kombe, Meenmutty, Thiruvananthapuram Dist., Kerala Peppara Wildlife Sanctuary, JNTBGRI campus, Palode, Thiruvananthapuram Dist., Kerala
Asterina hibisci	*Hibiscus* sp.		JNTBGRI campus, Palode, Thiruvananthapuram Dist., Kerala
Asterina triumfetticola	*Triumfetta* sp		JNTBGRI campus, Palode, Thiruvananthapuram Dist., Kerala
Irenopsis eriolaenae	*Eriolaena quinquelocularis*		Forest along the road from Painavu to Kulamavu, Idukki, Kerala
Irenopsis helicteridis	*Helicteres isora*		Peppara Wildlife Sanctuary, Thiruvananthapuram Dist., Kerala
Irenopsis kleinhoviae	*Kleinhovia hospita*		Karimancode, Thiruvananthapuram Dist., Kerala
Irenopsis madumalaiensis	*Kydia calycina*		Sampaje Ghat, Kodagu, Karnataka
Irenopsis molleriana	*Hibiscus furcatus* *Hibiscus hispidissimus* *Hibiscus* sp. *Hibiscus teliaceus* *Urena lobata*		Neyyar Wildlife Sanctuary, Kombe, Thiruvananthapuram Dist., Kerala JNTBGRI campus, Palode, Thiruvananthapuram Dist., Kerala JNTBGRI campus, Palode, Thiruvananthapuram Dist., Kerala Karimancode, Thiruvananthapuram Dist., Kerala
Irenopsis sidae var. *indica*	*Sida* sp.		JNTBGRI campus, Palode, Thiruvananthapuram Dist., Kerala
Asterina madikeriensis	*Memecylon* sp.		Nishanemotta, Madikeri, Kodagu, Karnataka
Asterina memecylonis	*Memecylon edule* *Memecylon* sp. *Memecylon sylvaticum*	Melastomataceae	JNTBGRI campus, Palode, Thiruvananthapuram Dist., Kerala Devarakadu, Hoddur, Karnataka

(*Continued*)

TABLE 2.1 (Continued)

Fungi	Host plant	Family of the host plant	Locality
Echidnodella memecyli	*Memecylon* sp.		Meenmutty, Neyyar Wildlife Sanctuary, JNTBGRI Campus, Palode, Thiruvananthapuram Dist., Kerala
Meliola attayarica	*Memecylon* sp.		Attayar, Neyyar Wildlife Sanctuary, Thiruvananthapuram Dist., Kerala
Meliola heudelotii	*Memecylon umbellatum*		JNTBGRI Arboretum, Palode, Thiruvananthapuram Dist., Kerala
Meliola memecyli	*Memecylon edule*		JNTBGRI Campus, Palode, Thiruvananthapuram Dist., Kerala Eanikara, Thiruvananthapuram Dist., Kerala Nedumangad, Thiruvananthapuram Dist., Kerala
Meliola memecyli var. *microspora*	*Memecylon* sp.		Kombe, Thiruvananthapuram Dist., Kerala Nedumangad, Thiruvananthapuram Dist., Kerala
Meliola memecylicola var. *longiseta*	*Memecylon umbellatum*		JNTBGRI Campus, Palode, Thiruvananthapuram Dist., Kerala
Rehmidothis osbeckiae	*Osbeckia wightiana*		Calvary Mount, Idukki, Kerala
Asterina aglaiae	*Aglaia* sp.	Meliaceae	Jodupal, Madikeri, Karnataka
Asterina cipadessae	*Cipadessa baccifera*		JNTBGRI Campus, Palode, Thiruvananthapuram Dist., Kerala Peppara Wildlife Sanctuary, Near Peppara Dam Bonacaud, Thiruvananthapuram Dist., Kerala Athirumala, Peppara Wildlife Sanctuary
Asterina chukrasiae	*Chukrasia tabularis*		Akane, Hoddur, Karnataka

Meliola aglaicola	*Aglaia minutiflora*		Pamba, Pathanamthitta Dist., Kerala
Meliola chukrasiae	*Chukrasia tabularis*		Akare, Hoddur, Karnataka Devarakadu, Hoddur, Karnataka
Sarcinella cipadessae	*Cipadessa baccifera*		Peppara Wildlife Sanctuary, Thiruvananthapuram Dist., Kerala
Schiffnerula azadirachtae	*Azadirachta indica*		Kallara, Trivandrum, Thiruvananthapuram Dist., Kerala
Maheshwaramyces pachygones	*Pachygone ovata*	Menispermaceae	JNTBGRI Campus, Palode, Thiruvananthapuram Dist., Kerala Bonacaud, Thiruvananthapuram Dist., Kerala
Meliola cycleae	*Cyclea peltata*		Palode Forest, Thiruvananthapuram Dist., Kerala Meenmutty, Idukki
Meliola subramanyaensis	*Cyclea peltata*		Palode Forest, Thiruvananthapuram Dist., Kerala
Prillieuxina anamirtae	*Anamirta cocculus*		JNTBGRI Campus, Palode, Thiruvananthapuram Dist., Kerala
Viegasia cissampeli	*Cissampelos pareira*		Kallara, Thiruvananthapuram Dist., Kerala
Asteridiella ficiicola	*Ficus microcarpa*	Moraceae	JNTBGRI campus, Palode, Thiruvananthapuram Dist., Kerala
Asteridiella strebli	*Streblus asper*		Peppara Wildlife Sanctuary, Thiruvananthapuram Dist., Kerala
Asterina hopeae	*Hopea ponga*		JNTBGRI campus, Palode, Thiruvananthapuram Dist., Kerala Peppara Wildlife Sanctuary, Thiruvananthapuram Dist., Kerala
Cerotelium fici	*Morus alba*		Vazhathope, Idukki, Kerala

(*Continued*)

TABLE 2.1 (Continued)

Fungi	Host plant	Family of the host plant	Locality
Cirsosia hopeae	*Hopea ponga*		Peppara Wildlife Sanctuary, Thiruvananthapuram Dist., Kerala
Cronartium fici	*Ficus asperrima*		Vazhathope, Idukki, Kerala
Hansfordiellopsis licheniicola	*Ficus* sp.		JNTBGRI Campus, Palode, Thiruvananthapuram Dist., Kerala
Irenopsis benguetensis	*Ficus exasperata* *Ficus* sp. *Ficus tinctoria* subsp. *gibbosa* *Ficus asperrima* *F. gibbosa* *F. hispida*		Forest near Bonacaud, Thiruvananthapuram Dist., Kerala Calvary Mount, Idukki, Kerala Forest along the road from Painavu to Kulamavu, Idukki, Kerala Calvary Mount, Idukki, Kerala Lakshmi Estate, Idukki, Kerala along the road from Painavu to Kulamavu, Idukki, Kerala
Sphaerodothis raoii	*Ficus religiosa*		Pattom, Thiruvananthapuram Dist., Kerala Nedumangad, Thiruvananthapuram Dist., Kerala Govt. Botanic Garden, Thiruvananthapuram Dist., Kerala JNTBGRI campus, Palode, Thiruvananthapuram Dist., Kerala
Uncinula fici-nervosae	*Ficus nervosa*		Oothu, Kodaikanal, Tamil Nadu
Uncinula fici-religiosae	*Ficus religiosa*		Myladumparai, Kodaikanal, Tamil Nadu
Cordana musae	*Musa paradisiaca*	Musaceae	Ayattawadam, Idukki, Kerala
Amazonia peregrine	*Maesa indica*	Myrsinaceae	Forest along the road from Painavu to Kulamavu, Idukki Periyakanal, Kodaikanal, Tamil Nadu Periyakanal, Kukkal shola forest, Kodaikanal, Tamil Nadu

Asterina ardisiae	*Ardisia solanacea*		Mannavan Shola, Idukki, Kerala
Meliola ardisiicola	*Ardisia solanacea*		Mandalpatti, Madikeri, Karnataka
Meliola groteana	*Maesa indica*		Periyakanal, Samikanal and Kuthiraiyar, Kukkal Shola forest, Kodaikanal, Tamil Nadu
Meliola groteana var. *maesae*	*Maesa indica*		Kuthiraiyar, Kodaikanal, Tamil Nadu
Questieriella ardisiae	*Ardesia solanacea*		Hoddur, Karnataka
Asteridiella myristicacearum	Myristicaceae member	Myristicaceae	Attayar, Neyyar Wildlife Sanctuary, JNTBGRI campus, Palode, Thiruvananthapuram Dist., Kerala
Asterina myristicacearum	*Myristica malabarica*		JNTBGRI Campus, Palode, Thiruvananthapuram Dist., Kerala
Asterina myristicae	*Myristica malabarica* *Myristica* sp.		JNTBGRI Campus, Palode, Thiruvananthapuram Dist., Kerala
Asteridiella brahmagiriensis	*Syzygium* sp.	Myrtaceae	Brahmagiri, Talacauveri, Kodagu, Karnataka
Amazonia syzygii	*Syzygium cumini* *Syzygium* sp.		Forest along the road from Painavu to Kulamavu, Idukki Vanachalu, Kodagu, Karnataka Sampaje Ghat, Kodagu, Karnataka
Asterina claviflori	*Syzygium cumini* *Syzygium cumini* *Syzygium mundagam*		JNTBGRI Campus, Palode, Thiruvananthapuram Dist., Kerala
Asterina jambolana	*Syzygium cumini* *Syzygium* sp.		JNTBGRI Campus, Palode, Thiruvananthapuram Dist., Kerala Mannavan shola, Munnar, Idukki, Kerala

(*Continued*)

TABLE 2.1 (Continued)

Fungi	Host plant	Family of the host plant	Locality
Asterina rhodomyrti	*Rhodomyrtus tomentosa*		Rajamala, Munnar, Idukki, Kerala
Hadronema verrucosae	*Syzygium cumini*		Panamkutty, Idukki, Kerala
Lembosia hosagoudarii	*Syzygium cumini* *Syzygium* sp. *Syzygium arnottianum*		JNTBGRI Campus, Palode, Thiruvananthapuram Dist., Kerala Rajamala, Munnar, Idukki, Kerala
Meliola cauveriana	*Syzygium mundagam*		Medikeri, Kodagu, Karnataka
Meliola densa	*Syzygium ram-varmae* *Syzygium* sp. *Syzygium munroni* *S. laetum*		Athirumala, Thiruvananthapuram Dist., Kerala JNTBGRI campus, Palode, Thiruvananthapuram Dist., Kerala Chemunji, Thiruvananthapuram Dist., Kerala Forest along the road from Painavu to Kulamavu, Idukki, Kerala
Meliolina pulcherrima	*Syzygium cumini*		JNTBGRI campus, Palode, Thiruvananthapuram Dist., Kerala Mylamood, Thiruvananthapuram Dist., Kerala Peringamala, Thiruvananthapuram Dist., Kerala Karimancode, Palode, Thiruvananthapuram Dist., Kerala, Attayar, Neyyar Wildlife Sanctuary Thiruvananthapuram Dist., Kerala
Phaeoseptoria eucalypti	*Eucalyptus globulus*		Meenmutty, Idukki, Kerala
Phyllachora ambigua	*Syzygium cumini*		Lakshmi Estate, Idukki, Kerala
Oidium nyctaginacearum	*Mirabilis jalapa*	Nyctaginaceae	Oothu, Kodaikanal, Tamil Nadu

Asterina erythropalicola	*Erythropalum scandens* *Erythropalum populifolium*	Olacaceae	Athirumala, Peppara Wildlife Sanctuary Thiruvananthapuram Dist., Kerala Hoddur, Karnataka
Meliola erythropali	*Erythropalum populifolium*		JNTBGRI Campus, Palode, Thiruvananthapuram Dist., Kerala Karimancode, Thiruvananthapuram Dist., Kerala Nedumangad, Kallara, Thiruvananthapuram Dist., Kerala Ponmudi, Thiruvananthapuram Dist., Kerala Near Peppara Dam Hoddur, Karnataka
Asteridiella americana	*Chionanthus mala-elengi*	Oleaceae	Forest along the road from Painavu to Kulamavu, Idukki, Kerala
Asterina erysiphoides	*Jasminum angustifolium* *Jasminum ritchiei*		Kusavoor, Thiruvananthapuram Dist., Kerala Nedumangad, Thiruvananthapuram Dist., Kerala JNTBGRI campus, Palode, Thiruvananthapuram Dist., Kerala
Asterina pongalaparensis	*Jasminum* sp.		Ponmudi, Thiruvananthapuram Dist., Kerala Chemunji, Thiruvananthapuram Dist., Kerala
Eupelte amicta	*Olea dioica*		Periyar Tiger Reserve, Idukki, Kerala
Hansfordiellopsis licheniicola	*Jasminum cordifolium*		JNTBGRI Campus, Palode, Thiruvananthapuram Dist., Kerala
Meliola daviesii var. *kodaikalensis*	*Jasminum brevilobum*		Periyakanal, Kukkal shola forest, Kodaikanal, Tamil Nadu
Meliola gemellipoda	*Jasminum ozoricum* *Jasminum* sp. *J. malabaricum*		Calvary Mount; Kanchiar forest; Lakshmi Estate, Idukki, Kerala Nishane motta, Madikeri, Karnataka Vanachalu, Hoddur, Karnataka

(*Continued*)

TABLE 2.1 (Continued)

Fungi	Host plant	Family of the host plant	Locality
Meliola jasmini	*Jasminum angustifolium* *J. angustifolium* *J. arborescence* *J. auriculatum* *J. azoricum* *J. didymum* *J. grandiflorum* *J. rottlerianum* *J. tortuosum* *J. angulare* *J. angustifolium* *J. cuspidatum* *J. frutescens* *J. multiflorum* *J. laurifolium* *J. pubescence* *J. rigidum* *J. rottlerianum* *J. sambac* *J. sinense* *J. simplicifolium* subsp. *le-ratii* *Jasminum multiflorum* *Jasminum* sp.		JNTBGRI Campus, Palode, Thiruvananthapuram Dist., Kerala Peppara Wildlife Sanctuary, Thiruvananthapuram Dist., Kerala Kombe, Meenmutty, Thiruvananthapuram Dist., Kerala Karimancode, Palode, Thiruvananthapuram Dist., Kerala

Meliola jasminicola var. *indica*	*Jasminum angustifloium* var. *angustifolium* *J. auriculatum* *Jasminum* sp.		Field Marshal Cariappa College Campus, Hoddur, Karnataka Sampaje Ghat, Kodagu, Karnataka Periyakanal, Kukkal shola forest, Kodaikanal, Tamil Nadu
Meliola linocierae-malabaricae	*Ligustrum perrottetii*		Samikanal and Kuthiraiyar, Kodaikanal, Tamil Nadu Thalavakanal and Boothakanal, Kukkal Shola Forest, Kodaikanal, Tamil Nadu
Meliola malabarensis	*Olea dioica*		Kombe, Meenmutty, Palode forest, Peppara and Neyyar Wildlife Sanctuaries, Thiruvananthapuram Dist., Kerala Bhagamandala, Kodagu, Karnataka
Meliola mayapeae	*Ligustrum perrottetii*		Kuthiraiyar, Samikanal and Thalavakanal, Kukkal Shola Forest, Kodaikanal, Tamil Nadu
Meliola mayapiicola var. *indica*	*Ligustrum perrottetii*		Thambiran Shola, Kodaikkanal, Tamil Nadu
Meliola oleacearum	*Olea dioica*		Peppara and JNTBGRI Campus, Palode, Thiruvananthapuram Dist., Kerala
Puccinia exhauriens	*Jasminum pubescens* *Jasminum* sp.		Kombe, Meenumutty, Thiruvananthapuram Dist., Kerala Athirumala, Neyyar, Thiruvananthapuram Dist., Kerala
Schiffnerula pulchra	*Ligustrum* sp.		Nishanemotta, Madikeri, Karnataka
Zhaghounia oleae	*Olea dioica*		Bonacaud Forest, Athirumala, Peppara Wildlife Sanctuary, Mylamood and Palode, Thiruvananthapuram Dist., Kerala
Meliola cansjericola	*Cansjera rheedii*	Opiliaceae	Calvary Mount, Idukki, Kerala
Asterina averrhoae	*Averrhoa carambola*	Oxalidaceae	JNTBGRI Campus, Palode, Thiruvananthapuram Dist., Kerala
			(*Continued*)

TABLE 2.1 (Continued)

Fungi	Host plant	Family of the host plant	Locality
Lembosia pandanacearum	*Pandanus* sp.	Pandanaceae	Hoddur, Karnataka
Meliola pandanacearum	*Pandanus odoratissimus* *Pandanus* sp.		JNTBGRI Campus, Palode, Thiruvananthapuram Dist., Kerala Peppara Wildlife Sanctuary, Thiruvananthapuram Dist., Kerala
Meliola kapoorii	*Pandanus thwaitesii*		Kodagu, Karnataka
Schiffnerula mirabilis	*Passiflora edulis*	Passifloraceae	Kallara, Thiruvananthapuram Dist., Kerala Karimancode, Palode, Thiruvananthapuram Dist., Kerala
Asterina aporusae	*Aporosa lindleyana* *Aporusa bourdillonii*	Phyllanthaceae	Kombe, Thiruvananthapuram Dist., Kerala Peppara Wildlife Sanctuary, JNTBGRI Campus, Palode, Thiruvananthapuram Dist., Kerala Mylamood, Palode, Thiruvananthapuram Dist., Kerala Neyyar Wildlife Sanctuary Thiruvananthapuram Dist., Kerala Hoddur, Kodagu, Karnataka
Meliolaster aporusae	*Aporosa cardiosperma* *Aporusa lindleyana*		JNTBGRI Campus, Palode, Thiruvananthapuram Dist., Kerala Hoddur, Kodagu, Karnataka
Phyllachora keralica	*Phyllanthus* sp.		In forest near Bonacaud, Thiruvananthapuram Dist., Kerala
Asterina piperina	*Piper* sp.	Piperaceae	JNTBGRI Campus, Palode, Thiruvananthapuram Dist., Kerala
Meliola stenospora	*Piper* sp.		Athirumala, Peppara Wildlife Sanctuary Thiruvananthapuram Dist., Kerala; Kottoor Reserve Forests, Peppara Wildlife Sanctuary, Thiruvananthapuram Dist., Kerala

Nigrocornus scleroticus	*Themeda cymbaria*	Poaceae	Forest along the road from Painavu to Kulamavu, Idukki Dist., Kerala
Kweilingia divina	*Ochlandra travancorica*		Idukki Dam site, Kerala
Balansia oryzae-sativae	*Eulalia trispicata*		The savanna along the road from Painavu to Kulamavu, Idukki, Kerala
Meliola cymbopogonis	*Cymbopogon flexuous*		Vazhathope, Idukki, Kerala Forest along the road from Painavu to Kulamavu, Idukki, Kerala
Meliola panici	*Cymbopogon* sp. *Crysopogon* sp.		Kombe, Meenmutty, Thiruvananthapuram, JNTBGRI Campus, Palode, Thiruvananthapuram Kuthiraiyar, Kukkal shola forest, Kodaikanal, Tamil Nadu
Meliola panici var. *vetiveriicola*	*Vetiveria zizanoides*		JNTBGRI Campus, Palode, Thiruvananthapuram Dist., Kerala
Meliola phyllostachydis var. *microspora*	*Sinarundinaria microphylla*		Thalavakanal, Kodaikanal, Tamil Nadu
Phakopsora apoda	*Pennisetum polystachyon*		Vazhathope, Idukki, Kerala
Tilletia themedicola	*Themeda triandra*		Meenmutty, Idukki, Kerala
Asterina xanthophylli	*Xanthophyllum arnottianum*	Polygalaceae	Chinnar, Munnar, Idukki, Kerala
Meliola thiyagesanii	*Polygala arillata*		Kuthiraiyar, Kukkal shola forest, Kodaikanal, Tamil Nadu
Phyllachora prataprajii	*Xanthophyllum flavescens*		Forest along the road from Painavu to Kulamavu, Idukki, Kerala

(*Continued*)

TABLE 2.1 (Continued)

Fungi	Host plant	Family of the host plant	Locality
Podoconis anacardii	*Xanthophyllum flavescens*		JNTBGRI campus, Palode, Thiruvananthapuram Dist., Kerala
Meliola groteana var. *maesae*	*Maesa indica*	Primulaceae	Ponmudi, Thiruvananthapuram Dist., Kerala
Asterina naraveliae	*Naravelia zeylanica*	Ranunculaceae	JNTBGRI Campus, Palode, Thiruvananthapuram Dist., Kerala
Meliola hoveniae	*Hovenia acerba*		Periyakanal, Kukkal shola forest, Kodaikanal, Tamil Nadu
Meliola ziziphi	*Ziziphus oenoplea* *Ziziphus rugosa*	Rhamnaceae	Peppara Wild Life Sanctuary Thiruvananthapuram Dist., Kerala Kombe, Thiruvananthapuram Dist., Kerala Mylamoodu, Palode, Thiruvananthapuram Dist., Kerala Forest area, Kusavoor, Thiruvananthapuram Dist., Kerala Near Peppara dam Ponmudi, Thiruvananthapuram Dist., Kerala
Phyllachora sageratiae	*Sageretia* sp.		Kukkal Shola, Kodaikanal, Tamil Nadu
Sarcinella caralliae	*Carallia brachiate*	Rhizophoraceae	Kaimada Field, Hoddur, Karnataka
Hamaspora longissima	*Rubus ellipticus*	Rosaceae	Rajmala Hills, Munnar, Idukki, Kerala
Appendiculella calostroma	*Rubus ellipticus*		Periyakanal and Kuthiraiyar, Kukkal Shola Forest, Kodaikanal, Tamil Nadu
Oidium rosacearum	*Rosa indica*		Oothu, Kodaikanal, Tamil Nadu
Prataprajella rubi	*Rubus ellipticus*		Periyakanal, Kukkal shola forest Kodaikanal, Tamil Nadu

Amazonia goosii	*Canthium* sp.	Rubiaceae	Agasthyamala Biosphere Reserve, Thiruvananthapuram Dist., Kerala
Amazonia psychotriae var. *macrospora*	Rubiaceous plant		Mannavan Shola, Munnar, Idukki, Kerala
Asteridiella ixorae	*Ixora coccinea*		Palode Forest area, Thiruvananthapuram Dist., Kerala
Asterostomula pavettae	*Pavetta* sp.		JNTBGRI Campus, Thiruvananthapuram Dist., Kerala
Balladynocallia glabra	*Canthium* sp. *Ixora* sp.		JNTBGRI campus, Palode, Athirumala, Peppara Wildlife Sanctuary Thiruvananthapuram Dist., Kerala
Hemileia vastatrix	*Coffea arabica*		Churuly Coffee Estate, Idukki, Kerala
Irenopsis xeromphidis	*Tamilnadia uliginosa*		JNTBGRI campus, Palode, Thiruvananthapuram Dist., Kerala
Monosporidium pavettae	*Pavetta indica*		Forest along the road from Painavu to Kulamavu, Idukki, Kerala
Meliola anceps	*Mussaenda arcuata* *Mussaenda belilla* *Mussaenda frondosa* *Mussaenda philippica* *Mussaenda* sp.		Forest along the road from Painavu to Kulamavu, Idukki, Kerala JNTBGRI campus and Karimancode, Palode, Thiruvananthapuram Dist., Kerala
Meliola canthii-angustifolii	*Canthium* sp.		Athirumala, Thiruvananthapuram Dist., Kerala
Meliola coorgiana	Rubiaceae member		Madikeri, Kodagu, Karnataka
Meliola ixorae var. *psychotriae*	*Ixora coccinea* *Psychotria macrocarpa*		Athirumala, Neyyar Wildlife Sanctuary, JNTBGRI Campus, Palode, Thiruvananthapuram Dist., Kerala
Meliola kanniyakumariana	*Hedyotis* sp.		Samikanal, Kodaikanal, Tamil Nadu

(*Continued*)

TABLE 2.1 (Continued)

Fungi	Host plant	Family of the host plant	Locality
Meliola kanniyakumariana var. *brahmagiriense*	*Knoxia sumatrensis*		Sampaje Ghat, Kodagu, Karnataka
Meliola luculiae	*Luculia grandiflora*		Periyakanal, Kukkal shola forest, Kodaikanal, Tamil Nadu
Meliola mitragynae	*Mitragyna parvifolia*		Meenmutty, Neyyar Wildlife Sanctuary, Thiruvananthapuram Dist., Kerala
Meliola mitragynae-tubulosae	*Mitragyna tubulosa*		forest near Kushavoor, Palode, Thiruvananthapuram Dist., Kerala
Meliola randiicola	*Ixora coccinea*		Nanniyode, Plavara, JNTBGRI Campus, Palode, and Peppara Wildlife Sanctuary, Thiruvananthapuram Dist., Kerala
Meliola weberae	*Tarenna flava*		Kuthiraiyar, Kukkal shola forest, Kodaikanal
Meliola wendlandiae	*Wendlandia* sp.		Kuthiraiyar, Kukkal shola forest, Kodaikanal Periyakanal, Kukkal shola forest, Kodaikanal
Prillieuxina ixorigena	*Ixora coccinea*		Palode, Thiruvananthapuram Dist., Kerala
Questieriella ophiorhizae	*Ophiorrhiza* sp.		In the campus of Bharatiya Vidyabhavan, Kodagu Vidyalaya, Madikeri, Karnataka
Schiffnerula braunii	*Morinda pubescence* *Morinda reticulata* *Morinda* sp. *Morinda tinctoria*		JNTBGRI campus, Palode, Thiruvananthapuram Dist., Kerala
Schiffnerula canthii	*Canthium* sp.		Ponmudi, Thiruvananthapuram Dist., Kerala

Uredo chasaliae	*Chasalia ophixyloides*		Forest along the road from Painavu to Kulamavu, Idukki, Kerala
Uredo ophiorrhizae	*Ophiorrhiza mungos*		Forest along the road from Painavu to Kulamavu, Idukki, Kerala
Asterina acronychiae	*Acronychia pedunculata*	Rutaceae	Vanachalu, Kodagu, Karnataka
Asterina atalantiae	*Atalantia rotundifolia* *Atalantia wightii*		Pandipathu, Thiruvananthapuram Dist., Kerala
Asterina melicopecola	*Melicope lunu-ankenda*		Neyyar Wildlife Sanctuary, Agasthiyar hill and Athirumala, Thiruvananthapuram Dist., Kerala
Asterina toddaliae	*Toddalia* sp.		Kuttikanam, Idukki, Kerala
Didymopsorella toddaliae	*Toddalia asiatica*		forest along the road from Painavu to Kulamavu, Idukki, Kerala
Hansfordiellopsis lichenicola	*Glycosmis pentaphylla*		JNTBGRI Campus, Palode, Thiruvananthapuram Dist., Kerala
Irenopsis murrayae	*Murraya koenigii*		Kallar; JNTBGRI campus, Palode, Karimancode, Bonacaud, Peppara Wildlife Sanctuary Thiruvananthapuram Dist., Kerala
Meliola butleri	*Citrus aurantifolia* *Citrus grandis* *Citrus limon* *Citrus* sp.		JNTBGRI campus, Palode, Thiruvananthapuram Dist., Kerala Narikal, Poovathoor, Nedumangad, Thiruvananthapuram Dist., Kerala
Meliola cadigensis var. *glycosmidis*	*Glycosmis mauritiana*		Peppara Wildlife Sanctuary, Thiruvananthapuram Dist., Kerala Painavu, Idukki, Kerala

(*Continued*)

TABLE 2.1 (Continued)

Fungi	Host plant	Family of the host plant	Locality
Meliola citricola	*Citrus grandis* *Citrus lemon* *Citrus* sp *Citrus medica* *Citrus aurantium*		JNTBGRI campus, Palode, Thiruvananthapuram Dist., Kerala Kusavoor, Palode, Thiruvananthapuram Dist., Kerala JNTBGRI campus, Palode, Thiruvananthapuram Dist., Kerala Kallara, Nedumangad, Thiruvananthapuram Dist., Kerala Calvary Mount, Idukki, Kerala Vazhathope, Idukki, Kerala
Meliola clausenae	*Clausena indica*		Kanchiar Forest, Idukki, Kerala
Meliola cranei	*Atalantia wightii* *A. rotundifolia*		Forest near Painavu, Idukki, Kerala Forest along the road from Painavu to Kulamavu, Idukki, Kerala Near Painavu, Idukki, Kerala
Meliola tecleae var. *toddaliae-asiaticae*	*Toddalia asiatica* *Toddalia asiatica* var. *floribunda*		Mannavan Shola, Munnar, Idukki, Kerala Periyakanal, Kukkal shola forest, Kodaikanal, Tamil Nadu
Meliola tenella	*Murraya paniculata*		Kuthiraiyar, Kukkal shola forest, Kodaikanal, Tamil Nadu Samikanal, Kukkal shola forest, Kodaikanal, Tamil Nadu
Meliola tenella var. *atalantiae*	*Atalantia wightii*		JNTBGRI Campus, Palode, Thiruvananthapuram Dist., Kerala
Meliola tenella var. *atalantiicola*	*Atalantia* sp.		JNTBGRI Campus, Palode, Thiruvananthapuram Dist., Kerala
Meliola toddaliae	*Pamburus missionis*		JNTBGRI Campus, Palode, Thiruvananthapuram Dist., Kerala

Meliola zanthoxyli	*Zanthoxylum* sp.		Bhagamandala, Kodagu, Karnataka Poombarai, Kukkal shola forest, Kodaikanal, Tamil Nadu
Oidium citri	*Aegle marmelos* *Citrus aurantium*		Ayyappan Kovil, Nedumangad, Thiruvananthapuram Dist., Kerala Kodaikanal, Tamil Nadu
Questieriella toddaliae	*Toddalia asiatica* var. *floribunda*		Periyakanal, Kukkal shola forest, Kodaikanal, Tamil Nadu Samikanal, Kukkal shola forest, Kodaikanal, Tamil Nadu
Questieriella zanthoxyli	*Zanthoxylum* sp.		Bhagamandala, Kodagu, Karnataka
Sarcinella limoniae	*Limonia acidissima*		JNTBGRI campus, Palode, Thiruvananthapuram Dist., Kerala
Schiffnerula girijae	*Aegle marmelos*		Karimancode, Thiruvananthapuram Dist., Kerala, JNTBGRI Campus, Palode, Thiruvananthapuram Dist., Kerala
Spiropes guareicola	*Citrus limon* *Citrus* sp. *Murraya paniculata*		Netta, Nedumangad, Thiruvananthapuram Dist., Kerala JNTBGRI Campus, Palode, Thiruvananthapuram Dist., Kerala
Asterina meliosmae-simplicifoliae	*Meliosma simplicifolia*	Sabiaceae	Hoddur, Karnataka
Asterina congesta	*Santalum album*	Santalaceae	JNTBGRI campus, Palode, Thiruvananthapuram Dist., Kerala Govt. Botanic Garden, Thiruvananthapuram Dist., Kerala
Oidium santalacearum	*Santalum album*		Rajamani Estate, Kodaikanal, Tamil Nadu
Meliola osyridicola	*Osyris wightiana*		Kuthiraiyar, Kodaikanal, Tamil Nadu Samikanal, Kodaikanal, Tamil Nadu
Meliola osyridicola var. *indica*	*Osyris wightiana*		Samikanal, Kukkal Shola forest, Kodaikanal, Tamil Nadu Kuthiraiyar, Kukkal shola forest, Kodaikanal, Tamil Nadu

(*Continued*)

TABLE 2.1 (Continued)

Fungi	Host plant	Family of the host plant	Locality
Meliola scleropyricola	*Scleropyrum pentandrum*		JNTBGRI Campus, Palode, Thiruvananthapuram Dist., Kerala
Meliola allophyligena	*Allophylus cobbe* *Allophylus* sp.	Sapindaceae	JNTBGRI campus, Palode, Thiruvananthapuram Dist., Kerala
Meliola allophyli-serrulati	*Allophylus cobbe*		Karadimala, Athirumala, Neyyar Wildlife Sanctuary, Thiruvananthapuram Dist., Kerala JNTBGRI campus, Palode, Thiruvananthapuram Dist., Kerala
Meliola capensis var. *allophylicola*	*Allophylus* sp.		Palode Forest, Thiruvananthapuram Dist., Kerala
Meliola capensis var. *malayensis*	*Dimocarpus longan* *Nephelium longan*		Kombe, Meenmutty, Thiruvananthapuram Dist., Kerala JNTBGRI campus, Palode, Thiruvananthapuram Dist., Kerala Peppara Wildlife Sanctuary, Thiruvananthapuram Dist., Kerala Painavu, Idukki, Kerala
Meliola commixta	*Nephelium longan*		Forest along the road from Painavu to Kulamavu, Idukki, Kerala
Meliola dimocarpi	*Dimocarpus longan*		Athirumala Forest, Neyyar Wildlife sanctuary, Thiruvananthapuram Dist., Kerala
Meliola filicii var. *indica*	*Filicium decipiens*		Ponmudi, Thiruvananthapuram Dist., Kerala

Meliola otophorae var. *indica*	*Sapindus laurifolius*		JNTBGRI Campus, Palode, Thiruvananthapuram Dist., Kerala Valicode, Nedumangad, Thiruvananthapuram Dist., Kerala
Meliola serjaniae var. *major*	*Sapindus emarginatus*		Hoddur, Kodagu, Karnataka
Asteridiella sapotacearum	*Madhuca longifolia*	Sapotaceae	Sampaje Forest Nursery, Kodagu, Karnataka
Asterina chrysophylligena	*Chrysophyllum roxburghii*		Kodagu, Karnataka
Asterina mimusopsidicola	*Mimusops elengi*		Palode, JNTBGRI Campus, Thiruvananthapuram Dist., Kerala
Echidnodella mimusopsidis	*Mimusops elengi*		Mandrane, Hoddur, Karnataka
Maravalia echinulata	*Madhuka longifolia*		Vazhathope, Idukki, Kerala
Symphaster mimusopsidis	*Mimusops elengi*		JNTBGRI Campus, Palode, Thiruvananthapuram Dist., Kerala
Meliola ailanthi	*Ailanthus excelsa* *Ailanthus* sp. *Ailanthus triphysa*	Simaroubaceae	JNTBGRI Arboretum, Palode, Thiruvananthapuram Dist., Kerala Peppara wildlife sanctuary Vazhayila, Nedumangad, Thiruvananthapuram Dist., Kerala
Meliola samaderae	*Quassia indica*		JNTBGRI Campus, Palode, Thiruvananthapuram Dist., Kerala Kallara, Nedumangad, Thiruvananthapuram Dist., Kerala
Meliola smilacis	*Smilax* sp.	Smilacaceae	Palode Forest, Thiruvananthapuram Dist., Kerala
Asteridiella solani kodaikalensis	*Solanum* sp.	Solanaceae	Periyakanal, Kukkal shola forest, Kodaikanal, Tamil Nadu

(*Continued*)

TABLE 2.1 (Continued)

Fungi	Host plant	Family of the host plant	Locality
Asterina diplopoda	*Solanum ferox*		Mannavan Shola, Idukki, Kerala
Golovinomyces orontii	*Solanum erianthum*		Manchur, Kodaikanal, Tamil Nadu
Leveillula taurica	*Solanum erianthum*		Perumalmalai, Kodaikanal, Dindigui Anna Dt
Fibroidium saeforthiani	*Solanum seaforthianum*		Near Bonacaud, Thiruvananthapuram Dist., Kerala
Schiffnerula palodensis	*Solanum* sp.		Chippanchira, Palode, Thiruvananthapuram Dist., Kerala, Kallara, Thiruvananthapuram Dist., Kerala, JNTBGRI campus, Palode, Thiruvananthapuram Dist., Kerala
Prataprajella turpiniicola	*Turpinia malabarica* *Turpinia nepalensis*	Staphyleaceae	Forest along the road from Painavu to Kulamavu, Idukki Dist., Kerala Mannavan Shola, Idukki Dist., Kerala Kuthiraiyar, Kodaikanal, Tamil Nadu
Asteridiella kodaikanalensis	*Symplocos anamallayana*		Shenbaganur Shola, Kodaikanal, Tamil Nadu
Asteridiella shenbaganurensis	*Symplocos anamallayana*		Shenbaganur Shola, Kodaikanal, Tamil Nadu
Asterina indica	*Symplocos laurina* *Symplocos* sp. *Symplocos cochinchinensis* ssp. *laurina* *S. torene*	Symplocaceae	Neyyar Wild Life Sanctuary, Pongalapara Agasthyamala (MPCA), Chemunji JNTBGRI Campus, Thiruvananthapuram Dist., Kerala Mannavan Shola, Munnar, Idukki, Kerala Periyar Tiger Reserve, Thekkadi, Munnar, Idukki, Kerala

Asterina suttonii	*Symplocos* sp.		Neyyar Wildlife sanctuary, Kombe, Thiruvananthapuram Dist., Kerala
Meliola symplocicola	*Symplocos cochinchinensis*		JNTBGRI Campus, Palode, Thiruvananthapuram Dist., Kerala
Amazonia gordoniicola	*Gordonia* sp.	Theaceae	Gravel Banks, Munnar, Idukki, Kerala
Appendiculella shettyi	*Gordonia obtusa*		Mannavan Shola, Munnar, Idukki, Kerala
Appendiculella vivekananthanii	*Gordonia obtusa*		Mannavan Shola, Munnar, Idukki, Kerala
Asterina euryae	*Eurya nitida* *E. japonica*		Mannavan Shola, Munnar, Idukki, Kerala Silent valley, Palghat, Kerala
Asterina theacearum	*Eurya* sp. *Gordonia* sp.		Rajamala Graval Bank, Idukki, Kerala Mannavan Shola, Munnar, Idukki, Kerala Vagapara, Munnar, Idukki, Kerala
Exobasidium vexans	*Camellia sinensis*		Munnar, Idukki, Kerala
Irenopsis gordoniae	*Gordonia* sp.		Mannavan Shola, Munnar, Idukki, Kerala
Meliola emespatilii	*Thea sinensis*		Kodaikanal, Tamil Nadu
Schiffnerula camelliae	*Camellia sinensis*		JNTBGRI campus, Palode, Thiruvananthapuram Dist., Kerala
Irenopsis coimbatorica	*Grewia serrulata*	Tiliaceae	Devarakadu, Hoddur, Karnataka
Irenopsis triumfettae var. *indica*	*Triumfetta rhomboidea*		FMC Compound, Madikeri, Karnataka
Oidium doidgei	*Triumfetta* sp.		Kodaikanal, Tamil Nadu

(*Continued*)

TABLE 2.1 (Continued)

Fungi	Host plant	Family of the host plant	Locality
Asterina elatostematis	*Elatostema lineolatum*	Urticaceae	Athirumala, Peppara Wildlife Sanctuary, Thiruvananthapuram Dist., Kerala
Sarcinella pouzolziae	*Pouzolzia* sp.		Nishane motta, Madikeri, Karnataka
Asterina girardiniae	*Girardinia diversifolia*		Mannavan Shola, Munnar, Idukki, Kerala
Appendiculella vacciniorum	*Vaccinium leschenaultia* *Vaccinium neilgherrense*	Vacciniaceae	Mannavan Shola, Munnar, Idukki, Kerala Kuthiraiyar, Kodaikanal, Tamil Nadu
Irenopsis vaccinia	*Vaccinium leschenaultia* *Vaccinium neilgherrense*		Rajamala, Munnar, Idukki, Kerala Kuthiraiyar, Kodaikanal, Tamil Nadu
Asteridiella callista	*Stachytarpheta jamaicensis*	Verbenaceae	Peppara Wildlife sanctuary, Palode, JNTBGRI, Near Peppara dam, Nettukaltheri, Neyyar Wild Life Sanctuary, Thiruvananthapuram Dist., Kerala Jodupal, Kodagu, Karnataka Nishane motta, Madikeri, Kodagu, Karnataka Field Marshal Cariappa College Campus, Madikeri Kodagu, Karnataka
Asteridiella depokensis	*Vitex negundo*		Hoddur, Karnataka
Asteridiella madikeriensis	*Premna* sp.		Nishane motta, Madikeri, Kodagu, Karnataka

Asteridiella viticis-negundoi	*Vitex negundo* L.		Hoddur, Karnataka
Schiffnerula hoddurensis	*Vitex negundo*		Hoddur, Karnataka
Schiffnerula tectonae	*Tectona grandis*		Jodupal, Kodagu, Karnataka
Asterina cissi	*Cissus* sp.	Vitaceae	Rajamala, Idukki, Kerala
Asterina vitacearum	*Cissus repens* Lam.		Hoddur, Karnataka
Meliola bakeri	Vitaceae member		Bhagamandala, Kodagu, Karnataka
Asteridiella amomi	*Amomum* sp.	Zingiberaceae	Neyyar Wildlife Sanctuary, Thiruvananthapuram Dist., Kerala

These studies reveal that Western Ghats is very rich in its fungal diversity. Environmental factors greatly influence on the growth of fungi and on their reproduction just as they do with other living beings. Not only in rates of growth but differences in types of growth occur depending on such factors. A fungi–host–host family–locality index showing Phylloplane fungi reported from the Southern Western Ghats is given in Table 2.1.

Scope of the study

Western Ghats is under increasing anthropogenic pressure. While there are several recommendations available for its conservation, the criteria for such recommendation are always criticized. Dr. Kasthurirangan report wisely demarcates ecologically sensitive areas in Western Ghats providing scope for sustainable development.

Systematic quantitative assessment of phylloplane fungi will help in identification of rare or new species. The outcome of the study will be highly useful for policy making, decision making for conservation priorities and efficient utilization of the resources.

Biodiversity is the greatest asset of a country and taxonomy is the base to understand the biodiversity. More we learn about biodiversity the more we realize how much the world dependent on it. There is an urgent need to protect the biodiversity since whole species of plants, animals, fungi and microscopic organisms are being lost at alarming rates (Figs. 2.1–2.3).

References

Hansford, C.G., 1961. The Meliolineae. A monograph. Sydowia, Beih 2, 1–806.

Hawksworth, D.L., Lucking, R., 2017. Fungal diversity revisited: 2.2 to 3.8 million species. Microbiol. Spectrum 5 (4), . Available from: https://doi.org/10.1128/microbiolspec.FUNK-0052-2016FUNK-0052-2016.

Hosagoudar, V.B., 1996. Meliolales of India. Botanical Survey of India, Calcutta, p. 363.

Hosagoudar, V.B., 2008. Meliolales of India, vol. II. Botanical Survey of India, Calcutta, p. 390.

Hosagoudar, V.B. 2010. Foliicolous Fungal Flora of Peppara and Neyyar Wildlife Sanctuaries in Kerala State, India. Balavidya Ganapathy Educational and Charitable Trust, Sundarakkottai, Mannargudi 614 016, India, pp. 124 +32 plates.

Hosagoudar, V.B., 2013. My contribution to the fungal knowledge of India. J. Threatened Taxa 5 (8), 4129–4344. Available from: https://doi.org/10.11609/JoTT.o3470.4129-344.

Hosagoudar, V.B., Agarwal, D.K., 2008. Taxonomic Studies of Meliolales. Identification Manual. International Book Distributors, Dehra Dun, p. 263.

Hosagoudar, V.B., Archana, G.R., 2012. Foliicolous Fungal Flora of Palode Forest Range, Thiruvananthapuram, Kerala. Sadguru Publications, Udaipur, p. 328.

Hosagoudar, VV.B., Riju, M.C., 2013. Foliicolous fungi of Silent Valley National Park, Kerala, India. J. Threatened Taxa 5 (3), 1–88.

Hosagoudar, V.B., Sabeena, A., 2014. Foliicolous fungi of Wayanad District in Kerala State, India. J. Threatened Taxa 6 (7), 5909–6052. Available from: https://doi.org/10.11609/JoTT.o3658.5909-6052.

Hosagoudar, V.B., Thomas, J., 2013. Meliolales in Peppara and Neyyar Wildlife Sanctuaries in Kerala State. Sadguru Publications, Udaipur, p. 254.

Hosagoudar, V.B., Biju, C.K., Agarwal, D.K., 2011a. Foliicolous Fungal Flora of Idukki District in Kerala State. International Book Distributors, Dehra Dun, Uttarakhand, pp. 1–436.

Hosagoudar, V.B., Chandra Prabha, A., Agarwal, D.K., 2011b. Asterinales of Kerala. Associated Publishing Company, p. 270, + pl.23.

Hosagoudar, V.B., Dhivaharan, VV. Nithyatharani, R., 2012. Foliicolous Fungi of Kodaikanal in Tamil Nadu. Published as a supplement to Vol: 2 (1) of Bulletin of Basic and Applied Plant Biology, pp. 94.

Manoharachary, C., Sridhar, K., Singh, R., Adholeya, A., Suryanarayanan, T.S., Rawat, S., et al., 2005. Fungal biodiversity: distribution, conservation and prospecting of fungi from India. Curr. Sci. 89, 1.

Robin, P.J., Hosagoudar, V.B. 2010. Foliar fungal parasites of Ponthanpuzha reserve forests of Kottayam district in Kerala state. National Conference on Plant Biodiversity on 23rd and 24th April. Dept. of Botany, Yashvantarao Chavan Institute of Science, Satara, Maharashtra, p. 37.

Sabeena, A., Hosagoudar, V.B., Divaharan, V., 2018. Foliicolous fungi on medicinal plants in Thiruvananthapuram District, Kerala, India. J. Threatened Taxa 10 (3), 11470–11479. Available from: http://doi.org/10.11609/jott.3761.10.3.11470-11479.

Thimmaiah, C.J., Hosagoudar, V.B., Jayashankar, M., 2013. Black mildews of Kodagu, Karnataka, India. J. Threatened Taxa 5 (16), 5021–5180. Available from: https://doi.org/10.11609/JoTT.o3657.5021-180.

Chapter 3

Fungal endophytes from seaweeds and bio-potential applications in agriculture

A. Noorjahan*, B. Aiyamperumal and P. Anantharaman
CAS in Marine Biology, Faculty of Marine Sciences, Annamalai University, Chidambaram, India
**Corresponding author.* Email: noorbiotek@gmail.com

Introduction

In recent years there is increasing research interest in exploring the field of marine microbial ecology, in part driven by the technological advances that allow for a comprehensive and detailed description of endophytic fungal diversity in seaweeds and its function.

Seaweeds

Seaweeds or marine macro algae are sessile multicellular photosynthetic eukaryotes that are differentiated from plants by their lack of specialized tissues (e.g. root system and vascular structures) (Graham and Wilcox, 1989). Due to difficult terrain and not easily accessible, seaweeds are the least explored in many countries. In some Asian countries, especially China and Japan utilize seaweed as a part of their dietary materials. However, people of western world are exploiting seaweed extract for application in food and in cosmetics industries. Seaweed preparations can be used as raw, dried or cooked form. Seaweeds have very low calorie value, they are valuable source of important minerals, vitamins, proteins, trace elements, oils, antioxidants and fats (Mouritsen, 2013).

Major pigments of green forms in seaweeds are chlorophyll *a*, chlorophyll *b*, α-, β-, and γ- carotene, and siphonoxanthin; brown seaweeds possess pigments such as chlorophyll *a*, c*1*, c*2*, fucoxanthin and β carotene. Important pigments harbored by red seaweeds are α-, β- carotene, allophycocyanin, c-phycoerythrin, chlorophyll a, and r-phycocyanin. Earlier, seaweeds

Fungi Bio-Prospects in Sustainable Agriculture, Environment and Nano-Technology.
DOI: https://doi.org/10.1016/B978-0-12-821394-0.00003-2

were utilized mainly as gelling substance and stabilizers in food and drug industry, but, presently, therapeutic roles of seaweed derived compounds as anticancer, anticoagulant, anti hyperlipidemic, immunomodulatory, thyroid stimulating, ant estrogenic, antibacterial, antiviral, ant obesity and antifungal agent have been demonstrated in disease management (Mohamed et al., 2012).

Endophytic fungi

Presently, there are several groups of fungi which have symbiotic relationships with plants have been discovered. These fungi are known as "endophytic fungi" and they live both inside and between plant cells. The alga which are the habitats of endophytic fungi provide nutrients and shelters which are suitable for the endophytic fungi while the endophytic fungi produce certain chemical substances enabling algae to protect them from environmental hazards and against pathogens, i.e., bacteria, fungi and virus. The bioactive substances which are produced by the endophytic fungi are known as "secondary metabolites" which play an important because, apart from protecting the plants from hazards, they can also repel the free radicals (Arunpanichlert et al., 2010).

Endophytic fungi may be diverse at an exceedingly small scale; a single conifer needle may harbor several dozen species. They are typically present as internal, unseen, microscopic hyphae and their presence is revealed externally only when they sporulate, usually in a seasonal and ephemeral event. Many endophytes are highly host- or tissue-specific. The regular usual conventional methods for sampling endophytic fungi are inadequate for accurately enumerating micro fungi. The details of distributions of even the most familiar taxa remain sketchy. Diagnostic and quantification of these endophytic fungi generally require selective isolation procedures.

Identification usually involves microscopic examination of host tissue and often requires a high degree of taxonomic expertise. That is especially true for isolates in pure culture that fail to produce spores or identifiable structures. Determination of growth conditions that induce sporulation is very important. Fungi that neither grow nor sporulate in culture must be detected and identified by molecular methods, such as comparisons of ribosomal DNA (rDNA) gene sequences, which also can elucidate phylogenetic position of the fungi (Guo et al., 2000). The absence or lack of basic taxonomic information is a major obstacle to ecological studies of endophytic fungi. In nature, fungal endophytes may represent themselves as obligate or facultative organisms (Hardoim et al., 2008).

Marine endophytic fungus derived from seaweeds has been received increased attention as sources for novel natural products (Bugni and Ireland 2004; Zhang et al., 2006; Jones et al., 2008; Raghukumar 2008; Schulz et al., 2008; Kjer et al., 2010). For example, the marine algal endophyte *Drechslera dematioidea* (Bubák et al. Wróblewski) Subram. *et* Jain produces several bioactive compounds including ten new sesquiterpenes (Bugni and Ireland 2004). Jones et al. (2008) stress the need for studying tropical marine algae as few of these seaweeds have been screened.

Diversity of molecules produced by seaweed endophytic fungi

Seaweeds comprise the second largest, source of diverse assemblage of marine fungi. A number of seaweeds encompassing genera mainly belonging to Phaeophyceae like *Adenocystis*, *Ascophyllum*, *Desmarestia*, *Dictyota*, *Fucus*, *Lobophora*, *Padina*, *Phaeurus*, *Stocheospermum*, *Sargassum*, and *Turbinaria* members of Rhodophyceae viz. *Gelidiella*, *Gracilaria*, *Grateloupia*, *Halymenia*, *Palmaria*, *Plocamium*, *Portieria*, *Pyropia* as well as Chlorophyceae members such as *Acrosiphonia*, *Caulerpa*, *Halimeda*, *Monostroma*, and *Ulva*, have been studied extensively for their fungal associations (Singh et al., 2015). Red and brown seaweeds have greater fungal species diversity whereas green seaweeds recorded lower fungal species diversity (Suryanarayanan et al., 2010). This poor diversity may be due to short life cycle of different green seaweeds and slow growth of the endosymbionts (Zuccaro and Mitchell, 2005).

Nearly 100 endophytic fungi from seaweeds have been reported by several workers from different regions Globally, 75% of which are known to occur in Baltic Sea, Canada, China, India, North Sea and United Kingdom. Out of 100 seaweeds collected, 41 host species belong to red algae followed by 32 species of brown algae and 19 species of green algae. In a survey by Guiry (2012), the total number of algae existing on earth is 72,500, out of which 7000 are red algae, 8000 are green algae and 2000 brown algae. Among them only about 100 have been reported to act as host for endophytic fungi which necessitates further investigation of newer host for endophytes which can be explored for novel compounds. Several fungal endophytes associated with variety of algal hosts are still unidentified.

Suryanarayanan et al. (2010) described the endophytic fungi of eleven brown algae, six green algae and eight red algae from Tamil Nadu coast, India. Alga group of phaeophyceae recorded higher diversity of endophytes as compared to chlorophyceae. endophytic fungi from ten different seaweeds from Mandapam and Pondicherry coastal region of the southern part of India. These have been screened for the production of bioactive compounds (Mathan et al., 2013). A total of 156 endophytic fungal isolates were reported to inhabit the seaweed. Among the collected sea algae, highest isolation frequency was represented by *Codium* species (80%) followed by *Ulva fasicata* (10%). Type and quantity of bioactive molecules synthesized by an endophyte depends on the host organisms and environmental factor. Based on detailed literature review, Sarasan et al. (2017) concluded that dominance of bioactive molecules derived from different endophytic fungi inhabiting within marine algae were found in the order of brown algae (39%) > red algae (28%) > green algae (23%).

The same species of endophytic fungi isolated from different algal host showed variation in their active metabolites profile thus indicating genetic variation among the endophytic fungi. Seymour et al. (2004) strongly commended that genetically similar endophytic fungi may have significant differences in their metabolic profiles. Thus, although species diversity of macroalgal endophytes may be low due to the occurrence of fungi with wide host ranges, the secondary metabolite diversity may be high.

Due to awareness of global warming and environmental pollution, microbial-based bio fertilizers and bio pesticides are considered to be among key agricultural components to improve crop productivity and contribute to sustainable environmental friendly agro-ecosystems. It is a component that aggregates a variety of microbial-based bio-products whose bioactivities are essential to stimulate and improve biological processes of the intricate plant–microbe–soil continuum (Singh et al., 2015). Agricultural microbial biotechnology through the integration of beneficial plant–microbes and micro biomeinteric actions may represent a promising sustainable solution to improve agricultural production (Timmusk et al., 2017).

Generally, microbial-based bio formulation may be classified into four types: (1) NF microbes, (2) P solubilizing/mobilizing microorganisms, (3) composting microorganisms, and (4) bio pesticides (Pathak and Kumar, 2016).

Role of endophytic fungi in plant growth

Endophytic fungi can influence plant growth differently among species and strains, not by a single mechanism but by several ways through which plant growth is promoted. Research have been mainly focused on studying the potential of various endophytic fungi to influence plant growth. Conceptually, endophytic fungi may influence plant growth either directly or indirectly. Direct influence of plant growth occurs either (i) the plant growth promoting endophytic fungi enable the attaining of resources from the environment including potassium, nitrogen, phosphorous, and iron; (ii) modify plant growth by providing or regulating various plant hormones including gibberellin, cytokinin, IAA, auxin or ethylene. Indirect influence of plant growth by endophytic fungi through production of metabolites, HCN, siderophore and antibiotics against pathogenic microorganisms (Fig. 3.1).

Seaweed endophytic fungi as phosphate solubilizers

Phosphorus is known to play an important physiological and bio-chemical role in plants. It regulates the crop growth and yield to the greater extent. Several studies have indicated that 49.3 per cent of cultivated lands are

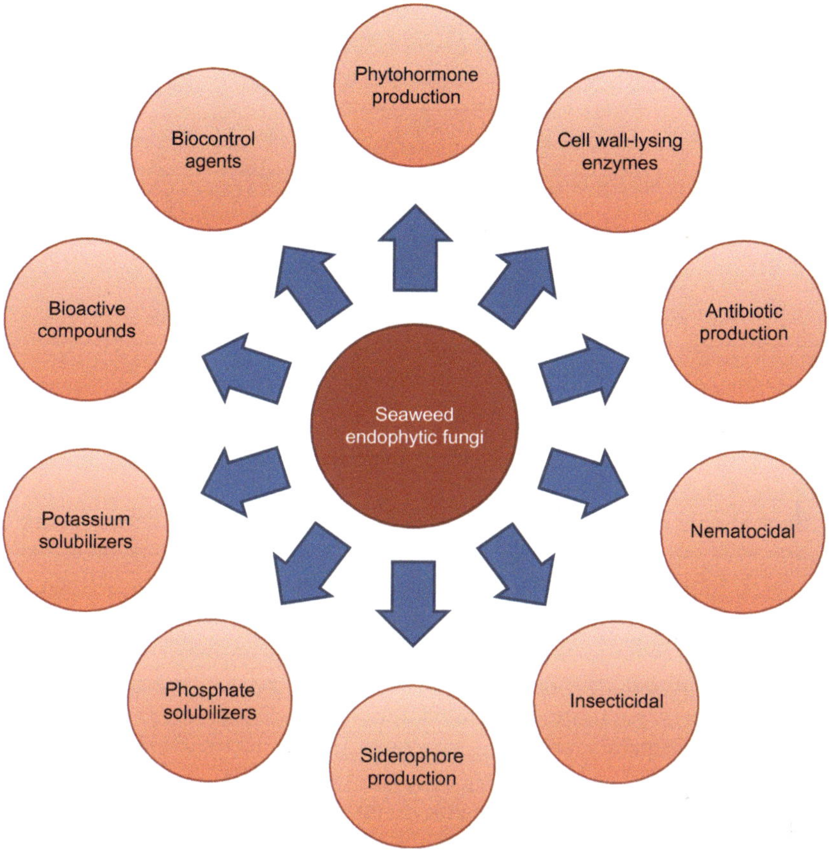

FIGURE 3.1 Different biological applications of seaweed endophytic fungi in agriculture and biotechnology.

deficient in available phosphorus (Zhang et al., 2018). The applied fertilizer phosphorus gets fixed in soil and it is not readily available to the plants Therefore, primary approach in agronomic management of phosphorus is to scavenge the native/fixed phosphorus and also to overcome the fixation of applied P-fertilizer. Phosphate-solubilizing fungi (PSF) generally enhance the availability of phosphorus (P) released from soil, which contributes to plants' P requirement, especially in P-limiting regions.

Noorjahan et al. (2019) described six endophytic fungi isolated from green and brown seaweeds from coastal regions of Tamil Nadu, India, and screened for Phosphate solubilization both qualitatively and quantitatively. *Penicillium oxalicum, P. citrinums* and A*spergillus* sp. showed maximum phosphate solubilizing activity (Fig. 3.2). Fungal endophyte assemblages of brown algae are more diverse than those of the green algae. Some endophytic fungi species have a wide host range. Algae endophytes are a good

FIGURE 3.2 Phosphate solubilization by endophytic fungi.

source of bioactive metabolites, such as antioxidants, antibiotics, anti-algal, antifungal and anti-insect compounds.

Phosphorus is an essential macro nutrient for growth and development of plants. Phosphorus plays an essential role in metabolic pathways like photosynthesis, biological oxidation, nutrient uptake and cell division (Illmer and Schinner, 1999). Worldwide soils are supplemented within organic P as chemical fertilizers to support crop production but repeated use of fertilizers deteriorates soil quality (Tewari et al., 2004). Therefore, the present scenario is shifting towards a more sustainable agriculture. Glass (1989) revealed that phosphorous is mainly absorbed by plants in two forms which are soluble forms: them monobasic (H_2PO_4) and the dibasic (HPO_4). Though, a large amount of phosphorous are present in insoluble forms and is readily not accessible for plant growth (McLaughlin et al., 2011).

Application of phosphate-solubilizing fungi increases soil fertility due to their ability to convert insoluble P to soluble P by releasing organic acids, chelation and ion exchange (Nahas, 1996). P solubilization efficiency depends more on quality rather than quantity of organic acids and P sources (Scervino et al., 2010). The latter authors, when comparing P solubilizing properties of two fungal strains (*Talaromyces flavus* and *Penicillium purpurogenum*), found that both strains exhibit equal solubilization potential, although *T. flavus* produced four times less organic acids than *P. purpurogenum*. This finding, which is in line with a study by Zhen et al. (2016), indicated a clear relationship between organic acids profile and the source of P regarding the bio-solubilization process. In addition to their positive contribution in enhancing P bio availability, soil microorganisms mediating P availability possess other substantial attributes of agronomic interests including production of phyto hormones, enhancing their salience to biotic and abiotic stress through the production of specific compounds (antifungal compounds), and the regulation of key metabolic pathways (Khan et al., 2007; Popavath et al., 2008; Sharma et al., 2013).

Seaweed endophytic fungi as potassium solubilizers

Potassium (K) is the third major essential macronutrient required for plant growth and development, and it is commonly applied as fertilizer for optimal nutrition of a crop production. The development of intensively managed agriculture resulted in an increased K removal from soil, and Globally soils showing K-deficiency, thus becoming one of the constrains in agricultural production (Sparks and Huang, 1985). In India, most of K-fertilizers is imported, costing a huge amount of foreign exchange, as there is no K-reserve in India for commercial manufacturing of K-fertilizers.

To evaluate the efficacy of endophytic fungi isolated from macroalgae *Caulerapa racemosa, Sargassum wightii* and *Turbinaria* sp. in Taminadu coastal regions which are agriculturally beneficial fungi for potash solubilization and to develop myco-potash cultures for use in crop growth. Five fungal cultures were utilized from the study, viz. *Paecilomyces lilacinus*, *Fomitopsis* sp., *Aspergillus* sp., and *Trichoderma viride and Penicillium* sp. Among them, *Penicillium* sp. and *Fomitopsis* sp. were found to produce maximum solubilization index around the colony after 7 days of incubation (DAI) on Aleksandrov medium supplemented with mica as potash source (Fig. 3.3 (Unpublished data)).

Endophytic fungus isolated from the tissues of *Caulerpa scalpelliformis*, *Caulerpa sertularioides*, *Gracilaria edulis*, *Sargassum ilicifolium* and *Ulva lactuca* also produced insecticidal compounds (Suryanarayanan et al., 2010).

Disease suppression by biocontrol agents is the sustained exposure of interactions among pathogen, bioagent, microbes inhabitated in the plant and in the environment (Handelsman and Stabb, 1996). In natural habitats, interactions between microbial cells can be as complex as they are dynamic. Each interaction is influenced by various factors such as the species, microbial diversity, various physicochemical parameters including temperature, water activity, chaotropicity and pH; motility and distance between competing

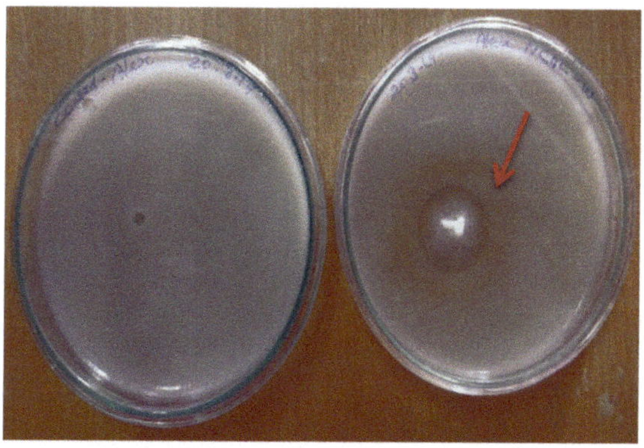

FIGURE 3.3 Efficacy of endophytic fungi as potassium solubilizer in Alekzendrov medium.

microorganisms and antimicrobials produced by the stress and other metabolic traits (Cray et al., 2013).

Fungal colonies developing in vicinity may impact significant changes in each other or they may require close hyphal connection. Variations include trophic stimulation, morphogenetic changes and different degrees of growth inhibition. Interactions between hyphae appear to be substantial in defining the pattern of colonization of fertilizer. In recent years, there has been a worldwide swing towards the use of eco-friendly methods for protecting crops from pests and diseases (Rao et al., 1998). Despite the fact that the quantity of biocontrol products is increasing, they still represent only about 1% of agricultural chemical sales (Fravel, 2005).

The efficacy of potential endophytic fungi from seaweed against *Macrophomina phaseolina*, a serious plant pathogen of economic importance under *in vitro* conditions were investigated. Among 25 isolates assayed for antagonism, seven different fungi such as *Penicillium* sp., *Aspergillus* sp., *Fomitopsis* sp. significantly inhibited colony growth of *M. phaseolina* in dual culture plates (Fig. 3.4 (unpublished data)).

Inhibition of *M. phaseolina* by *A. flavus* was supported by Cotty and Bayman (1993) indicated that atoxigenic strains of *A. flavus* will competitively exclude the aflatoxin producing strains of same fungal species. *A. terreus* as a biocontrol agent against *M. phaseolina* was similarly reported by Melo et al. (2006) specified sclerotial viability of the pathogen was found to be affected by treatment of sclerotia, with conidia of the hyperparasite *A. terreus*. *Penicillium oxalicum* slightly inhibits *M. phaseolina* supported by De Cal et al. (1997) identified disease reduction was observed when *P. oxalicum* was applied before or after inoculation with the pathogen.

Seaweed endophytic fungi also produce numerous novel bioactive metabolites that exhibit, antibacterial, anti-plasmodial, nematicidal, antiviral, and

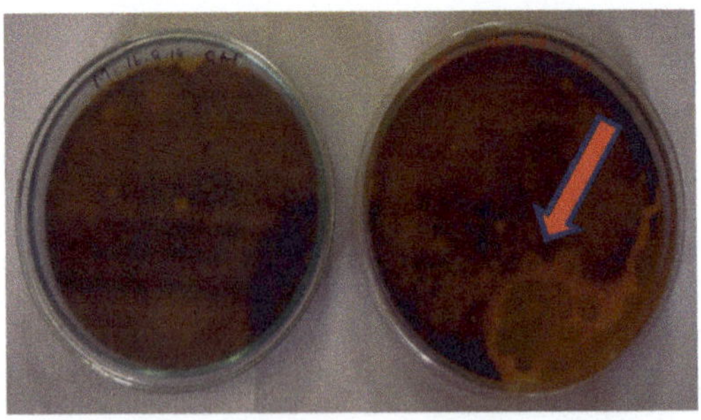

FIGURE 3.4 Dual culture and colony interactions of endophytic fungi and *M. phaseolina*.

antiangiogenic activities (Suryanarayanan et al., 2010; Oliveira et al., 2012; Flewelling et al., 2013; Godinho et al., 2013; Lee et al., 2013; Furbino et al., 2014; Suja et al., 2014). Suryanarayanan and coworkers (Suryanarayanan et al., 2010; Suryanarayanan, 2012) demonstrated that fungal isolates from green, red, and brown seaweeds produce anti-algal, anti-fungal, and insecticidal metabolites which may help in deterring colonization of algal thalli by other microbes.

Bioactive compounds

Mostly, bioactive compounds producing microorganisms are evolved through competitive environment due to nutrient and space limitation on their host surface that led them to produce allele-chemicals capable of preventing secondary colonization (Egan et al., 2001, 2008). Bio-activities of such secondary metabolites indicate that they possess pharmaceuticals, industrial, agricultural, and biotechnological applications (Armstrong et al., 2001; Penesyan et al., 2011).

Marine-derived fungi are currently being vigorously screened for bioactive compounds as they produce several novel chemicals (Liu et al., 2006). 27% of the new compounds obtained from the marine derived fungi come from those associated with algae (Bugni and Ireland, 2004). Some of these fungi produce compounds possessing totally novel carbon frameworks (Kjer et al., 2010).

Suryanarayanan et al. (2010) pointed out in his research that endophytes of marine algae are good potential source of antioxidants including phenolic acids, flavonoids, tannins, hydroxyuinones and terpenoids, which scavenge superoxide and hydroxyl free radicles. White and Torres (2010) suggest that endophytes protect their host plants from diseases, drought and heavy metal toxicity by producing antioxidant chemicals, thereby increasing tolerance of the host to oxidative stress. Further studies are necessary to determine whether endophytes of marine macro-algae also confer such benefits to their hosts by enhancing their ability to tolerate oxidative stress.

Additionally, it has been found that endophytes produce antioxidant chemicals that protect their host plants from diseases, drought, and heavy metal toxicity by increasing tolerance of the host to oxidative stress (White and Torres, 2010). Agriculturally beneficial fungi play a major role as biodegrader and biocontrol agent, but meagre efforts have been made till date to exploit endophytic fungi isolated from seaweeds as biofertilizers.

Future perspectives

a. To preserve marine biodiversity and greater organization are needed to collect and catalogue the endophytic fungi (depositories) throughout the world.

b. Research has to be done to characterize the host specificity of the endophytic fungi and its functional responsibility in the host. This will provide a deeper insight into symbiosis and the associated metabolites of the endophytic fungi and their host algae.
c. Bioactive natural products from endophytic fungi have enormous potential as the source of new Pharmaceutical, Agricultural and Industry, and methods to facilitate the identity of appropriate natural products from marine algae are required.
d. Studying different physiological and ecological conditions that have led to the activation of secondary metabolism gene clusters might also be useful in optimizing environmental condition for maximum production of metabolites.
e. Investigation of seaweed-associated microbes using new tools and techniques, such as those of high-throughput genomic and meta-genomic approaches, will lead to the discovery of more novel bioactive natural products in future, and will help in exploiting their biotechnological potential

Natural Marine-derived fungicides may be a source of new alternative active compounds, in particular with their microbial activity. Since several studies concerning the marine environment, seaweeds have emerged as a vast source of unique structures and bioactive metabolites. This research frontier, represents a new natural products source to be explored, implies that, in the future, the intriguing chemical structures already isolated from these microorganisms may provide high quality agricultural sustainability to improve environment.

Acknowledgments

We greatly thankful to Annamalai university and DBT-BIRAC for funding the project entitled "Development of integrated product with plant growth and defence potential through end to end utilization of marine biological resources". Ref BT/SBIRI1394/31/16.

References

Armstrong, E., Yan, L., Boyd, K.G., Wright, P.C., Burgess, J.G., 2001. The symbiotic role of marine microbes on living surfaces. Hydrobiologia 461, 37–40.

Arunpanichlert, J., Rukachaisirikul, V., Tadpetch, K., Phongpaichit, S., Hutadilok-Towatana, N., Supaphon, O., et al., 2010. A dimeric chromanone and a phthalide: metabolites from the seagrass derived fungus Bipolaris sp. PSU-ES64. J. Phytochem. Lett. 4, 604–608.

Bugni, T.S., Ireland, C.M., 2004. Marine-derived fungi: a chemically and biologically diverse group of microorganisms. Nat. Prod. Rep. 21, 143–163.

Cotty, P.J., Bayman, P., 1993. Competitive exclusion of a toxigenic strain of *Aspergillus flavus* by an atoxigenic strain. Am. Phytopathol. Soc. 83, 1283–1287.

Cray, J.A., Bell, A.N.W., Bhaganna, P., Mswaka, A.Y., Timson, D.J., Hallsworth, J.E., 2013. The biology of habitat dominance; can microbes behave as weeds? Microb. Biotechnol. 6, 453–492.

De Cal, A., Pascual, S., Melgarejo, P., 1997. Involvement of resistance induction by *Penicillium oxalicum* in the biocontrol of tomato wilt. Plant Pathol. 46, 72–79.

Egan, S., James, S., Holmstrom, C., Kjelleberg, S., 2001. Inhibition of algal spore germination by the marine bacterium *Pseudoalteromonas tunicata*. FEMS Microbiol. Ecol. 35, 67–73.

Egan, S., Thomas, T., Kjelleberg, S., 2008. Unlocking the diversity and biotechnological potential of marine surface associated microbial communities. Curr. Opin. Microbiol. 11, 219–225.

Flewelling, A.J., Johnson, J.A., Gray, C.A., 2013. Isolation and bioassay screening of fungal endophytes from North Atlantic marine macroalgae. Bot. Mar. 56, 287–297.

Fravel, D., 2005. Commercialization and implementation of biocontrol. Annu. Rev. Phytopathol. 43, 337–359.

Furbino, L.E., Godinho, V.M., Santiago, I.F., Pellizari, F.M., Alves, T.M.A., Zani, C.L., et al., 2014. Diversity patterns; ecology and biological activities of fungal communities associated with the endemic macroalgae across the Antarctic peninsula. Microb. Ecol. 67, 775–787.

Glass, A.D.M., 1989. Plant Nutrition: An Introduction to Current Concepts. Jones Bartlett Publishers, Boston, pp. 234–239.

Godinho, V.M., Furbino, L.E., Santiago, I.F., Pellizzari, F.M., Yokoya, N.S., Pupo, D., et al., 2013. Diversity and bioprospecting of fungal communities associated with endemic and cold-adapted macroalgae in Antarctica. ISME J. 7, 1434–1451.

Graham, L., Wilcox, L., 1989. Algae. Prentice-Hall, Upper Saddle River, NJ.

Guiry, M.D., 2012. How many species of algae are there? J. Phycol. 48 (5), 1057–1063.

Guo, B., Dai, J.R., Ng, S., 2000. Cytonic Acids A and B: novel tridepside inhibitors of hCMV protease from the endophytic fungus Cytonaema Species. J. Nat. Prod. 63, 602–604.

Handelsman, J., Stabb, E.V., 1996. Biocontrol of soilborne plant pathogens. Plant Cell 8, 1855–1869.

Hardoim, P.R., van Overbeek, L.S., van Elsas, J.D., 2008. Properties of bacterial endophytes and their proposed role in plant growth. Trends Microbiol. 16 (10), 463–471.

Illmer, P.A., Schinner, F., 1999. Solubilization of inorganic phosphates by microorganisms isolated from forest soil. Soil Biol. Biochem. 24, 389–395.

Jones, E.B.G., Stanley, S.J., Pinruan, U., 2008. Marine endophyte sources of new chemical natural products: a review. Bot. Mar. 51, 163–170.

Khan, M.S., Zaidi, A., Wani, P.A., 2007. Role of phosphate-solubilizing microorganisms in sustainable agriculture. Rev. Agron. Sustain. Dev. 27, 29–43.

Kjer, J., Debbab, A., Aly, A.H., Proksch, P., 2010. Methods for isolation of marine-derived endophytic fungi and their bioactive secondary products. Nat. Protoc. 5, 479–490.

Lee, Y.M., Kim, M.J., Li, H., Zhang, P., Bao, B., Lee, K.J., et al., 2013. Marine derived Aspergillus species as a source of bioactive secondary metabolites. Mar. Biotechnol. 15, 499–519.

Liu, R., Zhu, W., Zhang, Y., Zhu, T., Liu, H., Fang, Y., et al., 2006. A new diphenyl ether from marine-derived fungus *Aspergillus* sp. B-F-2. J. Antibiot. 59, 362–365.

Mathan, S., Subramanian, V., Nagamony, S., 2013. Optimization and antimicrobial metabolite production from endophytic fungi *Aspergillus terreus* KC 582297. Eur. J. Exp. Biol. 3, 138–144.

McLaughlin, M.J., McBeath, T.M., Smernik, R., Stacey, S.P., Ajiboye, B., Guppy, C., 2011. The chemical nature of P accumulation in agricultural soils—implications for fertilizer management and design: an Australian perspective. Plant Soil 349, 69–87.

Melo, I.S., Faull, J.L., Nascimento, R.S., 2006. Antagonism of *Aspergillus terreus* to *Sclerotinia sclerotiorum*. Braz. J. Microbiol. 37, 417–419.

Mohamed, S., Hashim, S.N., Rahman, H.A., 2012. Seaweeds: a sustainable functional food for complementary and alternative therapy. Trends Food Sci. Technol. 23 (2), 83–96.

Mouritsen, O.G., 2013. Seaweeds: edible, available, and sustainable. University of Chicago Press, London.

Nahas, E., 1996. Factors determining rock phosphate solu-bilization by microorganisms isolated from soil. W.J. Microbiol. Biotechnol. 12, 567–572.

Noorjahan, A., Aiyamperumal, B., Anantharaman, P., 2019. Isolation and characterisation of seaweed endophytic fungi as an efficient phosphate solubiizers. Biosci. Biotechnol. Res. Asia 16 (1), 33–39.

Oliveira, A.L.L., de Felício, R., Debonsi, H.M., 2012. Marine natural products: chemical and biological potential of seaweeds and their endophytic fungi. Braz. J. Pharmacogn. 22, 906–920.

Pathak, D.V., Kumar, M., 2016. Microbial inoculants as biofertilizers and biopesticides. In: Singh, D.P., Singh, H.B., Prabha, R. (Eds.), Microbial Inoculants in Sustainable Agricultural Productivity: Research Perspectives, vol. 1. Springer, Berlin, pp. 197–209.

Penesyan, A., Tebben, J., Lee, M., Thomas, T., Kjelleberg, S., Harder, T., et al., 2011. Identification of the antibacterial compound produced by the marine epiphytic bacterium Pseudovibrio sp. D323 and related sponge-associated bacteria. Mar. Drugs 9, 1391–1402.

Popavath, R.N., Gurusamy, R., Kannan, B.N., Natarajan, S., 2008. Assessment of genetic and functional diversity of phosphate solubilizing fluorescent Pseudomonads isolated from rhizospheric soil. BMC Microbiol.

Raghukumar, C., 2008. Marine fungal biotechnology: an ecological perspective. Fungal Divers. 31, 19–35.

Rao, M.S., Reddy, P.P., Nagesh, M., 1998. Evaluation of plant based formulations of *Trichoderma harzianum* for the management of *Meloidogyne incognita* on egg plant. Nematol. Mediterr. 26, 59–62.

Sarasan, M., Puthumana, J., Job, N., Han, J., Lee, J.S., Philip, R., 2017. Marine algicolous endophytic fungi—a promising drug resource of the era. J. Microbiol. Biotechnol. 27, 1039.

Scervino, J.M., Mesa, M.P., Monica, I.D., Recchi, M., Moreno, N.S., Godeas, A., 2010. Soil fungal isolates produce different organic acid patterns involved in phosphate salts solubilization. Biol. Fertil. Soil.

Schulz, B., Draeger, S., Cruz, T.E., Rheinheimer, J., Siems, K., Loesgen, S., et al., 2008. Screening strategies for obtaining novel, biologically active, fungal secondary metabolites from marine habitats. Bot. Mar. 51, 219–234.

Seymour, F.A., Cresswell, J.E., Fisher, P.J., Lappin-Scott, H.M., Haag, H., Talbot, N.J., 2004. The influence of genotypic variation of two endophytic fungal species. Fungal Genet. Biol. 41, 721–734.

Sharma, S.B., Sayyed, R.Z., Trivedi, M.H., Gobi, T.A., 2013. Phosphate solubilizing microbes: sustainable approach for managing phosphorus deficiency in agricultural soils. Springer Plus 2, 587.

Singh, R.P., Kumari, P., Reddy, C.R.K., 2015. Antimicrobial compounds from seaweeds-associated bacteria and fungi. Appl. Microbiol. Biotechnol. 99 (4), 1571–1586.

Sparks, D.L., Huang, P.M., 1985. Physical chemistry of soil potassium. In: Munson, R.D. (Ed.), Potassium in Agriculture. American Society of Agronomy, Madison, WI, USA, pp. 202–276.

Suja, M., Vasuki, S., Sajitha, N., 2014. Anticancer activities of compounds isolated from marine endophytic fungus *Aspergillus terreus*. World J. Pharm. Pharm. Sci. 3 (6), 661–672.

Suryanarayanan, T.S., 2012. Fungal endosymbionts of seaweeds. In: Raghu Kumar, C. (Ed.), Biology of marine fungi; progress in molecular and subcellular biology. Springer, Berlin, pp. 53–69.

Suryanarayanan, T.S., Venkatachalam, A., Thirunavukkarasu, N., Ravishankar, J.P., Doble, M., Geetha, V., 2010. Internal mycobiota of marine macroalgae from the Tamilnadu coast: distribution, diversity and biotechnological potential. Bot. Mar. 53 (5), 457–468.

Tewari, S.K., Das, B., Mehrotra, S., 2004. Cultivation of medicinal plants—tool for rural development. J. Rur. Technol. 3, 147–150.

Timmusk, S., Behers, L., Muthoni, J., Muraya, A., Aronsson, A.C., 2017. Perspectives and challenges of microbial application for crop improvement. Front. Plant. Sci. 8, 49.

White Jr, J.F., Torres, M.S., 2010. Is plant endophyte-mediated defensive mutualism the result of oxidative stress protection? Physiol. Plant 138, 440–446.

Zhang, W., Guo, Y.-W., Gu, R.X., 2006. Secondary metabolites from the South China Sea invertebrates: chemistry and biological activity. Curr. Med. Chem. 13, 2041–2090.

Zhang, Y., Chen, F.-S., Wu, X.-Q., Luan, F.-G., Zhang, L.-P., Fang, X.-M., et al., 2018. Isolation and characterization of two phosphate-solubilizing fungi from rhizosphere soil of moso bamboo and their functional capacities when exposed to different phosphorus sources and pH environments. PLoS One 13 (7), e0199625.

Zhen, L., Bai, T., Dai, L., Wang, F., Tao, J., Meng, S., 2016. A study of organic acid production in contrasts between two phosphate solubilizing fungi: Penicillium oxalicum and Aspergillus niger. Sci. Rep. 6, 25313.

Zuccaro, A., Mitchell, J.I., 2005. Fungal communities of seaweeds. In: 3rd ed. Digton, J., White, J.F., Oudemans, P. (Eds.), The fungal community: its organization and role in the ecosystem. Mycology Series, vol. 23. CRC Press, New York, pp. 533–579.

Chapter 4

Fungal endophytes: Entry, establishment, diversity, future prospects in agriculture

Nitin Swamy and Sardul Singh Sandhu*
Rani Durgavati University, Jabalpur, India
**Corresponding author.* Email: directordicrdvv@gmail.com

Introduction

With the advent of the 20th century, the population of the world has taken a gigantic leap. This increase in the size of the population had put a challenge in front of people who were associated with the agriculture sector to supply food requirements to fulfill the demand of this population. The world population is expected to reach 9.7 by 2050 (Desa, 2015), this increase would further increase the pressure on the agriculture sector. Along with this various environmental factors likes pollution, habitat loss, water scarcity, loss of biodiversity and climatic change are on the rise (Hodkinson et al., 2011) which further adds to the crisis of producing high crop yields. However, the objective was attained by the initiation of the Green revolution which advocates for the modern agricultural practices like using pesticides, insecticides, and fertilizers to enhance the productivity of the soil to yield more production per hectare. These practices were somehow very useful in attaining short time goals of providing people enough food, but these practices, in the long run, have left a major question on the health of the soil. With the excess use of these chemically composed insecticides, pesticides, and fertilizers the health of the soil has tainted to a level that was never imagined. The yielding capacity of the soil has reduced dramatically. This mismanagement of the chemical fertilizers has developed the resistance in the pathogens to these chemicals being used. These reasons have paved the way for the approaches which are natural and sustainable in increasing agricultural productivity to cope with the ever-increasing food demand of the population. Researches around the globe are working on exploiting natural tools instead of

Fungi Bio-Prospects in Sustainable Agriculture, Environment and Nano-Technology.
DOI: https://doi.org/10.1016/B978-0-12-821394-0.00004-4

chemicals to maintain the ecological balance. The microbial population represents one such approach, where microbes alone or in concoction with other microbes provide mutual benefits to the plants and microbes both. This approach is useful in providing protection to the plants. Among microbes, endophytes have emerged as the thrust area of research. Endophytes provide various avenues in which it can help us attain sustainable agriculture. Especially Fungal endophytes produce an array of secondary metabolites which are very useful for plant protection and their growth. Association of these fungal endophytes and plants also enriches the soil fertility.

What are endophytes?

Plants live in a dynamic association with large numbers of the microbial community, out of these there are some which enter and reside inside the plant tissue without causing any disease or negative impact. These closely associated microbes are given the name endophytes. The German botanist Heinrich Friedrich Link describes the endophytes for the first time in 1809. At that time they were called entophytae in 1809 which were partial parasitic fungi in plants (Hardoim et al., 2015; Link, 1809). The term endophyte was first used by the De Bary in 1866 to elucidate the fungi which colonize interior tissues of leaves and stems (Wilson, 1995). The word "endophytes" is made up of two Greek words Endon = within and "phyton" means plant (Chanway, 1996). Since then De Bary's definition has been modified various times, three of the most widely accepted definition follow: "Endophytes colonize symptomlessly the living, internal tissues of their host, even though the endophyte may, after an incubation or latency period, cause disease" (Petrini, 1991). "Endophytes are fungi or bacteria which, for all or part of their life cycle ... cause unapparent and asymptomatic infections entirely within plant tissues ..." (Wilson, 1995). But the most common definition of endophytes is given by Hallmann and coauthors that is derived from the practical description given in 1997 by Hallmann and coauthors, who stated that "endophytes are those (bacteria) that can be isolated from surface-disinfested plant tissue or extracted from within the plant, and that do not visibly harm the plant" (Hallmann et al., 1997). This definition has been well accepted for cultivable species around the globe but is not completely true for the endophytes which are non-cultured endophytes with the advent of molecular detection procedures in endophyte research (Garbeva et al., 2001). The endophytic-host interaction may possess a significant impact on the growth and development of the plant, and resistance from disease-causing pests and pathogens (Santoyo et al., 2016).

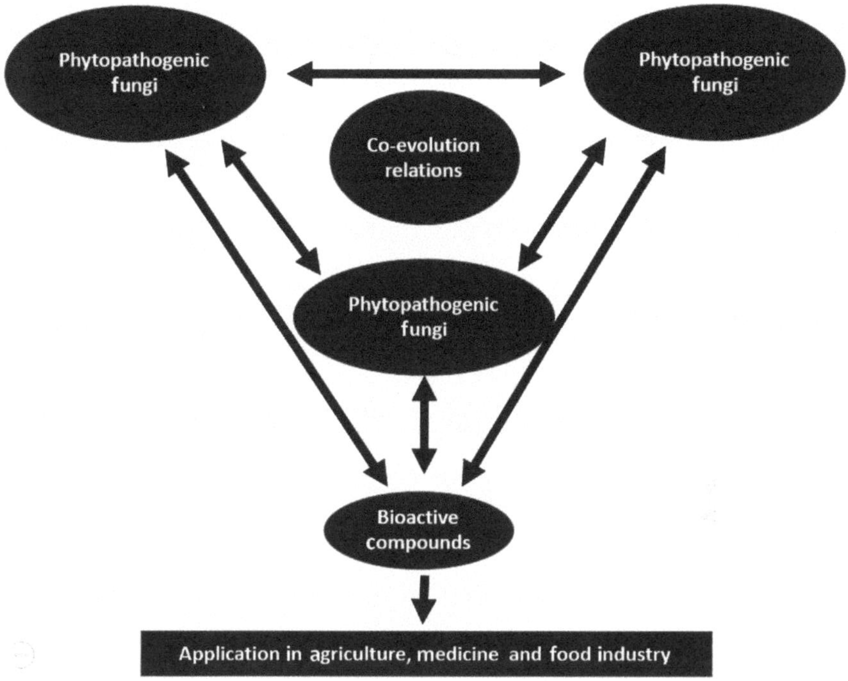

Co-evolution of fungi and bioactive compounds.

The co-evolution of fungal endophytes and host plant, has conferred fungal endophytes with an ability to synthesize the bioactive compounds which can be further used in various applications in medicine, agricultural and food industries. This symbiotic relationship has beneficial effects for both.

How do these endophytes enter and establish inside the plants?

Fungal endophytes originate from the environment of the host plants where they reside, it includes the microorganisms, fungal spores, and feeding habits and processes of insects and animals around (Rodriguez et al., 2009; Sasse et al., 2017). Fungal endophytes typically follow two transmission patters, vertical transmission and horizontal transmission. In vertical transmission, endophytes are transmitted from maternal plants to the progeny seed, this method is the widespread method of the infection in offspring (Gagic et al., 2018). Under favorable conditions, the seed germinates and the fungal endophytes present in the seed enter into the newly formed plants, thus establishing the transmission of fungal endophytes from the maternal plant to the newly formed offspring (Hodgson et al., 2014). Horizontal transmission is a preferred mode for the tissue above the ground, these fungal endophytes transmitted through spores, biotic factors like herbivores or insects and abiotic agents like wind or rain from one plant to another plant, thus establishing the transmission of fungal endophytes between different host plants (Wiewiora et al., 2015).

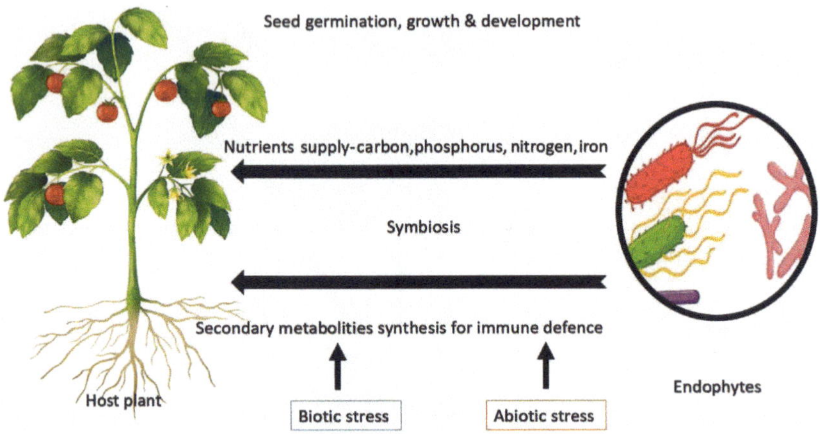

Interaction of host plant and endophytes.

These fungal endophytes create a symbiosis with the host plants and plants provides them the nourishment need for their growth and in return fungal endophytes produce diverse secondary metabolites which protect plants from various biotic and abiotic stress by regulating the various nutrients like carbon, phosphorus, nitrogen and iron etc. these secondary metabolites also helps in boosting the immune defense of the plants.

How diverse these endophytes are?

Fungal endophytes a megadiverse entity in the microbial world their utility and productivity are credited to its diverse nature. There are more than 300,000 plant species on earth. Each individual plant harbors one or more fungal endophytes. Each plant has a separate niche. It has been estimated that there are more than one million fungal endophytes making them a hyperdiverse family (Huang et al., 2007; Petrini, 1991). With reference to the evolutionary perspective, fungal endophytes have to be studied across the globe for the diversity and their functionality in the ecosystem. Researchers have studied the diversity and distribution of fungal endophytes on extensive geographical areas which include terrestrial region, semiaquatic region, freshwater region, etc. Along with it a wide number of phototrophs have investigated for fungal endophytes which include medicinal plants, mangroves, forest trees, grass. Till now the interest in the fungal endophytes is restricted towards the application perspective rather than taxonomic. Although extensive work is been done in this field but still the science to discover them is still in its naïve stage. Out of many geographical regions mentioned above the terrestrial ecosystem has been explored than the aquatic ecosystem due to the climate and environment feasibility. Under the terrestrial ecosystem, fungal endophytes of medicinal plants have been explored extensively for the secondary metabolites they produce in the symbiosis with medicinal plants.

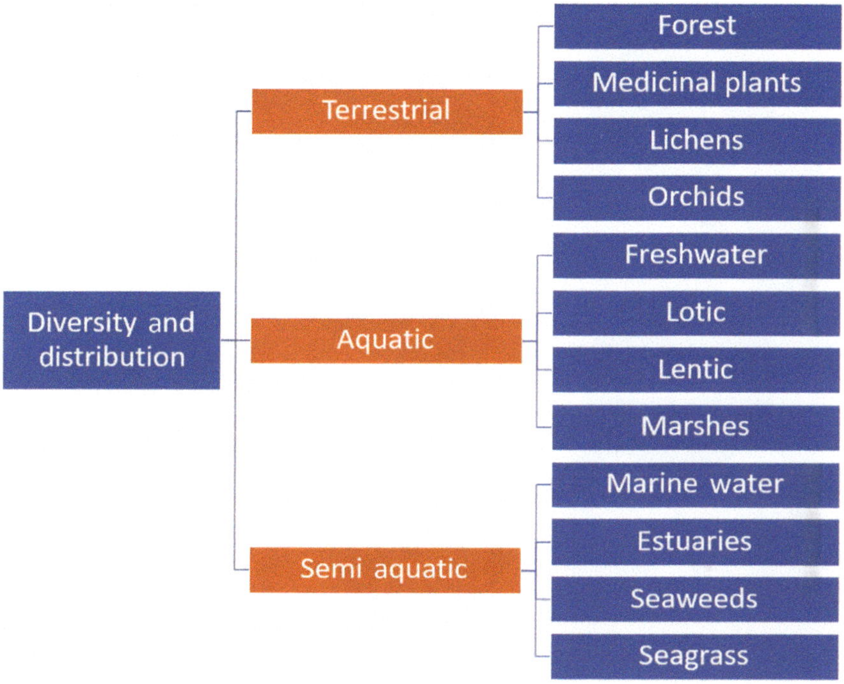

Geographical diversity of fungal endophytes.

How these fungal endophytes can contribute to sustainable agriculture?

Recently many researchers have advocated that fungal endophytes may be responsible for the adaptability in the plants to various abiotic stresses, such as attack by insect, herbivory or pathogens via the secondary metabolites production (Barz et al., 1988). Fungal endophytes provide various ways of contributing to sustainable agriculture, fungal endophytes can act like biofertilizers, antibiotic production (Kumar et al., 2014), siderophore production, phytohormone production, etc. they different ways contributes in the making host plant able to withstand various biotic and abiotic stress conditions and increase the production of the crop. Fungal endophytes also helps the host plants in the absorption of the various nutrients from the soil by increasing the uptaking ability of the nutrients and these nutrients in turn help the plant in boosting their immune defense.

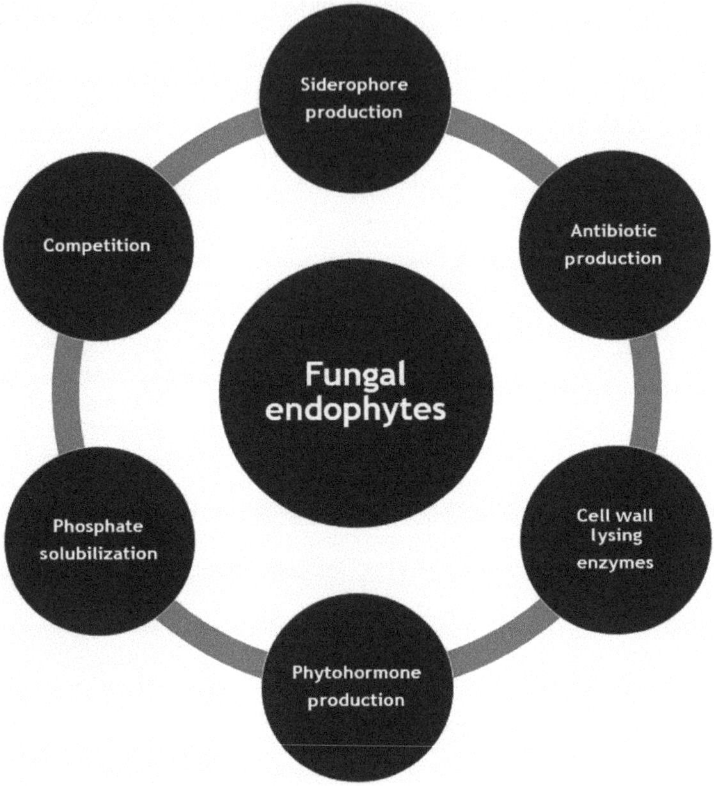

Benefits of fungal endophytes for agriculture.

Phosphate solubilization

Phosphate is the second most essential nutrient after nitrogen for the overall and growth and productivity of plants but its free availability is decreased due to its structural and chemical properties and making phosphate a limiting nutrient for its proper growth (Mehta et al., 2013a). Phosphate is abundantly present in the soil in inorganic and organic forms. In spite of being present in such a large quantity its availability is greatly reduced due to various fixation reactions occurred during biogeochemical cycles (Martinazzo et al., 2007). Around 75% of the phosphate is adsorbed when the phosphate is added externally. Soil type and pH strongly influences the adsorption-precipitation of phosphate (Lin et al., 2006).

Phosphate based fertilizers were used extensively to increase the productivity of the crop as it is the major nutrient element required by plants for growth (Sharma et al., 2015) but these chemical-based fertilizers have raised many concerns to the environment and human health.

An alternative approach to minimize the usage of these harmful chemical fertilizers is to find the solution from nature by utilizing the potential of microbes that are able to solubilize insoluble phosphate in the soil.

Pikovskaya (1948) testified for the first time that insoluble phosphate solubilization can be done by microorganisms. Followed him in the past few decades various phosphate solubilizing microbes have been discovered and reported. Numerous other microbes are being investigated for the potential of being Phosphate Solubilizing. Recent years have seen a leap in the usage of endophytic microbes as agricultural inoculants. These inoculants increase plant growth along with the production of various secondary metabolites (Chauhan et al., 2016). Phosphate solubilization contributes a lot in providing nutrition to plants in spite of being only 0.1–0.5% of total fungal populations (Kucey, 1983). These fungal endophytes have as edge over their bacterial counter part as their potential does not exhaust even after repeated culturing and synthesizes a greater amount of organic acids than bacteria (Venkateswarlu et al., 1984). Therefore they can act more efficiently in solubilizing phosphate in the soil. In recent times, many phosphate solubilizing endophytic fungi have been discovered including the genera *Curvularia*, *Penicillium*, *Aspergillus*, *Piriformospora*, etc. (Khan and Khan, 2002).

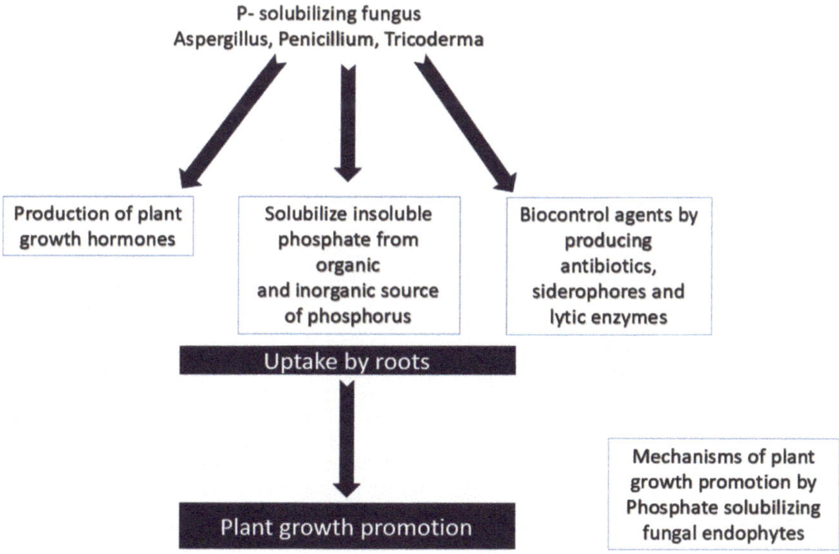

Conclusion

To fulfill the food-related demands the modern agricultural practices are extensively using the chemical fertilizers, pesticides and insecticides. These chemical based agricultural practices are ruining the ecological balance in myriad ways. Hence, researchers are focusing on nature to come up with a solution that is ecofriendly and not detrimental to nature itself. Fungal endophytes provide an excellent way to increase agricultural productivity without creating any harm to nature. Fungal endophytes with its ability to create the secondary metabolites which helps the host plants in strengthening its defense system and increase the yielding capacity. This has made researchers around the globe to focus on endophytic research to come up with the novel solution to the world's need for food without compromising with nature in a sustainable way.

References

Barz, W., Daniel, S., Hinderer, W., Jaques, U., Kessmann, H., Koster, J., et al., 1988. In: Pais, M., Mavituna, F., Novais, J. (Eds.), Plant Cell Biotechnology. Springer, Berlin, Heidelberg, New York, pp. 211–213. (NATO ASI series).

Chanway, C.P., 1996. Endophytes: they're not just fungi!. Can. J. Bot. 74, 321–322.

Chauhan, A., Guleria, S., Balgir, P., Walia, A., Mahajan, R., Mehta, P., et al., 2016. Tricalcium phosphate solubilization and nitrogen fixation by newly isolated Aneurinibacillus aneurinilyticus CKMV1 from rhizosphere of Valeriana jatamansi and its growth promotional effect. Braz. J. Microbiol. 48 (2), 294–304.

Desa, U., 2015. World Population Prospects: The 2015 Revision, Key Findings and Advance Tables. Working Paper No ESA/P/WP. 241, United Nations, Department of Economic and Social Affairs, Population Division, New York.

Gagic, M., Faville, M.J., Zhang, W., Forester, N.T., Rolston, M.P., Johnson, R.D., et al., 2018. Seed transmission of Epichloë endophytes in Lolium perenne is heavily influenced by host genetics. Front. Plant Sci. 9, 1580. Available from: https://doi.org/10.3389/Fpls.2018.01580.

Garbeva, P., van Overbeek, L.S., van Vuurde, J.W.L., van Elsas, J.D., 2001. Analysis of endophytic bacterial communities of potato by plating and denaturing gradient gel electrophoresis (DGGE) of 16S rDNA based PCR fragments. Microb. Ecol. 41, 369–383. Available from: http://dx.doi.org/10.1007/s002480000096.

Hallmann, J., Quadt-Hallmann, A., Mahaffee, W.F., Kloepper, J.W., 1997. Bacterial endophytes in agricultural crops. Can. J. Microbiol. 43, 895–914. Available from: http://dx.doi.org/10.1139/m97-131.

Hardoim, P.R., van Overbeek, L.S., Berg, G., Pirttilä, A.M., Compant, S., Campisano, A., et al., 2015. The hidden world within plants: ecological and evolutionary considerations for defining functioning of microbial endophytes. Microbiol. Mol. Biol. Rev. Available from: https://doi.org/10.1128/MMBR.00050-14July.

Hodgson, S., de Cates, C., Hodgson, J., Morley, N.J., Sutton, B.C., Gange, A.C., 2014. Vertical transmission of fungal endophytes is widespread in forbs. Ecol. Evol. 4 (8), 1199–1208. Available from: https://doi.org/10.1002/ece3.953.

Hodkinson, T.R., Jones, M.B., Waldren, S., Parnell, J.A.N. (Eds.), 2011. Climate Change Ecology and Systematics. Cambridge University Press, Cambridge.

Huang, W.Y., Cai, Y.Z., Xing, J., Corke, H., Sun, M., 2007. Potential antioxidant resource: endophytic fungi isolated from traditional Chinese medicinal plants. Econ. Bot. 61, 14–30.

Khan, M.R., Khan, S.M., 2002. Effects of root-dip treatment with certain phosphate solubilizing microorganism on the Fusarial wilt of tomato. Bioresour. Technol. 85, 213–215.

Kucey, R.M.N., 1983. Phosphate solubilizing bacteria and fungi in various cultivated and virgin Alberta soils. Can. J. Soil Sci. 63, 671–678.

Kumar, S., Aharwal, R.P., Shukla, H., Rajak, A., Sandhu, S.S., 2014. Endophytic fungi: as a source of antimicrobials bioactive compounds. World J. Pharm. Pharm Sci. 3 (2), 1179–1197.

Lin, T.F., Huang, H.I., Shen, F.T., Young, C.C., 2006. The protons of gluconic acid are the major factor responsible for the dissolution of tricalcium phosphate by Burkholderia cepacia CC-A174. Bioresour. Technol. 7, 957–960.

Link, H.F., 1809. Observationes in Ordines Plantarum Naturales, Dissertation Prima, Complectens Anandrarum Ordines Epiphytas, Mucedines, Gastromycos et Fungos. Der Gesellschaft Naturforschender Freunde zu Berlin, Berlin, Germany.

Martinazzo, R., Santos, D.R., Gatiboni, L.C., Brunetto, G., Kaminski, J., 2007. Fósforo microbiano no solo sob sistema plantio direto em resposta à adição de fosfato solúvel. Rev. Bras. Cienc. Solo 31, 563–570.

Mehta, P., Walia, A., Chauhan, A., Kulshrestha, S., Shirkot, C.K., 2013a. Phosphate solubilization and plant growth promoting potential by stress tolerant *Bacillus* sp. isolated from rhizosphere of apple orchards in trans Himalayan region of Himachal Pradesh. Ann. Appl. Biol. 163, 430–443.

Petrini, O., 1991. Fungal endophytes of tree leaves. In: Andrews, J., Hirano, S. (Eds.), Microbila Ecology of Leaves. Springer, Berlin Heidelberg New York, pp. 179–197.

Pikovskaya, R.I., 1948. Mobilization of phosphorus in soil in connection with the vital activity of some microbial species. Mikrobiologiya 7, 362–370.

Rodriguez Jr, R.J., White, J., Arnold, A.E., Redman, R.S., 2009. Fungal endophytes: diversity and functional roles. N. Phytol. 182 (2), 314–326. Available from: https://doi.org/10.1111/j.1469-8137.2009.02773.x.

Santoyo, G., Loeza, P.D., Villegas, J., Farias, R., Contreras, M., 2016. A glimpse of the endophytic bacterial diversity in roots of blackberry plants (Rubus fruticosus). Genet. Mol. Res. 15 (3), gmr.15038542.

Sasse, J., Martinoia, E., Northen, T., 2017. Feed your friends: doplant exudates shape the root microbiome? Trends Plant. Sci. 23 (1), 25–41. Available from: https://doi.org/10.1016/j.tplants.2017.09.003.

Sharma, R., Walia, A., Chauhan, A., Shirkot, C.K., 2015. Multi-trait plant growth promoting rhizobacteria from tomato rhizosphere and evaluation of their potential as bioinoculants. Appl. Biol. Res. 17 (2), 1–12.

Venkateswarlu, B., Rao, A.V., Raina, P., Ahmad, N., 1984. Evaluation of phosphorus solubilization by microorganisms isolated from arid soil. J. Indian Soc. Soil Sci. 32, 273–277.

Wiewiora, B., Zurek, G., Panka, D., 2015. Is the vertical transmission of Neotyphodium lolii in perennial ryegrass the only possible way to the spread of endophytes? PLoS One 10 (2), e0117231. Available from: https://doi.org/10.1371/journal.pone.0117231.

Wilson, D., 1995. Endophyte – the evolution of a term, and clarification of its use and definition. Oikos 73, 274–276.

Chapter 5

Fungal endophytes, biodiversity and biopotential applications

J. Immanuel Suresh* and N.M. Sona
Department of Microbiology, The American College, Madurai, India
**Corresponding author.* Email: immanuelsuresh1978@gmail.com

Introduction

The marine environment is currently well explored as one of the crucial sources regarding natural products in research. Since microorganisms from marine have exhibited exceptional biological, biochemical and biosynthetic potential. Similarly, microorganism's natural products represent a considerable area for novel therapeutic compounds search. Many reviews highlighted microbial metabolites as targets for discovery and development of latest drugs, especially anti-cancer, anti-biotic, anti-fungal and anti-parasitics among others.

Marine fungal endophytes are therefore virtually unlimited sources of novel compounds with numerous potential therapeutic applications thanks to their immense diversity and their ability to supply natural products of medicinal and pharmaceutical importance some of them produce metabolites with various biological activities against human pathogens. Which inspires researchers to study them further. The unique physicochemical properties of the marine environment are likely to possess conferred marine fungi with special physiological adaptations that would be exploited in Biotechnology. Endophytic fungi associated with marine invertebrates especially sponges in marine extreme environments.

History and evolution of endophytic fungi

Terrestrial fungi have long been known as a rich source of biologically active secondary metabolites. Since the invention of penicillin by Sir Fleming in 1928, which resulted during a breakthrough within the treatment of bacterial infections, fungi became a crucial source of drugs for the treatment of a various diseases. Antibiotics have saved millions of lives and played an important role in human history. Besides other well-known anti-microbial agents of

Fungi Bio-Prospects in Sustainable Agriculture, Environment and Nano-Technology.
DOI: https://doi.org/10.1016/B978-0-12-821394-0.00005-6

fungal origin like fusidic acid and Griseofulvin1, new semi-synthetic antifungal drugs like Anidulafungin, Caspofungin and Retapamulin are likewise derived from fungal metabolites. In 1971 the discovery of cyclosporine isolated from *Tolypocladium inflatum* an important step in immunopharmacology was taken, and improvements in the field of organ transplantation and treatment of autoimmune diseases are still in progress, as the introduction of substances such as the fungal-derived mycophenolic acid (Myfortic) to the market in 2004. In addition, pharmacologically important fungal metabolites include antilipidemic drugs collectively referred to as 'statins' with their root compounds mevastatin and lovastatin isolated from *Penicillium citrinum* and *Aspergillus terreus*, respectively. Statins reduce blood cholesterol levels and are used for the treatment of coronary diseases (Kjer et al., 2010). Fungal metabolites are, however, not only indispensable for medicine but also are important for plant protection, as demonstrated by the invention of the strobilurines that were first isolated from *Strobilurus* sp. and served as head compounds for synthetic fungicidal such as trifloxystrobin. However, the rediscovery of high numbers of previously described metabolites has to some extent precluded the study of traditional terrestrial sources of fungi and recently, the interest of natural product chemists and pharmacologists has turned to so far less investigated habitats and ecological niches such as the oceans. The oceans cover nearly three-quarters of the earth's surface and can be considered a 'soup' of essentially all imaginable types of microbes. Ecological niches, e.g. deep-sea hydrothermal vents, mangrove forests, algae, and sponges, provide distinct habitats for the isolation of particular microorganisms.

Marine micro-organisms including bacteria, cyanobacteria, microalgae, and fungi have become an essential source of new pharmacologically active metabolites. Recent reviews demonstrate the importance of these organisms as latent sources of pharmaceutical leads. Especially, marine fungi have shown promising potential as sources of new and bioactive natural products as suggested by the chemical diversity of their secondary metabolites, even though sample collection and preparation as the first step might be a challenge for marine-derived fungi than for terrestrial material because of difficulties inherent to the collection in marine environment, such as the need for scuba diving. In spite of the foremost famous group of bioactive compounds obtained from marine-derived fungi are still the cephalosporines with cephalosporine C, first isolated by G. Brotzu in 1945 from a marine strain of *Acremonium chrysogenum*, there are also more recent promising examples such as halovir and several naturally occurring analogs, which are potent inhibitors of Herpes simplex viruses 1 and 2 (Kjer et al., 2010). The new alkaloid sorbicillactone A isolated from the sponge-derived fungus *Penicillium chrysogenum* showed promising activity against leukemia cells without exhibiting notable cytotoxicity. Large-scale fermentation of the fungus yielded sufficient amounts of the compound for preclinical investigations. Overall, research on marine-derived fungi has led to the discovery of more than 270 new natural products mainly from the early 1990s to mid-2002 (with more than two-thirds of these compounds isolated from sponge- and plant-derived fungi), whereas more than 330 new metabolites were reported only in the 5-year period between 2002 and 2006, including numerous compounds with new carbon skeletons. This sharp rise in numbers intimates a growing interest in marine-derived fungi as sources of new bioactive metabolites.

Endophytic fungi

The endophytic word means “in the plant” (endon Gr. = within, phyton = -plant). The term endophyte coined by de Bary in 1866. To illustrate the fungi found in the internal tissues of stems and leaves. Endophytes are an important group of microorganisms linked with different organs and tissues of the aquatic and terrestrial plant, whose infections are unclear and the infected host tissues are symptomless rather than parasites which caused disease and reduce the fitness of their host plant (Khiralla et al., 2016). According to different author’s definitions of the term, endophyte differs. Two widely accepted definitions are

> Endophytes colonize symptomless the internal tissues of their living host, and after a period of time, they may or may not cause diseases.
> The endophytes fungi or bacteria are living a part or all of their life cycle and causing asymptomatic infections inside plant tissues.

The most significant group of eukaryotic microorganisms is fungi. The structure of fungi is branching colorless hypha which extends through the substrate where the fungus grows then; a vast number of the hyphae of each fungus are twisted to form a tangled web named mycelium. They have been isolated from dust, sand, soil, fresh and seawater, marine sediment, vertebrates, marine invertebrates and tissues of terrestrial and marine plants (Khiralla et al., 2016). They play important roles in nature, biotechnology, agriculture, and many living organisms. Endophytic fungi research has a long history and its diversity among plants is large. Each plant has been colonized by one or more endophytes. Lately, endophytes are determined to be a source of bioactive antimicrobial natural products and secondary metabolites.

Symbiosis is a widespread phenomenon in nature. Endophytes are described as all microorganisms that colonize asymptomatically within living healthy tissues. In general, endophytes are recognized as commensalistic symbionts, where they receive nutrients and shelter from the host. In return, they are thought to provide the host with chemical Recently, endophytic fungi have drawn particular attention, due to their considerable biodiversity, unparalleled metabolic pathways, and unique habitats. Therefore, they were recognized as an unusual source of novel secondary metabolites, exhibiting a variety of biological activity, which are in practice in modern agriculture, pharmaceutical, and biotechnological industry. They easily influence the physiological activities of their host without causing any harmful effects. Fungal endophytes enhance their host resistance against abiotic stress, extreme conditions, diseases by producing a broad range of fungal metabolites. Metabolites isolated from endophytic fungi refer to diverse chemical classes, including: alkaloids, steroids, flavonoids, terpenoids, quinones, and phenol.

Biodiversity of fungal endophytes in India

India has about 329 million hectares of the geographical area of which a coastline of over 7500 km. The ecosystem diversity of India is huge, ranging from sea level to the highest mountain ranges in the world; hot and arid conditions in the northwest to cold arid conditions in the trans-Himalayan region; tropical wet evergreen forests in Northeast India and the Western Ghats; mangroves of Sundarbans and freshwater aquatic to marine ecosystems (Mane and Vedamurthy, 2019). Although many studies illustrate endophyte diversity in several ecologies, there's no reliable estimation of the amount of endophytic species, of their host- and tissue-selectivity, since environmental factors have a complex effect on these features.

Unique features of marine environment and their relevance to marine fungi

A consideration of the unique properties of the marine environment is important for marine biotechnology for several reasons:

1. A good understanding of the ecosystem will help the prospect for novel genes
2. Biotechnological production processes are influenced by the special adaptations of organisms to their environment.

    The physical factors that influence the marine fungi most are

    1. salinity and pH,
    2. low water potential,
    3. high concentrations of sodium ions,
    4. low temperature,
    5. oligotrophic nutrient conditions and
    6. high hydrostatic pressure, the unique parameters of the deep-sea environment are the last three of the above.

The average salinity of seawater is 33–35 ppt. Freshwater in comparison contains less than 0.05% salts (0.5 ppt). Far more saline than the sea is hypersaline waters such as those in the Dead Sea, containing 5–10% salts or salinities of 50–100 ppt.

Low water potential is one of the major problems posed by seawater. Organisms living in it, therefore, need to maintain water potentials lower than that of seawater in their cells to enable water uptake (Zhang et al., 2007). Marine fungi maintain this gradient by accumulating osmolytes like glycerol, mannitol, polyol, and trehalose. Marine yeasts produce glycerol to maintain their internal osmotic response potential to increased salinity.

The presence of high levels of sodium ions in seawater also confers some unique properties to the cells of organisms living in the sea. Sodium, even in small concentrations is toxic to most of the living cells in the terrestrial and freshwater environments. Many marine fungi are known to reduce the toxicity of sodium ions by sequestering them in vacuoles or have a very efficient sodium efflux ATPase. The straminipilan fungi, the thraustochytrids, and labyrinthulids, on the other hand, there is a necessity for sodium for their growth and sporulation. Thus, these two groups are great tools for understanding the physiology of growth and enzyme production in the presence of sodium.

Terrestrial fungi generally grow best at pH 4.5–6.0, whereas facultative marine fungi were demonstrated to grow and produce many extracellular enzymes at pH 7–8 (Zhang et al., 2007). Lorenz and Molitoris demonstrated that salinity optimum for growth in some marine fungi shows an upward shift with rising incubation temperature. This interaction of salinity and temperature is termed Phoma pattern since it was first expressed in the marine species of *Phoma.*

Classification of endophytes

Endophytic fungi have a great effect on plant fitness, ecology, and evolution. Various groups of these microorganisms are able to produce the number of bioactive agents. In the past, endophytic fungi were classified into two groups: – Clavicipitaceous and non-Clavicipitaceous according to their taxonomy, evolution, ecological function and host specificity. After that, illustrated that there are four special groups depending on six properties which are tissue (s) colonized, plant colonization pattern, host range, plant biodiversity levels, ecological functions and mechanism of transmission between host generations. Class (1) is Clavicipitaceous endophytes while; non-Clavicipitaceous endophytes are divided into three special groups class (2), class (3) and class (4). Most of them are belong to the class (2) non-clavicipitaceous endophytes (Khiralla et al., 2016).

Sponge associated fungal endophytes

Sponges are exclusively aquatic and mostly marine. They are found from the deepest oceans to the edge of the sea. Approximately there are 15,000 species of sponges in the world. Out of this 150 occur in freshwater, but only 17 have commercial values. Nearly 486 species of sponges have been identified in India (Indira et al., 2015). In the Gulf of Mannar and Palk Bay, a maximum number of 275 species of sponges have been recorded. Among this *Callyspongia diffusa, Echinodictyum gorgonoides, Callyspongia reticutis, Gelliodes cellaria*, and *Thalysias vulpine* are the common sponge species present in the southeast coastal region *Aspergillus* sp. and *Penicillium* sp. are common associates of sponge species found in different geographical locations (Thirunavukkarasu et al., 2011). Among the group of marine organisms, sponges are the most diverse and abundant, due to their soft bodies and sedentary lifestyle (Chandrakala and Priya, 2017; Issac Dhinakaran et al., 2012). These marine invertebrates have evolved a chemical defense mechanism against other invading organisms, which involves the production of primary and secondary metabolites along with fungal endophytes. The secondary metabolites do not occur randomly but are connected with ecological factors. Nevertheless, a growing number of metabolites from sponge derived fungal strains have been reported in the last years. It provides an summary of sponge species investigated taxonomy of isolated fungi and reported metabolites. The secondary metabolites of the 18,000 marine natural products represented over 30% are from sponges and of the antitumor natural products patent registrations in recent years over 75% are from sponges (Samuel et al., 2011).

Technologies are developed to produced novel products from marine sponges associated endophytic fungi. A variety of natural products from the marine sponges have been found to exhibit remarkable activities. To understand their coevolution, lifestyle, life history, diversity, ecology, natural products, biological control bio-prospects, stress tolerance, and bioremediation. Which stimulate researchers to pay more attention towards a mutualistic association between sponge species and endophytic fungi

Bioactive compounds and their biopotential applications

Primary and secondary metabolites

Metabolites that are produced by endophytic fungi associated with sponge species at their primary level of metabolism are said to be primary metabolites. These are the substance isolated from several biosynthetic pathways such as isoprenoids, polyketide, amino acid derivatives, belonged to diverse structural groups: terpenoids, steroids, xanthones, chinones, phenols, isocumarines, benzopyranones, tetralones, cytochalasines, and enniatine.

The secondary metabolites of the fungal strains of the genus *Aspergillus* sp., both terrestrial and marine origin, have been extensively studied. Like other members in this genus, marine isolates of *Aspergillus niger* has been proven to be a prolific source of new secondary metabolites unknown from those of the terrestrial ones. For example, some strains of *A. niger* from marine sponges and ascidians produced novel metabolites including asperazine, aspernigrins A and B, yanuthones AE, nafuredin, and so on. Other examples such as Anthraquinones and Betaenones from the sponge-derived fungus *Microsphaeropsis* sp. which act as a inhibitor of protein kinase (Raghu Kumar, 2008).

Fungal Endophytes are an untapped, unappreciated microbial society. Endophytic fungal biology has appeared in many clinical applications with molecular approaches.

- They mimic the characteristics of the host and their behavior and are able to produce secondary metabolites as novel compounds.
- Fungal endophytes formulation based bio-fertilizers may be used to increase soil fertility and crop yield.
- Fungal endophytes can be used in the degradation of plastics, electrical materials, etc. by taking help from rDNA technology.
- Fungal endophytes can be used to cure cancer, tuberculosis, Malaria, Diabetes Mellitus, etc.

Other biopotential applications including, The fungal endophytes can also be used for different fermentation procedures. Molecular study of fungal endophytes may help to enhance the activity of the drug research area (Mane and Vedamurthy, 2019). However, over time researchers should address questions about endophyte biology exploring the procedures to confirm fungal endophytes and differentiate fungal endophytes from fungal Epiphytes.

Factors affecting the production of metabolites

Fungal endophytes require the optimum conditions for the production of bioactive metabolites. Parameters like incubation temperature and pH are found to be optimum for maximum mycelial growth and bioactive metabolite production.

Physical factors: temperature, salinity, pH, and light.

Chemical factors: media components, precursors and inhibitors (Suja et al., 2013) (for the growth of microbes).

Future prospects

From the foregoing discussion, the following areas are recommended to be strengthened in the future.

We need to investigate whether endophytic fungi living in algae and other marine plants and endosymbionts in marine invertebrates induce the host to produce structurally as well as activity-wise novel bioactive molecules.

A systematic search should be initiated for fungi from the special extreme environments such as deep-sea, hypoxic zones (with low oxygen levels) and hydrothermal vents for enzymes, degradation of xenobiotics and bio-remediation applications.

Salt-tolerant fungi and their salt-tolerant enzymes are used as a bio remedial tool to control pollutants (Raghu Kumar, 2008).

The future of anti-microbial discovery is based on the genome mining of marine microorganisms. It is a useful method for screening biopotential endophytes, by using the genomic sequence of the microorganisms stored in the databank.

Further imminent prospects may involve solving questions related to how endophytes communicate with each other in the view of their pathogenicity, the biodiversity of Fungal Endophytes functional classes across environmental gradients, their survival mechanism, biogeographic patterns, evolutionary origins, and habitat adaptations and can fungal endophytes be used by rDNA technology successfully.

Fungal endophytes are the future research of microbial science, due to their unique adaptation in nature. Fungal endophytes have an efficient activity to produce bioactive secondary metabolites with antimicrobial, anticancer, antioxidants, insecticidal and cytotoxic activity. They have already shown to be a potent source for the discovery of bioactive compounds, but still, new and innovative ideas are needed for natural product-based drug discovery to overcome the drug-resistant problem. For such an innovative initiation, well-organized approaches are essential for the study of fungal endophytes.

References

Chandrakala, N., Priya, S., 2017. Sponge of South East Coast: A Review Article.

Indira, K., Jayaprabha, N., Srinivasan, M., 2015. Antimicrobial activity of endophytic fungi isolated and identified from salt marsh plant in vellar estuary. J. Microbiol. Antimicrob. 13–20.

Issac Dhinakaran, D., Manohari, V., Tamilselvi, K., Lipton, A.P., 2012. Antifungal and cytotoxic activities of some marine sponges collection from the southeast coast of India. J. Appl. Pharm. Sci. 02 (01), 52–55.

Khiralla, A., Yagi, S., Rosella, S., Laurain-Mattar, D., 2016. Endophytic fungi: Occurrence, Classification, Function and Natural Products. NOVA Publishers, New York.

Kjer, J., debbab, A., Aly, A.H., Proksch, P., 2010. Methods for isolation of Marine-derived fungi and their bioactive secondary products. Nat. Protoc. 5 (3).

Mane, R.S., Vedamurthy, A.B., 2019. The fungal endophytes: source and future prospects. Artic. J. Med. Plants 6 (2), 121–126.

Raghu Kumar, C., 2008. Marine fungal biotechnology: an ecological perspective. Fungal Divers. 31, 19–35.

Samuel, P., Prince, L., Prabakaran, P., 2011. Antibacterial activity of marine derived fungi collected from southeast coast of Tamil Nadu, India. J. Microbiol. Biotechnol. Res. 1 (4), 86–94.

Suja, M., Vasuki, S., Sajitha, N., 2013. Optimization and antimicrobial metabolite production from endophytic fungi *Aspergillus terreus* KC 582297. Euro. J. Exp. Biol. 3 (4), 138–144.

Thirunavukkarasu, N., Suryanarayanan, T.S., Venkatachelam, A., Geetha, V., Jagadesan, P.R., Muresh, D., 2011. Fungal symbionts of marine sponges from Rameshwaram Southern India: species composition and bioactive metabolites.

Zhang, Y., Li, X.-M., Wang, B.-G., 2007. Nigerasperones A∼C, new monomeric and dimeric naptho-pyrones from a marine alga-derived endophytic fungus *Aspergillus niger* EN-13. J. Antibiot. 60 (3), 204–210.

Chapter 6

The role of fungi in abiotic stress tolerance of plants

Sushma[1], Rajnish Kumar Verma[2], Shivani Thakur[1], Harpreet Singh[2] and Dhiraj Kapur[1,*]

[1]*Department of Botany, Dolphin PG College of Science and Agriculture, Fatehgarh Sahib, India,* [2]*Department of Agriculture, Swami Vivekanand Group of Institutes, Banur, India*

**Corresponding author.* Email: dhiraj1543@gmail.com

Introduction

Abiotic stresses, which include heavy metal, drought, salinity, cold and heat stress are a severe threat to agricultural ecosystems which greatly hampers the crop productivity and also deteriorates the environment (Wang et al., 2003). It was estimated that worldwide, these stresses are responsible for reduction of average yield of various major crops by more than 50% (Boyer, 1982; Bray et al., 2000). These stresses cause many morphological, physiological, biochemical and molecular alterations that finally leads to loss of productivity and degradation of ecosystem (Wang et al., 2001). Climatic changes and excessive use of fertilizers and pesticides in modern agricultural practices to increase production worsen the effects of these abiotic stresses. Only 0.1% of the applied pesticides reach the target organism, while the residual contaminates the soil and decreases growth, productivity of crop and also alters the biological and chemical properties of the soil (Carriger et al., 2006). Furthermore, increase and expansion of human population with time has caused degradation of habitats, increased pollution and increased salinization of agricultural fields. All the stresses mentioned above are often interconnected and these are known to initiate similar cellular damage (Wang et al., 2003), e.g. both, drought and salinity stress which are principally shown as osmotic stress, brings about the change in cell homeostasis and ion dispersal in the cell (Serrano et al., 1999; Zhu, 2001). High temperature causes denaturation and accumulation of deoxyribonucleic acid (DNA), cellular proteins, causes induction of thermo tolerance and accumulation of heat shock proteins (hsps) (Krishna, 2003) whereas low temperature impairs many metabolic processes, modifies membrane properties, changes structure of proteins and inhibits enzymatic activity (Heino and Tapio, 2003). Heavy

Fungi Bio-Prospects in Sustainable Agriculture, Environment and Nano-Technology.
DOI: https://doi.org/10.1016/B978-0-12-821394-0.00006-8

metals (HMs) like arsenic (As), cadmium (Cd), chromium (Cr), copper (Cu), lead (Pb), mercury (Hg) and nickel (Ni) interferes with photosynthesis, water and mineral uptake and also causes modification of amino acids, proteins, deoxyribonucleic acid (DNA) and cell membrane damage (Yadav, 2010; Rascio and Navari-Izzo, 2011; Pierart et al., 2015). All plants are considered capable of recognizing and responding to these stresses (Bohnert et al., 1995; Bartels and Sunkar, 2005). Common plant responses against these stresses include altering water transport, production of different osmolytes and scavenging of harmful reactive oxygen species (ROS) by action of different enzymatic and non-enzymatic antioxidants (Leone et al., 2003; Maggio et al., 2003; Tuberosa et al., 2003).

Plants are considered autonomous organisms however, they provide a unique ecological niche to various communities of symbiotic microorganisms which often benefits plants with increased photosynthetic efficiency, nutrient and water uptake (Barrow et al., 2008). In addition, these beneficial microbes, mainly bacteria and fungi are useful in overcoming the harmful effects and to enhance plant's performance under various environmental stresses (Levy et al., 1983; Barrow et al., 2008). Fossil records have verified that fungi are linked to plants since atleast 400 million years ago (Redecker et al., 2000; Krings et al., 2007) and fungal symbionts are thought to be accountable for the movement of plants onto land (Pirozynski and Malloch, 1975). Present day research suggests that all plants in natural environments are symbiotic with fungi and other microorganisms such as bacteria, nematodes and yeast on their leaves, root, in rhizosphere and internal tissues that affects their performance (Ganley et al., 2004; van der Heijden et al., 2006). Growing of high yielding varieties to increase production, rigorous cropping system, uneven and excessive use of chemical fertilizers are the main reasons which depletes soil nutrients and develop nutrient deficiency, lessens yield, hampers soil fertility and quality of food. Hence, to solve these problems it has become a necessity to develop sustainable strategies to alleviate the negative effect of intensive practices adopted by the farmers (Elkoca et al., 2010). Scientists around the globe are working to develop novel options to increase agricultural productivity with sustainability, but it undoubtedly is a huge challenge for them. To use beneficial microbial symbionts of plants with an objective of enhancing productivity of various agricultural crops is one of the most vital sustainable practices (Freitas et al., 2007).

Fungal symbiosis

Symbiosis is defined as "the close relationship between two organisms of different species, with benefits to one or both the individuals" (Yuan et al., 2010). Symbioses may be 'obligate', in which the two individuals are so interdependent, that they can't survive without the other or 'facultative', in which the two individual forms an association through choice, and can even survive individually. Many animals and plants are known to form associations with a number of microorganisms. Evolutionarily, in order to sustain in a particular habitat and continue their normal growth and development plants need specialized microbial associates (Yuan et al., 2010). Various types of symbionts like actinorhizal, rhizobial and mycorrhizal have long been studied and are considered as the primary mutualistic microbes associated with plant roots (Pawlowski and Bisseling, 1996; Diouf et al., 2003), aboveground many plant-fungal associations are known to exist in temperate and tropical grasses and many endophytic fungi inhabiting inner leaf tissue of grass are known to increase the resistance of the hosts against herbivores, pathogens and multiple stresses (Kuldau and Bacon, 2008). In plant-fungal relation the outcome of the association to host plants can be positive (mutualism), negative (amensalism, competition and parasitism) or neutral (commensalism and neutralism) (Lewis, 1985; Rodriguez et al., 2008). Successful colonization of plant root by fungus is completed in three main steps (1) fungus penetration into plant tissues (2) colonization of plant tissues by the entering fungus (3) expression of a fungus in symbiotic association. However, in order to establish a compatible interaction with the host symbiotic as well as pathogenic, fungus needs to overcome or manipulate defense system of host plant (Pieterse and Dicke, 2007; Robert-Seilaniantz et al., 2007). It is anticipated that some type of biochemical and/or genetic communication take place among the companions (symbiont and host), but exactly what type of communication occurs which result in the expression of different symbiotic lifestyles, is still not known (Redman et al., 2001). However, in different cases, different strategies have been utilized by the symbiont, such as disturbing plant's defense signaling pathways or even reconstructing host metabolism for example alterations in hormonal homeostasis and antioxidant contents (Robert-Seilaniantz et al., 2007; Göhre and Robatzek, 2008). Earlier, it was believed that specific fungus is restricted to specific symbiotic relationships which can be either mutualistic or commensalistic or parasitic (Lewis, 1985) but later studies suggest that fungi can express different symbiotic relationship depending upon the host genotypes and various environmental factors e.g. depending on the functional status of plants, some mycorrhizal fungi may behave as mutualistic or parasitic (Francis and Read, 1995; Johnson et al., 1997; Graham and Eissenstat, 1998).

Types of fungal symbionts

The fungal symbionts associated with plants are categorized into two major classes: (1) Endophytic fungi, which dwell entirely inside plant tissues and are found associated with roots, stems and leaves (2) Mycorrhizal fungi, are the fungi found in roots only but also extend out into the rhizosphere (Sherwood and Carroll, 1974; Carroll, 1988; Stone et al., 2004; Smith and Read 2008). The term "endophyte" which was first introduced by de Bary in

1866 is derived from the Greek words, *endon* = within and *phyte* = plant. The word Mycorrhiza is combination of two Greek words "mycos = fungus" and "rhiza = root" which means "fungus root" (Smith and Read, 2008).

Endophytic fungi are divided into two major groups based on differences in their taxonomy, host range, colonization patterns, tissue specificity and ecological function. Group one consists of the Clavicipitaceous endophytes (C-endophytes or class-1 endophytes) which infect some grasses. Group two consists of the Nonclavicipitaceous endophytes (NC-endophytes or class-2 endophytes) which can be found in tissues of nonvascular plants, ferns and fern-allies, conifers and angiosperms (Schaechter, 2011). On the other hand, Rodriguez et al. (2009) classified fungal endophytes into four classes. Two major endophytic groups (Clavicipitaceous and Nonclavicipitaceous) based on phylogeny and life history characters. Furthermore, they categorized NC-endophytes into three functional classes based on host colonization and transmission and fitness benefits granted to the hosts. The C-endophytes include free living and symbiotic species which are found associated with insects, grasses and sedges (Bacon and White, 2000; Bischoff and White, 2005). Mycelium of C-endophytes can be found in intercellular spaces of leaf sheaths, culms, rhizomes and rarely on the surface of leaf (White et al., 1996; Moy et al., 2000; Dugan et al., 2002; Bacon and de Battista, 1991). C-endophytes enhance plant biomass, increases drought tolerance and produce compounds that are poisonous to animals hence decrease herbivory (Clay, 1988). The NC-endophytes include the endophytic fungi which colonize leaves of tropical trees (Fröhlich and Hyde, 1999; Arnold et al., 2000; Gamboa and Bayman, 2001), over the ground parts of nonvascular plants, seedless vascular plants, woody and herbaceous angiosperms (Carroll and Carroll, 1978; Petrini, 1991; Stone, 1988). A very unique feature of NC-endophytes is their ability to asymptomatically colonize and provide habitat-adapted fitness benefits to the genetically distant host plants (Rodriguez et al., 2008).

Harley and Smith (1983) described seven types of mycorrhizal associations which still is the most accepted classification. The different groups of mycorrhizal associations are classified on the presence of extraradical or intraradical mycelia (Bonfante and Perotto, 2000). These are named ectomycorrhizae, endomycorrhizae, ectendomycorrhizae, arbutoid mycorrhizae, monotropoid mycorrhizae and orchid mycorrhizae (Brundrett, 2004; Smith and Read, 2008). Arbuscular mycorrhiza (AM) which is included in endomycorrhizae (Harley and Smith, 1983), is the most widespread of all mycorrhizal associations in terms of geographical distribution and of great ecological and economic importance (Brundrett et al., 1996; Jeffries et al., 2003; Fitter, 2005; Smith et al., 2009; Gutjahr and Parniske, 2013). Arbuscular mycorrhizal fungus (AMF) is an obligate biotroph, which obtains carbohydrates required to sustain its life from the host plant, in return of which it provides the plant with various nutrients such as nitrogen, phosphate, sulfur and also water (Smith and Read, 2008; Smith and Smith, 2011).

Fungal symbionts express various types of associations with the host including commensalism, mutualism and parasitism (Lewis, 1985). In mutualistic interaction host is conferred with fitness benefits such as drought and metal tolerance (Read, 1999; Ahanger et al., 2014), disease resistance (Kawaguchi and Minamisawa, 2010; Cui et al., 2017) thermotolerance (Redman et al., 2001), enhanced growth and development (Marks and Clay, 1990; Varma et al., 1999; Baslam et al., 2011), herbivore resistance (Latch, 1993) and improved nutrient status (Jiang et al., 2017; Luginbuehl et al., 2017), while, fungi obtains nutrients from hosts and distribution by seed broadcast (Schardl et al., 2004; Schulz, 2006). In commensal association there are no beneficial or deleterious effects on hosts whereas in parasitic association the organism negatively impacts the growth and development of host or induces disease which may result in plant death (Redman et al., 2001). Various studies on host-fungus association have revealed that some fungal species could show mutualistic, pathogenic or commensalistic lifestyle depending upon the host plant e.g. *Colletotrichum* spp. are considered as infectious pathogen, but some species can form mutualistic association also. In mutualism *Colletotrichum* spp. increase disease resistance, enhances growth and development and also confers drought tolerance (Redman et al., 2001). *C. gloeosporioides* is a pathogen of strawberry plant, whereas, it increased plant biomass and drought tolerance of tomato plants and was therefore called a mutualist (Rodriguez and Redman, 2008). Fungi assist plants in different ways, such as by improving nutrient uptake, protection against pathogens and production of beneficial phytohormones (Barea et al., 2005). Furthermore, endophytic (Table 6.1) and mycorrhizal (Table 6.2) fungi are also known to confer tolerance to host plants under various abiotic stresses such as salt, drought, heavy metal and temperature stress (Redman et al., 2001; Mahajan and Tuteja, 2005; Baltruschat et al., 2008; Hajiboland et al., 2010; Latef and Chaoxing, 2011; Li et al., 2011; Aroca et al., 2012; Khan et al., 2012b; Wang et al., 2012; Hashem et al., 2014; Wu et al., 2017, Sabra et al., 2018).

Role of fungus in various abiotic stress tolerance

Salinity stress

Soil salinization is a serious environmental problem which is increasing steadily around the globe, particularly in arid and semiarid areas (Giri et al., 2003; Al-Karaki, 2006). Increased salinization of cultivable land has resulted in 30% loss of the arable land which will reach up to 50% by the mid of 21st century (Wang et al., 2003). Salinity stress suppress the growth of plants by altering the vegetative growth and net assimilation rate leading to reduced

TABLE 6.1 Examples of endophytic fungi that confers tolerance in different abiotic stresses.

Abiotic Stress	Fungi	Host	References
Salinity	*Piriformospora indica*	Barley *Medicago truncatula* *Brassica compestris*	Waller et al. (2005), Baltruschat et al. (2008), Li et al. (2017), Khalid et al. (2018)
	Fusarium culmorum	*Leymus mollis*	Rodriguez et al. (2008)
	Paecilomyces formosus	Cucumber	Khan et al. (2012a)
	Aspergillus fumigatus *Penicillium minioluteum* *Penicillium funiculosum*	Soybean	Khan et al. (2011a), Khan et al. (2011c), Khan et al. (2011b)
	Epichloe spp.	*Festuca* spp. *Lolium* spp.	Schardl et al. (1994), Tsai et al. (1994)
	Aspergillus flavus	Soybean	Lubna et al. (2018)
Drought	*Piriformospora indica*	*Arabidopsis* Chinese cabbage	Sherameti et al. (2008), Sun et al. (2010)
	Trichoderma spp.	*Theobroma cacao*	Bailey et al. (2006)
	Trichoderma harzianum	Tomato	Mona et al. (2017)
	Neotyphodium	*Lolium perenne* *Elymus dahuricus*	Hahn et al. (2008), Zhang and Nan (2007)
Heavy metal	*Piriformospora indica*	Wheat	Shahabivand et al. (2012)
	Serendipita indica	*Ocimum basilicum*	Sabra et al. (2018)
	Exophiala pisciphila	Maize	Li et al. (2011)
	Phialocephala fortinii *Rhizodermea veluwensis* *Rhizoscyphus* sp.	*Clethra barbinervis*	Yamaji et al. (2016)
	Lasiodiplodia theobromae	*Boswellia ovalifoliolata*	Aishwarya et al. (2017)

(*Continued*)

TABLE 6.1 (Continued)

Abiotic Stress	Fungi	Host	References
	Trichoderma atroviride	*Brassica juncea*	Cao et al. (2008)
	Neotyphodium caenophialum	*Lolium arundinaceum, Festuca arundinacea*	Ren et al. (2011), Soleimani et al. (2010)
	Neotyphodium uncinatum	*Festuca pratensis*	Soleimani et al. (2010)
	Mucor sp. *M. circinelloides*	*Brassica compestris;* Oil seed rape	Zahoor et al. (2017), Zhu et al. (2015)
	Penicillium funiculosum	Soybean	Khan and Lee (2013)
	Fusarium tricinctum	*Solanum nigrum*	Khan et al. (2017)
Heat	*Aspergillus japonicus*	Soybean and Sunflower	Ismail et al. (2018)
	Exophiala dermatitidis Thermomcyes spp. *Exophiala* sp. LHL8	Cucumber	Matos et al. (2002), Ali et al. (2017), Khan et al. (2012b)
	Paecilomyces formosus	Japonica rice	Waqas et al. (2015)
	Penicillium resedanum	*Capsicum annuum*	Khan et al. (2013)
	Curvularia sp.	*Dichanthelium lanuginosum*	Redman et al. (2002)

productivity and yield (Paul, 2012; Hasanuzzaman et al., 2013; Ahanger et al., 2017). Increased salinity of soil also leads to both hyperionic and hyperosmotic stress which finally can cause the death of plant. Soil salinity of a given soil depends upon various variables like, amount of evaporation which increases salt levels or the quantity of precipitation which decreases in salt levels (Mahajan and Tuteja, 2005). High salt levels in the soil can alter the basic soil texture resulting in reduced soil porosity and aeration causing decreased water holding capacity and imbalanced acquisition of nutrients (van Hoorn et al., 2001). As the basic physiology of salt and drought stress overlaps with one another, therefore, salt stress results in water shortage

TABLE 6.2 Examples of mycorrhizal fungi that confers tolerance in different abiotic stresses.

Abiotic Stress	Fungi	Host	References
Salinity	*Glomus* sp. *Acaulospora* sp.	Date Palm	Ait-El-Mokhtar et al. (2019)
	Glomus intraradices *G. mosseae*	*Capsicum annuum*, Lotus glaber Tomato Maize	Cekic et al. (2012), Sannazzaro et al. (2007), Hajiboland et al. (2010), Feng et al. (2002), Sheng et al. (2008)
	Glomus deserticola	Sweet basil	Elhindi et al. (2017)
	Glomus fasciculatum	*Allium sativum* *Phragmites australis*	Borde et al. (2010), Al-Garni (2006)
	Claroideoglomus claroideum	Lettuce	Santander et al. (2019)
	Funneliformis mosseae *Diversispora versiformis*	*Chrysanthemum morifolium*	Wang et al. (2018)
	Funneliformis mosseae *Rhizophagus intraradices* *Claroideoglomus etunicatum*	*Ephedra aphylla*	Alqarawi et al. (2014)
	Glomus clarum	Moongbean	Rabie (2005)
	Glomus mosseae *G. intraradices* *G. claroideum*	Olive	Porras-Soriano et al. (2009)
Drought	*Acaulospora* *Claroideoglomus etunicatum*	*Cynophalla flexuosa*	Barros et al. (2018)
	Funneliformis mosseae *Paraglomus occultum*	*Poncirus trifoliata*	Wu et al. (2017)
	Glomus intraradices	Zucchini *Lotus glaber*	Colla et al. (2008), Sannazzaro et al. (2006)

(*Continued*)

TABLE 6.2 (Continued)

Abiotic Stress	Fungi	Host	References
	Glomus macrocarpum	*Sesbania aegyptiaca S. grandiflora*	Giri and Mukerji (2004)
	Glomus coronatum *G. constrictum* *G. intraradices* *G. claroideum* *G. mosseae* *G. geosporum*	*Lactuca sativa*	Marulanda et al. (2003)
	Glomus fasciculatum	*Capsicum annuum*	Mena-Violante et al. (2006)
	Rhizophagus irregularis	Lettuce and Tomato	Ruiz-Lozano et al. (2016)
Heavy metal	*Glomus* isolate Br1	Maize	Kaldorf et al. (1999)
	Glomus mosseae	*Trifolium repens*	Medina et al. (2005)
	Glomus fasciculatum (BEG 53)	*Plantago lanceolata*	Parádi et al. (2003)
	Gigaspora intraradices	*Pisum sativum* Zinc violet *Medicago sativa* Wheat	Rivera-Becerril et al. (2002), Hildebrandt et al. (1999), Wang et al. (2012), Heydarian et al. (2018)
	Rhizophagus intraradices *Funneliformis mosseae*	Rice	Li et al. (2016)
	Rhizophagus irregularis	Pigeonpea	Garg and Kashyap (2019)
Heat	*Glomus fasciculatum*	*Cyclamen persicum*	Maya and Matsubara (2013)
	Glomus mosseae	Strawberry	Matsubara (2010)
Cold	*Funneliformis mosseae*	*Cucumis sativus*	Liu et al. (2014a)
	Glomus mosseae	Tomato	Latef and Chaoxing (2011)
	Glomus hoi	*Plantago lanceolata*	Barrett et al. (2011)

(*Continued*)

TABLE 6.2 (Continued)

Abiotic Stress	Fungi	Host	References
	Glomus intraradices	Rice *Phaseolus vulgaris*	Liu et al. (2014b), Aroca et al. (2007)
	Glomus fasciculatum *Glomus macrocarpum*	Sorghum	Raju et al. (1990)
	Rhizophagus irregularis	*Medicago truncatula*	Hu et al. (2015)

condition in plant which appears as a physiological drought (Mahajan and Tuteja, 2005).

Endophytes

When barley plants were exposed to different NaCl levels in hydroponic culture, they showed leaf chlorosis and reduced growth, but the deleterious effects of salt stress were fully eradicated with the colonization by endophytic fungus *Piriformospora indica* and the colonized plants showed higher biomass compared to control plants. The possible mechanism for this might be the composition of a cellular environment with enhanced antioxidative capabilities of the plant (Waller et al., 2005). In another study, there was an increase in metabolic activity rate and ascorbic acid content while lipid-peroxidation (LPO) decreased in the barley plant colonized by *P. indica* under salt stress, hence, indicating the role of fungus to compensate the inhibited leaf metabolic activity (Baltruschat et al., 2008). *Medicago truncatula* plants inoculated with the fungus *P. indica* showed, strongly promoted roots development and biomass under all the concentrations of salt stress (Li et al., 2017). Redman et al. (2011), revealed that the rice plants infected by *Fusarium culmorum* strain FcRed1 showed higher biomass and increased tolerance to saline conditions. *Fusarium culmorum* (FcRed1) was also isolated from the tissues and seed coat of *Leymusmollis* spp. and was found to confer salt tolerance to the plant (Rodriguez et al., 2008). Endophyte *Paecilomyces formosus* LHL10 inoculation greatly enhanced growth, reduced electrolyte leakage, membrane damage, abscisic acid (ABA) of host plant and also mitigated the adverse effects of salinity stress (Khan et al., 2012a). Similarly in different studies, colonization of soybean by *Aspergillus fumigatus* LH02 and *Penicillium funiculosum* LHL06 and *Penicillium minioluteum* LHL09 caused a significant increase in plant biomass i.e. fresh/dry weights and also improved plant growth under moderate as well as high salinity stress. Whereas, the non-colonized plants showed symptoms like leaf necrosis, stunted growth and decreased photosynthesis rates (Khan et al., 2011a, 2011b, 2011c). *Epichloe* spp. is endophytic fungi found in many cold climate grasses like *Festuca* spp. and *Lolium* spp. (Schardl et al., 1994; Tsai et al., 1994).

Epichloe spp. have been reported to play a positive part in increasing host plants resistance to abiotic stress, including salinity (Yin et al., 2014), Cd stress (Ren et al., 2011) and Ni stress (Mirzahossini et al., 2015). Soybean plants introduced with fungus, *Aspergillus flavus* CHS1 revealed significant increase in hormones level (IAA and gibberllic acid) and antioxidant enzyme activities under salinity stress, in comparison to control plants, therefore, fungus introduction helped plants in tolerating stress by decreasing ROS level through increased APX, CAT, and SOD activities (Lubna et al., 2018). Additionally, many other studies have also reported that plants with endophyte presence under salt stress showed increased activities of antioxidant enzymes which helped the plants to tolerate stress and diminish H_2O_2 formation (Hashem et al., 2014; Ahmad et al., 2015; Yasmeen and Siddiqui, 2017). ABA which is also known as stress hormone is very important for plant growth as it decreases the stress damage by mediating the stomatal closure to lessen water loss and also regulates some stress related genes thereby induce stress tolerance in plants (Zhang et al., 2006). Khalid et al. (2018), reported that *P. indica* assisted to cope up the adverse effects of salinity stress via regulation of cation transporter genes, phytohormones (salicylic acid and gibberellic acid) and antioxidant enzymes in *Brassica campestris* ssp. *Chinensis*.

Mycorrhiza

Mycorrhizal symbiosis is considered a key factor in helping plants persist under adverse conditions (Augé et al., 1992) and AMF is broadly found in saline conditions (Wang and Liu, 2001). It had been reported that AMF symbiosis leads to increased nutrient uptake, osmoregulators accumulation and increased photosynthesis and water use efficiency (WUE) in stress, however, these positive effects are AMF dependent (Sylvia and Williams, 1992; Marulanda et al., 2003; Wu et al., 2007). Recently, in a study conducted on date palm under salinity stress, AMF colonization proved beneficial in improving the physiological parameters including photosynthetic rate, stomatal conductance, water relations and decreasing Na^+ uptake (Ait-El-Mokhtar et al., 2019). Studies have confirmed that plants inoculated with AMF under NaCl stress showed increased uptake of P, K^+ and Ca^{2+} and reduced uptake of Na^+ (Sallaku et al., 2019). Cekic et al. (2012) have found that under long term saline conditions, colonization of *Capsicum annuum* by mycorrhiza exhibited increased chlorophyll concentration and increased Mg^{2+} and N uptake whereas reduced Na^+ transport. In another study, Santander et al. (2019) have found that the lettuce plants colonized by mycorrhiza showed increased biomass production, enhanced proline synthesis and N uptake and changes in ionic relations, mainly reduced Na^+ accumulation, in comparison to non-mycorrhizal plants under salt stress. Furthermore, Wang et al. (2018) revealed significant increase in fresh and dry weight and also N levels in shoot and root of *Chrysanthemum morifolium* in presence of mycorrhizal colonization when subjected to moderate salinity levels. Enhanced growth of mycorrhiza colonized date palm grown in saline soils could be attributed to, mycorrhiza-mediated enhanced P nutrition (Alqarawi et al., 2014). Various investigators have found that AMF colonized plants show increased fitness and better growth compared to non-colonized plants under salt stress (Cantrell and Linderman, 2001; Giri et al., 2007; Sannazzaro et al., 2007; Zuccarini and Okurowska, 2008). In another study on tomato plant, Hajiboland et al. (2010) revealed that high salinity levels decreased dry matter production in two cultivars of tomato, but all mycorrhiza colonized plants showed better growth compared to non-mycorrhizal plants.

AMF introduced olive plants showed a better survival under salinity stress in terms of improved root-shoot growth, improved nutrient acquisition and lesser biomass loss in comparison to non-introduced plants (Porras-Soriano et al., 2009). Free polyamine (putrescine, spermidine and spermine) levels and root-shoot length declined under salinity stress condition; whereas, the introduction of AMF to host plants enhanced free polyamine levels and also restored the decline in root-shoot length (Sannazzaro et al., 2007). Mycorrhizal inoculation prominently showed an improvement in chlorophyll content, photosynthetic rate, gas exchange traits and WUE of sweet basil plants subjected to saline conditions (Elhindi et al., 2017). It has been observed that AMF colonized plants show increase in levels of cytokinin which results in enhanced translocation of photosynthate in salinity stress (Hameed et al., 2014; Talaat and Shawky, 2014). An improvement in growth parameters which included leaf area index, fresh and dry biomass was recorded in AMF inoculated *Allium sativum* plants under saline conditions (Borde et al., 2010). AMF-colonized pepper plants showed a decrease in oxidative stress as there was a decrease in lipid membrane peroxidation under saline conditions (Latef and Chaoxing, 2014). Salinity stress decreased plant biomass and also disturbed the ionic status of two pigeon pea genotypes, but application of AMF lessened the negative effects of the stress and amongst all AMF used, *Rhizophagus irregularis* displayed highest efficiency (Garg and Pandey, 2015). Another tolerance mechanism adopted by plants involves buildup of soluble sugars which helps to lower down the osmotic potential of the plant (Thanna and Nawar, 1994). The accumulation of soluble sugars in AMF colonized plants is a positive response to salt stress because sugars can avoid stress induced structural changes in soluble protein, maintain the osmotic balance in plant cells and protect membrane structure (Evelin et al., 2013). Al-Garni (2006), reported that AMF infected *Phragmites australis* plants showed to higher accumulation of soluble sugars compared to non-mycorrhizal plants. Many other studies, have also reported accumulation of soluble sugars in AMF inoculated plants such as mungbean (Rabie, 2005), maize (Feng et al., 2002; Sheng et al., 2008) and citrus (Wu et al., 2009), under salinity stress compared to non-inoculated plants.

Drought stress

Drought stress is known to induces a wide range of physiological and biochemical changes in plants which includes stomatal closure (Schroeder et al., 2001; Roelfsema and Hedrich, 2005), decreased growth and photosynthesis (Flexas et al., 2004; Ruiz-Lozano et al., 2016), reduced transpiration (Impa et al., 2012) and activation of respiration (Rennenberg et al., 2006). Furthermore, drought stress leads to osmotic stress which causes turgor loss, finally, causing inhibition of plant growth and development (Selmar and Kleinwächter, 2013).

Endophyte

When *Arabidopsis* plants were subjected to drought conditions, fungus (*Piriformospora indica*) inoculated seedlings showed good growth, whereas, those without fungus showed symptoms of withering. Furthermore, level of proteins involved in drought enhanced faster in leaves of *P. indica* colonized seedlings in comparison to the uncolonized controls (Sherameti et al., 2008). Many species of genus *Trichoderma* which are best known for its role as biocontrol fungi also invade and colonize plant roots, stem, twigs and enhances plant resistance in various abiotic stresses. A study on *Theobroma cacao* exposed to drought stress revealed that, *Trichoderma* spp. increase plant resistance by delaying changes in net photosynthesis, stomatal conductance and green fluorescence emissions (Bailey et al., 2006). This improved tolerance for drought stress in cacao plant was attributed to the enhanced root growth due to colonization by *Trichoderma* spp. which improved acquisition of water resulting in improved water status of the plant (Bae et al., 2009). The capability of some *Trichoderma* spp. to adapt to extreme environmental conditions facilitates their existence in diverse geographical areas (Hermosa et al., 2004). It has been reported that, endophytic fungi *Penicillium citrinum* produce secondary metabolites which positively affect plant growth by enhancing physiological characteristics including chlorophyll and carotenoids content, photosynthetic rate and tolerance in drought and high-light field condition (Syamsia et al., 2018). Tomato plants inoculated with *Trichoderma harzianum*, showed an improved root-shoot growth, photosynthetic pigments and also enhanced plant resistance in comparison to control plants uninoculated with fungi under drought conditions (Mona et al., 2017). Chinese cabbage when inoculated with *P. indica* under drought conditions, showed reduction in drought induced degradation of chlorophylls and thylakoid proteins and decrease in photosynthetic activity, whereas the activities of oxidative stress enzymes such as POX, CAT and SOD enhanced (Sun et al., 2010).

It is recognized that, enhanced drought tolerance of the plants because of colonization by fungi is associated with change in water relations and stimulation of osmotic adjustments (amino acids, free sugars, sugar alcohols) leading to tissue survival and growth of colonized plants in comparison to non-colonized plants (Richardson et al., 1992; Elmi and West, 1995; Elbersen and West, 1996; Hahn et al., 2008). Many other reports have also shown that the deleterious effects of drought stress were mitigated by endophytes colonization via up-regulating the activity of antioxidant and accumulation of carbohydrates, mannitol, phenolics and proline which protects plant cell membrane from oxidative damage under stressed conditions (Malinowski and Belesky, 2000; Rodriguez et al., 2008; Gundel et al., 2012; Hamilton et al., 2012). In another study with endophyte *Neotyphodium*, which infected *Lolium perenne*, it was observed that it can affect the growth of perennial ryegrass in a non-uniform way and different growth parameters such as shoot dry weight, seed yield, vegetative and reproductive tillers increased under water deficit conditions (Hahn et al., 2008). Depending on the availability of water in the habitats, certain genotypes of *L. perenne* may get benefitted by endophyte infection while recovering from drought conditions and produce higher dry weight compared to non-infected plants (Hesse et al., 2003). Zhang and Nan (2007) reported an increased antioxidative enzymes activity, biomass, plant height and tiller numbers in *Elymus dahuricus* infected by *Neotyphodium* exposed to low water conditions. West (1994) found that the tall fescue grasses which were infected with the endophyte were efficient in prolonged net carbon assimilation or in recovering their growth after the drought is over.

Mycorrhiza

Many mycorrhizal fungi are known to adapt to adverse environments, hence, they can prove beneficial to plants under stressed conditions (Mosse et al., 1981). Interestingly, when drought stress is perceived by most of the plants, they ask AMF for help, by secreting a signaling molecule into the rhizosphere called as "strigolactone" a class of phytohormones (Oldroyd, 2013). At the same time, mycorrhizal hyphae extensively improves soil structure as well as its water holding capacity (Miller and Jastrow, 2000), hence, improves overall growth of the plants in comparison to their non-mycorrhizal counterparts (Augé, 2004). In addition, AMF can also form hyphal fusion (anastomosis) which is a vital mechanism for AMF persistence and to enhances their chances of survival under water deficit conditions (Giovannetti et al., 2015). Previous studies have found that, host plant's association with mycorrhiza increases the water status of the plant, as shown by relative leaf water content (RLWC) (Wu et al., 2017; Barros et al., 2018). Plants colonized by AMF show enhanced biosynthesis of osmolytes such as sugars and proline, which lowers the water potential of drought stressed plants (Latef et al., 2016; Ruiz-Lozano, 2003), and this lowered water potential allow plants to maintain turgor and carry on various physiological activities such as growth, photosynthesis and stomatal opening (Smith et al., 2010). There are many studies which have reported that inoculation of AMF enhances water content in the leaves of Zucchini plants leaves (Colla et al., 2008), water potential and nutrients status in maize plants (Feng et al., 2000) and chlorophyll content of plants, for example, *Sesbania aegyptica, S. grandiflora* and *Lotus glaber* (Colla et al., 2008; Giri and Mukerji, 2004; Sannazzaro et al., 2006).

Ouledali et al. (2018) revealed that AMF colonization enhances root growth and architecture which overall develops a greatly functional root system and this leads to the uptake of water and nutrients. In a study on *Lactuca sativa* it was reported that plants infected by AM fungi showed higher content of water in comparison to non-infected control plants under water deficit conditions (Marulanda et al., 2003). Root-shoot dry weights, fruit fresh weight, chlorophyll content, carotenoids content showed a significant increase under drought stress in *Capsicum annuum* plants colonized by AMF in comparison to non-AMF plants (Mena-Violante et al., 2006). Li et al. (2019) found that AMF colonization sufficiently reduced the adverse effects of water deficit conditions on plant growth and enhanced plant biomass, photosynthetic rate, stomatal conductance, intrinsic water use efficiency (iWUE), superoxide dismutase (SOD) activity and reduced malondialdehyde (MDA) levels in *Leymus chinensis* (C_3 plant) under light and moderate drought and in *Hemarthria altissima* (C_4 plant) AMF colonization increased the aboveground biomass and CAT activity in light and moderate drought stress. In another experiment, when the effects of three different water regimes on lettuce and tomato plant were examined in AMF and non-AMF plants it was found that AMF infected plants showed a better growth and efficiency of photosystem II (PS II) in comparison to non-AMF plants from very early stages of plant colonization (Ruiz-Lozano et al., 2016). Regarding its effect on drought, in different crops studied, AMF colonization alleviates the deleterious effects of the stress by making the host plant more tolerant to drought (Aroca et al., 2012; Bárzana et al., 2014; Augé et al., 2015; Moradtalab et al., 2019).

Heavy metal stress

Heavy metals are defined as naturally occurring elements which have atomic weight and density at least 5 times higher than water (Fergusson, 1990).

Multiple application of these elements in industries, homes, agriculture and health sector have led to their wide distribution in different ecosystems, hence, raising a great concerns over their possible outcomes on human and the environment (Tchounwou et al., 2012). Commercial fertilizers and pesticides are very essential inputs to enhance agricultural production (Zhang and Zhang, 2007; Zhang et al., 2011), but their long-term and excessive use has caused contamination of agricultural soils as they contain heavy metals such as As, Cd, Cr, Cu, Hg, Ni and Pb (Arao et al., 2010). Soils containing high levels of these heavy metals, either from natural origin or from anthropogenic action, may pose great challenge to exposed biota (Colpaert et al., 2011). Once these heavy metals get accumulated in crops, fruits, vegetables and/or soils, they leads to various health risks via dietary exposure (Liu et al., 2013; Yousaf et al., 2016). The absorption of these heavy metals by plants takes place via the exposed parts of the plants to the polluted air and/or from contaminated soils (Khairiah et al., 2004; Kachenko and Singh, 2006). HM toxicity is one of the major abiotic stress which is responsible for productivity loss ranging from 25% to 80% of different cultivated crops (Eid et al., 2019). It is recognized that in acidic soils, the low productivity and decreased fertility of the soil are primarily because of toxicity of manganese and aluminum at the same time K, Mg, P and Ca deficiencies (Singh et al., 2011). Various fungal hyphae of internal as well as external origin can immobilize these heavy metals (Ouziad et al., 2005).

Endophyte

When the effect of endophytic fungus (*Serendipita indica*) and AMF (*Rhizophagus irregularis*) was examined on sweet basil (*Ocimum basilicum*) grown in pots having soil contaminated with lead and copper, both fungi caused an increase in shoot-root dry weight of the plants in all treatments and also decreased the translocation of lead and copper to shoots individually as well as in combination (Sabra et al., 2018). Successful colonization of maize roots by *Exophiala pisciphila* H93 in Cd, Pb and Zn contaminated soils mitigated the harmful effects of these HMs, promoted root-shoot growth and tolerance of plants (Li et al., 2011). Fungal endophytes like *Aspergillus*, *Cladosporium*, *Exophiala*, *Fusarium*, *Gaeumannomyce*, *Glomerella*, *Humicola*, *Lasiodiplodia*, *Nannizzia*, *Paecilomyces*, *Penicillium*, *Phomopsis* and *Rhizopus* have been known for their ability to enhance growth and counteract metal stress in soil contaminated with heavy metals such as Al, Cd, Cu, Pb and Zn (Ezzouhri et al., 2009; Li et al., 2011; Valix et al., 2001; Ban et al., 2012; Iram et al., 2013; Deng et al., 2014, 2016). Inoculation of three root fungal endophytes namely, *Phialocephala fortinii*, *Rhizodermea veluwensis* and *Rhizoscyphus* sp. to *Clethra barbinervis* seedlings growing in mine soil significantly enhanced the growth of seedlings, K uptake to shoots and reduced the levels of Cd, Cu, Ni, Pb and Zn in roots, whereas, seedlings without endophytes could hardly grow in the heavy-metal contaminated soil (Yamaji et al., 2016). Aishwarya et al. (2017), reported that *Lasiodiplodia theobromae* isolated from *Boswellia ovalifoliolata* exhibited resistance to four heavy metals namely Cd, Co, Cu and Zn, when evaluated for its tolerance ability. Ni uptake and extraction efficiency were positively influenced when *Brassica juncea* was inoculated with *Trichoderma atroviride* strain F6 and also the inoculated plants showed a greater tolerance to Ni toxicity (Cao et al., 2008).

Endophyte *Neotyphodium coenophialum*, when infected tall fescue (*Lolium arundinaceum*) it conferred tolerance against metal Cd which was noticeable from enhanced growth, biomass and photosynthetic rate in comparison to non-infected plants (Ren et al., 2011). In another study, soil applied with different levels of Cd negatively affected the growth of two grass species i.e. *Festuca arundinacea* and *F. pratensis*, but this loss was lessened when the grass species were inoculated with *Neotyphodium coenophialum* and *N. uncinatum* respectively, as parameters like photosynthetic parameters and plant growth improved (Soleimani et al., 2010). Zahoor et al. (2017) found that, application of *Mucor* sp. MHR-7 seized various HMs in the mycelium which made them less available to plant root thereby reduced HMs uptake and toxicity in *Brassica campestris*. In addition, Zhu et al. (2015) also reported that two species belonging to the genus *Mucor* namely *M. circinelloides* (Z4) and *M. racemosus* (Z8), improved plant height, plant biomass and increased the accumulation of Cd and Pb in roots of Guizhou oilseed rape. Soybean plants infected by *Penicillium funiculosum* reported more production of reduced glutathione under Cu stress than control i.e. non-infected plants (Khan and Lee, 2013). Many researchers have reported that, the plants colonized by endophytic fungi under metal stress produces more biomass than the non-colonized plants (Zhu et al., 2015; Wang et al., 2016; Rozpądek et al., 2018).

Solanum nigrum infected by *Fusarium tricinctum* showed enhanced production of IAA and shoots and roots growth stimulation under Cd toxicity (Khan et al., 2017). In a study by Gong et al. (2017), it was revealed that *Purpureocillium* sp. A5 protects *Kandelia candel* plants under Cu stress, as metal uptake was reduced, soil pH characterization changed and parameters like chlorophyll a, b and RWC improved. *Paecilomyces formosus* inoculation to soybean played a positive role to counteract Ni induced stress, enhance chlorophyll level and enhance plant growth as secretion of plant growth regulators like GA and IAA was stimulated and this could be attributed to decrease in levels of H_2O_2 and MDA in comparison to non-inoculated plants (Bilal et al., 2017). Babu et al. (2014), reported enhanced chlorophyll levels in maize plants subjected to Pb stress following introduction of endophytic fungus *Trichoderma virens* strains. When indole acetic acid (IAA) producing endophytic fungi *Penicillium ruqueforti* was introduced into maize plants grown on heavy metals (Cd, Cu, Ni, Pb and Zn) polluted site, it restricted the transfer of these heavy metals from soil into the plants by secreting IAA and the plants had increased root-shoot length, chlorophyll contents, fresh and dry weight compared to non-introduced plants (Ikram et al., 2018). *Piriformospora indica* inoculation to wheat plant under Cd toxicity enhanced chlorophyll content and performance Index (PI) compared to non-inoculated plants, indicating the positive effect of this plant-microbe interaction on photosynthetic apparatus (Shahabivand et al., 2012).

Mycorrhiza

AMF are widely known to support plant establishment in soils contaminated with heavy metals, as they have potential to strengthen plants defense system and promote growth and development. Mycorrhizal fungi are also known to enhance plant growth by improving its nutrition status by increasing water and nutrients uptake and decreased soil compaction (Gaur and Adholeya, 2004). Results of many studies have proved that, AM fungi are present in soils contaminated with HMs such as As (Sun et al., 2016), Cd (González-Chávez et al., 2009), Cu (Chen et al., 2005), Pb (Long et al., 2010; Zarei et al., 2010) and Zn (Zarei et al., 2010). Furthermore, in many multi-metal contaminated sites, 59–79% of the plant roots have been reported to be colonized by AMF (Al-Ghamdi and Jais, 2012). It is believed that fixation or chelation of heavy metals in the rhizospheric soil by fungi can cause metal dilution in plant tissues (Kapoor et al., 2013; Audet, 2014). Researchers have demonstrated that mycorrhizal presence can alter the uptake of different metals from soil into the plant and further their translocation to the aerial parts (Dong et al., 2008; Li et al., 2015). AMF have evolved various strategies which can lessen heavy metal induced negative effects and also decrease their entry into the food chains (Hall, 2002; Kramer, 2005). For example, AMF is known to bind heavy metals by releasing an insoluble glycoprotein called glomalin beyond the plant rhizosphere (González-Chávez et al., 2004; Göhre and Paszkowski, 2006). In addition, fungal hyphae having internal and/or external origin can immobilize and fix heavy metals in their cell wall and store them in vacuole or chelate them with various compounds in the cytoplasm (Ouziad et al., 2005; Hildebrandt et al., 2007; Punamiya et al., 2010) and thereby confer metal tolerance to the plants. Moreover, mycorrhizal fungi take part in nontoxic symbiotic association with plants to accumulate heavy metals in plant root (Joner and Leyval, 1997).

Kaldorf et al. (1999) reported that, in maize heavy metals were selectively retained in the inner parenchyma cells overlapping with fungal structures. In various cropping patterns which involves growing of legumes such as clover, mycorrhizal buffer to the stress imposed by Cd is provided by AMF (Rivera-Becerril et al., 2002, 2005; Medina et al., 2005) or tolerance for Cd is enhanced by changing polyamine metabolism, therefore, fixing Cd in the roots of colonized plants (Parádi et al., 2003). (Li et al., 2016) reported that in rice, colonization by AMF was highly successful in lessening Cd levels in vacuoles and cell wall, which brought about Cd detoxification. It has been observed in alfalfa (*Medicago sativa*), that AMF infected plants showed improved tolerance to Cd which was possibly due to alteration of chemical forms of Cd in various plant tissues (Wang et al., 2012). Wheat plants grown in Ni polluted soil showed toxicity symptoms like chlorosis, growth inhibition and root tips browning; decreased chlorophyll levels, seed number per spike and seed yield per plant when no AMF was applied, whereas, AMF application ameliorated negative effect of Ni (Heydarian et al., 2018). In another study, pigeonpea plants subjected to As stress showed decreased plant growth and productivity, root to biomass ratio and oxidative burst, whereas, AMF inoculations improved plant growth as the uptake of As was significantly reduced and also the oxidative burst was attenuated (Garg and Kashyap, 2019). Studies have demonstrated that plants colonized by AMF had higher chlorophyll levels, lower MDA levels, lower ascorbate peroxidase (APX) and guaiacol peroxidase (GPX) activity compared to the non-AMF plants under Cr contamination (Davies et al., 2001; Rahmaty and Khara, 2011).

AMF performance in alleviation of heavy metals induced stress in contaminated soils has been documented by many studies (Hall, 2002; Hildebrandt et al., 2007; Latef, 2011; Miransari, 2011; Latef, 2013; Heydarian et al., 2018). Colonization of plant roots by AMF significantly affected the expression of several plant genes which codes for proteins that are apparently involved in heavy metal tolerance (Repetto et al., 2003; Rivera-Becerril et al., 2005). AMF isolate *Glomus intraradices* Br1, consistently conferred heavy metal tolerance to plants grown on various heavy metal contaminated soils under optimum fertilization, furthermore, the isolate Br1 was more effective in conferring metal tolerance to tomato, maize or alfalfa plant compared to *G. intraradices* Sy167 isolate (Hildebrandt et al., 1999; Kaldorf et al., 1999). Furthermore, Kaldorf et al. (1999) found that AMF infected maize grown on soils contaminated with two different heavy metals had enhanced uptake of essential elements like Mg, K and P and reduced uptake of heavy metals such as Ni, Fe, Zn or Cu when compared to non-AMF colonized roots.

Temperature (heat and cold stress)

Temperature is a very crucial environmental factor which determine the crops productivity across the globe. Many biochemical and physiological processes of plant are influenced by high and low temperature stresses (Zhu et al., 2011). Now days, extreme temperature occurrence have become very frequent along with global climate change. Generally, temperature ranging 0–15 °C (low but not freezing) is considered as low temperature stress (Theocharis et al., 2012), and a 10–15 °C increase above ambient temperature is considered as high temperature stress (Wahid et al., 2007). Severely high temperatures cause severe cellular injury and even cell death within minutes, which could be accredited to a disastrous collapse of cellular organization (Schöffl et al., 1999). Cellular injuries which are directly related to high temperatures include increased fluidity of bio-membranes, protein denaturation and aggregation, whereas, indirect injuries include inactivation of enzymes, microtubules formation, loss of membrane integrity and inhibition of protein synthesis (Howarth, 2005). These injuries ultimately cause inhibition of growth, production of toxic compounds and reactive oxygen species (ROS) (Schöffl et al., 1999; Howarth, 2005). Low temperatures i.e. chilling stress often leads to reduction in growth and photosynthesis, leaf chlorosis and wilting or even necrosis, cell membrane damage and oxidative stress in plants (Suzuki and Mittler, 2006). Furthermore, temperatures below 0 °C cause freezing stress, which leads to formation of ice crystals within the cell, mechanical damages and metabolic disturbances in plants (Xin and Browse, 2000). Destabilization of the membranes, with an increased permeability, is the main result produced by chilling stress (Raison and Chapman, 1976; Thomashow, 1999; Ma et al., 2018).

Endophyte

In soybean and sunflower plants *Aspergillus japonicus* colonization proved helpful in improving plant biomass and other growth parameters under high temperature stress (40 °C) compared to non-colonized plants (Ismail et al., 2018). Khan et al. (2012b) reveled that, *Exophiala* sp. LHL8 was effective in lessening high-temperature induced oxidative burst in cucumber plants, as shown by the increased levels of total polyphenol and reduced levels of glutathione, superoxide anion, lipid peroxidation and electrolyte leakage. In another study, Matos et al. (2002) have also reported that *Exophiala* sp. can survive and grow in environments with temperature higher than 42 °C, hence, confers heat tolerance to the plants under high temperature stress. The fungi of genera *Aspergillus*, *Candida*, *Curvularia*, *Myceliophthora*, *Thermomucor* and *Thermomyces* are thermo-tolerant fungi which can survive in extreme temperatures up to 62 °C with increased enzymatic activities of cellulases and xylanases (Márquez et al., 2007; Moretti et al., 2012). Inoculation of cucumber seeds with *Thermomcyes* spp. under high temperature stress alleviated the harmful effects of stress by maintaining quantum efficiency (QE) of PS II, photosynthetic rate, WUE and root length compared to untreated control plants. Furthermore, significant increase in flavonoids, saponins, total sugars, soluble proteins and antioxidant enzyme activities was also reported in inoculated plants (Ali et al., 2017). Hubbard et al. (2013), reported that the introduction of endophytic fungi SMCD 2206 to wheat plants raised heat stress tolerance of plants by lowering photosynthetic stress and improving PS II maximum efficiency compared to untreated control plants. Symbiotic association of fungus *Paecilomyces formosus* LWL1 conferred heat tolerance to japonica rice cv. Dongjin as there was significant improvement in plant growth variables like plant height, fresh weight, dry weight, chlorophyll content and decreased levels of stress-signaling compounds like ABA and jasmonic acid (Waqas et al., 2015). *Penicillium resedanum* LK6 infected *Capsicum annuum* plant had higher number of leaves and biomass per plant and reduced leaf damage, electrolyte leakage and lipid peroxidation in comparison to non-infected plants during 7 and 15 days of heat stress (Khan et al., 2013). Redman et al. (2002) revealed that, *Curvularia* sp. inoculated host plant *Dichanthelium lanuginosum* showed thermal stress tolerance compared to non-inoculated controls.

Mycorrhiza

With increase in soil and air temperatures, the response of plant community may depend on symbiotic interactions for sustainable yield and productivity (Bunn et al., 2009). Gavito et al. (2005) stated that, AMF-colonized plants show superior growth and survival compared to non-AMF colonized ones under thermal stress. Maya and Matsubara (2013) have reported that, AMF *Glomus fasciculatum* interaction with *Cyclamen persicum* positively influenced the growth and development of the plant under high temperature conditions. Additionally, Matsubara (2010) testified that polyphenol and ascorbic acid production enhanced in the leaves of AMF colonized strawberry in comparison to non-AMF plants under temperature stress conditions. Various other studies have reported, AMF symbiosis improve temperature stress tolerance in plants, for example, *Glomus fasciculatum* with asparagus (Matsubara and Harada, 1996), pepper (Martin and Stutz, 2004) and maize (Zhu et al., 2010). Latef and Chaoxing (2011), found that tomato plants infected by *Glomus mosseae* under low temperature stress improved plant growth, chlorophyll levels and soluble sugar; increased activity of APX, POD and SOD. In another study on *Medicago truncatula*, it was found that plants colonized by *Rhizophagus irregularis* reported increased seed number and plant biomass, when there was a constant increase in night temperature by 1.53 °C (Hu et al., 2015). Inoculation of *Plantago lanceolata* with *Glomus hoi* proved beneficial for host under low temperature stress as it improved N and ^{15}N content, hence, conferred nutritional benefit to its host (Barrett et al., 2011). Liu et al. (2014a) reported that *Funneliformis mosseae* was effective in improving total fresh and dry weights, root activity, ATPase activity, level of plasma membrane protein, expression of ATPase-related genes and decrease the levels of H_2O_2 and NADPH oxidase activity in *Cucumis sativus* under low temperature. Furthermore, Liu et al. (2014b) revealed, inoculation of *Oryza sativa* by *Glomus intraradices* under low temperature conditions improved relative water content (RWC), root length and trehalose content. It was observed that, AMF colonized *Dichanthelium lanuginosum* plants showed improved root length and root diameter in comparison to the non-AMF plants when exposed to high temperature stress (Bunn et al., 2009). *Phaseolus vulgaris* plants subjected to cold stress were inoculated with *Glomus intraradices* which resulted in improved transpiration rate and water relations (Aroca et al., 2007). Raju et al. (1990) reported that when sorghum plants were infected by *Glomus fasciculatum* and *Glomus macrocarpum* mixtures, the levels of elements like Ca, Cu, Fe, K, Mg, Mn, S and Zn increased in the plants.

Conclusion

Various studies have documented the importance of fungi (endophytic and mycorrhizal) in alleviating different abiotic stresses; improving plant nutrition and defense. This review comprises the existing information related to the beneficial role of symbiotic fungi under various stress conditions. Earlier, symbiotic fungus were mainly discussed for their role in nutrient uptake from soil; but now results of various studies have clearly proved that the fungi inoculated plants are well efficient to counteract various environmental stresses such as salinity, drought, heavy metal and temperature (cold and hot). Fungi colonized plants not only efficiently counteract stress, but also show improved growth and development and thereby increased per hectare yield of a large number of crops. Isolation of native and adapted fungi from polluted soils can be a useful tool for successful restoration of degraded environments. Certainly, utilization of fungi for improving agricultural output can significantly reduce the usage of synthetic fertilizers and pesticides. Furthermore, fungi mediated environmental friendly growth of different crops can be very helpful to meet increasing demand of feeding the increasing population across the globe.

References

Ahanger, M.A., Tittal, M., Mir, R.A., Agarwal, R.M., 2017. Alleviation of water and osmotic stress-induced changes in nitrogen metabolizing enzymes in *Triticum aestivum* L. cultivars by potassium. Protoplasma 254 (5), 1953–1963.

Ahanger, M.A., Tyagi, S.R., Wani, M.R., Ahmad, P., 2014. Drought tolerance: role of organic osmolytes, growth regulators, and mineral nutrients. In: Ahmad, P., Wani, M.R. (Eds.), Physiological Mechanisms and Adaptation Strategies in Plants Under Changing Environment, 1. Springer, New York, NY, pp. 25–55.

Ahmad, P., Hashem, A., Abd-Allah, E.F., Alqarawi, A., John, R., Egamberdieva, D., et al., 2015. Role of *Trichoderma harzianum* in mitigating NaCl stress in Indian mustard (*Brassica juncea* L) through antioxidative defense system. Front. Plant. Sci. 6, 868. Available from: https://doi.org/10.3389/fpls.2015.00868.

Aishwarya, S., Venkateswarulunagam, Netala, V.R., Tartte, V., 2017. Screening and identification of heavy metal-tolerant endophytic fungi *Lasiodiplodia theobromae* from *Boswellia ovalifoliolata* an endemic plant of tirumala hills. Asian J. Pharm. Clin. Reserch 10 (3), 488–491.

Ait-El-Mokhtar, M., Laouane, R.B., Anli, M., Boutasknit, A., Wahbi, S., Meddich, A., 2019. Use of mycorrhizal fungi in improving tolerance of the date palm (*Phoenix dactylifera* L.) seedlings to salt stress. Sci. Horticulturae 253, 429–438.

Al-Garni, S.M.S., 2006. Increasing NaCl-salt tolerance of a halophytic plant *Phragmites australis* by mycorrhizal symbiosis. American-Eurasian J. Agric. Environ. Sci. 1, 119–126.

Al-Ghamdi, A.A.M., Jais, H.M., 2012. Interaction between arbuscular mycorrhiza and heavy metals in the rhizosphere and roots of *Juniperus procera*. Int. J. Agric. Biol. 14, 69–74.

Ali, A.H., Abdelrahman, M., Radwana, U., El-Zayata, S., El-Sayeda, M.A., 2017. Effect of *Thermomyces* fungal endophyte isolated from extreme hot desert-adapted plant on heat stress tolerance of cucumber. Appl. Soil. Ecol. 124, 155–162.

Al-Karaki, G.N., 2006. Nursery inoculation of tomato with arbuscular mycorrhizal fungi and subsequent performance under irrigation with saline water. Sci. Horticult. 109, 1–7.

Alqarawi, A.A., Abd Allah, E.F., Hashem, A., 2014. Alleviation of salt-induced adverse impact via Mycorrhizal fungi in *Ephedra aphylla* Forssk. J. Plant. Interact. 9, 802–810.

Arao, T., Ishikawa, S., Murakam, I.M., 2010. Heavy metal contamination of agricultural soil and counter measures in Japan. Paddy Water Environ. 8 (3), 247–257.

Arnold, A.E., Maynard, Z., Gilbert, G., Coley, P.D., Kursar, T.A., 2000. Are tropical endophytic fungi hyper diverse. Ecol. Lett. 3, 267–274.

Aroca, R., Porcel, R., Ruiz-Lozano, J.M., 2007. How does arbuscular mycorrhizal symbiosis regulate root hydraulic properties and plasma membrane aquaporins in *Phaseolus vulgaris* under drought, cold or salinity stresses? N. Phytol. 173 (4), 808–816.

Aroca, R., Porcel, R., Ruiz-Lozano, J.M., 2012. Regulation of root water uptake under abiotic stress conditions. J. Exp. Bot. 63 (1), 43–57.

Audet, P., 2014. Arbuscular mycorrhizal fungi and metal phytoremediation: ecophysiological complementarity in relation to environmental stress. In: Ahmad, P., Rasool, S. (Eds.), Emerging Technologies And Management Of Crop Stress Tolerance. Academic Press, San Diego, pp. 133–160.

Augé, R.M., 2004. Arbuscular mycorrhizae and soil/plant water relations. Can. J. Soil. Sci. 84 (4), 373–381.

Augé, R.M., Foster, J.G., Loescher, W.H., Stodola, A.J.W., 1992. Symplastic sugar and free amino acid molality of *Rosa* roots with regard to mycorrhizal colonization and drought. Symbiosis 12, 1–17.

Augé, R.M., Toler, H.D., Saxton, A.M., 2015. Arbuscular mycorrhizal symbiosis alters stomatal conductance of host plants more under drought than under amply watered conditions: a meta-analysis. Mycorrhiza 25, 13–24.

Babu, A.G., Shim, J., Bang, K.S., Shea, P.J., Oh, B.T., 2014. *Trichoderma virens* PDR-28: a heavy metal-tolerant and plant growth-promoting fungus for remediation and bioenergy crop production on mine tailing soil. J. Environ. Manag. 132, 129–134.

Bacon, C.W., de Battista, J., 1991. Endophytic fungi of grasses. Handb. Appl. Mycol. 1, 231–256.

Bacon, C.W., White, J.F., 2000. Physiological adaptations in the evolution of endophytism in the Clavicipitaceae. Microbial Endophytes. CRC Press, pp. 251–276.

Bae, H., Sicher, R.C., Kim, M.S., Kim, S.H., Strem, M.D., Melnick, R.L., et al., 2009. The beneficial endophyte *Trichoderma hamatum* isolate DIS 219b promotes growth and delays the onset of the drought response in *Theobroma cacao*. J. Exp. Bot. 60 (11), 3279–3295.

Bailey, B.A., Bae, H., Strem, M.D., Roberts, D.P., Thomas, S.E., Crozier, J., et al., 2006. Fungal and plant gene expression during the colonization of cacao seedlings by endophytic isolates of four *Trichoderma* species. Planta 224 (6), 1449–1464.

Baltruschat, H., Fodor, J., Harrach, B.D., Niemczyk, E., Barna, B., Gullner, G., et al., 2008. Salt tolerance of barley induced by the root endophyte *Piriformospora indica* is associated with a strong increase in antioxidants. N. Phytol. 180 (2), 501–510.

Ban, Y., Tang, M., Chen, H., Xu, Z., Zhang, H., Yang, Y., 2012. The response of dark septate endophytes (DSE) to heavy metals in pure culture. PLoS One 7, 10.

Barea, J.M., Pozo, M.J., Azcon, R., Azcon-Aguilar, C., 2005. Microbial co-operation in the rhizosphere. J. Exp. Bot. 56 (417), 1761–1778.

Barrett, G., Campbell, C.D., Fitter, A.H., Hodge, A., 2011. The arbuscular mycorrhizal fungus *Glomus hoi* can capture and transfer nitrogen from organic patches to its associated host plant at low temperature. Appl. Soil. Ecol. 48 (1), 102–105.

Barros, V., Frosi, G., Santos, M., Ramos, D.G., Falcao, H.M., Santos, M.G., 2018. Arbuscular mycorrhizal fungi improve photosynthetic energy use efficiency and decrease foliar construction cost under recurrent water deficit in woody evergreen species. Plant. Physiol. Biochem. 127, 469–477.

Barrow, J.R., Lucero, M.E., Reyes-Vera, I., Havstad, K.M., 2008. Do symbiotic microbes have a role in plant evolution, performance and response to stress? Commun. Integr. Biol. 1 (1), 69–73.

Bartels, D., Sunkar, R., 2005. Drought and salt tolerance in plants. Crit. Rev. Plant. Sci. 24 (1), 23–58.

Bárzana, G., Aroca, R., Bienert, G.P., Chaumont, F., Ruiz-Lozano, J.M., 2014. New insights into the regulation of aquaporins by the arbuscular mycorrhizal symbiosis in maize plants under drought stress and possible implications for plant performance. Mol. Plant-Microbe Interact. 27, 349–363.

Baslam, M., Garmendia, I., Goicoechea, N., 2011. Arbuscular mycorrhizal fungi (AMF) improved growth and nutritional quality of greenhouse grown lettuce. J. Agric. Food Chem. 59 (10), 5504–5515.

Bilal, S., Khan, A.L., Shahzad, R., Asaf, S., Kang, S.M., Lee, I.J., 2017. Endophytic *Paecilomyces formosus* LHL10 Augments *Glycine max* L. adaptation to Ni-contamination through affecting endogenous phytohormones and oxidative stress. Front. Plant. Sci. 8, 870.

Bischoff, J.F., White, J.F., 2005. Evolutionary development of the Clavicipitaceae. In: Dighton, J., White, J.F., Oudemans, P. (Eds.), The Fungal Community: Its Organization and Role in the Ecosystem. Taylor & Francis, Boca Raton FL, USA, pp. 505–518.

Bohnert, H.J., Nelson, D.E., Jensen, R.G., 1995. Adaptations to environmental stresses. Plant. Cell 7, 1099–1111.

Bonfante, P., Perotto, S., 2000. Outside and inside the roots: cell to-cell interactions among arbuscular mycorrhizal fungi, bacteria and host plants. In: Podila, G.K., Douds, D.D. (Eds.), Current Advances in Mycorrhizae Research. APS Press, St. Paul, MN, pp. 141–155.

Borde, M., Dudhane, M., Jite, P.K., 2010. AM fungi influences the photosynthetic activity, growth and antioxidant enzymes in *Allium sativum* L. under salinity condition. Not. Sci. Biol. 2 (4), 64–71.

Boyer, J.S., 1982. Plant productivity and environment. Science 218 (4571), 443–448.

Bray, E.A., Bailey-Serres, J., Weretilnyk, E., 2000. Responses to abiotic stresses. In: Gruissem, W., Buchannan, B., Jones, R. (Eds.), Biochemistry and Molecular Biology of Plants. American Society of Plant Physiologists, Rockville MD, pp. 1158–1203.

Brundrett, M., Bougher, N., Dell, B., Grove, T., Malajczuk, N., 1996. Working with mycorrhizas in forestry and agriculture, ACIAR Monograph 32. Australian Centre for International Agricultural Research, Canberra, Australia, pp. 173–196.

Brundrett, M.C., 2004. Diversity and classification of mycorrhizal associations. Biol. Rev. 79 (3), 473–495.

Bunn, R., Lekberg, Y., Zabinski, C., 2009. Arbuscular mycorrhizal fungi ameliorate temperature stress in thermophilic plants. Ecology 90 (5), 1378–1388.

Cantrell, I.C., Linderman, R.G., 2001. Preinoculation of lettuce and onion with VA mycorrhizal fungi reduces deleterious effects of soil salinity. Plant. Soil. 233 (2), 269–281.

Cao, L., Jiang, M., Zeng, Z., Du, A., Tan, H., Liu, Y., 2008. *Trichoderma atroviride* F6 improves phytoextraction efficiency of mustard (*Brassica juncea* (L.) Coss. var. foliosa Bailey) in Cd, Ni contaminated soils. Chemosphere 71, 1769–1773.

Carriger, J.F., Rand, G.M., Gardinali, P.R., Perry, W.B., Tompkins, M.S., Fernandez, A.M., 2006. Pesticides of potential ecological concern in sediment from South Florida Canals: an ecological risk prioritization for aquatic arthropods. Soil. Sediment. Contamination 15 (1), 21–45.

Carroll, G., 1988. Fungal endophytes in stems and leaves from latent pathogen to mutualistic symbiont. Ecology 69, 2–9.

Carroll, G.C., Carroll, F.E., 1978. Studies on the incidence of coniferous needle endophytes in the pacific northwest, USA. Can. J. Bot. 56 (24), 3034–3043.

Cekic, F.O., Unyayar, S., Ortas, I., 2012. Effects of arbuscular mycorrhizal inoculation on biochemical parameters in *Capsicum annuum* grown under long term salt stress. Turkish J. Bot. 36 (1), 63–72.

Chen, B., Tang, X., Zhu, Y., Christie, P., 2005. Metal concentrations and mycorrhizal status of plants colonizing copper mine tailings: potential for revegetation. Sci. China Ser. C. 48, 156–164.

Clay, K., 1988. Fungal endophytes of grasses: a defensive mutualism between plants and fungi. Ecology 69, 10–16.

Colla, G., Rouphae, Y., Cardarelli, M., Tulio, M., Rivera, C.M., Rea, E., 2008. Alleviation of salt stress by arbuscular mycorrhiza in zucchini plants grown at low and high phosphorus concentration. Biol. Fertil. Soils 44 (3), 501–509.

Colpaert, J.V., Wevers, J.H., Krznaric, E., Adriaensen, K., 2011. How metal-tolerant ecotypes of ectomycorrhizal fungi protect plants from heavy metal pollution. Ann. For. Sci. 68 (1), 17–24.

Cui, J.L., Wang, Y.N., Jiao, J., Gong, Y., Wang, J.H., Wang, M.L., 2017. Fungal endophyte-induced salidroside and tyrosol biosynthesis combined with signal cross-talk and the mechanism of enzyme gene expression in *Rhodiola crenulata*. Sci. Rep. 7 (1), 12540.

Davies Jr, F.T., Puryear, J.D., Newton, R.J., Egilla, J.N., Saraiva, G.J.A., 2001. Mycorrhizal fungi enhance accumulation and tolerance of chromium in sunflower (*Helianthus annuus*). J. Plant. Physiol. 158, 777–786.

Deng, B., Yang, K., Zhang, Y., Li, Z., 2016. Can heavy metal pollution defend seed germination against heat stress? Effect of heavy metals (Cu(2 +), Cd(2 +) and Hg(2 +)) on maize seed germination under high temperature. Environ. Pollut. 216, 46–52.

Deng, Z., Zhang, R., Shi, Y., Tan, H., Cao, L., 2014. Characterization of Cd-, Pb-, Zn-resistant endophytic *Lasiodiplodia* sp. MXSF31 from metal accumulating *Portulaca oleracea* and its potential in promoting the growth of rape in metal-contaminated soils. Environ. Sci. Pollut. Res. 21 (3), 2346–2357.

Diouf, D., Diop, T.A., Ndoye, I., 2003. Actinorhizal, mycorrhizal and rhizobial symbioses: how much do we know? Afr. J. Biotechnol. 2 (1), 1–7.

Dong, Y., Zhu, Y.G., Smith, F.A., Wang, Y., Chen, B., 2008. Arbuscular mycorrhiza enhanced arsenic resistance of both white clover *Trifolium repens* L. and ryegrass *Lolium perenne* L. plants in an arsenic-contaminated soil. Environ. Pollut. 155, 174–181.

Dugan, F.M., Sitton, J.W., Sullivan, R.F., White, J.F., 2002. The *Neotyphodium* endophyte of wild barley (*Hordeum brevisubulatum* sub sp. *violaceum*) grows and sporulates on leaf surfaces of the host. Symbiosis 32 (2), 147–159.

Eid, A.M., Salim, S.S., Hassan, S.E.D., Ismail, M.A., Fouda, A., 2019. Role of endophytes in plant health and abiotic stress management. Microbiome in Plant Health and Disease. Springer, Singapore, pp. 119–144.

Elbersen, H.W., West, C.P., 1996. Growth and water relations of field-grown tall fescue as influenced by drought and endophyte. Grass Forage Sci. 51 (4), 333–342.

Elhindi, K.M., El-Din, S.A., Elgorban, A.M., 2017. The impact of arbuscular mycorrhizal fungi in mitigating salt-induced adverse effects in sweet basil (*Ocimum basilicum* L.). Saudi J. Biol. Sci. 24 (1), 170–179.

Elkoca, E., Turan, M., Donmez, M.F., 2010. Effects of single, dual and triple inoculations with *Bacillus subtilis*, *Bacillus megaterium* and *Rhizobium leguminosarum* bv. *phaseoli* on nodulation, nutrient uptake, yield and yield parameters of common bean (*Phaseolus vulgaris* l. cv. 'Elkoca-05′). J. Plant. Nutr. 33 (14), 2104–2119.

Elmi, A.A., West, C.P., 1995. Endophyte infection effects on stomatal conductance, osmotic adjustment and drought recovery of tall fescue. N. Phytol. 131, 61–67.

Evelin, H., Giri, B., Kapoor, R., 2013. Ultrastructural evidence for AMF mediated salt stress mitigation in *Trigonella foenum-graecum*. Mycorrhiza 23, 71–86. Available from: https://doi.org/10.1007/s00572-012-0449-8.

Ezzouhri, L., Castro, E., Moya, M., Espinola, F., Lairini, K., 2009. Heavy metal tolerance of filamentous fungi isolated from polluted sites in Tangier, Morocco. Afr. J. Microbiol. Res. 3 (2), 35–48.

Feng, G., Li, X.L., Zhang, F.S., Li, S.X., 2000. Effect of AM fungi on water and nutrition status of corn plants under salt stress. Chin. J. Appl. Ecol. 11 (4), 595–598.

Feng, G., Zhang, F.S., Li, X.L., Tian, C.Y., Tang, C., Rengel, Z., 2002. Improved tolerance of maize plants to salt stress by arbuscular mycorrhiza is related to higher accumulation of soluble sugars in roots. Mycorrhiza 12, 185–190.

Fergusson, J.E., 1990. The Heavy Elements: Chemistry, Environmental Impact and Health Effects. Pergamon Press, Oxford.

Fitter, A.H., 2005. Darkness visible: reflections on underground ecology. J. Ecol. 93 (2), 231–243.

Flexas, J., Bota, J., Loreto, F., Cornic, G., Sharkey, T.D., 2004. Diffusive and metabolic limitations to photosynthesis under drought and salinity in C_3 plants. Plant. Biol. 6 (3), 269–279.

Francis, R., Read, D.J., 1995. Mutualism and antagonism in the mycorrhizal symbiosis, with special reference to impacts on plant community structure. Can. J. Bot. 73 (1), 1301–1309.

Freitas, A.D.S., Vieira, C.L., de Rosália, C.E., Santos, S., Stamford, N.P., 2007. Caracterização de rizóbios isolados de Jacatupé cultivado em solo salino do estado de Pernambuco, Brasil. Bragantia 66 (3), 497–504.

Fröhlich, J., Hyde, K.D., 1999. Biodiversity of palm fungi in the tropics: are global fungal diversity estimates realistic? Biodivers. Conserv. 8 (7), 977–1004.

Gamboa, M.A., Bayman, P., 2001. Communities of endophytic fungi in leaves of a tropical timber tree (*Guarea guidonia*: Meliaceae). Biotropica 33 (2), 352–360.

Ganley, R.J., Brunsfeld, S.J., Newcombe, G., 2004. A community of unknown, endophytic fungi in western white pine. Proc. Natl Acad. Sci. 101 (27), 10107–10112.

Garg, N., Kashyap, L., 2019. Joint effects of Si and mycorrhiza on the antioxidant metabolism of two pigeonpea genotypes under As (III) and (V) stress. Environ. Sci. Pollut. Res. 26, 7821–7839.

Garg, N., Pandey, R., 2015. Effectiveness of native and exotic arbuscular mycorrhizal fungi on nutrient uptake and ion homeostasis in salt-stressed *Cajanus cajan* L. (Millsp.) genotypes. Mycorrhiza 25, 165. Available from: https://doi.org/10.1007/s00572-014-0600-9.

Gaur, A., Adholeya, A., 2004. Prospects of arbuscular mycorrhizal fungi in phytoremediation of heavy metal contaminated soils. Curr. Sci. 86, 528–534.

Gavito, M.E., Olsson, P.A., Rouhier, H., Medinapeñafel, A., Jakobsen, I., Bago, A., 2005. Temperature constraints on the growth and functioning of root organ cultures with arbuscular mycorrhizal fungi. N. Phytol. 168, 179–188. Available from: https://doi.org/10.1111/j.1469-8137.2005.01481.x.

Giovannetti, M., Avio, L., Sbrana, C., 2015. Functional significance of anastomosis in arbuscular mycorrhizal networks. Mycorrhizal Networks. Springer, Dordrecht, pp. 41–67.

Giri, B., Kapoor, R., Mukerji, K.G., 2003. Influence of arbuscular mycorrhizal fungi and salinity on growth, biomass, and mineral nutrition of *Acacia auriculiformis*. Biol. Fertil. Soils 38 (3), 170–175.

Giri, B., Kapoor, R., Mukerji, K.G., 2007. Improved tolerance of *Acacia nilotica* to salt stress by arbuscular mycorrhiza, *Glomus fasciculatum* may be partly related to elevated K/Na ratios in root and shoot tissues. Microb. Ecol. 54 (4), 753–760.

Giri, B., Mukerji, K.G., 2004. Mycorrhizal inoculant alleviates salt stress in *Sesbania aegyptica* and *Sesbania grandiflora* under field conditions: evidence for reduced sodium and improved magnesium uptake. Mycorrhiza 14 (5), 307–312.

Göhre, V., Paszkowski, U., 2006. Contribution of arbuscular mycorrhizal symbiosis to heavy metal phytoremediation. Planta 223, 1115–1122.

Göhre, V., Robatzek, S., 2008. Breaking the barriers: Microbial effector molecules subvert plant immunity. Annu. Rev. Phytopathology 46, 189–215.

Gong, B., Liu, G., Liao, R., Song, J., Zhang, H., 2017. Endophytic fungus *Purpureocillium* sp. A5 protect mangrove plant *Kandelia candel* under copper stress. Braz. J. Microbiol. 48, 530–536.

González-Chávez, M.C., Carrillo-González, R., Gutiérrez-Castorena, M.C., 2009. Natural attenuation in a slag heap contaminated with cadmium: the role of plants and arbuscular mycorrhizal fungi. J. Hazard. Mater. 161, 1288–1298.

González-Chávez, M.C., Carrillo-González, R., Wright, S.F., Nichols, K.A., 2004. The role of glomalin, a protein produced by arbuscular mycorrhizal fungi, in sequestering potentially toxic elements. Environ. Pollut. 130 (3), 317–323.

Graham, J.H., Eissenstat, D.M., 1998. Field evidence for the carbon cost of citrus mycorrhizas. N. Phytol. 140 (1), 103–110.

Gundel, P.E., Hamilton, C.E., Seal, C.E., Helander, M., Martínez-Ghersa, M.A., Ghersa, C.M., et al., 2012. Antioxidants in *Festuca rubra* L. seeds affected by the fungal symbiont *Epichloë festucae*. Symbiosis 58 (1–3), 73–80.

Gutjahr, C., Parniske, M., 2013. Cell and developmental biology of arbuscular mycorrhiza symbiosis. Annu. Rev. Cell Dev. Biol. 29, 593–617.

Hahn, H., McManus, M.T., Warnstorff, K., Monahan, B.J., Young, C.A., Davies, E., et al., 2008. *Neotyphodium* fungal endophytes confer physiological protection to perennial ryegrass (*Lolium perenne* L.) subjected to a water deficit. Environ. Exp. Botany 63, 183–199.

Hajiboland, R., Aliasgharzadeh, N., Laiegh, S.F., Poschenrieder, C., 2010. Colonization with arbuscular mycorrhizal fungi improves salinity tolerance of tomato (*Solanum lycopersicum* L.) plants. Plant. Soil. 331, 313–327.

Hall, J.L., 2002. Cellular mechanism of heavy metal detoxification and tolerance. J. Exp. Bot. 53 (366), 1–11.

Hameed, A., Dilfuza, E., Abd-Allah, E.F., Hashem, A., Kumar, A., Ahmad, P., 2014. Salinity stress and arbuscular mycorrhizal symbiosis in plants, Use of microbes for the alleviation of soil stresses, 1. Springer, New York, NY, pp. 139–159.

Hamilton, C.E., Gundel, P.E., Helander, M., Saikkonen, K., 2012. Endophytic mediation of reactive oxygen species and antioxidant activity in plants: a review. Fungal Divers. 54 (1), 1–10.

Harley, J.L., Smith, S.E., 1983. Mycorrhizal Symbiosis. Academic, London.

Hasanuzzaman, M., Gill, S.S., Fujita, M., 2013. Physiological role of nitric oxide in plants grown under adverse environmental conditions. In: Tuteja, N., Gill, S.S. (Eds.), Plant Acclimation to Environmental Stress. Springer, New York, NY, pp. 269–322.

Hashem, A., Abd-Allah, E., Alqarawi, A., Al Huqail, A.A., Egamberdieva, D., 2014. Alleviation of abiotic salt stress in *Ochradenus baccatus* (Del.) by *Trichoderma hamatum* (bonord.) bainier. J. Plant. Interact. 9, 857–868.

Heino, P., Tapio, P.E., 2003. Signal transduction in plant cold acclimation. In: Hirt, H., Shinozaki, K. (Eds.), Plant Responses to Abiotic Stress, 4. Springer-Verlag, Berlin Heidelberg, pp. 151–186.

Hermosa, M.R., Emma, K.E.C.K., Chamorro, I., Rubio, B., Luis, S.A.N.Z., Vizcaíno, J.A., et al., 2004. Genetic diversity shown in *Trichoderma* biocontrol isolates. Mycol. Res. 108 (8), 897–906.

Hesse, U., Schoberlein, W., Wittenmayer, L., Forster, K., Warnstorff, K., Diepenbrock, W., et al., 2003. Effects of *Neotyphodium* endophytes on growth, reproduction and drought-stress tolerance of three *Lolium perenne* L. genotypes. Grass Forage Sci. 58 (4), 407–415.

Heydarian, A., Tohidi, M.H.R., Donath, T.W., Sohrabi, M., 2018. Study of effect of arbuscular mycorrhiza (*Glomus intraradices*) fungus on wheat under nickel stress. Agron. Res. 16 (4), 1660–1667.

Hildebrandt, U., Kaldorf, M., Bothe, H., 1999. The zinc violet and its colonization by arbuscular mycorrhizal fungi. J. Plant. Physiol. 154 (5–6), 709–711.

Hildebrandt, U., Regvar, M., Bothe, H., 2007. Arbuscular mycorrhiza and heavy metal tolerance. Phytochemistry 68, 139–146.

Howarth, C.J., 2005. Genetic improvements of tolerance to high temperature. In: Ashraf, M., Harris, P.J.C. (Eds.), Abiotic Stresses: Plant Resistance Through Breeding and Molecular Approaches. Howarth Press Inc, New York.

Hu, Y., Wu, S., Sun, Y., Li, T., Zhang, X., Chen, C., et al., 2015. Arbuscular mycorrhizal symbiosis can mitigate the negative effects of night warming on physiological traits of *Medicago truncatula* L. Mycorrhiza 25 (2), 131–142.

Hubbard, M., Germida, J.J., Vujanovic, V., 2013. Fungal endophytes enhance wheat heat and drought tolerance in terms of grain yield and second-generation seed viability. J. Appl. Microbiol. 116 (1), 109–112.

Ikram, M., Ali, N., Jan, G., Jan, F.G., Rahman, I.U., Iqbal, A., et al., 2018. IAA producing fungal endophyte *Penicillium roqueforti* Thom., enhances stress tolerance and nutrients uptake in wheat plants grown on heavy metal contaminated soils. PLoS One 13 (11), e0208150.

Impa, S.M., Nadaradjan, S., Jagadish, S.V.K., 2012. Drought stress induced reactive oxygen species and anti-oxidants in plants. In: Ahmad, P., Prasad, M.N.V. (Eds.), Abiotic Stress

Responses in Plants: Metabolism, Productivity and Sustainability. LLC: Springer Science + Business Media, pp. 131–147.

Iram, S., Zaman, A., Iqbal, Z., Shabbir, R., 2013. Heavy metal tolerance of fungus isolated from soil contaminated with sewage and industrial wastewater. Pol. J. Environ. Stud. 22 (3), 691–697.

Ismail, H.M., Hussain, A., Iqbal, A., Khan, S.A., Lee, I.J., 2018. Endophytic fungus *Aspergillus japonicus* mediates host plant growth under normal and heat stress conditions. Biomed. Res. Int. 3, 1–11.

Jeffries, P., Gianinazzi, S., Perotto, S., Turnau, K., Barea, J.M., 2003. The contribution of arbuscular mycorrhizal fungi in sustainable maintenance of plant health and soil fertility. Biol. Fertil. Soils 37 (1), 1–16.

Jiang, Y., Wang, W., Xie, Q., Liu, N., Liu, L., Wang, D., et al., 2017. Plants transfer lipids to sustain colonization by mutualistic mycorrhizal and parasitic fungi. Science 356 (6343), 1172–1175.

Johnson, N.C., Graham, J.H., Smith, F., 1997. Functioning of mycorrhizal associations along the mutualism-parasitism continuum. N. Phytol. 135 (4), 575–586.

Joner, E., Leyval, C., 1997. Uptake of 109 Cd by roots and hyphae of a *Glomus mosseae/Trifolium subterraneum* mycorrhiza from soil amended with high and low concentrations of cadmium. N. Phytol. 135, 353–360.

Kachenko, A.G., Singh, B., 2006. Heavy metals contamination in vegetables grown in urban and metal smelter contaminated sites in Australia. Water, Air Soil. Pollut. 169 (1–4), 101–123.

Kaldorf, M., Kuhn, A.J., Schröder, W.H., Hildebrandt, U., Bothe, H., 1999. Selective element deposits in maize colonized by a heavy metal tolerance conferring arbuscular mycorrhizal fungus. J. Plant. Physiol. 154 (5–6), 718–728.

Kapoor, R., Evelin, H., Mathur, P., Giri, B., 2013. Arbuscular mycorrhiza: approaches for abiotic stress tolerance in crop plants for sustainable agriculture. In: Tuteja, N., Gill, S.S. (Eds.), Plant Acclimation to Environmental Stress. LLC: Springer Science + Business Media, pp. 359–401.

Kawaguchi, M., Minamisawa, K., 2010. Plant–Microbe Communications for Symbiosis. Plant. Cell Physiol. 51 (9), 1377–1380.

Khairiah, T., Zalifah, M.K., Yin, Y.H., Aminath, A., 2004. The uptake of heavy metals by fruit type vegetables, grown in selected agricultural areas. Pak. J. Biol. Sci. 7, 1438–1442.

Khalid, M., Hassani, D., Liao, J., Xiong, X., Bilal, M., Huang, D., 2018. An endosymbiont *Piriformospora indica* reduces adverse effects of salinity by regulating cation transporter genes, phytohormones, and antioxidants in *Brassica campestris* ssp. *Chinensis*. Environ. Exp. Bot. 153, 89–99.

Khan, A.L., Hamayun, M., Ahmad, N., Hussain, J., Kang, S.M., Kim, Y.H., et al., 2011c. Salinity stress resistance offered by endophytic fungal interaction between *Penicillium minioluteum* LHL09 and *Glycine max* L. J. Microbiol. Biotechnol. 21 (9), 893–902.

Khan, A.L., Hamayun, M., Kang, S.M., Kim, Y.H., Jung, H.Y., Lee, J.H., et al., 2012a. Endophytic fungal association via gibberellins and indole acetic acid can improve plant growth potential in abiotic stress: an example of *Paecilomyces formosus* LHL10. BMC Microbiol. 12, 3.

Khan, A.L., Hamayun, M., Kim, Y.H., Kang, S.M., Lee, I.J., 2011b. Ameliorative symbiosis of endophyte (*Penicillium funiculosum* LHL06) under salt stress elevated plant growth of *Glycine max* L. Plant. Physiol. Biochem. 49 (8), 852–861.

Khan, A.L., Hamayun, M., Kim, Y.H., Kang, S.M., Lee, J.H., Lee, I.J., 2011a. Gibberellins producing endophytic *Aspergillus fumigatus* sp. LH02 influenced endogenous phytohormonal levels, isoflavonoids production and plant growth in salinity stress. Process. Biochem. 46 (2), 440–447.

Khan, A.L., Hamayun, M., Waqas, M., Kang, S.M., Kim, Y.H., Kim, D.H., et al., 2012b. *Exophiala* sp. LHL08 association gives heat stress tolerance by avoiding oxidative damage to cucumber plants. Biol. Fertil. Soils 48 (5), 519–529.

Khan, A.L., Kang, S.M., Dhakal, K.H., Hussain, J., Adnan, M., Kim, J.G., et al., 2013. Flavonoids and amino acid regulation in *Capsicum annuum* L. by endophytic fungi under different heat stress regimes. Sci. Horticult. 155, 1–7.

Khan, A.L., Lee, I.J., 2013. Endophytic *Penicillium funiculosum* LHL06 secretes gibberellin that reprograms *Glycine max* L. growth during copper stress. BMC Plant. Biol. 13, 86.

Khan, A.R., Ullah, I., Waqas, M., Park, G.-S., Khan, A.L., Hong, S.J., et al., 2017. Host plant growth promotion and cadmium detoxification in *Solanum nigrum*, mediated by endophytic fungi. Ecotoxicol. Environ. Saf. 136, 180–188.

Kramer, U., 2005. Phytoremediation: novel approaches to cleaning up polluted soils. Curr. Opin. Biotechnol. 16 (2), 133–141.

Krings, M., Taylor, T.N., Hass, H., Kerp, H., Dotzler, N., Hermsen, E.J., 2007. Fungal endophytes in a 400 million-yr-old land plant: infection pathways, spatial distribution and host responses. N. Phytol. 174 (3), 648–657.

Krishna, P., 2003. Plant responses to heat stress. In: Hirt, H., Shinozaki, K. (Eds.), Plant Responses to Abiotic Stress, 4. Springer-Verlag, Berlin, Heidelberg, pp. 73–101.

Kuldau, G., Bacon, C., 2008. Clavicipitaceous endophytes: their ability to enhance resistance of grasses to multiple stresses. Biol. Control. 46 (1), 57–71.

Latch, G.C.M., 1993. Physiological interactions of endophytic fungi and their hosts. Biotic stress tolerance imparted to grasses by endophytes. Agriculture, Ecosyst. Environ. 44 (1–4), 143–156.

Latef, A.A.H.A., 2011. Influence of arbuscular mycorrhizal fungi and copper on growth, accumulation of osmolyte, mineral nutrition and antioxidant enzyme activity of pepper (*Capsicum annuum* L.). Mycorrhiza 21, 495–503.

Latef, A.A.H.A., 2013. Growth and some physiological activities of pepper (*Capsicum annuum* L.) in response to cadmium stress and mycorrhizal symbiosis. J. Agric. Sci. Technol. 15, 1437–1448.

Latef, A.A.H.A., Chaoxing, H., 2011. Arbuscular mycorrhizal influence on growth, photosynthetic pigments, osmotic adjustment and oxidative stress in tomato plants subjected to low temperature stress. Acta Physiol. Plant. 33 (4), 1217–1225.

Latef, A.A.H.A., Chaoxing, H., 2014. Does the inoculation with *Glomus mosseae* improve salt tolerance in pepper plants? J. Plant. Growth Regul. 33 (3), 644–653. Available from: https://doi.org/10.1007/s00344-014-9414-4.

Latef, A.A.H.A., Hashem, A., Rasool, S., Abd-Allah, E.F., Alqarawi, A.A., Egamberdieva, D., et al., 2016. Arbuscular mycorrhizal symbiosis and abiotic stress in plants: a review. J. Plant. Biol. 59 (5), 407–426.

Leone, A., Perrotta, C., Maresca, B., 2003. Plant tolerance to heat stress: current strategies and new emergent insight. In: di Toppi, L.S., Pawlik-Skowronska, B. (Eds.), Abiotic Stresses in Plants. Kluwer Academic Pub, London, pp. 1–22.

Levy, Y., Dodd, J., Krikun, J., 1983. Effect of irrigation, water salinity and rootstock on the vertical distribution of vesicular-arbuscular mycorrhiza in citrus roots. N. Phytol. 95 (3), 397–403.

Lewis, D.H., 1985. Symbiosis and mutualism: crisp concepts and soggy semantics. In: Boucher, D.H. (Ed.), The Biology of Mutualism. Croom Helm Ltd, London, pp. 29–39.

Li, H., Chen, X.W., Wong, M.H., 2015. Arbuscular mycorrhizal fungi reduced the ratios of inorganic/organic arsenic in rice grains. Chemosphere 145, 224–230.

Li, H., Luo, N., Zhang, L.J., Zhao, H.M., Li, Y.W., Cai, Q.Y., et al., 2016. Do arbuscular mycorrhizal fungi affect cadmium uptake kinetics, subcellular distribution and chemical forms in rice? Sci. Total. Environ. 571, 1183–1190.

Li, J., Meng, B., Chai, H., Yang, X., Song, W., Li, S., et al., 2019. Arbuscular mycorrhizal fungi alleviate drought stress in C_3 (*Leymus chinensis*) and C_4 (*Hemarthria altissima*) grasses via altering antioxidant enzyme activities and photosynthesis. Front. Plant. Sci. 10, 499.

Li, L., Wang, X., Zhu, P., Wu, H., Qi, S., 2017. Plant growth-promoting endophyte *Piriformospora indica* alleviates salinity stress in *Medicago truncatula*. Plant. Physiol. Biochem. 119, 211–223.

Li, T., Liu, M.J., Zhang, X.T., Zhang, H.B., Sha, T., Zhao, Z.W., 2011. Improved tolerance of maize (*Zea mays* L.) to heavy metals by colonization of a dark septate endophyte (DSE) *Exophiala pisciphila*. Sci. Total. Environ. 409, 1069–1074.

Liu, A., Chen, S., Chang, R., Liu, D., Chen, H., Ahammed, G.J., et al., 2014a. Arbuscular mycorrhizae improve low temperature tolerance in cucumber via alterations in H_2O_2 accumulation and ATPase activity. J. Plant. Res. 127 (6), 775–785.

Liu, X., Song, Q., Tang, Y., Li, W., Xu, J., Wu, J., et al., 2013. Human health risk assessment of heavy metals in soil-vegetable system: a multi-medium analysis. Sci. Total. Environ. 463, 530–540.

Liu, Z., Ma, L., He, X., Tian, C., 2014b. Water strategy of mycorrhizal rice at low temperature through the regulation of PIP aquaporins with the involvement of trehalose. Appl. Soil. Ecol. 84, 185–191.

Long, L.K., Yao, Q., Guo, J., Yang, R.H., Huang, Y.H., Zhu, H.H., 2010. Molecular community analysis of arbuscular mycorrhizal fungi associated with five selected plant species from heavy metal polluted soils. Eur. J. Soil. Biol. 46, 288–294.

Lubna, Asaf, S., Hamayun, M., Khan, A.L., Waqas, M., Khan, M.A., et al., 2018. Salt tolerance of *Glycine max* L. induced by endophytic fungus *Aspergillus flavus* CSH1, via regulating its endogenous hormones and antioxidative system. Plant. Physiol. Biochem. 128, 13–23.

Luginbuehl, L.H., Menard, G.N., Kurup, S., Van Erp, H., Radhakrishnan, G.V., Breakspear, A., et al., 2017. Fatty acids in arbuscular mycorrhizal fungi are synthesized by the host plant. Science 356 (6343), 1175–1178.

Ma, X., Chen, C., Yang, M., Dong, X., Lv, W., Meng, Q., 2018. Cold-regulated protein (SlCOR413IM1) confers chilling stress tolerance in tomato plants. Plant. Physiol. Biochem. 124, 29–39.

Maggio, A., Bressan, R.A., Ruggiero, C., Xiong, L., Grillo, S., 2003. Salt tolerance: placing advances in molecular genetics into a physiological and agronomic contextIn: di Toppi, L. S., Pawlik-Skowronska, B. (Eds.), Kluwer Academic Pub, London, pp. 53–70.

Mahajan, S., Tuteja, N., 2005. Cold, salinity and drought stresses: an overview. Arch. Biochem. Biophys. 444 (2), 139–158.

Malinowski, D.P., Belesky, D.P., 2000. Adaptations of endophyte-infected cool-season grasses to environmental stresses: mechanisms of drought and mineral stress tolerance. Crop. Sci. 40 (4), 923–940.

Marks, S., Clay, K., 1990. Effects of CO_2 enrichment, nutrient addition and fungal endophyte infection on the growth of two grasses. Oecologia 84, 207–214.

Márquez, L.M., Redman, R.S., Rodriguez, R.J., Roossinck, M.J., 2007. A virus in a fungus in a plant-three way symbiosis required for thermal tolerance. Science 315, 513–515.

Martin, C.A., Stutz, J.C., 2004. Interactive effects of temperature and arbuscular mycorrhizal fungi on growth, P uptake and root respiration of *Capsicum annum* L. Mycorrhiza 14, 241–244.

Marulanda, A., Azcon, R., Ruiz-Lozano, J.M., 2003. Contribution of six arbuscular mycorrhizal fungal isolates to water uptake by *Lactuca sativa* plants under drought stress. Physiol. Plant. 119 (4), 526–533.

Matos, T., de-Hoog, G.S., de Boer, A.G., de Crom, I., Haase, G., 2002. High prevalence of the neurotrope *Exophiala dermatitidis* and related oligotrophic black yeasts in sauna facilities. Mycoses 45, 373–377.

Matsubara, Y., 2010. High Temperature Stress Tolerance and the Changes in Antioxidative Ability in Mycorrhizal Strawberry Plants. Horticulture in the 21st century, Nova Science, New York, pp. 179–192.

Matsubara, Y., Harada, T., 1996. Effect of constant and diurnally fluctuating temperatures on arbuscular mycorrhizal fungus infection and the growth of infected asparagus (*Asparagus officinalis* L.) seedlings. Soc. Horticult. Sci. 65 (3), 565–570.

Maya, M.A., Matsubara, Y., 2013. Influence of arbuscular mycorrhiza on the growth and antioxidative activity in *Cyclamen* under heat stress. Mycorrhiza 23 (5), 381–390. Available from: https://doi.org/10.1007/s00572-013-0477-z.

Medina, A., Vassilev, N., Barea, J.M., Azcon, R., 2005. Application of *Aspergillus niger* treated agrowaste residue and *Glomus mosseae* for improving growth and nutrition of *Trifolium repens* in a Cd contaminated soil. J. Biotechnol. 116 (4), 369–378.

Mena-Violante, H.G., Ocampo-Jimenez, O., Dendooven, L., Martinez-Soto, G., Gonzalez-Castafeda, J., Davies, F.T., et al., 2006. Arbuscular mycorrhizal fungi enhance fruit growth and quality of chile ancho *Capsicum annuum* L. cv San Luis plants exposed to drought. Mycorrhiza 16 (4), 261–267.

Miller, R.M., Jastrow, J.D., 2000. Mycorrhizal fungi influence soil structure. In: Kapulnik, Y., Douds, D.D. (Eds.), Arbuscular Mycorrhizas: Physiology and Function. Kluwer Academic Publishers, Dordrecht, pp. 3–18.

Miransari, M., 2011. Hyper-accumulators, arbuscular mycorrhizal fungi and stress of heavy metals. Biotechnol. Adv. 29, 645–653.

Mirzahossini, Z., Shabani, L., Sabzalian, M.R., Sharifi-Tehrani, M., 2015. ABC transporter and metallothionein expression affected by NI and *Epichloe* endophyte infection in tall fescue. Ecotoxicol. Environ. Saf. 120, 13–19.

Mona, S.A., Hashem, A., Abd-Allah, E.F., Alqarawi, A.A., Soliman, D.W.K., Wirth, S., et al., 2017. Increased Resistance of drought by *Trichoderma harzianum* fungal treatment correlates with increased secondary metabolites and proline content. J. Integr. Agric. 16 (8), 1751–1757.

Moradtalab, N., Roghieh, H., Nasser, A., Tobias, E.H., Günter, N., 2019. Silicon and the association with an arbuscular-mycorrhizal fungus (*Rhizophagus clarus*) mitigate the adverse effects of drought stress on strawberry. Agronomy 9, 41.

Moretti, M.M.S., Bocchini-Martins, D.A., Da Silva, R., Rodrigues, A., Sette, L.D., Gomes, E., 2012. Selection of thermophilic and thermotolerant fungi for the production of cellulases and xylanases under solid-state fermentation. Braz. J. Microbiol. 43 (3), 1062–1071.

Mosse, B., Stribley, D.P., Le Tacon, F., 1981. The ecology of mycorrhizae and mycorrhizal fungi. In: Alexender, M. (Ed.), Advances in Microbial Biology. Plenum Press, New York, pp. 137–210.

Moy, M., Belanger, F., Duncan, R., Freehoff, A., Leary, C., Meyer, W., et al., 2000. Identification of epiphyllous mycelial nets on leaves of grasses infected by clavicipitaceous endophytes. Symbiosis 28 (4), 291–302.

Oldroyd, G.E.D., 2013. Speak, friend, and enter: signalling systems that promote beneficial symbiotic associations in plants. Nat. Rev. Microbiol. 11 (4), 252–263.

Ouledali, S., Ennajeh, M., Zrig, A., Gianinazzi, S., Khemira, H., 2018. Estimating the contribution of arbuscular mycorrhizal fungi to drought tolerance of potted olive trees (*Olea europaea*). Acta Physiol. Plant. 40 (5), 81.

Ouziad, F., Hildebrandt, U., Schmelzer, E., Bothe, H., 2005. Differential gene expressions in arbuscular mycorrhizal-colonized tomato grown under heavy metal stress. J. Plant. Physiol. 162 (6), 634–649.

Parádi, I., Bratek, Z., Láng, F., 2003. Influence of arbuscular mycorrhiza and phosphorus supply on polyamine content, growth and photosynthesis of *Plantago lanceolata*. Biol. Plant. 46 (4), 563–569.

Paul, D., 2012. Osmotic stress adaptations in rhizobacteria. J. Basic. Microbiol. 53 (2), 1–10.

Pawlowski, K., Bisseling, T., 1996. Rhizobial and actinorhizal symbioses: what are the shared features? Plant. Cell 8, 1899–1913.

Petrini, O., 1991. Fungal endophytes of tree leaves. In: Andrews, J.A., Hirano, S.S. (Eds.), Microbial Ecology of Leaves. Springer-Verlag, New York, NY.

Pierart, A., Shahid, M., Séjalon-Delmas, N., Dumat, C., 2015. Antimony bioavailability: knowledge and research perspectives for sustainable agricultures. J. Hazard. Mater. 289, 219–234.

Pieterse, C.M.J., Dicke, M., 2007. Plant interactions with microbes and insects: from molecular mechanisms to ecology. Trends Plant. Sci. 12 (12), 564–569.

Pirozynski, K.A., Malloch, D.W., 1975. The origin of land plants a matter of mycotrophism. Biosystems 6, 153–164.

Porras-Soriano, A., Soriano-Martin, M.L., Porras-Piedra, A., Azcon, R., 2009. Arbuscular mycorrhizal fungi increased growth, nutrient uptake and tolerance to salinity in olive trees under nursery conditions. J. Plant. Physiol. 166 (13), 1350–1359.

Punamiya, P., Datta, R., Sarkar, D., Barber, S., Patel, M., Da, P., 2010. Symbiotic role of *Glomus mosseae* in phytoextraction of lead in vetiver grass *Chrysopogon zizanioides* L. J. Hazard. Mater. 177 (1–3), 465–474.

Rabie, G.H., 2005. Influence of VA-mycorrhizal fungi and kinetin on the response of moongbean plants to irrigation with sea water. Mycorrhiza 15, 225–230.

Rahmaty, R., Khara, J., 2011. Effects of vesicular arbuscular mycorrhiza *Glomus intraradices* on photosynthetic pigments, antioxidant enzymes, lipid peroxidation, and chromium accumulation in maize plants treated with chromium. Turkish J. Biol. 35, 51–58.

Raison, J.K., Chapman, E.A., 1976. Membrane phase changes in chilling-sensitive vigna radiate and their significance to growth. Funct. Plant. Biol. 3 (3), 291–299.

Raju, P.S., Clark, R.B., Ellis, J.R., Maranville, J.W., 1990. Effects of species of VA-mycorrhizal fungi on growth and mineral uptake of sorghum at different temperatures. Plant. Soil. 121, 165–170.

Rascio, N., Navari-Izzo, F., 2011. Heavy metal hyperaccumulating plants: how and why do they do it? And what makes them so interesting? Plant. Sci. 180 (2), 169–181.

Read, D.J., 1999. Mycorrhiza–the State of the Art. In Mycorrhiza. Springer, Berlin, Heidelberg.

Redecker, D., Kodner, R., Graham, L.E., 2000. Glomalean fungi from the Ordovician. Science 289, 1920–1921.

Redman, R.S., Dunigan, D.D., Rodriguez, R.J., 2001. Fungal symbiosis: from mutualism to parasitism, who controls the outcome, host or invader? N. Phytol. 151, 705–716.

Redman, R.S., Kim, Y.O., Woodward, C.J.D.A., Greer, C., Espino, L., Doty, S.L., et al., 2011. Increased fitness of rice plants to abiotic stress via habitat adapted symbiosis: a strategy for mitigating impacts of climate change. PLoS One 6 (7), 14823.

Redman, R.S., Sheehan, K.B., Stout, R.G., Rodriguez, R.J., Henson, J.M., 2002. Thermotolerance conferred to plant host and fungal endophyte during mutualistic symbiosis. Science 298 (5598), 1581.

Ren, A., Li, C., Gao, Y., 2011. Endophytic fungus improves growth and metal uptake of *Lolium arundinaceum* Darbyshire ex. Schreb. Int. J. Phytoremediat. 13 (3), 233–243.

Rennenberg, H., Loreto, F., Polle, A., Brilli, F., Fares, S., Beniwal, R.S., et al., 2006. Physiological responses of forest trees to heat and drought. Plant. Biol. 8 (5), 556–571.

Repetto, O., Bestel-corre, G., Dumas-gaudot, E., Berta, G., Gianinazzi-pearson, V., Gianinazzi, S., 2003. Targeted proteomics to identify cadmium-induced protein modifications in *Glomus mosseae*-inoculated pea roots. N. Phytologist 157, 555–567.

Richardson, M.D., Chapman, G.W., Hoveland, C.S., Bacon, C.W., 1992. Sugar alcohols in endophyte-infected tall fescue under drought. Crop. Sci. 32 (4), 1060–1061.

Rivera-Becerril, F., Calantzis, C., Turnau, K., Caussanel, J.P., Belimov, A.A., Gianinazzi, S., et al., 2002. Cadmium accumulation and buffering of cadmium induced stress by arbuscular mycorrhiza in three *Pisum sativum* L. genotypes. J. Exp. Bot. 53 (371), 1177–1185.

Rivera-Becerril, F., van Tuinen, D., Martin-Laurent, F., Metwally, A., Dietz, K.J., Gianinazzi, S., et al., 2005. Molecular changes in *Pisum sativum* L. roots during arbuscular mycorrhiza buffering of cadmium stress. Mycorrhiza 16, 51–60. Available from: https://doi.org/10.1007/s00572-005-0016-7.

Robert-Seilaniantz, A., Navarro, L., Bari, R., Jones, J.D., 2007. Pathological hormone imbalances. Curr. Opin. Plant. Biol. 10 (4), 372–379.

Rodriguez, R., Redman, R., 2008. More than 400 million years of evolution and some plants still can't make it on their own: plant stress tolerance via fungal symbiosis. J. Exp. Bot. 59 (5), 1109–1114.

Rodriguez, R.J., Henson, J., Van Volkenburgh, E., Hoy, M., Wright, L., Beckwith, F., et al., 2008. Stress tolerance in plants via habitat-adapted symbiosis. ISME J. 2 (4), 404.

Rodriguez, R.J., White Jr., J.F., Arnold, A.E., Redman, A.R.A., 2009. Fungal endophytes: diversity and functional roles. N. Phytol. 182 (2), 314–330.

Roelfsema, M.R., Hedrich, R., 2005. In the light of stomatal opening: new insights into 'the Watergate'. N. Phytol. 167 (3), 665–691.

Rozpądek, P., Domka, A., Ważny, R., Nosek, M., Jędrzejczyk, R., Tokarz, K., et al., 2018. How does the endophytic fungus *Mucor* sp. improve *Arabidopsis arenosa* vegetation in the degraded environment of a mine dump? Environ. Exp. Bot. 147, 31–42.

Ruiz-Lozano, J.M., 2003. Arbuscular mycorrhizal symbiosis and alleviation of osmotic stress. N. Perspect. Mol. Studies. Mycorrhiza 13, 309–317.

Ruiz-Lozano, J.M., Aroca, R., Zamarreno, A.M., Molina, S., Andreo-Jimenez, B., Porcel, R., et al., 2016. Arbuscular mycorrhizal symbiosis induces strigolactone biosynthesis under drought and improves drought tolerance in lettuce and tomato. Plant, Cell Environ. 39 (2), 441–452.

Sabra, M., Aboulnasr, A., Franken, P., Perreca, E., Wright, L.P., Camehl, I., 2018. Beneficial root endophytic fungi increase growth and quality parameters of sweet basil in heavy metal contaminated soil. Front. Plant. Sci. 9, 1726.

Sallaku, G., Sandén, H., Babaj, I., Kaciu, S., Balliu, A., Rewald, B., 2019. Specific nutrient absorption rates of transplanted cucumber seedlings are highly related to RGR and influenced by grafting method, AMF inoculation and salinity. Sci. Horticult. 243, 177–188.

Sannazzaro, A.I., Echeverria, M., Alberto, E.O., Ruiz, O.A., Menéndez, A.B., 2007. Modulation of polyamine balance in *Lotus glaber* by salinity and arbuscular mycorrhiza. Physiol. Biochem. 45 (1), 39–46.

Sannazzaro, A.I., Oscar, R., Edgardo, A., Ana, M., 2006. Alleviation of salt stress in *Lotus glaber* by *Glomus intraradices*. Plant. Soil. 285, 279–287.

Santander, C., Sanhueza, M., Olave, J., Borie, F., Valentine, C., Cornejo, P., 2019. Arbuscular mycorrhizal colonization promotes the tolerance to salt stress in lettuce plants through an efficient modification of ionic balance. J. Soil. Sci. Plant. Nutr. 19 (2), 321–331.

Schaechter, M., 2011. Eukaryotic Microbes. Elsevier, San Diego, CA.

Schardl, C.L., Leuchtmann, A., Tsai, H.F., Collett, M.A., Watt, D.M., Scott, D.B., 1994. Origin of a fungal symbiont of perennial ryegrass by interspecific hybridization of a mutualist with the ryegrass choke pathogen *Epichloe typhina*. Genetics 136, 1307–1317.

Schardl, S.L., Leuchtmann, A., Spiering, M.J., 2004. Symbioses of grasses with seedborne fungal endophytes. Annu. Rev. Plant. Biol. 55, 315–340.

Schöffl, F., Prandl, R., Reindl, A., 1999. Molecular responses to heat stress. In: Shinozaki, K., Yamaguchi-Shinozaki, K. (Eds.), Molecular Responses to Cold, Drought, Heat and Salt Stress in Higher Plants. R.G. Landes Co., Austin, Texas, pp. 81–98.

Schroeder, J.I., Kwak, J.M., Allen, G.J., 2001. Guard cell abscisic acid signalling and engineering drought hardiness in plants. Nature 410, 327–330.

Schulz, B.J.E., 2006. Mutualistic interactions with fungal root endophytes. In: Schulz, B.J.E., Boyle, C.J.C., Sieber, T.N. (Eds.), Microbial Root Endophytes. Springer-Verlag, Berlin, pp. 261–280.

Selmar, D., Kleinwächter, M., 2013. Influencing the product quality by deliberately applying drought stress during the cultivation of medicinal plants. Ind. Crop. Products 42, 558–566.

Serrano, R., Mulet, J.M., Rios, G., Marquez, J.A., de Larrinoa, I.F., Leube, M.P., et al., 1999. A glimpse of the mechanisms of ion homeostasis during salt stress. J. Exp. Bot. 50, 1023–1036.

Shahabivand, S., Maivan, H.Z., Goltapeh, E.M., Sharifi, M., Aliloo, A.A., 2012. The effects of root endophyte and arbuscular mycorrhizal fungi on growth and cadmium accumulation in wheat under cadmium toxicity. Plant. Physiol. Biochem. 60, 53–58.

Sheng, M., Tang, M., Chen, H., Yang, B., Zhang, F., Huang, Y., 2008. Influence of arbuscular mycorrhizae on photosynthesis and water status of maize plants under salt stress. Mycorrhiza 18, 287–296.

Sherameti, I., Tripathi, S., Varma, A., Oelmüller, R., 2008. The root-colonizing endophyte *Piriformospora indica* confers drought tolerance in Arabidopsis by stimulating the expression of drought stress related genes in leaves. Mol. Plant-Microbe Interact. 21 (6), 799–807.

Sherwood, M., Carroll, G., 1974. Fungal succession on needles and young twigs of old-growth Douglas fir. Mycologia 66 (3), 499–506.

Singh, D., Singh, N.P., Chauhan, S.K., Singh, P., 2011. Developing aluminium tolerant crop plants using biotechnological tools. Curr. Sci. 100 (12), 1807–1814.

Smith, J.L., De Moraes, C.M., Mescher, M.C., 2009. Jasmonate- and salicylate-mediated plant defense responses to insect herbivores, pathogens and parasitic plants. Pest Management Science: formerly. Pesticide Sci. 65 (5), 497–503.

Smith, S.E., Facelli, E., Pope, S., Smith, F.A., 2010. Plant performance in stressful environments: interpreting new and established knowledge of the roles of arbuscular mycorrhizas. Plant. Soil. 326, 3–20.

Smith, S.E., Read, D.J., 2008. Mycorrhizal Symbiosis. Academic, London.

Smith, S.E., Smith, F.A., 2011. Roles of arbuscular mycorrhizas in plant nutrition and growth: new paradigms from cellular to ecosystem scales. Annu. Rev. Plant. Biol. 62, 227–250.

Soleimani, M., Hajabbasi, M.H., Afyuni, M., Mirlohi, A., Borggaard, O.K., Holm, P.E., 2010. Effect of endophytic fungi on cadmium tolerance and bioaccumulation by *Festuca arundinacea* and *Festuca pratensis*. Int. J. Phytoremediat. 12 (6), 535–549.

Stone, J.K., 1988. Fine structure of latent infection by *Rhabdocline parkeri* on Douglas fir, with observation on uninfected epidermal cells. Can. J. Bot. 66 (1), 45–54.

Stone, J.K., Polishook, J.D., White, J.R.J., 2004. Endophytic fungi. In: Mueller, G., Bills, G.F., Foster, M.S. (Eds.), Biodiversity of Fungi: Inventory and Monitoring Methods. Elsevier, Burlington MA, USA, pp. 241–270.

Sun, C., Johnson, J.M., Cai, D., Sherameti, I., Oelmüller, R., Lou, B., 2010. *Piriformospora indica* confers drought tolerance in Chinese cabbage leaves by stimulating antioxidant enzymes, the expression of drought-related genes and the plastid-localized CAS protein. J. Plant. Physiol. 167 (12), 1009–1017.

Sun, Y., Zhang, X., Wu, Z., Hu, Y., Wu, S., Chen, B., 2016. The molecular diversity of arbuscular mycorrhizal fungi in the arsenic mining impacted sites in Hunan Province of China. J. Environ. Sci. 39, 110–118.

Suzuki, N., Mittler, R., 2006. Reactive oxygen species and temperature stresses: a delicate balance between signaling and destruction. Physiol. Plant. 126, 45–51. Available from: https://doi.org/10.1111/j.0031-9317.2005.00582.x.

Syamsia, Idhan, A., Noerfitryani, Nadir, M., Reta, Kadir, M., 2018. Paddy chlorophyll concentrations in drought stress condition and endophytic fungi application. IOP Conf. Series: Earth Environ. Sci. 156 (1), 012040.

Sylvia, D.M., Williams, S.E., 1992. Vesicular arbuscular mycorrhizae and environmental stress. In: Bethlenfalvay, G.T., Linderman, R.D. (Eds.), Mycorrhiza in Sustainable Agriculture. USA, Special Publication, Madison, pp. 101–124.

Talaat, N.B., Shawky, B.T., 2014. Protective effects of arbuscular mycorrhizal fungi on wheat (*Triticum aestivum* L.) plants exposed to salinity. Environ. Exp. Botany 98, 20–31.

Tchounwou, P.B., Yedjou, C.G., Patlolla, A.K., Sutton, D.J., 2012. Heavy metals toxicity and the environment. Molecular, Clinical and Environmental Toxicology. Springer, Basel, pp. 133–164.

Thanna, E., Nawar, A., 1994. Salinity and mycorrhizal association in relation to carbohydrate status, leaf chlorophyll and activity of peroxidase and polyphenol oxidase enzymes in sour orange seedlings. Alex. J. Agric. Res. 39, 263–280.

Theocharis, A., Clement, C., Barka, E.A., 2012. Physiological and molecular changes in plants grown at low temperature. Planta 235, 1091–1105.

Thomashow, M.F., 1999. Plant cold acclimation: freezing tolerance genes and regulatory mechanisms. Annu. Rev. Plant. Biol. 50 (1), 571–599.

Tsai, H.F., Liu, J.S., Staben, C., Christensen, M.J., Latch, G.C., Siegel, M.R., et al., 1994. Evolutionary diversification of fungal endophytes of tall fescue grass by hybridization with *Epichloë* species. Proc. Natl Acad. Sci. 91 (7), 2542–2546.

Tuberosa, R., Grillo, S., Ellis, R.P., 2003. In: di Toppi, L.S., Pawlik-Skowronska, B. (Eds.), Unravelling The Genetic Basis of Drought Tolerance in Crops. Kluwer Academic Pub, London, pp. 71–122.

Valix, M., Tang, J., Malik, R., 2001. Heavy metal tolerance of fungi. Miner. Eng. 14 (5), 499–505.

van der Heijden, M.G., Bakker, R., Verwaal, J., Scheublin, T.R., Rutten, M., Van Logtestijn, R., et al., 2006. Symbiotic bacteria as a determinant of plant community structure and plant productivity in dune grassland. FEMS Microbiol. Ecol. 56 (2), 178–187.

van Hoorn, J.W., Katerji, N., Hamdy, A., Mastrorilli, M., 2001. Effect of salinity on yield and nitrogen uptake of four grain legumes and on biological nitrogen contribution from the soil. Agric. Water Manag. 51 (2), 87–98.

Varma, A., Verma, S., Sudha, Sahay, N., Butehorn, B., Franken, P., 1999. *Piriformospora indica*, a cultivable plant-growth-promoting root endophyte. Appl. Environ. Microbiol. 65 (6), 2741–2744.

Wahid, A., Gelani, S., Ashraf, M., Foolad, M., 2007. Heat tolerance in plants: an overview. Environ. Exp. Bot. 61, 199–223. Available from: https://doi.org/10.1016/j.envexpbot.2007.05.011.

Waller, F., Achatz, B., Baltruschat, H., Fodor, J., Becker, K., Fischer, M., et al., 2005. The endophytic fungus *Piriformospora indica* reprograms barley to salt-stress tolerance, disease resistance, and higher yield. Proc. Natl Acad. Sci. 102 (38), 13386–13391.

Wang, F.Y., Liu, R.J., 2001. A preliminary survey of arbuscular mycorrhizal fungi in saline alkaline soils of the Yellow River Delta. Chin. Biodivers. 9 (4), 389–392.

Wang, J., Li, T., Liu, G., Smith, J.M., Zhao, Z., 2016. Unraveling the role of dark septate endophyte (DSE) colonizing maize (*Zea mays*) under cadmium stress: physiological, cytological and genic aspects. Sci. Rep. 6, 22028.

Wang, W.X., Vinocur, B., Altman, A., 2003. Plant responses to drought, salinity and extreme temperatures: towards genetic engineering for stress tolerance. Planta 218, 1–14.

Wang, W.X., Vinocur, B., Shoseyov, O., Altman, A., 2001. Biotechnology of plant osmotic stress tolerance: physiological and molecular considerations. IV Int. Symp. Vitro Cult. Horticult. Breed. 560, 285–292.

Wang, Y., Jing, H., Gao, Y., 2012. Arbuscular mycorrhizal colonization alters subcellular distribution and chemical forms of cadmium in *Medicago sativa* L. and resists cadmium toxicity. PLoS One 7, 3161–3164.

Wang, Y., Wang, M., Li, Y., Wu, A., Huang, J., 2018. Effects of arbuscular mycorrhizal fungi on growth and nitrogen uptake of *Chrysanthemum morifolium* under salt stress. PLoS One 13 (4), e0196408.

Waqas, M., Khan, A.L., Shahzad, R., Ullah, I., Khan, A.R., Lee, I.J., 2015. Mutualistic fungal endophytes produce phytohormones and organic acids that promote japonica rice plant growth under prolonged heat stress. J. Zhejiang Univ.-Sci. B 16 (12), 1011–1018.

West, C.P., 1994. Physiology and drought tolerance of endophyte-infected grasses. In: Bacon, C. W., White, J.F. (Eds.), Biotechnology of Endophytic Fungi of Grasses. CRC Press, Boca Raton, FL, pp. 87–99.

White, J.F.J., Martin, T.I., Cabral, D., 1996. Endophyte-host associations in grasses. III. Conidia formation by *Acremonium* endophytes in the phylloplanes of *Agrostis hiemalis* and *Poa rigidifolia*. Mycologia 88, 174–178.

Wu, H.H., Zou, Y.N., Rahman, M.M., Ni, Q.D., Wu, Q.S., 2017. Mycorrhizas alter sucrose and proline metabolism in trifoliate orange exposed to drought stress. Sci. Rep. 7, 42389.

Wu, Q.S., Zou, Y.N., He, X.H., 2009. Contributions of arbuscular mycorrhizal fungi to growth, photosynthesis, root morphology and ionic balance of citrus seedlings under salt stress. Acta Physiol. Plant. 32, 297–304. Available from: https://doi.org/10.1007/s11738-009-0407-z.

Wu, Q.S., Zou, Y.N., Xia, R.X., Wang, M.Y., 2007. Five *Glomus* species affect water relations of *Citrus tangerine* during drought stress. Bot. Stud. 48 (2), 147–154.

Xin, Z., Browse, J., 2000. Cold comfort farm: the acclimation of plants to freezing temperatures. Plant. Cell Environ. 23, 893–902.

Yadav, S.K., 2010. Heavy metals toxicity in plants: an overview on the role of glutathione and phytochelatins in heavy metal stress tolerance of plants. South. Afr. J. Bot. 76, 167–179.

Yamaji, K., Watanabe, Y., Masuya, H., Shigeto, A., Yui, H., Haruma, T., 2016. Root fungal endophytes enhance heavy-metal stress tolerance of *Clethra barbinervis* growing naturally at mining sites via growth enhancement, promotion of nutrient uptake and decrease of heavy-metal concentration. PLoS One 11 (12), e0169089.

Yasmeen, R., Siddiqui, Z., 2017. Physiological responses of crop plants against *Trichoderma harzianum* in saline environment. Acta Bot. Croatica 76 (2), 154–162.

Yin, L., Ren, A., Wei, M., Wu, L., Zhou, Y., Li, X., et al., 2014. *Neotyphodium coenophialum*-infected tall fescue and its potential application in the phytoremediation of saline soils. Int. J. Phytoremediat. 16 (3), 235–246.

Yousaf, B., Liu, G., Wang, R., Imtiaz, M., Zia-ur-Rehman, M., Munir, M.A.M., et al., 2016. Bioavailability evaluation, uptake of heavy metals and potential health risks via dietary exposure in urban-industrial areas. Environ. Sci. Pollut. Res. 23 (22), 22443–22453.

Yuan, Z., Zhang, C., Lin, F.L., 2010. Role of diverse non-systemic fungal endophytes in plant performance and response to stress: progress and approaches. J. Plant. Growth Regul. 29 (1), 116–126.

Zahoor, M., Irshad, M., Rahman, H., Qasim, M., Afridi, S.G., Qadir, M., et al., 2017. Alleviation of heavy metal toxicity and phytostimulation of *Brassica campestris* L. by endophytic Mucor sp. MHR-7. Ecotoxicol. Environ. Saf. 142, 139–149.

Zarei, M., Hempel, S., Wubet, T., Schäfer, T., Savaghebi, G., Jouzani, G.S., et al., 2010. Molecular diversity of arbuscular mycorrhizal fungi in relation to soil chemical properties and heavy metal contamination. Environ. Pollut. 158, 2757–2765.

Zhang, J., Jia, W., Yang, J., Ismail, A.M., 2006. Role of ABA in integrating plant responses to drought and salt stresses. Field Crop. Res. 97, 111–119.

Zhang, W.J., Jiang, F.B., Ou, J.F., 2011. Global pesticide consumption and pollution: with China as a focus. Proc. Int. Acad. Ecol. Environ. Sci. 1 (2), 125–144.

Zhang, W.J., Zhang, X.Y., 2007. A forecast analysis on fertilizers consumption worldwide. Environ. Monit. Assess. 133, 427–434.

Zhang, Y.P., Nan, Z.B., 2007. Growth and anti-oxidative systems changes in *Elymus dahuricusis* affected by *Neotyphodium* endophyte under contrasting water availability. J. Agron. Crop. Sci. 193 (6), 377–386.

Zhu, J.K., 2001. Plant salt tolerance. Trends Plant. Sci. 6 (2), 66–71.

Zhu, S., Tang, J., Zeng, X., Wei, B., Yang, S., Huang, B., 2015. Isolation of *Mucor circinelloides* Z4 and *Mucor racemosus* Z8 from heavy metal-contaminated soil and their potential in promoting phytoextraction with Guizhou oilseed rap. J. Cent. South. Univ. 22, 88–94.

Zhu, X., Song, F., Xu, H., 2010. Influence of arbuscular mycorrhiza on lipid peroxidation and antioxidant enzyme activity of maize plants under temperature stress. Mycorrhiza 20, 325–332.

Zhu, X.C., Song, F.B., Liu, S.Q., Liu, T.D., 2011. Effects of arbuscular mycorrhizal fungus on photosynthesis and water status of maize under high temperature stress. Plant. Soil. 346, 189–199.

Zuccarini, P., Okurowska, P., 2008. Effects of mycorrhizal colonization and fertilization on growth and photosynthesis of sweet basil under salt stress. J. Plant. Nutr. 31 (3), 497–513.

Chapter 7

Phytopathogenic fungi and their biocontrol applications

Indu Sharma*
Department of Biotechnology, Maharishi Markandeshwar Deemed to be University, Ambala, India
**Corresponding author.* Email: endusharma@gmail.com

Introduction

Plant pathology (phytopathology) is the scientific study of diseases in plants caused by pathogens (infectious organisms) and environmental conditions (physiological factors). Infectious organisms that cause disease include fungi, bacteria, viruses, viroids, protozoa, nematodes and parasitic plants. Ectoparasites like insects, mites, vertebrate, or other pests that affect plant health by eating of plant tissues are not included. Plant pathology also involves the study of pathogen identification, disease etiology, disease cycles, economic impact, plant disease epidemiology, plant disease resistance and management of plant diseases.

Plant pathology

Plant pathology (gr., path -"suffering"- "ology", the science of) is the study of plant diseases and the abnormal conditions that create plant disorders.

Etiology is the purpose and study of the cause of disease. A pathogen can be living or non-living. A pathogen is an organism which causes a disease. Pathogenic is having the characteristics of a pathogen and pathogenicity is the capability of a pathogen to cause a disease.

Definition of plant pathology

The science which deals with plant disease is known as Plant Pathology or Phytopathology

Fungi Bio-Prospects in Sustainable Agriculture, Environment and Nano-Technology.
DOI: https://doi.org/10.1016/B978-0-12-821394-0.00007-X

Plant pathologist

One who specifies in the study of plant disease is called a Plant Pathologist or Phytopathologist. He is chiefly concerned with the life histories of disease causing organisms. His final goal is to prevent or control plant disease.

Definition of disease

> Any deviation from the healthy condition which interferes with the normal structure and performance of vital functions of a plant is called a disease. The diseased plant can be noticeable by changes in their structure or physiological processes

The deviation from the normal structure and function often develop to such an extent that signs of such effects become observable. They are characteristic of a given disease. The external signs on the plant which indicates disease are called symptoms. The injurious effects of the disease may lower its value or even threaten its existence. The disease may effects the state of the whole plant. It is called a systemic disease. Some disease, however, effect certain organism of the host or part of those organs. These are called the localized disease.

The disease triangle

The amount of disease that develops in a plant community is determined by the host, the pathogen and the environment and can be depicted in the form of a disease triangle (Fig. 7.1). A fourth factor, namely 'human interference' (making a disease square) can be added, but, as the other three aspects have

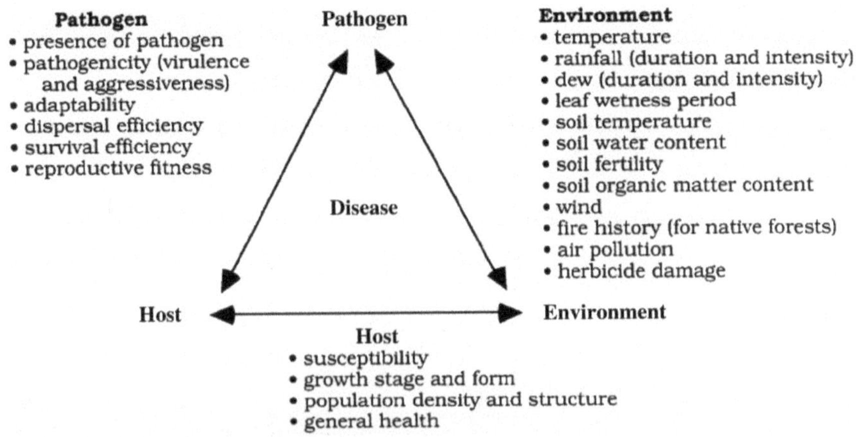

FIGURE 7.1 Disease triangle showing interaction of host, pathogen and environment and factors affecting the occurrence of plant disease.

a degree of human influence, the disease triangle is necessary as a framework for discussing the various factors that affect disease. It is important to note the arrows indicating interactions between the various factors in the disease triangle. It is the balance of these interactions that determines whether or not disease develops to destructive levels in a particular situation. Development of epidemics requires the interaction of a highly virulent pathogen and a susceptible plant host in an environment that favors the development of disease. The environment can affect both the susceptibility of the host and the activity of the pathogen. The pathogen can affect the host and the host can influence the pathogen (secreting chemical factors). Similarly, the host can influence the environment (influencing the microclimate within the canopy). An understanding of these factors and their interactions for a particular disease in a particular locality allows prediction of disease outbreaks and intervention to reduce the amount of disease.

Factors affecting disease development

Pathogen factors

The main factor which determines whether or not disease occurs at all is the presence or absence of a pathogenic strain of the pathogen. The amount of disease that develops is often determined by the pathogenicity of the prevalent population of the pathogen. The term pathogenicity comprises both the virulence of the pathogen and its aggressiveness.

Pathogenicity is one aspect of the reproductive fitness of a pathogen, which also includes aspects of dispersal and survival fitness. Control of diseases often involves monitoring the pathogenicity of the prevailing pathogen population as a guide to the breeding of resistant host cultivars. The adaptability of a pathogen is very important in determining its ability to overcome resistance in newly released cultivars or to adapt to changed environmental conditions.

Adaptability is governed by the genetic flexibility of the pathogen population and its reproductive efficiency. *Puccinia graminis* has high reproductive efficiency and, in spite of the absence of a sexual reproductive cycle it, has the potential to develop new pathotypes by asexual methods such as mutation, heterokaryosis and parasexualism.

Introduction of new pathogen

Some pathogens, epidemic in certain area, may become quite aggressive and outbreak as epidemic when introduced to new area. For example late blight of potato caused by *Phytophthora infestans* was epidemic in South America. This disease became epidemic when the infected tubers were introduced in

Europe (in 1843–45). Fire blight (*Erwinia amylovora* in North America is endemic).

Presence of aggressive strain of the pathogen

All the strains of a pathogen are not aggressive. Only the aggressive strains are capable of causing infectious diseases which spread as epidemic. They are characterized by rapid cycle of infection and causing successful infection in new hosts. Rapid cycle of infection is essential for successful infection and it happens only by aggressive strain of the pathogen. e.g., *Puccinia graminis tritici* (wheat black rust) in India, stripe rust, bunt and loose smut of wheat in the U.S.A. and Europe.

High birth rate of the pathogen

Pathogen with high reproductive capacity and capable of rapid dissemination over wide areas mostly causes epidemics. The fungal members causing powdery mildews, downy mildews, rusts, blasts, blights, etc., produce enormous amount of spores. These spores are easily dispersed by using water or insects and cause infections to new plants.

Low death rate

Epiphytotics may also be caused by low death rate diseases. These diseases are caused by agents of systemic nature which are protected by plant tissues. As they are protected by plant tissues the chances of high mortality is reduced to the minimum. In these diseases the chief source for accumulation of inoculum for epiphytotics is the diseased plant organ used for vegetative propagation (corms, setts, tubers, etc.).

Easy and rapid dispersal of the pathogen

The ability of the pathogen to cause epidemic depends both on the high birth rate and dispersal. The propagules of the pathogen produced should be dispersed for development of an epidemic. It may happen by external agencies like wind, water, insects, mites and nematodes. Fungal spores/conidia are minute and light and resistant to adverse conditions.

Adaptability of the pathogen

Pathogens have the capacity to adapt to adverse conditions. Fungi produce different types of spores like oospores, ascospores and smut spores (chlamydospores) which help in tiding over adverse conditions. Bacteria survive in diseased plant parts. Viruses and phytoplasmas live in collateral hosts or insect vectors in the absence of the suitable crop hosts.

Host factors

The main host factor affecting disease development is the occurrence of individuals in the host population that are susceptible to the particular pathogen. For a disease epidemic to occur, the host plant population must be largely susceptible to attack by the pathotypes of the pathogen in the vicinity.

Susceptibility of the host

Plants have ability to contest disease which manifests itself as susceptibility or resistance. Plants are susceptible to the attack depending on their nature, environment and stage of growth. Presence of susceptible varieties in an area may act as one of the causes of epidemic. For example, late maturing varieties of groundnut are more susceptible to early leaf spot (*Cercospora arachidicola*) and late leaf spot (*Phaeoisariopsis*) than the early maturing varieties. Similarly late maturing varieties of wheat are susceptible to loose smut (*Ustilago nuda tritici*) than the early maturing varieties. Early sown sugarcane varieties of sugarcane are more susceptible to leaf rust in Deccan canals in Bombay area than the late sown varieties. Wheat plant becomes susceptible to black rust (*Puccinia graminis tritici*) at the boot stage but is resistant when young. Susceptibility of rice plants to blast disease (*Pyricularia oryzae*) increases with application of heavy doses of nitrogenous fertilizers.

Aggregation and distribution of susceptible hosts

Abundance of susceptible hosts in an area is one of the major causes of the spread of epidemics. Continuous cultivation of susceptible variety or varieties in an area, that too in a large contiguous area help in the buildup of inoculum and improve the chances of epidemics. Under the above conditions the pathogen increases the rate of multiplication of its propagules and repeats the disease cycles in a short span.

Introduction of new host(s)

Disease proneness in the host is induced by environment and other factors. The host is liable to vigorous attack and successful infection by the pathogen. A resistant or moderately resistant variety may become susceptible or highly susceptible. A susceptible variety may become highly susceptible when conditions favoring proneness are existing and cause severe damage. Under the above conditions the pathogen multiplies faster, cause infection and produces more propagules for secondary spread.

Introduction of new collateral or alternate hosts

Alternate hosts are those plants on which the heteroecious pathogens pass part of their life cycles. Similarly, collateral hosts are some wild plants in which the pathogen survives when primary host is not available. Both alternate and collateral hosts are important in building up the primary inoculum to the next crop. They determine the course and intensity of an epidemic. Grass hosts (collateral hosts) of *Sclerospora sacchari, S. philippinensis* (downy mildews), *Pyricularia oryzae* (rice blast), *Ustilago scitaminea* (sugarcane smut) may produce abundant inoculum which aid in building up of epidemics

Environmental factors

Environmental factors have traditionally been considered to have the major impact on disease development. The environmental conditions such as temperature, relative humidity, rainfall, duration and intensity of light, etc. play very important role in causing epidemics. These are actually the deciding factors and influence almost all the stages of disease cycle. Favorable environmental conditions are needed for sporulation, liberation of spores, and dissemination of pathogen, germination, infection and establishment of pathogen in the host.

The term 'moisture' covers rainfall (duration and intensity), relative humidity, dew and leaf wetness (duration and intensity). The influence of rain splash and running water on dispersal of pathogen propagules is also important. Often the combination of leaf wetness (due to rainfall and dew deposition) and ambient temperature is critical in determining the proportion of pathogen propagules that infect the host. Most air-borne pathogens such as rusts and late blight of potato require a period of free water on the leaf surface for spore germination and infection.

Based on the mode of multiplication of pathogen, the diseases are classified of two types:

1. Simple interest diseases
2. Compound interest diseases

Simple interest diseases

In simple interest diseases the increase is mathematically analogous to simple interest in money. There is only one generation of the pathogen in the life of the crop. The primary inoculum is seed-borne or soil-borne.

Compound interest diseases

In compound interest diseases the rate of increase is mathematically analogous to compound interest in money. The pathogen produces enormous amount of spores at a very rapid rate. These spores are disseminated rapidly by wind and infect the other plants. Both the inoculation and sporulation period are short so that the pathogen spreads from plant to plant during the same growing season.

The edaphic environment

Soil-borne diseases are affected mainly by environmental conditions in the soil (the edaphic environment). As with the aerial environment, wetness is often a critical factor. Some pathogens such as the pythiaceous fungi causing damping off and eucalypt dieback require a period of soil saturation to allow germination of survival propagules and infection of roots. The zoospores of pythiaceous fungi require free water for their release and mobility. They also require a soil temperature above a critical level for optimum activity. On the other hand, nematodes and many pathogenic fungi are adversely affected by prolonged periods of soil saturation, which reduce the concentration of oxygen in the soil. Soil fertility can affect development of both soil- and airborne disease. Some facultative pathogens cause much more disease when plants are growing poorly under conditions of nutrient deficiency than when they are growing vigorously. The defenses of the plant that exclude minor pathogens from healthy plants are weakened by nutrient deficiency. On the other hand, biotrophic pathogens such as rusts and powdery mildews are often more common on vigorous well-fertilized plants than on poorly growing plants. The organic matter content of soil has an important influence on some soil borne pathogens. For example, eucalypt dieback is common in sandy or gravelly soils low in organic matter. Successful control of *Phytophthora cinnamomi* in avocado groves has been achieved by greatly increasing the organic matter content of the soils. In southern Australia, bare patch disease of cereals caused by *Rhizoctonia solani* is more common in poorer, sandy soils than in more fertile soils. In some instances, soils high in organic matter contain larger populations of antagonistic microorganisms which reduce the survival time of plant pathogens and the subsequent incidence of disease. As with the aerial environment, pollutants (e.g. herbicide residues) in soil can affect disease, or even directly damage plants.

Interactions among factors

Although the disease triangle can be described as three sets of two-way interactions (pathogen-environment, host-environment and host-pathogen), this is an over-simplification. All three groups of factors interact in any field situation, often in non-linear ways that are difficult to quantify and predict. Attempts have been made to model the complexity of these interactions using computers. Examples of such programs are EPIDEM, EPIMAY and EPMEN for *Alternaria solani, Bipolaris magdis* and *Venturia inaequalis*, respectively. Much research is needed to provide data on all the interactions involved. Often, from a practical point of view, it is sufficient to concentrate on the few key interactions that govern disease development (e.g. temperature and duration of leaf wetness for foliar pathogens).

Plant pathogens

Diseases in plants caused by pathogens (infectious organisms)and environmental conditions (physiological factors).

Classification of plant disease

The plant diseases are classified by different scientist. The plant disease is based on the nature of the causal agent, living or non-living. On the basis of the nature of the causal agent, the plant disease may be classified into two main categories as follows:

1. Non-parasitic or abiotic diseases

   These induced by unfavorable environmental conditions of soil or air such as mineral deficiencies or excesses in the soil, low and high temperature, improper water, oxygen and light relations. The non-parasitic disease may also be caused by mechanical injuries. Common examples of the non-parasitic disease are blossom and rot of potato, black heart of potatoes, scald of apples, tip burn of lettuce and heat canker of flax.
2. Parasitic diseases

These are caused by the attack of some other living agents called the causal agents. The causal agents may be a plant or an animal or a virus it is called the parasite. In the case of a disease caused by a parasitic organism the diseased plant is called a host and the causal agent is called pathogen. The parasite may subsist in whole or in part upon the living tissue of the host. The host is more appropriately called the suscept. It is the term applied to the plant which is prone to a given disease caused by a given causal agency. The parasitic disease of plant s may also be caused by the animal parasites such as the nematodes, molluscs (snails and snugs) and insects. Mostly the plant disease are caused by following pathogens:

Parasitic fungi: They produce a great majority of the parasitic plant disease called the fungal diseases. The fungal diseases are of first importance because they sometimes cause much damage and destruction.

Parasitic bacteria: They are also the causative agents of many plant diseases. The bacterial diseases of plants are second in importance to the fungal diseases.

Parasitic seed plants: There are a very few phanerogamic parasitic which cause plant disease the most common parasitic forms are the mistletoes (*Viscum*), Dodders (*Cuscuta*), Cystanche, *Orobanche,* etc.

Parasitic algae: There are a very few parasitic algae that cause plant diseases. The most important ones are: *Cephaleuros virescens* which causes red rust of tea, *C. coffeae* is parasitic on coffee, *C. parasitica* incites Orange rust of tea.

Virus: The viruses are pathogens all of which are the most extreme parasites being almost completely dependent upon their host for their growth and replication. None of them has ever been cultivated outside the bodies of plants or animals. They cause serious plant and animal (including man) diseases. All disease caused by the viruses are of systemic type. The common example of virus disease of plant are tomato and tobacco mosaic, potato and cucumber mosaic, curly top of sugar beet, spotted wilt of various economic plants such as pea, potato, tomato, tobacco and spinach and yellow disease of peach and aster.

Fungi

Fungi are among the dominant causal agents of plant diseases. Most phytopathogenic fungi belong to the Ascomycetes and the Basidiomycetes. Fungi are eukaryotic, achlorophyllous organisms that may reproduce sexually and asexually and whose filamentous branched somatic structures are typically surrounded by cell walls containing chitin or cellulose. Most phytopathogenic fungi belong to the Ascomycetes and the Basidiomycetes. The fungi reproduce both sexually and asexually via the production of spores and other structures. Spores may be spread long distances by air or water, or they may be soil borne. Many soil inhabiting fungi are capable of living saprotrophically, carrying out the part of their life cycle in the soil. These are facultative saprotrophs. Fungal diseases may be controlled through the use of fungicides and other agriculture practices. However, new races of fungi often evolve that are resistant to various fungicides. Biotrophic fungal pathogens colonize living plant tissue and obtain nutrients from living host cells. Necrotrophic fungal pathogens infect and kill host tissue and extract nutrients from the dead host cells. Significant fungal plant pathogens include strategies to colonization.

Phenomenon of infection – pre-penetration, penetration and post penetration

The "infection process" can be divided into three phases: **pre-entry**, **entry** and **colonization**. It encompasses the germination or multiplication of an infective propagule in or on a potential host through to the establishment of a parasitic relationship between the pathogen and the host. The process of infection is influenced by properties of the pathogen, the host and the external environment. If any of the stages of the infection process is inhibited by any of these factors, the pathogen will not cause disease in the host. While some parasites colonize the outside of the plant (ectoparasites), pathogens may also enter the host plant by penetration, through a natural opening (like a stomatal pore) or via a wound (endoparasites). The symptoms of the diseases produced by these pathogens result from the disruption of respiration, photosynthesis, translocation of nutrients, transpiration, and other aspects of growth and development of plant.

Phytopathogenic fungi use various Infection strategies to colonize in plants and cause disease

Pre entry (pre-penetration)

Before a pathogen can penetrate a host tissue, a spore must germinate and grow on the surface of the plant. In the case of motile pathogens, they must find the host and negotiate its surface before entering the host. Some pathogens develop specialized penetration structures, such as appressoria, while others utilize pre-existing openings in the plant's surface, such as wounds or stomatal pores. Plant viruses are often transported and introduced into the plant via vectors such as fungi or insects. The initial contact between infective propagules of a parasite and a potential host plant is called inoculation. Pathogens use a variety of stimuli to identify a suitable entry point. Several fungi use topographical cues on the plant surface to guide them towards a likely stomatal site. Once the hypha reaches a stoma, volatile compounds escaping from the pore appear to provide a signal for the formation of a specialized penetration structure, the appressorium. Sugars, amino acids and minerals secreted by plants at the leaf surface can non-specifically trigger spore germination or provide nutrition for the pathogen. Some pathogenic spores will not germinate in the absence of these substances. Pathogen development is influenced by temperature, moisture, light, aeration, nutrient availability and pH. The conditions necessary for survival and successful infection differ between pathogens.

Entry (penetration)

Pathogens exploit every possible pathway to enter their host, although individual species of pathogen tend to have a preferred method. Fungal pathogens often use direct penetration of the plant surface to enter the host. This requires adhesion to the plant surface, followed by the application of pressure and then enzymatic degradation of the cuticle and cell wall, in order to overcome the physical barriers presented by the plant's surface. During the degradation of the cuticle and wall, a succession of genes are switched on and off in the pathogen, so that cutinase, followed by cellulase, then pectinase and protease are produced, attacking the cuticle, cell wall, and middle lamella in the order that they are encountered. The pressure needed for the hypha to penetrate the cell wall is achieved by first firmly attaching the appressorium to the plant surface with a proteinaceous glue. The cell wall of the appressorium then becomes impregnated with melanin, making it watertight, and capable of containing the high turgor pressure that builds up within the appressorium. The point of the appressorium that is in contact with the cuticle is called the penetration pore, and the wall is thinnest at this point. The increasing turgor pressure causes the pore to herniate, forming a penetration peg, which applies huge pressure to the host cuticle and cell wall. The alternative pathway for pathogen entry is via a pre-existing opening in the plant surface. This can be a natural opening or a wound. Pathogenic bacteria and nematodes often enter through stomatal pores when there is a film of moisture on the leaf surface. Fungi can also penetrate open stomata without the formation of any specialized structures. Some fungi form a swollen appressorium over the stomatal aperture and a fine penetration hypha enters the airspace inside the leaf, where it forms a sub-stomatal vesicle, from which infection hyphae emerge and form haustoria in surrounding cells. Also vulnerable to pathogen invasion are hydathodes, pores at the leaf margin that are continuous with the xylem. Under particularly humid conditions, droplets of xylem fluid (guttation droplets) can emerge at the surface of the leaf where they can be exposed to pathogenic bacteria, which then enter the plant when the droplet retreats back into the hydathode as the humidity decreases. Lenticels are raised pores that allow gas exchange across the bark of woody plants. They exclude most pathogens, but some are able to enter the plant via this route. Some specialized pathogens can also use more unusual openings, such as nectaries, styles and ectodesmata. Entry through a wound does not require the formation of specialized structures, and many of the pathogens that utilize wounds to enter the plant are unable to penetrate the plant surface otherwise. Most plant viruses enter through wounds, such as those made by their insect vectors.

Colonization (post penetration)

A successful infection requires the establishment of a parasitic relationship between the pathogen and the host, once the host has gained entry to the plant. There are two broad categories of pathogens are **biotrophs** (those that establish an infection in living tissue) and **necrotrophs** (those that kill cells before colonizing them, by secreting toxins that diffuse ahead of the advancing pathogen). These two kinds of pathogens are also sometimes known as ‘sneaks’ and ‘thugs’, because of the tactics they use to acquire nutrients from their hosts. The toxins produced by necrotrophs can be specific to the host or non-specific. Non-specific toxins are involved in a broad range of plant-fungus or plant-bacterial interactions, and will therefore not usually determine the host range of the pathogen producing them. Necrotrophs often enter the plant through wounds and cause immediate and severe symptoms. An intermediate category of parasite is the **hemibiotrophs**, which start off as biotrophs and eventually become necrotrophic, employing tactics from both classes of pathogen. Pathogens that colonize the surface of plants, extracting nutrients through haustoria in epidermal or mesophyll cells are termed **ecto-parasites**. The haustoria are the only structures that penetrate the host cells. Some parasites colonize the area between the cuticle and the outer wall of the epidermal cells, penetrating host epidermal and mesophyll cells with haustoria. These are called **sub-cuticular infections**. Pathogens can also form colonies deeper in the plant tissues. These are **mesophyll and parenchyma infections**, and can be necrotrophic, hemibiotrophic or biotrophic relationships. Necrotrophs do not produce specialized penetration structures. Instead, they kill host cells by secreting toxins, then degrade the cell wall and middle lamella, allowing their hyphae to penetrate the plant cell walls and the cells themselves. In hemibiotrophic infections, intercellular hyphae can form haustoria in living mesophyll cells, but as the lesion expands under favorable conditions, those heavily parasitized cells at the inner, older part of the colony collapse and die. A similar sequence of events can take place in plants infected by burrowing nematodes. Viruses, mildews and rusts develop specialized biotrophic relationships with their hosts. Intercellular hyphae of downy mildew colonize host mesophyll cells and form haustoria. The mildew sporulates and the infected cells eventually die, although necrosis is delayed and contained, compared to that caused by necrotrophic pathogens. Rust fungi can also delay senescence in infected cells while they sporulate. **Vascular infections** usually cause wilting and discoloration as a result of the physical blockage of infected xylem vessels. True vascular wilt pathogens colonize the vascular tissue exclusively, although other pathogens can cause the same symptoms if they infect the vascular system as well as other tissues. There are a few pathogens that manage to achieve **systemic infection** of their host. For example, many viruses can spread to most parts of the plant, although not necessarily all tissues. Some downy mildews can also systemically infect their host by invading the vascular tissue and growing throughout the host, causing deformation, rather than necrosis. Finally, there are some pathogens that complete their entire life cycle within the cells of their host, and may spread from cell to cell during cytokinesis. These are **endobiotic infections**.

Pathogenesis – role of enzymes, toxins, growth regulators and polysaccharides

The term pathogenesis means step by step development of a disease and the chain of events leading to that disease due to a series of changes in the structure and/or function of a cell/tissue/organ being caused by a microbial, chemical or physical agent. The **pathogenesis** of a disease is the mechanism by which an etiological factor causes the disease. The term can also be used to describe the development of the disease, such as acute, chronic and recurrent.

The word comes from the Greek *pathos*, "disease", and *genesis*, "creation". There are several chemical weapons secreted by pathogens that are utilized as they carry out their activities. These weapons include enzymes, toxins, growth regulators and polysaccharides.

Enzymes

Cutinases, cellulases, pectinases and lignases are often secreted by the pathogenic organism. Fungi, nematodes and bacteria are all known to produce one or more of the above enzymes in specific pathogen-host combinations. Viruses and viroids are generally not considered to secrete enzymes, although some viruses may encapsidate an enzyme in their particle. Pathogenic organisms either continually secrete enzymes or upon contact with the host plant. As we mentioned in class on Tuesday, the first surface an organism comes into contact with is cuticle and the cell wall of the plant. As we also mentioned, the cuticle is comprised of a complex wax, cutin, which impregnates the cellulose wall. The cell wall is comprised of cellulose, which makes up the structural framework of the wall, along with the matrix molecules hemicellulose, glycoproteins, pectin and lignin. Thus, penetration into living parenchymatous tissues and degradation of middle lamella is due to the action of one or more enzymes which degrade these chemical substances. Cutinases degrade the cutin on the cuticle layer presoftening the tissue for mechanical penetration or as a first step in tissue degradation. Studies have shown that several fungi and at least one bacterial species produce cutinases. Further, evidence indicates that cutinases are continually produced, albeit in low concentrations, with degradation products often inducing even higher levels of cutinase secretion. Studies have shown that in some organisms, cutinase production may be linked to virulence. Pectic substances comprise the middle lamella and also form an amorphous gel between the cellulose microfibrils in the primary cell wall. Pectin degrading substances often termed pectinases or pectolytic enzymes include pectin methyl esterases (PME). polygalacturonases (PG) and pectin lyases or transeliminases. Pectin methyl esterases remove small groups such as methyl groups (CH_3) often altering solubility and thus affecting the rate of chain splitting by polygalacaturonase and pectin lyase. Polygalacturonases split chains by adding a molecule of water, while pectin lyases split chains by removing a water molecule from the linkage. Pectin degrading enzymes are involved in a wide range of plant diseases particularly in the soft rot diseases (Examples: bacterial soft rot due to *Erwinia carotovora*, leaf drop of lettuce and water soft rot of crucifers (*Sclerotinia sclerotiorum*) and damping off in various seedlings due to *Rhizoctonia solani*). Organisms such as these, as well as, elaborate pectic enzymes lead to tissue maceration. In fact, these enzymes are sometimes referred to as macerating enzymes.

Cellulases

Cellulose is the major framework molecule of the plant cell wall existing as microfibrils with matrix molecules (glycoproteins, hemicelluloses, pectins, lignins) filling the spaces between the microfibrils and cellulose chains. Cellulases have been shown to be produced by many pathogenic fungi, bacteria and nematodes. Cellulolytic enzymes play a role in softening and disintegration of cell walls. No doubt cellulolytic enzymes are involved in the invasion and spread of the pathogen, but also are instrumental in the collapse of cells and tissues. As pointed out in Agrios text, indirectly cellulolytic enzymes participate indirectly in disease development by releasing soluble sugars that may be used as nutrients by pathogens and also may be involved in the release of materials in the vascular system interfering with transport or translocation of water.

Hemicellulases

Hemicelluloses are complex polysaccharide polymers that link the ends of pectic compounds to cellulose microfibrils. Since hemicelluloses are such a diverse group of polymers such as xyloglucans, glucomannans, glactomannans, arabinoglucans, etc., several hemicellulases have been identified in many plant pathogenic fungi. The mechanism by which they participate in cell wall breakdown is not clear, nor is it known how they contribute to pathogenesis.

Ligninases

Lignin is a phenylpropanoid which is found in the middle lamella and secondary cell wall of plants. More than anything else, lignin confers the tough, woody nature to woody tissues. According to Agrios text, only about 500 species of fungi are capable of decomposing wood. Most lignin degradation is by basidiomycetes known as white-rot fungi. These fungi produce ligninases that enable the fungi to utilize lignin.

Toxins as chemical weapons of pathogens

Toxins have been implicated in plant disease as far back as deBary who advanced a theory of plant disease often termed the "toxin theory". A primary tenant of the toxin theory is that a toxin elaborated by a pathogen may produce all of the symptoms of the disease. As more information was developed the theory was largely discarded. As we will see a little later in this discussion, the discovery of the toxin victorin, a host specific toxin, revived interest in the toxin theory of plant disease. Toxins may act directly on living host cells, damaging or even killing the host. Some toxins are active on a wide range of plant species (non-host-specific) or in some cases, as with the toxin victorin (host-specific).

Non-Host Specific Toxins

Tabtoxin – *Pseudomonas syringae* p.v. tabaci

Phaseolotoxin – *Pseudomonas syringae* p.v. phaseolicola

Tentoxin – *Alternaria alternata*

Tabtoxin

This is the toxin involved in the "wildfire disease of tobacco". In this disease, leaves exhibit necrotic spots surrounded by a yellow halo. Identical symptoms of the disease may be induced by culture filtrates of the organism or purified toxins with symptoms identical to that of wildfire of tobacco. Similar effects may also be observed on a relatively wide range of hosts, thus making the toxin non-host specific. Tabtoxin chemically is a dipeptide composed of the amino acids threonine and tabtoxinine. In the cell tabtoxin is cleaved releasing the tabtoxinine moiety which is the active toxin. The inhibition of the enzyme glutamine synthetase is the primary mode of action of the toxin.

Phaseolotoxin

This is the toxin involved in one of the bacterial bean blights called "halo blight". Symptoms of the disease incited by the bacterium can be produced by the toxin alone. Chemically the toxin is a tripeptide of ornithine-alanine-arginine with a phosphosulfinyl group. Within cells the toxin is enzymatically cleaved releasing phosphosulfinylornithine which is the toxic moiety. Cellular affects are a result of the inactivation of the enzyme ornithine carbamoyltransferase.

Tentoxin

This is the toxin produced by *Alternaria alternata*. The disease induced by this organism is primarily a seedling disease in a wide range of plant species. Seedling death results when greater than one-third of the leaf area become chlorotic, and reduce vigor with less than that amount of leaf chlorosis. The toxin is a cyclic tripeptide that binds to and inactivates a chloroplast-coupling factor protein involved in energy transfer and also the inhibition of light dependent phosphorylation of ADP to form ATP.

Host-specific toxins

Discovery of host-specific toxins when one reduces the toxin theory to its most elementary form, it can be stated that all of the symptoms of a given disease result from the direct action of a toxic product of the pathogen in that disease. Toxin can be tested by the following criteria: (a) the toxin, applied at concentrations which could be reasonably expected in or around the diseased plant, produces in a susceptible host all the symptoms characteristic of the disease; (b) the pathogen and the toxin exhibit similar suspect (host) specificity (c) the ability of the pathogen to produce the toxin varies directly with its ability to cause disease (d) a single toxin is involved. The first toxin isolated that fulfilled the above criteria was that produced by *Cochliobolus* (*Helminthosporium*) victoriae. "*Helminthosporium*" leaf blight of oats appeared in 1945 with the introduction and widespread use of the oat variety Victoria' and its derivatives which contained the Vb gene for resistance to the crown rust disease. In the victoria lines, the infects the basal portions of plants, and as it infects, the fungus elaborates a toxin that is carried to the leaves, causing a severe leaf blight, and death of the plant. Chemically, the toxin is a complex chlorinated cyclic pentapeptide. Ultrastructure studies demonstrate the primary target of the toxin is the plasma membrane. Here, it apparently binds to proteins and in some yet to be determined mechanism alters the metabolism of susceptible cells

T-toxin

This is another good example of a host-specific toxin. The toxin is produced in common corn disease "Southern Corn Leaf Blight" incited by the fungus *Cochliobolus heterostrophus* formerly known as *Helminthosporium maidis*. T-toxin is produced by race T of the fungus that first appeared in the United States in 1968. Corn varieties with normal cytoplasm were found to be resistant to the fungus and also to the toxin. The toxin, chemically, is a mixture of linear, long polyketols with 35–45 carbon atoms. The toxin specifically affects mitochondria of susceptible cells where ATP synthesis is inhibited.

HC-toxin

One other example of a host-specific toxin in the HC-Toxin produced by the fungus *Cochliobolus* (*Helminthosporium*) carbonum which incites a leaf spot disease in corn. The toxin is specific to only certain maize (corn) lines. The mechanism of action of the toxin is poorly understood at the present time, although the molecular and biochemical basis for resistance is well understood.

Growth regulators as weapons

As you know, plant growth is the result of three factors: cell divisions, cell elongation and cell differentiation. Plant growth is controlled by a group of naturally occurring compounds with hormonal action that are often referred to as plant growth regulators, plant growth hormones, etc. The major groups of plant growth regulators are the auxins, gibberellins, kinins and ethylene. Plant growth regulators are not unlike any hormone, in that, they act is very small concentrations, with slight changes in the normal level resulting in profound changes in the growth pattern of the plant. Pathogens often cause an imbalance in the hormonal system by causing the infected plant to produce more or less hormone, or in some cases the pathogen itself elaborates hormone thus changing the hormone level. Some of the commonly observed symptoms related to effects on plant growth regulation are: stunting, overgrowths, galling, root branching, adventitious root formation, defoliation, rosetting, leaf epinasty, etc.

Auxins

Auxins are the major growth regulator in plants and are defined as growth regulators whose major mode of action is cell elongation; they resemble indole-3-acetic acid in their activity. The naturally occurring auxin is indole acetic acid (IAA). Increased levels of auxin occur in plants infected with fungi, bacteria, viruses, mollicutes and nematodes. Following are a few important diseases directly related to altered auxin levels: Corn smut (*Ustilago maydis*), club root of crucifers (*Plasmodiophora brassicae*), and southern bacterial wilt (*Rolstonia solanacearum* formerly *Pseudomonas solanacearum*), crown gall (*Agrobacterium tumefaciens*) good examples of diseases where altered auxin levels have been implicated in disease.

Gibberellins

A century ago, rice farmer in Asia noticed some exceptionally tall seedlings growing in their paddies. Before these rice seedlings could mature and flower, they grew so tall and spindly that they toppled over. In Japan, this aberration in growth pattern became known as *bakanae* ('foolish seedling disease') disease of rice. In 1926, Kurosawa, a Japanese scientist discovered that the disease was caused by a fungal pathogen, *Gibberella fujikuroi*. By the 1930s, Japanese scientists had determined that fungus produced hyperelongation of rice stems by secreting a chemical, which was given the name gibberellin. Gibberellins are normal constituents of green plants and also produced by several other microorganisms. The best known gibberellin is gibberellic acid. In the past years, scientists have identified more than 80 different gibberellins, many of them occurring naturally in plants. Spraying of diseased plants with gibberellin overcomes some of the symptoms (stunding) caused by several virus- or mollicute-pathogens indicating that gibberellin involves in disease development.

Ethylene

Ethylene production in infected tissues can be dramatically induced. This induction is largely dependent on activation of the ethylene biosynthetic pathway in plant tissues. Genes encoding several key enzymes involved in the ethylene biosynthesis are highly activated at the transcriptional level. It has not been shown that ethylene is produced directly by plant pathogenic fungi and bacteria. Ethylene has been considered as a signal in plant for wounding and senesces responses. Recent studies show that ethylene together with another signal component jasmonic acid may plays an essential role in plant defense responses of several pathosystems.

- Induced ethylene production in infected tissues.
- Ethylene induces expression of certain types of PR-genes and some signal components involved in defense signaling.
- Ethylene insensitive mutants block several PR-gene-expression.
- Ethylene insensitive mutants can be beneficial against some pathogens but deleterious to resistance against other pathogens in a specific a gene-for-gene manner.

Other chemical weapons include polysaccharides, plant defense suppressors, transporters etc.

Plant viruses and viroids are not known to produce any substances themselves, but they adapt, induce and manipulate the host metabolism to replicate themselves

General principles of plant diseases management – importance, general principles: avoidance, exclusion, eradication, protection and therapy, immunization

Information on etiology, symptoms, pathogenesis and epidemiology of plant diseases are intellectually interesting and scientifically justified but most important of all they are useful as they help in formulation of methods developed for successful management of disease and thereby increasing the quantity and improving the quality of plant and plant products. Practices disease management vary considerably from one disease to another depending upon the type of pathogen, the host and the biotic and abiotic factors involved. Contrary to management of human and animal diseases where every individual is attended, the plants are generally treated as populations and measures used as preventive rather than curative.

Biocontrol of plant disease

Fungal plant pathogens are among the most important factors that cause serious losses to agricultural products every year. Biological control of plant diseases including fungal pathogens has been considered a viable alternative method to chemical control. In plant pathology, the term biocontrol applies to the use of microbial antagonists to suppress diseases. Throughout their lifecycle, plants and pathogens interact with a wide variety of organisms. These interactions can significantly affect plant health in various ways. Different mode of actions of biocontrol-active microorganisms in controlling fungal plant diseases include hyperparasitism, predation, antibiosis, cross protection, competition for site and nutrient and induced resistance.

Plant pests (harmful insects, parasitic weeds and pathogens) are among the most important biotic agents causing serious losses and damages to agricultural products. Plant pests need to be controlled to ensure food, feed and fiber production quantitatively and qualitatively.

A number of different strategies are currently being employed to manage and control plant pests (Agrios, 1988; Baker, 1987; Cook, 1993; Bargabus et al., 2002, 2004; Benhamou, 2004; Chisholm et al., 2006; Heydari, 2007). Beyond good agronomic and cultural practices, growers often rely heavily on chemical pesticide application.

However, the environmental pollution caused by excessive use of agrochemicals has led to considerable changes in people's attitudes towards the use of pesticides in agriculture. Today, there are strict regulations on chemical pesticide use and there is political pressure to remove the most hazardous chemicals from the market. In addition to the above-mentioned issues, the spread of plant diseases in natural ecosystems may preclude successful application of chemicals, because of the scale to which such applications might have to be applied.

The need for the development of non-chemical alternative methods to control plant diseases is therefore clear. Biological control of plant diseases has been considered a viable alternative method to manage plant diseases. Biological control is the inhibition of growth, infection or reproduction of one organism using another organism. Biocontrol is environmentally safe and in some cases is the only option available to protect plants against pathogens. Biological control employs natural enemies of pests or pathogens to eradicate or control their population. This can involve the introduction of exotic species or its, can be a matter of harnessing whatever form of biological control exists naturally in the ecosystem. The induction of plant resistance using non-pathogenic or incompatible microorganisms is also a form of biological control (Cook, 1993; Schouten et al., 2004).

Fungal plant diseases are considered the most important microbial agents causing serious losses in the agriculture annually. Some fungal diseases that have successfully been controlled using biological agents are pathogens of pruning wounds and other cut surfaces, diseases of leaves and flowers, such as powdery mildew, diseases of fruits and vegetables, such as *Botrytis* and fungal pathogens in the soil.

A variety of biological controls are available for use, but further development and effective adoption will require a greater understanding of the complex interactions among plants, people and the environment.

In plant pathology, the term biocontrol applies to the use of microbial antagonists to suppress diseases as well as the use of host-specific pathogens to control weed populations. In both fields, the organism that suppresses the pest or pathogen is referred to as the Biological Control Agent (BCA). More broadly, the term biological control also has been applied to the use of the natural products extracted or fermented from various sources. These formulations may be very simple mixtures of natural ingredients with specific activities or complex mixtures with multiple effects on the host as well as the target pest or pathogen. While such inputs may mimic the activities of living organisms, non-living inputs should more properly be referred to as biopesticides or biofertilizers, depending on the primary benefit provided to the host plant (Cook, 1993; Agrios, 1988; Baker, 1987; Heydari, 2007; Heydari et al., 2004, 2007; Heydari and Misaghi, 1998, 1999, 2003).

Interactions between plants and beneficial microbes

Throughout their lifecycle, plants and pathogens interact with a wide variety of organisms. These interactions can significantly affect plant health in various ways. In order to understand the mechanisms of biological control, it is helpful to appreciate the different ways that organisms interact. In order to interact, organisms must have some form of direct or indirect contact. The types of interactions between plants and microorganisms have been referred to as mutualism, protocooperation, commensalisms, neutralism, competition, amensalism, parasitism and predation (Bankhead et al., 2004; Bull et al., 2002; Katska, 1994; Chisholm et al., 2006; Fitter and Grabaye, 1994; Hoitink and Boehm, 1999).

While the terminology has been developed for macroecology, examples of all of these types of interactions can be found in the natural world at both the macroscopic and microscopic level. And, because the development of plant diseases involves both plants and microbes, the interactions that lead to biological control take place at multiple levels of scale From the plant's point of view, biological control may be considered a positive result arising from different specific and non-specific interactions (Cook, 1993; Bull et al., 2002; Katska, 1994; Chisholm et al., 2006; Fitter and Grabaye, 1994; Weller et al., 2002).

Mutualism is an association among several species where all of them are benefited from this association. Sometimes, it can be an obligatory relation involving close physical and biochemical contact between two organisms, such as those between plants and mycorrhizal fungi. However, they are generally facultative and opportunistic. For example, *Rhizobium* bacteria reproduce either in the soil or, to a much greater degree, through their mutualistic association with legume plants. These types of mutualism can contribute to biological control, by providing plant with improved nutrition and/or by stimulating host defense mechanism and ability (Bull et al., 2002; Katska, 1994; Chisholm et al., 2006; Fitter and Grabaye, 1994).

Many of the microorganisms isolated and classified as biocontrol agents (BCA) can be considered facultative mutualists, because host and disease suppression by them will vary depending on the prevailing environmental conditions.

Commensalism is also a symbiotic interaction between two living organisms, where one organism benefits and the other is neither harmed nor benefited (Fitter and Grabaye, 1994). Most plant-associated microorganisms are assumed to be commensals with regards to the host plant, because their presence, individually or in total, rarely results in positive or negative consequences to the plant. While the presence of these microorganisms may present a variety of challenges to an infecting pathogen, their absence decreases pathogen infection or disease severity and is indicative of commensal interactions (Cook, 1993). Biological interactions in which the population density of one species has absolutely no effect on the other are called neutralism (Katska, 1994; Berg et al., 2005; Chisholm et al., 2006).

Related to biological control, an inability to associate the population dynamics of pathogen with that of another organism would indicate neutralism. In contrast, antagonism between organisms results in a negative outcome for one or both. Competition within and between species caused a decreased growth, activity, and/or fecundity of the interacting organisms.

Biocontrol can occur when non-pathogens compete with pathogens for nutrients and sites in host plant. Direct interactions that benefit one population at the expense of another also affect our understanding of biological control (Cook, 1993). Parasitism is also a symbiotic relation in which two organisms coexist over a prolonged period of time. In this type of interaction, one organism, usually the physically smaller (parasite) benefits and the other (host) is harmed. The activities of various hyperparasites, for example those agents that parasitize plant pathogens, can result in biocontrol (Lo et al., 1997). Another interesting contribution to biocontrol is when host infection and parasitism by relatively avirulent pathogens may lead to biocontrol of more virulent pathogens through the stimulation of host defense systems. Finally predation refers to the hunting and killing of one organism by another for consumption and sustenance. While the term predator typically refers to animals that feed at higher trophic levels in the macroscopic world, it has also been applied to the actions of microorganisms such as protists and mesofauna, e.g. fungal feeding nematodes and microarthropods, which consume pathogen biomass for sustenance (Chisholm et al., 2006; Lo et al., 1997).

Biological control and PGPR

Plant growth promoting rhizobacteria Biological control is defined as the reduction of inoculum density or disease producing activities of a pathogen or parasite in its active or dormant stage by one or more organisms accomplished naturally or through manipulation of the environment, host or by introduction of one or more antagonists or by mass introduction of one or more antagonists. Biological control is but control of plant diseases using living microorganisms. Root rot disease (*Macrophomina phaseolina*) is a major disease in pulses, oilseeds, cofton, etc., and the most common method of control is using fungicides. So, the biological control can be very efficacy used for the root rot disease management as the biological agent multiply in soil and offer protection throughout the crop growth. The four main mechanisms involved in the biocontrol are (i) the biological agent (antagonist), may parasite the other organism, (ii) antagonist may secrete metabolites (antibiotics) harmful to the pathogens (Antibiosis), (iii) antagonist may compete with the pathogens for nutrients or space (Competition) and (iv) may cause death of the parasite by producing enzymes (Lysis).

Parasitism and lysis

The biocontrol against parasitizes the pathogen by coiling around the hyphae, e.g., *Trichoderma* viride; various bacteria and fungi secrete hydrolytic about the degradation of cell wall of pathogens. e.g. (i) *Bacillus* sp. causes hyphal lysis of *Gaeumanornyces graminis*. (ii) The chitnolytic enzymes of *Serratia marcescens* caused cell wall lysis of *Scierotium rolfsii*. (iii) *Trichoderma* sp. produces chitinases and β-1,3 glucanases which lyses the cell wall of *Rhizoctonia solani*.

Antibiosis

The antibiotic compounds secreted by the biocontrol agent suppress the growth of the pathogen. e.g. Phenazine-L-carboxylic acid produced by *P. fluorescens* plays an important role in suppressing the take all disease of wheat.

Competition

The biocontrol bacteria and fungi compete for food and essential elements with the pathogen thereby displacing and suppressing the growth of pathogen. e.g. (i) the competition for nutrients between *Pythium aphanidermatum, P. ultimum* and bacteria suppress the damping off disease in cucumbers. (ii) Fluorescent siderophores (iron chelaters) such as pseudobactinis &

pyoverdins produced by *P. fluorescons* chelates iron available in the soil, thereby depriving the pathogen of Its Fe requirements.

Trichoderma viride

The fungus, *Trichoderma viride* is one such biocontrol agent, mainly used for the control of root rot diseases of pulses and oil seeds in Tamil Nadu. A mass production technology for *T. viride* has been developed by Tamil Nadu Agricultural University, Coimbatore.

Bacillus subtilis

This bacterium is widely used for the control of soil-bome plant pathogens like *Macrophomina phaseolina, Rhizoctonia solani, Fusarium* spp., etc. This treatment also considerably improves the plant growth and yield. *Bacillus subtilis* is a rod shaped, thermophilic gram positive, aerobic bacterium. Roots may be formed in chains. It is 5–6 mm in length and 2–3 mm in width. It forms endospores during adverse conditions.

Pseudomonas fluorescens

This is another bacterium effectively used in controlling sheath blight and blast of paddy, wilt diseases of redgram, and banana. *Pseudomonas fluorescens* is a gram negative, rod shaped nonspore forming bacteria which may be mono or lopotrichous or non-motile. It produces greenish, fluorescent and water soluble pigment, pyoverdin. The direct influence of pseudomonas on plant growth is mediated either by release of auxin-like substances or through improved uptake of nutrients in the environment. The indirect promotion of plant growth is achieved when fluorescent *Pseudomonas* decreases or prevents the deleterious influence of phytopathogens.

Plant products and antiviral principles in plant disease management

Plant products play an important role in evolving an ecologically sound and environmentally acceptable disease management system. Plant products have been found to have fungicidal, bactericidal and antiviral properties. It is well established that about 346 plant products have fungicidal properties, 92 have bactericidal and 90 have antiviral properties. This clearly indicates that the plant kingdom is a vast storehouse of chemicals that can check several plant pathogens. As many of them have more than one type of activity there is a less chance for development of resistance and moreover, the plant products are safer to non-target organisms.

Neem products

Among the plant products, the neem derivatives are reported to be effective in controlling several diseases. The neem tree (*Azadirachta indica*), popularly called as china berry, crackjack, Nim, Indian lilac, margosa and paradise tree, contains several active principles in various parts. The important active principles are Azadirachtin, Nimbin, Nimbidin, Nimbinene, Nimbridic acid and Azadirone which have antifungal and insecticidal properties.

Neem seed Kernel extract (NSKE)

It is prepared by soaking 5 kg of powdered neem seed kernel (in a gunny bag) in 100 L of water for 8 hours. The gunny bag is then removed after thorough shaking. Then, 100 mL of teepol is mixed thoroughly, before spraying. The quantity of extract required for a hectare is 500 L.

Neem oil solution

One hundred nil of teepol is mixed first with 100 L of water. Then, 3 L of neem oil is slowly added to this solution with constant shaking. The milky solution formed is ready for spray. The spray volume is 500 L/ha.

Neem cake extract

Ten kg of powdered neem cake in a gunny bag is soaked in 100 L of water for 8 hours. The gunny bag is removed after thorough shaking. Then, 100 mL of sticker is added and mixed well. The quantity of spray fluid required is 500 L/ha.

Neem cake

Powdered neem cake is directly applied to the field at the time of last ploughing. The quantity applied is 150 kg/ha.

Other plant products

In addition to the neem products, products from several other plant species are also found to be effective in disease management. The leaf extract of tuisi (*Ocimum sanctum*) is found effective against *Helminthosporium oryzae* (paddy brown spot). The leaf and pollen extracts of vilvam (*Aegle marmelos*) effectively reduced early blight of tomato (*Altenaria solani*) and blight of onion (*A. porri*). *A. solani* is also effectively checked by flower extract of periwinkle (*Catharanthus roseus*) and bulb extract of garlic (*Allium sativum*) Rice discoloration caused by *Drechslera oryzae* is effectively reduced by leaf extract of mint (*Mentha piperita*). The bulb extract of garlic is also effective in reducing leaf blight of finger millet (*H. nodulosum*) and blast of paddy (*Pyricularia oryzae*). The root exudates of kolinji and rhizome extract of banana are effectively used against *Ganoderma lucidum, the* pathogen of Thanjavur wilt of coconut. The seed oil of pinnai (*Calophyllum inophyllum*) is effective against *Puccinia arachidis* causing groundnut rust. Leaf extract of nochi (*Vitex negundo*) effectively reduced, Rice Tungro viruses by checking the vector, *Nephotettix virescens.*

PGPR

Plant growth promoting rhizobacteria are bacteria that colonize plant roots, and in doing so, they promote plant growth and/or reduce disease or insect damage. There has been much research interest in PGPR and there is now an increasing number of PGPR being commercialized for crops. Organic growers may have been promoting these bacteria without knowing it. The addition of compost and compost teas promote existing PGPR and may introduce additional helpful bacteria to the field. The absence of pesticides and the more complex organic rotations likely promote existing populations of these beneficial bacteria. However, it is also possible to inoculate seeds with bacteria that increase the availability of nutrients, including solubilizing phosphate, potassium, oxidizing sulfur, fixing nitrogen, chelating iron and copper. Phosphorus (P) frequently limits crop growth in organic production. Nitrogen fixing bacteria are miniature of urea factories, turning N^2 gas from the atmosphere into plant available amines and ammonium via a specific and unique enzyme they possess called nitrogenase. Although there are many bacteria in the soil that 'cycle' nitrogen from organic material, it is only this small group of specialized nitrogen fixing bacteria that can 'fix' atmospheric nitrogen in the soil. Arbuscular mycorrhizal fungi (AMF) are root symbiotic fungi improving plant stress resistance to abiotic factors such as phosphorus deficiency or deshydratation. The fourth major plant nutrient after N, P and K is sulfur (S). Although elemental sulfur, gypsum and other sulfur bearing mined minerals are approved for organic production, the sulfur must be transformed (or oxidized) by bacteria into sulfate before it is available for plants. Special groups of microorganisms can make sulfur more available, and do occur naturally in most soils. One of the most common ways that PGPR improve nutrient uptake for plants is by altering plant hormone levels. This changes root growth and shape by increasing root branching, root mass, root length, and/or the amount of root hairs. This leads to greater root surface area, which in turn, helps it to absorb more nutrients.

Disease control

PGPR have attracted much attention in their role in reducing plant diseases. Although the full potential has not been reached yet, the work to date is very promising and may offer organic growers some of their first effective control of serious plant diseases. Some PGPR, especially if they are inoculated on the seed before planting, are able to establish themselves on the crop roots. They use scarce resources, and thereby prevent or limit the growth of pathogenic microorganisms. Even if nutrients are not limiting, the establishment of benign or beneficial organisms on the roots limits the chance that a pathogenic organism that arrives later will find space to become established. Numerous rhizosphere organisms are capable of producing compounds that are toxic to pathogens like HCN.

Ways that PGPR promote plant growth

- Increasing nitrogen fixation in legumes
- Promoting free-living nitrogen-fixing bacteria
- Increasing supply of other nutrients, such as phosphorus, sulfur, iron and copper
- Producing plant hormones
- Enhancing other beneficial bacteria or fungi
- Controlling fungal diseases
- Controlling bacterial diseases
- Controlling insect pests

Anti-viral principle (AVP)

Plants are also known to contain some compounds which are inhibitory to virus. They are called Anti-Viral Principles (AVP) or AntiViral Factors (AVF). The leaf extracts of sorghum, coconut, bougainvillea, *Prosopis juliflora and Cyanodon dactylon* are known to contain virus inhibiting principles.

Biotechnological developments

Modern biotechnological developments and referred to biological control as the use of natural or modified organisms, genes, or gene products, to reduce the effects of undesirable organisms and to favor desirable organisms such as crops, beneficial insects and microorganisms, but this definition encouraged much subsequent examinations and it was frequently considered too broad by many scientists who worked in the field. Published definitions of biocontrol differ depending on the target of suppression; number, type and source of biological agents and the degree and timing of human intervention. Most broadly, biological control is the suppression of damaging activities of one organism by one or more other organisms, often referred to as natural enemies. With regards to plant diseases, suppression can be accomplished in many ways. If grower's activities are considered relevant, cultural practices such as the use of rotations and planting of disease resistant cultivars (whether naturally selected or genetically engineered) would be included in the definition. Because the plant host responds to numerous biological factors, pathogenic and non-pathogenic, induced host resistance might be considered a form of biological control. More narrowly, biological control refers to the purposeful utilization of introduced or resident living organisms, other than disease resistant host plants, to suppress the activities and populations of one or more plant pathogens. This may involve the use of microbial inoculants to suppress a single type or class of plant diseases. This may also involve managing soils to promote the combined activities of native soil and plant-associated organisms that contribute to general suppression. Most narrowly, biological control refers to the suppression of a single pathogen by a single antagonist, in a single cropping system. Most specialists in the field would concur with one of the narrower definitions presented above.

Disease management by biotechnological methods

The use of genetically modified organisms and or modern techniques (genetic engineering, tissue culture, etc.) with biological systems for disease control is known as biotechnology. Genetic engineering or Genetic manipulation is the deliberate alteration of the composition of a genome by man. A growth of cells in a laboratory nutrient medium is known as tissue culture i.e. the technique of growing of plants in vitro. Cells of plants can be cultured in special nutrient medium and whole plants can be regenerated from cultured cells. Plant biotechnology is used for rapid clonal propagation of plants. It can help to produce industrial plant products under tissue culture conditions Biotechnological methods are employed to control important plant diseases which are not amenable to control by usual methods.

Genetic engineering

Genetic Engineering is the technology by which a particular gene is isolated from one organism and inserted into the genome of another organism and made to express at the right time.

Vectors for transfer of genes

Genetic engineering has been used to manage plant virus diseases. For transfer of genes to plants vectors are needed in which the gene to be transferred will multiply several folds. The most effective gene vector developed is the Tumor inducing plasmid of *Agrobacterium tumefaciens* from which the Tumor inducing genes have been removed *A. tumefaciens* induces tumors (crown galls) through di-plasmid (tumor-inducing) which is a circular double stranded DNA molecule containing up to 2,00,000 base pairs organized into several genes. The Ti-plasmid is transferred from the bacterium into the cell. A specific region of the plasmid, the T-DNA, is transferred from the plasmid to the nucleus of the plant cell. It becomes integrated into the plant nuclear genome, and is transcribed. Cauliflower mosaic virus (CaMV) is the only plant virus with double-stranded DNA genome. As it has DNA genome, it is used as a possible vector in introducing foreign genes into plant. It is possible to insert a non-viral gene into CaMV genome and obtain expression of the gene in the infected plant. The viral promoter regions from CaMV are effective for obtaining expression of other genes in plant cells. The genes to be expressed are now fused to a promoter element from CaMV and a gene of *A. tumefaciens*. They are then introduced into the plants using *A. tumefaciens* Ti-DNA transformation.

DNA construction

Messenger RNA is extracted and exposed to an enzyme reverse transcriptase which synthesizes a complimentary single stranded DNA. The complimentary DNA (cDNA) is exposed to another enzyme, DNA polymerase, which produces the double stranded cDNA. The cDNAs are inserted into the plasmids of *A. tumefaciens*.

Coat-protein expression in transgenic plants

Example: Transgenic tobacco plants expressing coat protein gene protected the plants against TMV. Transgenic tobacco plants showing resistance to alfalfa mosaic virus and tobacco rattle virus have also been developed. Transformation using a gene encoding the viral nucleocapsid protein of tomato spotted wilt virus (TSWV) has yielded transgenic tobacco plants that are resistant to TSWV. The expression of the viral genome in transgenic plants gives resistance to virus infection. Transgenic tobacco plants transformed with a DNA copy of the satellite RNA of cucumber mosaic virus (CMV) are shown to produce large amounts or satellite RNA following inoculation with CMV and symptom development is greatly reduced.

Satellite RNA expression in transgenic plants

Satellite RNAs are associated with several viruses. They are packaged into virus particles along with the genomic RNAs of the helper virus. They are not part of the viral genome and have no obvious sequence relationships with the helper virus. The presence of the satellite RNA suppresses the disease severity in many hosts. Hence transgenic plants which express satellite RNA have been produced to manage virus diseases. e.g., Transgenic plants of tobacco expressed the synthesis of satellite tobacco ring spot virus and reduce the virus disease incidence. Satellite RNA expressing tobacco plants against Cucumber Mosaic Virus (CMV) and Tobacco aspermy virus have been synthesized.

MIC RNA expression in transgenic plants

A DNA copy is made of one or more sections of the viral genome that include the initiation codon for proteins vital to virus replication. The DNA copy is inserted in the host-cell genome, Cells then produce an 'antisera RNA' called mic RNA (mRNA-interfering complementary to 5′ end of the gene). The mic RNA hybridizes in vivo with the viral mRNA blacking translation. The mic RNA is inserted into the plants using the Ti plasmid of *A. tumefaciens* . Plants regenerated from the transformed cells will be resistant to the particular virus. This possibility is also being exploited for the control of virus diseases.

Use of RFLP markers for cloning resistance genes

Molecular markers *viz.*, isozymes and DNA markers (Restriction Fragment Length Polymorphisms – RFLPs; Random Amplified Polymorphic DNA – RAPD and others) are being used in several areas relevant to identification of disease resistance genes. Some of the disease resistance genes using random DNA markers have been identified.

Disease resistance genes mapped using RFLP markers detoxification of pathotoxin

Pathogens that produce pathogenesis-related phytotoxins usually also have the capacity to metabolize i.e. detoxify, these compounds. The search for genes encoding the enzyme(s) performing the key catabolic step(s) should thus lead to a convenient source of resistance, which can be engineered into plants to protect them from the effects of the toxin. A gene encoding a tabtoxin acetyltransferase from the pathogen, *Pseudomonas syringae* pv. *tabaci* which causes wild fire disease of tobacco was isolated and transferred into tobacco under a strong constitutive promotor. The transgenic plants expressed this gene and, when treated with either the pathogen or its toxin, did not produce the chlorotic lesions typical of wild fire disease.

Activation of plant defense mechanism - phytoalexins

Phytoalexins have long been known to accumulate in certain plants upon infection by pathogens. The production of phytoalexins is also triggered by mechanical stimulation, ultraviolet (UV) irradiation, stress and a variety of chemical elicitors. Phytoalexins are part of the localized hypersensitive response at the site of damage or **Plant Pathogen** Tomato *Fusarium* *oxysporum* Citrus *Phytophthora* spp. pathogen ingress, which involves cell trauma and death. The importance of phytoalexins in the defense response is underscored by experiments and pathogenicity in *Nectria haematococca* was correlated to its ability to detoxify the phytoalexin, pisatin, by way of demethylation. By transferring the demethylase gene from Nectria, *Aspergillus nidulans*, a non-pathogen on peas, was rendered insensitive to pisatin.

Defense related genes

a. Single gene defense mechanism

There are some defense proteins which do not require any intermediate step both for their synthesis and their expression require only few steps and those genes encoding such proteins are called single gene defense mechanism. Chitinases and glucanases are those proteins belonging to single gene defense mechanism.

Chitinases and glucanases

Chitinases are abundant proteins found in wide variety of plants. Although the physiological function of chitinases is not known, there is strong correlative evidence that they are defense proteins with antifungal activity. Chitin is a major structural component of cell walls of many fungi. The low constitutive activity of chitinase found in many plants can be dramatically induced by wounding or by infection of the tissue with fungal pathogens. Chitinase in concrete with ß-1,3-glucanase (capable of degrading glucans present in fungal cell wall), degrades fungal cell walls and inhibits fungal growth at hyphal tips and has been shown to associate with hyphal walls in plants. The chitinase and glucanase enzymes are having direct action against several fungal pathogens compared to other defense related proteins. Since lytic enzymes are encoded by single genes, these defense should be high amenable to manipulation by gene transfer. The first reports of success with this approach was the expression of bean vacuolar chitinase gene under the control of the strong constitutive gene under the control of the strong constitutive promoter of the cauliflower mosaic virus (CaMV) 35 S transcript in tobacco and *Brassica napus*, which resulted in decreased symptom development by *Rhizoctonia solani*, the causative agent of post-emergence damping off. An endochitinase gene (from genomic tomato DNA library) was introduced into *Brassica napus*. var. *oleifera*. The transgenic *Brassica* showed enhanced resistance against several fungal pathogens like *Cylindrosporium concentricum, Phoma lingam* and *Sclerotinia sclerotiorum* under field conditions when compared to non-transgenic plants. More recently, chitinase gene from *Manduca sexta*, tobacco horn worm, has been cloned into *P. fluorescens* to increase their antagonistic potential against *R. solani*.

b. Multigenic defense mechanism

Defense responses such as phytoalexin biosynthesis or lignin deposition in the cell wall require the action of many genes.

Peroxidases

Anionic peroxidases in the cell wall catalyze the production of phenolic radicals for the oxidative polymerization of lignin from cinnamyl alcohols. In tomato, there is a marked induction of two linked genes encoding highly anionic peroxidases in an incompatible interaction with an avirulent form of *Verticillium albo-atrum*, with only weak induction in the compatible interaction with a virulent form of this vascular pathogen. Expression of one of these genes in transgenic tobacco under the control of either its own promoter or the CaMV 35 s promoter resulted in massive increase in anionic peroxidase activity and these plants apparently showed a significant increase in resistance to *Peronospora parasitica* as judged by symptom development and fungal sporulation.

Activation of defense genes by chemicals

Several classes of compounds have the potential to act as inducers of natural resistance mechanisms in horticultural crops and chemicals with such indirect modes of action may offer attractive alternatives or supplement to existing contact/systemic fungicides in integrated disease management. Increase was found to occur in response to salicylic acid treatment as well as oligosaccharides and glycoproteins originating from either fungal cell wall or host cell walls, the so called elicitors. Recently, chitosan seed treatment has been found to induce defense related genes like chitinase and glucanase in tomato and consequently the Fusarium crown and root rot diseases were significantly reduced. Pre-treatment with 2, 6-dichloroisonicotinic acid was highly effective in significantly reducing both anthracnose (caused by *Colletotrichum lindemuthianum*) and rust (caused by *Uromyces appendiculatus*) diseases in bean plants.

Cell and tissue culture

Tissue culture approach is one of the oldest techniques in the field of molecular biology and it is applied in several ways for the development of disease resistance cultivars in agriculture and horticulture.

Somaclonal variation

In the past two decades, several advances have been made in culturing of isolated plant cells and tissue under controlled conditions in vitro. When plants are regenerated from cultured cells, they exhibit new phenotypes, sometimes at high frequencies. If these are heritable and affecting desirable traits, such "somaclonal variation" can be incorporated into regular breeding programs. However, the finding of specific traits by these methods is largely left to chance and hence inefficient. Rather than relying on this undirected process, selection in vitro aims to specific traits by subjecting large populations of cultured cells to the action of a selective agent in the petri dish. For purpose of disease resistance, this selection can be done by fungal pathogens, culture filtrates of pathogens or isolated phytotoxins that are known to have a role in pathogenesis. The selection will allow only those cells to survive and proliferate that are resistant to the challenge. Plants regenerated from resistant cells often display a resistant phenotype when evaluated with either the toxin or the pathogen itself.

Disease resistant plants from tissue culture

Plant Culture System Selection Resistance to Pathogen Potato Protoplasts SCV *Phytophthora infestans, Alternaria solani*, Callus CF *Fusarium oxysporum*, Tomato Callus Fusaric *Fusarium oxysporum*, Protoplasts acid Banana Meristem SCV *Fusarium oxysporum* Strawberry Callus SCV *Fusarium oxysporum* (SCV – plant regeneration without selection; CF crude culture filtrate).

Although this method has obviously yielded some impressive results, it also has its drawbacks; *viz.*, (i) Many pathogens do not produce pathogenesis specific toxins useful for selection. (ii) Culture filtrates are rather artificial and neither pathogens nor plant cells grown together in vitro behave quite as they would in a natural environment. (iii) The selection approach can only detect mutations in plant genes that are expressed at the time that selection is applied. In order to be useful, new resistance traits, whether selected or not, must be heritable sexually or in the case of vegetatively propagated crops must be transmitted through vegetative propagules. The pathogens produced toxins can be used to screen calluses (cultured cells) which may regenerate resistant plants. The toxins will kill the calluses, but the mutant toxin resistant calluses will survive. The toxin-resistant calluses yield disease resistance plants. Vidhyasekaran obtained brown spot resistant rice plants using *Helminthosporium oryzae* toxin. Similarly, *H. maydis* resistant maize plants, *H. sacchari* resistant sugarcane plants and *Phytophthora infestans* resistant tobacco plants have been evolved.

Anther culture

In this method, the plants are produced directly from microspores (immature pollen grains). Through anther or microspore culture, one has immediate access to unique and rare combinations of genes representing the recombination of the genetic material contributed by the parents of the cross. Through anther culture, followed by chromosome doubling, such gene combinations can be fixed in their homozygous state as instant inbreds in a single step. Over the past two decades, anther culture has become widely accepted as a tool in cultivar development. This technique can be particularly useful for producing plants with novel combinations of resistance genes for managing fungal diseases.

Protoplasmic fusion

This generates hybrid cells by merging the total cellular components of somatic cells from which the cell walls have been removed to produce protoplasts. The incompatibility preventing sexual fertilization between species is thus avoided and viable hybrids have been created, even between unrelated distance species. Disease resistance genes have thus been transferred by protoplasts fusion from wild species into potato. biotechnology and nanotechnology in improvement of biocontrol mechanisms, manage plant diseases, increase yield, protect the environment and biological resources and approach a sustainable agricultural.

Reference

Agrios, N.A., 1988. Plant Pathology., 3rd Ed. Academic Press, USA, pp. 220–222.

Baker, K.F., 1987. Evolving concepts of biological control of plant pathogens. Annu. Rev. Phytopathol. 25, 67–85.

Bankhead, S.B., Landa, B.B., Lutton, E., Weller, D.M., Gardener, B.B., 2004. Minimal changes in rhizosphere population structure following root colonization by wild type and transgenic biocontrol strains. FEMS Microb. Ecol. 49, 307–318.

Bargabus, R.L., Zidack, N.K., Sherwood, J.E., Jacobsen, B.J., 2002. Characterization of systemic resistance in sugar beet elicited by a non-pathogenic, phylosphere-colonizing Bacillus mycoides, biological control agent. Physiol. Mol. Plant Pathol. 61, 289–298.

Bargabus, R.L., Zidack, N.K., Sherwood, J.E., Jacobsen, B.J., 2004. Screening for the identification of potential biological control agents that induce systemic acquired resistance in sugar beet. Biol. Control. 30, 342–350.

Benhamou, N., 2004. Potential of the mycoparasite, Verticillium lecanii, to protect citrus fruitagainst Penicillium digitatum, the causal agent of green mold: a comparison with the effect of chitosan. Phytopathology 94, 693–705.

Berg, G., Krechel, A., Ditz, M., Sikora, R.A., Ulrich, A., Hallmann, J., 2005. Endophytic and ectophytic potato associated bacterial communities differ in structure and antagonistic function against plant pathogenic fungi. FEMS Microbiol. Ecol. 51, 215–229.

Bull, C.T., Shetty, K.G., Subbarao, K.V., 2002. Interactions between Myxobacteria, plant pathogenic fungi and biocontrol agents. Plant Dis. 86, 889–896.

Chisholm, S.T., Coaker, G., Day, B., Staskawicz, B.J., 2006. Host-microbe interactions: shaping the evolution of the plant immune response. Cell 124, 803–814.

Cook, R.J., 1993. Making greater use of introduced microorganisms for biological control of plant pathogens. Annu. digitatum, the causal agent of green mold: a comparison with the effect of chitosan. Phytopathology 94, 693–705.

Fitter, A.H., Grabaye, J., 1994. Interactions between mycorrhizal fungi and other soil organisms. Plant Soil 159, 123–132.

Heydari, A., Misaghi, I.J., 1998. Biocontrol activity of Burkholderia cepacia against Rhizoctonia solani in herbicide-treated soils. Plant Soil 202, 109–116.

Heydari, A., Misaghi, I.J., 1999. Herbicide-mediated changes in the populations and activity of root associated microorganisms: a potential cause of plant stress. In: Pessarakli, M. (Ed.), Hand Book Plant Crop Stress, 2nd Ed. Marcel Dekker Press, New York.

Heydari, A., Misaghi, I.J., 2003. The role of rhizosphere bacteria in herbicide-mediated increase in *Rhizoctonia solani*-induced cotton seedling damping-off. Plant Soil 257, 391–396.

Heydari, A., 2007. Biological control of turfgrass fungal diseases. In: Pessarakli, M. (Ed.), Turfgrass Management Physiology. CRC Press, Florida, USA.

Heydari, A., Fattahi, H., Zamanizadeh, H.R., Zadeh, N.H., Naraghi, L., 2004. Investigation on the possibility of using bacterial antagonists for biological control of cotton seedling damping-off in greenhouse. Appl. Entomol. Phytopathol. 72, 51–68.

Heydari, A., Misaghi, I.J., Balestra, G.M., 2007. Pre-emergence herbicides influence the efficacy of fungicides in controlling cotton seedling damping-off in the field. Int. J. Agric. Res. 2, 1049–1053.

Hoitink, H.A.J., Boehm, M.J., 1999. Biocontrol within the context of soil microbial communities: a substrate dependent phenomenon. Annu. Rev. Phytopathol. 37, 427–446.

Katska, V., 1994. Interrelationship between vesicular-arbuscular mycorrhiza and rhizosphere microflora in apple replant disease. Biol. Plant. 36, 99–104.

Lo, C.T., Nelson, E.B., Harman, G.E., 1997. Biological control of *Pythium, Rhizoctonia* and *Sclerotinia* infected diseases of turfgrass with *Trichoderma harzianum*. Phytopathology 84, 1372–1379.

Schouten, A., van den Berg, G., Edel-Hermann, V., Steinberg, C., Gautheron, N., et al., 2004. Defense responses of *Fusarium oxysporum* to 2,4-DAPG, a broad spectrum antibiotic produced by *Pseudomonas fluorescens*. Mol. Plant-Microbe Interact. 17, 1201–1211.

Weller, D.M., Raaijmakers, J.M., Gardener, B.B.M., Thomashow, L.S., 2002. Microbial populations responsible for specific soil suppressiveness to plant pathogens. Annu. Rev. Phytopathol. 40, 309–348.

Chapter 8

Impact of fungal inoculants on sustainable agriculture

Kavita Rambal*
Department of Botany, M. D. College, Parel, Mumbai, India
**Corresponding author.* Email: kavita.rambal@rediffmail.com

Introduction

Population world over is increasing which needs to be taken care of as far as food is concerned, which means that there has to be more agricultural production. This agricultural production has to and needs to be without compromising ecology, environment and health of soil and humans i.e. sustainable manner.

An increase in agricultural production has previously also been carried out by virtue of chemical fertilizers, pesticides, herbicides etc i.e. conventional farming. The higher yields have been consequence of excessive use of fertilizers (Ju et al., 2009). Excessive use results in wastage and later on may result in depletion of these essential nutrients/elements. (Cordell et al., 2009; Schlesinger, 2009). The wastage or loss can result in limiting these factors or nutrients which have other implications. (Cordell et al., 2009). The wastage or loss includes leaching and gas emissions (Smil, 1999; Liu et al., 2010).

Not only are these lost but also result in severe environmental pollution (soil, air and water), eutrophication, global warming and reduction in biodiversity in ecosystems. (Galloway et al., 2003; Schlesinger, 2009).

Thus, there is a need for new technologies and methods to overcome all these problems in a sustainable manner which means there is a great need of sustainable agricultural practices. Sustainable agricultural practices should be such that these should help in enhancing yield and reducing environmental impacts (Schlesinger, 2009).

For sustenance of people the people should have access to nutritious healthy food. Healthy food can be produced from healthy soil, because health of soil is an indicator of healthy crops. (Robertson and Groffman, 2007; Ingham et al., 1985).

Soil fertility or in other words soil health is influenced greatly by soil micro-organisms (Gianinazzi and Schüepp, 1994).

Fungi Bio-Prospects in Sustainable Agriculture, Environment and Nano-Technology.
DOI: https://doi.org/10.1016/B978-0-12-821394-0.00008-1

Soil health determines the productivity of the crops, the two are directly proportional i.e., more fertile the soil more will be the crop production. For sustainable agriculture since there is requirement for more produce, thus there is a need for making the soil fertile.

Plants (roots) produce exudates especially in the rhizosphere region stimulates the microbial growth. Microbes in turn mobilize or provide nutrients in soluble form to the plants. Thus, soil ecology enhances the agricultural productivity (Grayston et al., 1997). Some plants are mycorrhizal because of which instead of rhizosphere zone it is called mycorrhizosphere (Rambelli, 1973).

Many nutrient transformations in soil are performed by soil organisms. These organisms drive the nutrient cycling and play an important role in making these nutrients available to plants and or store in the soil which can be lost also (Robertson and Groffman, 2007; van der Heijden et al., 2008). Earlier studies have proved that soil biota is responsible for nutrient mineralization and plant nutrition (Ingham et al., 1985; Setala and Huhta, 1991; Setala et al., 1990).

It is this microbial diversity which silently interacts with the roots and help in increasing the crop productivity. This interaction is very crucial and needs to be studied thoroughly for proper functioning of soil ecosystems (Johansson et al., 2004).

Soil is a complex ecosystem with diverse group of organisms. These organisms convert the dead decaying matter of plant or animal origin into reusable form where it can be recycled by the plants. Not only do they recycle nutrients, influence soil color, texture, but also make the soil environment congenial for other inhabitants of the soil which include flora and fauna. It also contributes substantially by supporting plant nutrient uptake and yield and also by reducing N leaching losses. (Bender et al., 2014). Besides N and P contents strongly increased with enriched soil life. Enhanced diversity increases nutrient use efficiency which reduces fertilizer amounts needed for agricultural production besides reducing nutrient losses due to leeching (Gabriel et al., 2013; Rockström et al., 2009).

Indiscriminate use of chemical fertilizers, herbicides and weedicides contain toxic heavy metals which lead to heavy metal pollution. Plant uptake is the most important gateway of toxic and heavy metals into the food chain and ultimately humans (Guerra et al., 2012). Lead being one of the heavy metals is not readily absorbed by the plant roots, but inhibits plant metabolic activities such as root growth, photosynthesis and water uptake. (Sharma and Dubey, 2005; Fahr et al., 2013). It also results in food contamination and accumulation of toxic compounds in soil as well as plants. Thus, there is a negative impact on environment and human health. All this needs to be taken care of without jeopardizing human and environment health (Alori et al., 2017a, 2017b).

With all this backdrop there is a need for sustainable agriculture. Microorganisms which form an essential part of soil ecosystem maintain health of the soil and ultimately the crops and higher up organisms along the food chain or food webs. There is a need of using micro-organisms instead of chemical fertilizers, herbicides, pesticides and weedicides etc. Lot of work has been done with regard to microbes like bacteria and few fungi and lots need to be done still. Presently various bacterial and fungal inoculants are being used in sustainable agriculture as these can help in improvement of crop production without any impact on environmental and human health that is in an eco-friendly manner.

Soil and soil organisms and nutrients

Soil can be defined as a dynamic natural body composed of minerals and organic solids, gases, liquids and living organisms which can serve as medium for plant growth, formed as a result of integrated effects of climate and living organisms over periods of time. (Brady and Weil, 2012). It is no doubt a precious invaluable resource that sustains life as also pointed out by Franklin Delano Roosevelt "The nation that destroys its soil, destroys itself."

The plants grow in the soil, the root is the main organ which remains in the soil and provides and supplies nutrients and water for all metabolic activities of the plant. The zone of soil immediately surrounding the living roots called rhizosphere is full of micro-organisms; full of diverse group of micro-organisms. These thrive on the exudations of the roots, organic substances exuding from the roots which stimulates the growth of these organisms and in turn increase the fertility of the soil. (Grayston et al, 1997; Gianinazzi and Schüepp, 1994). These include a milieu of micro-organisms along with symbiotic arbuscular mycorrhizal fungi. These naturally occurring micro-organisms are quite beneficial for the healthy crop production. The changes and cycling of nutrients take place in soil. All these changes are also possible because of the soil organisms. For the successful functioning of agricultural ecosystems, the soil organisms living in the soil play a vital role. The organisms include bacteria, fungi, protozoa, nematodes, earthworms, arthropods and other organisms. These influence plant growth and uptake of nutrients (Rambelli, 1973). Besides these, there are other fungi which form an important group of trophic level in the soil ecosystem called decomposers. The role of these decomposers is known since time immemorial as they decompose the dead organic matter and release the bound nutrients back into the nutrient pool.

Although both harmful and beneficial organisms reside both in rhizosphere and surrounding it yet beneficial organisms outnumber harmful ones. (Scow et al., 1994). Bacteria and fungi help in mineralization. Bacteria make nitrogen available to the plants in soluble form and mycorrhizal fungi; make phosphorous iron sulfur and other elements along with other organisms accessible to plants. Many biodegrading bacteria and fungi breakdown pesticides and other toxic chemicals. Besides, fungi also are useful in the formation of soil humus, soil aggregate formation and stability etc. The humus is important for the maintenance of the organic fraction of soil. (Paul and Clark, 1989; Killham, 1994). Mycorrhiza fungi associated plants show resistance to fungal and nematode pathogens. (Perrin, 1990; Hussey and Roncadori, 1982). It is because of the fine network of AM fungi that phosphorous from deeper regions are mobilized to the plant (Chiariello et al., 1982). Many nutrients are made available for the plant uptake through the agency of micro-organisms. (Robertson and Groffman, 2007).

The number of these naturally occurring diverse group of helpful organisms have decreased as a result of fertilizers and biocontrol agents of chemical origin (Gianinazzi and Schüepp, 1994; Gianinazzi et al., 2002; Galloway et al., 2003; Schlesinger, 2009). The excessive use of these has no doubt been a key factor in higher yields at present (Ju et al., 2009) but at the same time have also resulted in environmental pollution, loss of soil fertility, replacement of soil nutrients with toxic elements. These toxic elements are being taken up by the plants and later higher up organisms along the food chains and food webs.

More biodiversity of micro-organisms favors good plant growth while as decline in soil biodiversity can negatively affect many ecosystem functions (Wagg et al., 2014). The decline indeed has been due to the excessive use of chemical fertilizers, herbicides, pesticides etc.

Since time immemorial be it farmers or agriculturists various methods and technologies have been tried besides conventional methods of farming. More and more techniques eco-friendly for sustainable agriculture are being tried. Interactions between soil microbial communities and crops are being exploited for increasing food production for growing population.

The scenario because of conventional agriculture and genetically modified organisms (GMOs) with the help of chemical fertilizers, pesticides, herbicides has and is changing. Pests are becoming resistant to chemical pesticides, toxic elements etc being added to soil are becoming a nuisance and are being taken up by the plants thus resulting in bioaccumulation and being transferred to higher up the food chains and food webs, thus causing diseases and climate change, agricultural wash off is resulting in environmental pollution (Quarles, 2007). The situation is alarming and there is a public reaction and use of few pesticides have been banned in Europe (O′Callaghan, 2016).

Herbicides have destroyed habitats of butterflies, Nitrogen and Phosphorous fertilizers are contaminating streams resulting in toxic algal blooms. (Dodds et al., 2009). Carbon content necessary to support microbial populations that is beneficial to plant health has been drastically reduced (Quarles, 2018).

All these problems have generated solution in the form of microbial inoculants which are being marketed. These microbial inoculants include mycorrhizae, phosphorus solubilising microbes, nitrogen fixing bacteria, microbes, nitrogen fixing bacteria, microbial antagonists to plant pathogens, and microbial insecticides and pesticides. These inoculants have been found to enhance yield without compromising ecology and environment and also help in reducing pesticide and fertilizer applications. Inoculants are being tested and many have been commercialized (Callaghan, 2016).

Use of microbes

Use of microbes in farming is not new. Previously cover crops were used which encourage mycorrhizae. Composts and compost teas used act as biocontrol agents for pathogens.

For increasing yield mycorrhizal fungi are being inoculated on to the plants, to prevent soil borne diseases. Compost with microbial antagonists are being used. (Quarles, 2007; Ingham, 2005). For organic production two hundred inoculant products have been approved. (Quarles, 2018).

Methods have been developed to use inoculants with ease. These are being welcomed as these are in harmony with organic methods of farming and sustainable agriculture. These do not pollute the environment and can increase yields in an eco-friendly manner. (Amaranthus, 2013; Quarles, 1999).

Microbial inoculants

Microbial inoculants refer to formulations composed of beneficial microorganisms that play an important role in soil ecosystems for sustainable agriculture. They are eco-friendly and are a potential alternative to chemical fertilizers, herbicides and pesticides. (Babalola and Glick, 2012). They help in improving soil fertility and health and in turn human health with no impact on ecosystem. They could be phyto stimulants, bio fertilizers or microbial biocontrol agents (Alori et al., 2017a, 2017b) and include bacteria, fungi and algae (Alori and Babalola, 2018).

Microbial inoculants can reduce the negative impact of agrochemicals and improve the soil health and ultimately the health of the human, higher up the food chain. Microbial inoculants play a critical role in biocontrol technology as they are antifungal and antibacterial in nature (Rani et al., 2018) because of their ability to produce hydrolytic enzymes they stimulate defence responses. In short, they have been found to be eco-friendly.

Fungal inoculants

More work has been done on bacterial inoculants but now attention is being paid to fungal inoculants or myco-inoculants. The three major or macronutrients required by plants for growth and metabolism are NPK; Nitrogen, Phosphorus and Potassium respectively. These limit the growth and yield of plants (Gyaneshwar et al., 2002; Vance et al., 2003). Phosphorus is found in limited quantities in soil in available form for plant uptake, as it gets fixed and immobilised easily (Yadav and Tarafdar, 2010). In order to be taken up by plants it needs to undergo acidification and chelation and it is only by the agency of fungi which produce enough phytase enzymes that it's dissolution and hydrolysis take place (Oberson et al., 2001; Hedley et al., 1982; Richardson, 1994; Yadav and Tarafdar, 2007). The group of fungi mostly involved include *Aspergillus, Gliocladium, Penicillium, Emericella, Chaetomium* and *Trichoderma* besides AM fungi like *Glomus* and *Gigaspora*. (Tarafdar and Gharu, 2006; Yadav et al., 2010; Tarafdar,2019). More than 50% of the total organic phosphorus comprises of phytin and phytate (Turner et al., 2002). Rate of solubilization is more in presence of ammonium salts as compared to nitrate salts. Thus, solubilization is dependent or determined by the nitrogen source (Roos and Luckner, 1984).

AM fungi are associated with solubilization and mobilization of phosphorus in soil ecosystems (Yang et al., 2018). The solubilization is possible from the bed rocks of the soil because of fine network of AM fungi which live in symbiotic relationship with plants and help in making phosphorus available (Harrison, 1999).

At present these fungal inoculants are mainly used as seed inoculants and are commercially also available. Besides AM fungi free living organisms like *Penicillium* in the rhizosphere region have been found to indirectly mobilize phosphorus to the plants by producing phytohormones which increase root density and function (Harvey et al., 2009). Along with phosphorus, potassium, which is the third major essential macronutrient is required by plants for their growth and metabolism. Large agricultural lands in the world are potassium deficient. (Meena et al., 2014). It is present in low quantities naturally in soil but is made available by the use of fertilizer (Kinekar, 2011). It binds with other minerals and is therefore not available for plant uptake (Goldstein, 1994; Prajapati et al., 2013). Fertilizers have to be applied continuously in order to make it available to plants because it is transformed to insoluble form. (Kang et al., 2002). It has to be made available to plants in soluble form. Various potassium solubilising micro-organisms few bacteria and fungus Aspergillus have been isolate and employed for Potassium mobilization. (Bagyalakshmi et al., 2012; Prajapati et al., 2013; Zhang and Kong, 2014).

Three fungal isolates, isolated and purified from arid environment of Jodhpur have shown Potassium solubilization (Kasana et al., 2017). Potassium solubilization isolates from various rhizospheric soil samples have also been reported (Maurya et al., 2014). Genra of *Aspergillus* and *Fomitopsis* can solubilize potassium. These inoculants although after trials can be used.

Penicillium bilaii has been recommended as an inoculant for wheat and canola crops as increase in yields have been observed (Bhatt et al., 2016; Callaghan, 2016). Besides many fungi like *Trichoderma* can be used as inoculant for bio control.

Seed treatments with *Trichoderma* have been available commercially (Quarles, 1993). Not only does it supress soil borne pathogen but also secretes growth promoting substances (Dandurand and Knudsen, 1993; Windham et al., 1986). Increase in yields have also been reported. (Harman, 1991).

Fungus *Metarhizium anisopliae* which can colonize the rhizosphere can control insects when applied as a seed treatment. (Kabaluk and Ericsson, 2007).

Besides, mushrooms (mostly basidiomycetes) can be used for bioremediation as excessive use of chemical fertilizers and pesticides etc. have been responsible for pollution of soil with toxic pollutants. These can not only be taken up by mushrooms, but mushrooms can degrade toxic pollutants in an efficient and eco-friendly manner.

The enzymes produced by these fungi mostly include lignin peroxidases, manganese peroxidases and laccase etc, thus can be used to degrade agricultural waste, which is a cause of worry, as it is burnt indiscriminately leading to pollution. These mushrooms include *Phanerochaete chrysosporium, Agaricus bisporus, Trametes versicolor* and *Pleurotus ostreatus. Lantinus* spp and *Pleurotus tuberregium* have been employed in bioremediation of contaminated soil Nigeria (Adenipekun and Lawal, 2012). Thus, one can kill two birds with one stone i.e., first polluted soil can be decontaminated at the same time mushrooms can be produced which can take care of malnutrition. Various entomopathogenic fungal inoculants can be developed and used for controlling insect pests of crops. Some of the most widely used species include *Beauveria bassiana, Metarhizium anisopliae, Nomuraea rileyi, Paecilomyces farinosus* and *Verticillium lecanii.*

The discovery of endophytes that have beneficial application to agricultural crops need to be studied in detail and their inoculum can also be exploited in near future. More and more research need to be carried out for sustainable agriculture with the help of fungal inoculants as they are eco-friendly.

Future prospects and limitations

Lots of work has been carried out so far with bacterial inoculants since more attention was paid to Nitrogen (plant macronutrient) as compared to Phosphorus and Potassium. Although these two are also important yet less attention was paid so far.

Fungi besides performing the role of decomposers, because of which all ecosystems function can be utilized for other purposes, since they have been found to produce antibiotics, enzymes and other metabolites. These already exist in the soil along with other microorganisms and help in maintaining soil health in turn health of crops and humans. People now a days have become health conscious and want more good quality, nutritious food not at the cost of health and environment.

It is because of indiscriminate use of chemically based products which have been used so far in conventional farming many problems like loss of soil health, fertility and accumulation of toxic materials has taken place. These get accumulated (bio accumulation) and magnified (bio magnification) along the food chains, webs and get ultimately passed on to humans. Thus, become the cause of concern, as health of the population is at risk. It is not only the health of people who consume the produce but also the farm workers and poultry and animals raised on this produce. (Ojo, 2016; United States Environmental Protection Agency (USEPA), 2016; Alori and Babalola, 2018).

In future the population boom is quite expected, has to be balanced with sufficient food production. Search is on for various improved methods of farming Which no doubt the need of the hour. All these methods need to be safe and environment friendly. Thus, it has become imperative to use natural ways which can protect the natural resources, soil being one of them. It is because of this that naturally occurring micro-organisms found to be beneficial are being exploited.

Fungal or mycoinoculums on the basis of various functions which fungi have been found to perform like bioremediation, contaminated soil can be decontaminated. The toxic and other harmful materials in the soil can be degraded into nontoxic materials (Rhodes, 2014). Not only bioremediation and degradation, there inocula has been used to control viral diseases of honey bees which are of commercial value as they bring about pollination of economically important plants world over (Stamets et al., 2018). Stamets has carried out lot of work on mushrooms and reported their ability to not only taking up toxic material but also biodegrading them into harmless material.

Studies have revealed the role of fungal inocula facilitating mobilization, absorption of essential nutrients by plants. Their use has been found to lessen the burden on chemically based preparations so far being used (Alori and Babalola, 2018). Active strains of fungi directly or indirectly stimulate microbial activity which results in soil health improvement (Suyal et al., 2016).

Mycoinoculants being quite beneficial in future seem to replace all other chemically based ingredients being used in the conventional agriculture. With more and more research in future search for new beneficial characteristics can be carried out. Although these are eco-friendly and future looks bright, but there are certain limitations regarding their use. One of the major limitations so far being their beneficial activities have been established in closed systems i.e. laboratory or small-scale units. Agricultural farms where they'll be required in huge quantities their performance has not been tested. As reported earlier (Callaghan, 2016; Alori and Babalola, 2018; Timmusk et al., 2017). Moreover, there is a doubt as these are target oriented and not broad spectrum. Although after due pathogenicity testing these will be introduced into the soil ecosystem, yet their functioning can't be predicted. It's only after their introduction one can understand their behavior towards the soil flora and fauna. The inoculant should be such that it shouldn't disturb the homeostasis of the ecosystem. Local beneficial fungal population although scanty need to be isolated, multiplied and then introduced in the soil, so as not to disturb the other beneficial organisms in the soil.

Conclusion

Mycoinoculants have been associated with increase in yields without harming health and environment (Pathak and Kumar, 2016; Alori et al., 2017b). Their impact has been reported to improve the nutritional value of the produce in sustainable manner. (Rodrı́guez-Morató et al., 2015; Khalid et al., 2017). The impact has mostly been positive. They're also being utilized in food processing industries as they have been found to improve the nutritional value of food (Vitorino and Bessa, 2017).

In future Mycoinoculants seem to replace agrochemicals because of their beneficial activities. It seems like a quite useful alternative for increase in yield of crops with beneficial nutritional and health promoting values in a sustainable way.

References

Adenipekun C.O., Lawal R., 2012. Biotechnol. Mol. Biol. Rev. 7(3), 62–68.

Amaranthus, M., 2013. Getting back to our roots—agriculture discovers the living soil. Acres USA 43 (10), 6.

Alori, E.T., Babalola, O.O., 2018. Microbial inoculants for improving crop quality and human health in Africa. Front. Microbiol. 9 (2213), 1–12.

Alori, E.T., Dare, M.O., Babalola, O.O., 2017a. Microbial inoculants for soil quality and plant fitness. In: Lichtfouse, E. (Ed.), Sustainable Agriculture Review. Springer, Berlin, pp. 181–308.

Alori, E.T., Glick, B.R., Babalola, O.O., 2017b. Microbial phosphorus solubilization and its potential for use in sustainable agriculture. Front. Microbiol. 8, 971.

Babalola, O.O., Glick, B.R., 2012. The use of microbial inoculants in African agriculture: current practice and future prospects. J. Food Agric. Environ. 10, 540–549.

Bagyalakshmi, B., Ponmurugan, P., Marimuthu, S., 2012. Influence of potassium solubilizing bacteria on crop productivity and quality of tea *(Camellia sinensis)*. Afr. J. Agric. Res. 7 (30), 4250–4259.

Bardgett, R.D., Chan, K.F., 1999. Experimental evidence that soil fauna enhances nutrient mineralization and plant nutrient uptake in montane grassland ecosystems. Soil. Biol. Biochem. 31, 1007–1014.

Bender, S.F., Plantenga, F., Neftel, A., Jocher, M., Oberholzer, H.-R., Koehl, L., et al., 2014. Symbiotic relationships between soil fungi and plants reduce N_2O emissions from soil. ISME J. 8, 1336–1345.

Bhatt, S.C., Debnath, S., Pareek, N., 2016. Assessment of Penicillium bilaii inoculation in wheat for improving growth, yield, and phosphorus availability in Mollisols of India. J. Appl. Nat. Sci. 8 (1), 358–367.

Brady, N.C., Weil, R.C., 2012. Nature and Properties of Soils, The Indian Edition Published by Dorling Kindersley India Pvt Ltd.

Callaghan, M.O., 2016. Inoculation of seed for improved crop performance: issues and opportunities. Appl. Microbiol. Biotechnol. 100, 5729–5746.

Chiariello, N., Hickman, J.C., Mooney, H.A., 1982. Endomycorrhizal role for interspecific transfer of phosphorous in a community of annual plants. Science 217 (3), 941–943.

Cordell, D., Drangert, J.-O., White, S., 2009. The story of phosphorus: global food security and food for thought. Glob. Environ. Change 19, 292–305.

Dandurand, L.M., Knudsen, G.R., 1993. Influence of Pseudomonas fluorescens on hyphal growth and biocontrol activity of Trichoderma harzianum in the spermosphere and rhizosphere of pea. Phytopathology 83, 265–270.

Dodds, W.K., Bouska, W.W., Eitzmann, J.L., et al., 2009. Eutrophication of U.S. freshwaters: analysis of potential economic damages. Environ. Sci. Technol. 43 (1), 12–19.

Fahr, M., Laplaze, L., Bendaou, N., Hocher, V., El Mzibri, M., Bogusz, D., et al., 2013. Front. Plant. Sci. 4, 175.

Gabriel, D., Sait, S.M., Kunin, W.E., Benton, T.G., 2013. Food production vs. biodiversity: comparing organic and conventional agriculture. J. Appl. Ecol. 50, 355–364.

Galloway, J.N., Aber, J.D., Erisman, J.W., Seitzinger, S.P., Howarth, R.W., Cowling, E.B., et al., 2003. The nitrogen cascade. Bioscience 53, 341–356.

Gianinazzi, S., Schüepp, H., 1994. Impact of Arbuscular Mycorrhizas on Sustainable Agriculture and Natural Ecosystems. Birkhäuser, Verlag, Basel, p. 226.

Gianinazzi, S., Schüepp, H., Barea, J.M., Haselwandter, K., 2002. Mycorrhizal Technology in Agriculture – From Genes to Bioproducts. Birkhäuser, Basel.

Gianinazzi-Pearson, V., Gollotte, A., Dumas-Gaudot, E., Franken, P., Gianinazzi, S., 1994. Gene expression and molecular modifications associated with plant responses to infection by arbuscular mycorrhizal fungi. In: Daniels, M., Downic, J.A., Osbourn, A.E. (Eds.), Advances in Molecular Genetics of Plant–Microbe Interactions. Kluwer, Dordrecht, pp. 179–186.

Goldstein, A.H., 1994. Involvement of the quinoprotein glucose dehydrogenase in the solubilization of exogenous mineral phosphates by gram-negative bacteria. In: Torriani-Gorni, A., Yagil, E., Silver, S. (Eds.), Phosphate in Microorganisms: Cellular and Molecular Biology. ASM Press, Washington, pp. 197–203.

Grayston, S.J., Vaughan, D., Jones, D., 1997. Rhizosphere carbon flow in trees, in comparison with annual plants: the importance of root exudation and its impact on microbial activity and nutrient availability. Appl. Soil. Ecol. 5 (1), 29–56.

Guerra, F., Trevizam, A.R., Muraoka, T., Marcante, N.C., Canniatti-Brazaca, S.G., 2012. Heavy metals in vegetables and potential risk for human health. Sci. Agricola 69, 54–60.

Gyaneshwar, P., Kumar, G.N., Parekh, L.J., Poole, P.S., 2002. Role of soil microorganisms in improving P nutrition of plants. Plant. Soil 245, 83–93.

Harman, G.E., 1991. Seed treatments for biological control of plant disease. Crop. Prot. 10, 166–171.

Harrison, M.J., 1999. Molecular and cellular aspects of arbuscular mycorrhizal symbiosis. Ann. Rev. Plant. Physiol. Plant Mol. Biol. 50, 361–389.

Harvey, P.R., Warren, R.A., Wakelin, S.A., 2009. Potential to improve root access to phosphorus: the role of non-symbiotic microbial inoculants in the rhizosphere. Crop Pasture Sci 60, 144–151.

Hedley, M.J., Nye, P.H., White, R.E., 1982. Plant induced changes in the rhizospheres of rape (Brassica napus var. Emerald) seedlings. II. Origin of the pH change. New Phytol. 91, 31–44.

Hiltner, L., 1904. Über neuere Erfahrungen und Probleme auf dem Gebiet der Bodenbakteriologie unter besonderer Berüksichtigung der Gründüngung und Brache (On recent insights and problems in the area of soil bacteriology under special consideration of the use of green manure and fallowing). Arb. Dtsch. Landwirt. Ges. 98, 59–78.

Hussey, R.S., Roncadori, R.W., 1982. Vesicular-arbuscular mycorrhizae may limit nematode activity and improve plant growth. Plant. Dis. 66 (1), 9–14.

Ingham, R.E., Trofymow, J.A., Ingham, E.R., Coleman, D.C., 1985. Interactions of bacteria, fungi, and their nematode grazers: effects on nutrient cycling and plant growth. Ecol. Monogr. 55, 119–140.

Ingham, E., 2005. The Compost Tea Brewing Manual; Latest Methods and Research. Soil Food Web., Corvallis, O.R.

Johansson, J.F., Paul, L.R., Finlay, R.D., 2004. Microbial interactions in the mycorrhizosphere and their significance for sustainable agriculture. FEMS Microbiol. Ecol. 48 (1), 1–13.

Ju, X.T., Xing, G.X., Chen, X.P., Zhang, S.L., Zhang, L.J., Liu, X.J., et al., 2009. Reducing environmental risk by improving N management in intensive Chinese agricultural systems. Proc. Natl Acad. Sci. USA 106, 3041–3046.

Kabaluk, J.T., Ericsson, J.D., 2007. Metarhizium anisopliae seed treatment increases yield of field corn when applied for wireworm control. Agron. J. 99, 1377–1381.

Kang, S.C., Ha, C.G., Lee, T.G., Maheshwari, D.K., 2002. Solubilization of insoluble inorganic phosphates by a soil-inhabiting fungus Fomitopsis sp. PS 102. Curr. Sci. 82 (4), 439–442.

Kasana, R.C., Panwar, N.R., Burman, U., Pandey, C.B., Kumar, P., 2017. Isolation and identification of two potassium solubilizing fungi from arid soil. Int. J. Curr. Microbiol. App. Sci. 6 (3), 1752–1762.

Khalid, M., Hassani, D., Bilal, M., Asad, F., Huang, D., 2017. Influence of bio-fertilizer containing beneficial fungi and rhizospheric bacteria on health promoting compounds and antioxidant activity of Spinacia oleracea L. Bot. Stud. 58, 35.

Killham, K., 1994. Soil Ecology. Cambridge University Press, Cambridge.

Kinekar, B.K., 2011. Potassium fertilizer situation in India: current use and perspectives. Karnataka J. Agric. Sci. 24 (1), 1–6.

Liu, J., You, L., Amini, M., Obersteiner, M., Herrero, M., Zehnder, A.J.B., et al., 2010. A high-resolution assessment on global nitrogen flows in cropland. Proc. Natl Acad. Sci. USA 107, 8035–8040.

Maurya, B.R., Meena, V.S., Meena, O.P., 2014. Influence of inceptisol and alfisol's potassium solubilizing bacteria. KSB. Isolates Rel. K. Waste Mica VEGETOS 27 (1), 181–187.

Meena, V.S., Maurya, B.R., Verma, J.P., 2014. Does a rhizospheric microorganism enhance K + availability in agricultural soils? Microbiol. Res. 169, 337–347.

Oberson, A., Friesen, D.K., Rao, I.M., Buhler, S., Frossard, E., 2001. Phosphorus transformations in an oxisol under contrastingland-use system: the role of the microbial biomass. Plant. Soil 237, 197–210.

Ojo, J., 2016. Pesticides use and health in Nigeria. Ife J. Sci. 8, 981–991.

Pathak, D., Kumar, M., 2016. Microbial inoculants as biofertilizers and biopesticides. In: Pratap Singh, D., Bahadur Singh, H., Prabha, R. (Eds.), Microbial Inoculants in Sustainable Agricultural Productivity. Springer, Berlin, pp. 197–209.

Paul, E.A., Clark, F.E., 1989. Soil Microbiology and Biochemistry. Academic Press, San Diego, 272p.

Perrin, R., 1990. Interactions between mycorrhizae and diseases caused by soilborne fungi. Soil Use Manage. 6 (4), 189–195.

Prajapati, K., Sharma, M.C., Modi, H.A., 2013. Growth promoting effect of potassium solubilizing Microorganisms on okra. Abelmoscus Esculantus). Int. J. Agric. Sci. Res. 3 (1), 181–188.

Quarles, W., 1993. Alternatives to methyl bromide—Trichoderma seed treatments. IPM Practitioner 15 (9), 1–7.

Quarles, W., 1999. Plant disease biocontrol and VAM Fungi. IPM Practitioner 21 (4), 1–9.

Quarles, W., 2007. Global warming means more pests. IPM Practitioner 29 (9/10), 1–7.

Quarles, W., 2018. Regenerative agriculture can reduce global warming. IPM Practitioner 36 (1/2), 1–8.

Rambelli, A., 1973. The rhizosphere of mycorrhizae. In: Marks, G.L., Koslowski, T.T. (Eds.), Ectomycorrhizae. Academic Press, New York, pp. 299–343.

Rhodes, C.J., 2014. Mycoremediation (bioremediation with fungi) – growing mushrooms to clean the earth. Chem. Speciation Bioavailability 26 (3), 196–198.

Richardson, A.E., 1994. Soil microrganisms and phosphate availability. In: Pankhurst, C.E., Doulse, B.M., Gupta, V.V.S.R., Grace, P.R. (Eds.), Soil Biota Anagement in Sustainable Farming Systems. CSIRO, Melbourne, pp. 50–62.

Robertson, G.P., Groffman, P.M., 2007. Nitrogen transformations. Soil Microbiology. In: Paul, E.A. (Ed.), Biochemistry and Ecology. Springer, New York, NY, USA, pp. 341–364.

Rockström, J., Steffen, W., Noone, K., Persson, A., Chapin, F.S., Lambin, E.F., et al., 2009. A safe operating space for humanity. Nature 461, 472–475.

Rodríguez-Morató, J., Xicota, L., Fitó, M., Farré, M., Dierssen, M., de la Torre, R., 2015. Potential role of olive oil phenolic compounds in the prevention of neurodegenerative diseases. Molecules 20, 4655–4680.

Roos, W., Luckner, M., 1984. Relationships between proton extrusion and fluxes of ammonium ions and organic acid in Penicillium cyclopium. J. Gen. Microb. 130, 1007–1014.

Schlesinger, W.H., 2009. On the fate of anthropogenic nitrogen. Proc. Natl Acad. Sci. USA 106, 203–208.

Scow, K.M., Somasco, O., Gunapala, N., Lau, S., Venette, R., Ferris, H., et al., 1994. Transition from conventional to low-input agriculture changes soil fertility and biology. Calif. Agriculture 48, 20–2631.

Setala, H., Huhta, V., 1991. Soil fauna increase Betula pendula growth: laboratory experiments with coniferous forest floor. Ecology 72, 665–671.

Setala, H., Martikainen, E., Tyynismaa, M., Huhta, V., 1990. Effects of soil fauna on leaching of nitrogen and phosphorus from experimental systems simulating coniferous forest floor. Biol. Fertil. Soils 10, 170–177.

Sharma, P., Dubey, R.S., 2005. Lead toxicity in plants. Braz. J. Plant. Physiol. 17, 1–19.

Smil, V., 1999. Nitrogen in crop production: an account of global flows. Glob. Biogeochem. Cycles 13, 647–662.

Stamets, P.E., Naeger, N.L., Evans, J.D., Han, J.O., Hopkins, B.K., Lopez, D., et al., 2018. Extracts of polypore mushroom mycelia reduce viruses in honeybees. Sci. Rep. 8, 13936.

Suyal, D.C., Soni, R., Sai, S., Goel, R., 2016. Microbial inoculants as biofertilizer. In: Singh, D. P. (Ed.), Microbial Inoculants in Sustainable Agricultural Productivity. Springer India, New Delhi, pp. 311–318.

Tarafdar, J.C., Gharu, A., 2006. Mobilization of organic and poorly soluble phosphates by *Chaetomium globossum*. Appl. Soil. Ecol. 32, 273–283.

Tarafdar, J.C., 2019. Fungal inoculants for native phosphorus mobilization. In: Giri, B., Prasad, R., Wu, Q.S., Varma, A. (Eds.), Biofertilizers for Sustainable Agriculture and Environment. Soil Biology, vol. 55. Springer, Cham.

Timmusk, S., Behers, L., Muthoni, J., Muraya, A., Aronsson, A.-C., 2017. Perspectives and challenges of microbial application for crop improvement, Front. Plant Sci., 8. p. 49.

Turner, B.L., Paphazy, M.J., Haygarth, P.M., McKelive, I.D., 2002. Inositol phosphates in the environment. Philos. Trans. R. Soc. Lond. B 357, 449–469.

United States Environmental Protection Agency [USEPA], 2016. Melathion Human Health Risk Assessment. USEPA, Washington, DC, p. 258.

Vance, C.P., Uhde-Stone, C., Allan, D.L., 2003. Phosphorous acquisition and use: critical adaptations by plants for securing a nonrenewable resource. N. Phytol. 157, 423–447.

van der Heijden, M.G.A., Bardgett, R.D., van Straalen, N.M., 2008. The unseen majority: soil microbes as drivers of plant diversity and productivity in terrestrial ecosystems. Ecol. Lett. 11, 296–310.

Vitorino, L.C., Bessa, L.A., 2017. Technological microbiology: development and applications. Front. Microbiol. 8, 827.

Wagg, C., Bender, S.F., Widmer, F., Van der Heijden, M.G., 2014. Soil biodiversity and soil community composition determine ecosystem multifunctionality. Proc. Natl Acad. Sci. 111 (14).

Windham, M.T., Elad, Y., Baker, R., 1986. A mechanism for increased plant growth induced by Trichoderma spp. Phytopathology 76, 518521.

Yadav, B.K., Tarafdar, J.C., 2007. Ability of Emericella rugulosa to mobilize unavailable P compounds during Pearl millet (Pennisetum glaucum (L.) R. Br.) crop under arid condition. Ind. J. Microb. 47, 57–63.

Yadav, B.K., Tarafdar, J.C., 2010. Studies on phosphatases activity and clusterbean production as influenced by the P mobilization organisms Emericella rugulosa. Legume Pes. 33, 114–118.

Yadav, B.K., Tarafdar, J.C., Panwar, J., Yadav, R.S., 2010. Phytase producing fungi and their efficiency in hydrolyzing phytin-P compounds. Ann. Arid. Zone 49, 85–90.

Yang, H., Schroeder-Moreno, M., Giri, B., Hu, S., 2018. Arbuscular mycorrhizal fungi and their responses to nutrient management. In: Giri, B., et al., (Eds.), Root Biology, Soil Biology, vol. 5. pp. 429–449.

Zhang, C., Kong, F., 2014. Isolation and identification of potassium-solubilizing bacteria from tobacco rhizospheric soil and their effect on tobacco plants. Appl. Soil. Ecol. 82, 18–25.

Further reading

Aliyu, I.A., Yusuf, A.A., Uyovbisere, E.O., Masso, C., Sanders, IR, 2019. Effect of co-application of phosphorus fertilizer and in vitro-produced mycorrhizal fungal inoculants on yield and leaf nutrient concentration of cassava. PLoS One 14 (6), 1–19.

Alori, E.T., Fawole, O.B., 2017a. Impact of chemical inputs on arbuscular mycorrhiza spores in soil: response of AM Spores to fertilizer and herbicides. Alban J. Agric. Sci. 16, 10–13.

Alori, E.T., Fawole, O.B., 2017b. Microbial inoculants-assisted phytoremediation for sustainable soil management. In: Ansari, A.A., Gill, S.S., Lanza, G.R., Newman, L. (Eds.), Phytoremediation: Management of Environmental Contaminants, Switzerland. Springer International Publishing, Berlin.

Antunes, P.M., Koyama, A., 2017. Mycorrhizas as nutrient and energy pumps of soil food webs: multitrophic interactions and feedbacks. In: Johnson, N., Gehring, C., Jansa, J. (Eds.), Mycorrhizal Mediation of Soil. Elsevier, Amsterdam, pp. 149–173.

Burkle, L.A., Belote, R.T., 2015. Soil mutualists modify priority effects on plant productivity, diversity, and composition. Appl. Veget. Sci. 18, 332–342.

Chapin, F.S., 1980. The mineral nutrition of wild plants. Annu. Rev. Ecol. Syst. 11, 233–260.

Emam, T., 2016. Local soil, but not commercial AMF inoculum, increases native and non-native grass growth at a mine restoration site. Restor. Ecol. 24, 35–44.

Gharu, A., Tarafdar, J.C., 2016. Efficiency of phosphatases in mobilization of native phosphorus fractions under different vegetation. Agric. Res. 5, 335–345.

Hart, M.M., Antunes, P.M., Chaudhary, V.B., Abbott, L.K., 2018. Fungal inoculants in the field: is the reward greater than the risk? Funct. Ecol. 32, 126–135.

Harvey, P.R., Warren, R.A., Wakelin, S.A., 2019. Potential to improve root access to phosphorus: the role of non-symbiotic microbial inoculants in the rhizosphere. Crop. Pasture Sci. 60 (2), 144–151.

Hodge, A., Fitter, A.H., 2010. Substantial nitrogen acquisition by arbuscular mycorrhizal fungi from organic material has implications for N cycling. Proc. Natl Acad. Sci. USA 107, 13754–13759.

Johansen, A., Jakobsen, I., Jensen, E.S., 1992. Hyphal transport of 15N-labelled Nitrogen by a vesicular-arbuscular mycorrizal fungus and its effect on depletion of inorganic soil N. N. Phytologist 122, 281–288.

Kamel, L., Keller-Pearson, M., Roux, C., Ané, J.M., 2016. Biology and evolution of arbuscular mycorrhizal symbiosis in the light of genomics. N. Phytol. 213, 531–536.

Kasana, R.C., Salwan, R., Dhar, H., Dutt, S., Gulati, A., 2008. A rapid and easy method for the detection of microbial cellulases on agar plates using gram's iodine. Curr. Microbiol. 57, 503–507.

Koziol, L., Bever, J.D., Hawkes, C.V., 2015. Mycorrhizal response trades off with plant growth rate and increases with plant successional status. Ecology 96, 1768–1774.

Middleton, E.L., Rihardson, S., Koziol, L., Palmer, C.E., Yermakov, Z.Y., Henning, J.A., et al., 2015. Locally adapted arbuscular mycorrhizal fungi improve vigor and resistance to herbivory of native prairie plant species. Ecosphere 6, 1–16.

Quarles, W., 2013. New biopesticides for IPM and organic production. IPM Practitioner 33 (7/8), 19.

Rodriguez, A., Sanders, IR, 2015. The role of community and population ecology in applying mycorrhizal fungi for improved food security. ISME J. 9, 1053–1061.

Segaran, G., Sathiavelu, M., 2019. Fungal endophytes: a potent biocontrol agent and a bioactive metabolites reservoir. Biocatal. Agric. Biotechnol. 21, 101284.

Sturz, A.V., Carter, M.R., Johnston, H.W., 1997. A review of plant disease, pathogen interactions and microbial antagonism under conservation tillage in temperate humid agriculture. Soil. Tillage Res. 41, 169–189.

Suryanarayanan, T.S., Govinda Rajulu, M.B., Vidal, S., 2018. Biological control through fungal endophytes: gaps. In: Govinda Rajulu, M.B., Vidal, S. (Eds.), Knowledge Hindering Success Current Biotechnology, vol. 7. pp. 185–198. (3).

Torrez, V., Ceulemans, T., Mergeay, J., de Meester, L., Honnay, O., 2016. Effects of adding an arbuscular mycorrhizal fungi inoculum and of distance to donor sites on plant species recolonization following topsoil removal. Appl. Veget. Sci. 19, 7–19.

Chapter 9

Arbuscular mycorrhizal (AM) fungi: Potential role in sustainable agriculture

Khirood Doley and Mahesh Borde*
Department of Botany, Savitribai Phule Pune University, Pune, India
**Corresponding author.* Email: maheshmyco@gmail.com

Introduction

In nature, there exist several interactions where associations are established among various microorganisms that are considered to be favorable for a symbiotic association. In particular, the "mycorrhiza" coined by Frank (1885) (from Greek, mykos: fungus and Rhiza: root) is a symbiotic mutualistic association between a higher plant and a fungus which is found in the roots of plant that extends into the surrounding rhizospheric soil. And there are basically two type of mycorrhiza along with orchid and ericoid mycorrhizas viz., arbuscular (AM) or ectomycorrhizal (EM) where AM are characterized by unique tree-like structure known as arbuscules (Luginbuehl et al., 2017) and EM are characterized by fungal structures such as Hartig nets respectively (Strullu-Derrien et al., 2018). However, interestingly, there is still confusion over their classification due to absence of clear explanation as to which constitutes an arbuscular mycorrhizal and which one is ectomycorrhizal plants (Teste et al., 2019).

Nevertheless, this symbiotic association where there is involvement of interaction between fungi and plants are well acknowledged as arbuscular mycorrhizal fungi (AMF) (Oldroyd, 2013). Arbuscular mycorrhizal fungi belong to the fungal phylum of Glomeromycota (Tedersoo et al., 2018). It is reported that arbuscular mycorrhizal fungi are said to be in association with more than 200,000 species of plants which may include herbs, grasses, several trees and hornworts that represents more than 80% of terrestrial vascular plant species and 90% of agricultural crop plant (Smith and Read, 2008; van der Heijden et al., 2015). As per associations it establishes between plant and fungus, the mycorrhizal associations may be divided into three groups viz., ectomycorrhiza, endomycorrhiza, and ectoendomycorrhizae (transitional

Fungi Bio-Prospects in Sustainable Agriculture, Environment and Nano-Technology.
DOI: https://doi.org/10.1016/B978-0-12-821394-0.00009-3

existence between ectomycorrhiza and endomycorrhizae) (Roth-Bejerano et al., 2014). In this mutuality, the fungus receives 10–20% of total photosynthates and lipids from the host plant, whereas the plant is enhanced through uptake of water and mineral nutrient by the mycorrhizal partner (Gutjahr et al., 2013). Besides, the arbuscular mycorrhizal fungi (AMF) association is being given importance, as it can improve growth in host plants by increased nutrient uptake, production of growth promoting substances, tolerance to abiotic or biotic stresses and by synergistically interact with other beneficial microbes (Akhtar and Siddiqui, 2008). Since, AMF is ubiquitous in its distribution it may play a very crucial role in agricultural ecosystems, since they can play multiple influential roles. In addition, the current practice of using hazardous chemical based agro-products in farmland are being discouraged globally, hence among other options available at our disposal, the AMF can establish itself as suitable alternative to conventional agricultural which mostly employs chemicals. At present, several scientific investigations are undergoing via multidisciplinary approaches to understand mechanism of arbuscular mycorrhizal fungi (AMF) for their symbiotic associations which implies with their potential employment in development of sustainable agricultural system for the future in the most cultivated crops worldwide.

Role of AMF for improving agricultural yield and quality

The AMF symbiosis have exhibited its significant role in enhancing yield in various crop plants in terms of production which would play crucial part for our food security objective in near future (Fiorilli et al., 2018). In this regard, AMF is significantly playing role in increaseing the yield of crop plants is well documented which is beneficial for securing food for our ever increasing population. Already, the various crop plants when in association with AMF have been shown to increase the yield potential in several pot experiments. Adeyemi et al. (2019) showed significant increase in yield of soybean that were inoculated with commercially available AMF inoculum under controlled environment in terms of higher number of pods, dry pod weight and seed yield. Likewise, several other athors have reported significant increase in yield due to AMF inoculation in several crop plants (Eke et al., 2016; Begum et al., 2019). Also, the AMF has remarkably been showed to possess potential in increasing yield in field conditions. For example, the soybean and cotton showed marked increase in vegetal tissue as well as N and P content in higher amount which showed the effectiveness of AMF equivalent of fertilizers (Cely et al., 2016). Thus, it suggests AMF as better alternative to currently employed chemical based fertilizers for future agricultural practices. In addition, besides increase in yield of crop plant the AMF association also enhances quality of host plants such as nutritional quality in lettuce, tomato quality, increase in contents of flavonoids, sugars, vitamin C and organic acids in citrus plant (Baslam et al., 2011; Bona et al., 2017).

In recent investigation of Torres et al. (2018), it was found that in diverse climatic conditions, the quality of grapevine fruit was well maintained due AMF colonization which is very significant in context of our constantly changing climatic conditions. Also, AMF colonization can induce increase in anthocyanins, carotenoids, chlorophyll, tocopherols, total phenolics and several other important minerals (Baslam et al., 2011).

Role of AMF in biotic stress management

Arbuscular mycorrhizal fungi associations can significantly increase plant defenses against several pathogens and earlier reports suggested various beneficial activities of mycorrhizal induced protection against several plant pathogens such as *Alternaria*, *Sclerotium*, *Fusarium*, *Phytophthora*, *Pythium*, *Rhizoctonia*, *Thielaviopsis*, *Pseudomonas*, *Meloidogyne*, *Tetranychus urticae* and *Xiphinema index* (Pozo et al., 2002; Diedhiou et al., 2003; Akhtar and Siddiqui, 2008; Doley et al., 2017; Song et al., 2011; Hao et al., 2012; Doley et al., 2014). Therefore, AM fungi and their associated interactions with plants in reducing the damage caused by plant pathogens have received great attention by several investigators worldwide (Chen et al., 2018). Moreover, as the present day agricultural system makes use of enormous hazardous chemical based fertilizers which is causing more harm than good to our ecosystem and public health, rather it has raised serious concerns regarding the phenomenon of pathogen resistance to chemical based pesticides as a whole. In that scenario, the candidature of AM fungi may be of great importance which would be more suitable and environmental friendly acceptable alternative for sustainable agriculture. Several mechanisms have been proposed to explain the role of AMF in augmenting plant tolerance to against several disease causing plant pathogen which includes higher uptake of plant nutrients, increased cell wall lignification, competition with pathogens for nutrient and space, alteration in roots, induced systemic resistance (Cameron et al., 2013; Schouteden et al., 2015). Furthermore, the research on AMF, as how it confers bio-control abilities in plants by undertaking "omics" and systems biological approaches seems to add further interest in the coming years which will definitely lead to further insights comprehensively and may unravel the complex mechanisms involved in it. These insights might in turn lead to the successful application of large scale AMF in the field (Salvioli and Bonfante, 2013).

Viral diseases

Nowadays, viral diseases in plants are of significant importance for securing food globally and this problem is aggravating only due to change in climate, international trade and several management practices (Kreuze and Valkonen, 2017). Generally, plant viruses are obligate parasitic microorganisms that have distinct life cycles which depend on host plant machinery. And the plants virus management has been made possible by genetically resistant plants where gene-for-gene interaction occurs and resistance is acquired mainly by employing genetic engineering (Palukaitis et al., 2017). However, resistance genes still has not been able to fully provide complete resistance which has been referred to as tolerance or partial resistance genes (De Ronde et al., 2014). In addition, the management practices where agrochemicals which are not environment friendly for sustainable agricultural system are been made use of to target viral life cycle have often resulted into ineffective or not been fully developed for controlling the plant viruses (Hashimoto et al., 2016). Therefore, the inhibition of plant viruses via environment friendly microbes is need of the hour where AMF could provide significant resistance. In that context, several investigations have reported the beneficial effects of AMF colonization against plant viruses even though the effects have been found to be variable. For instance, the AMF colonization showed marked inhibition in the rate of reproduction and gall formation on roots against nematode vector *Xiphinema index* in soil which resulted into protection in grapevine against grapevine fanleaf virus (GFLV) (Hao et al., 2018). The pathogen *Alternaria solani* Sorauer causing the disease known as early blight in tomato (*Solanum lycopersicum* Mill.) has been shown to be decreased upon preinocluation by the mycorrhizal fungi (*Funneliformis mosseae*) significantly by increase in defense related activities of β-1,3-glucanase, chitinase, phenylalanineammonia-lyase (PAL) and lipoxygenase (LOX) in the leaves of tomato. And the stimulation of defense responses was found to be more when preinoculated with AMF when compared with uninoculated control ones in presence of pathogen (Song et al., 2015). Similarly, in tomato plants, a serious geminivirus known as **Tomato yellow leaf curl Sardinia virus** (TYLCSV) was tested for their efficacy as biocontrol agent. In results, it was shown that the virus were not able to alter the mycorrhiza- responsive genes which point towards hindrance in spreading of virus. Thus, symptoms due to virus were lower due to presence of mycorrhiza (*F. mosseae*) (Maffei et al., 2014). Contradictory to this, the AMF colonization have also been reported to increase the multiplication or severity of disease related symptoms which were already present in AMF colonized plants. In tomato plant infected with **Tomato spotted wilt virus** (TSWV), it was reported that the *Glomus mosseae* inoculation restricted induction of defense responses. Thus, the AMF inoculation suggested increase in susceptibility rather than recovery (Miozzi et al., 2011). Moreover, mycorrhizal colonization is being suggested as a main reason for enhancing multiplication of virus and even Miozzi et al. (2019) has proposed a term known as "mycorrhiza-induced susceptibility" (MIS) in view of detrimental consequences AMF colonization has on host plant on being infected by the virus. Hence, the various investigations were indicative of positive as well as negative consequences when host plants were colonized by AMF in presence of virus (Rùa et al., 2013; Aseel et al.,

2019). Moreover, Deja-Sikora et al. (2019) in his reviewed have mentioned that mycorrhizal colonization's contribution against **Potato virus Y** (PVY) in solanaceous crops has been dual in its nature in terms of both positive and negative effects. Furthermore, complete understanding of interactions of AMF against virus in host plant is still not been established yet. Therefore, there is need for advent in novel techniques and explorations which ultimately will help in increasing the knowledge regarding exact mechanisms that are involved for the biocontrol of plant pathogens.

Nematodes

Globally, plant parasitic nematodes comprises of widely distributed group in various habitats (Sawadogo et al., 2009). The plant-parasitic nematodes can cause enormous losses as pests on several important agricultural crop plants. They are classified according to their feeding habits (Perry and Moens, 2011). The most damaging pests of agricultural crops worldwide comprises of cyst and root-knot nematodes (Shah et al., 2017). The plant parasitic nematodes can cause significant yield losses (Nicol et al., 2011) which makes it an economically important pest. Moreover, the limitations have encountered while managing nematodes by use of chemical pesticides as they are posing serious threat to our ecosystem which has called upon for an alternative strategy (Bernard et al., 2017). Hence, AMF can be an environmentally friendly option for effective the management of nematodes as biological control agent. So, the AMF has showed its ability in suppression of nematode and have been previously reported (Alguacil et al, 2011; Nyaku et al., 2017). Even in the review of Schouteden et al. (2015), it was mentioned about mechanisms employed for the inhibition of nematodes been suggested were alteration in root morphology, enhancement in nutrient uptake, competition for space and nutrients, alteration in rhizospheric interactions and induced systemic resistance.

Role of AMF against abiotic stress management

Drought stress

The stress due to drought incurs upon one of the major challenges that has the potential to make already vulnerable global food towards scarcity (Zhang et al., 2018). Looking at the present scenario, it is relevant as the problem related to shortage of water condition is likely to be increased only in the future. Therefore, it has now become inevitable for the further emphasis on novel innovations for sustaining our agricultural systems for the future crop production requirements (Elliott et al., 2014). The stress of drought can be detrimental in growth of a plant as it may induce oxidative damages, osmotic adjustments, affects enzyme activity, adjustments in stomatal and photosynthesis conductance, nutrient assimilation, crops productivity (Hasanuzzaman et al., 2013; Li et al., 2015; Quiroga et al., 2017). In addition, the upregulation in osmotic root hydraulic conductivity in AMF colonized plants have been related due to increase in the expression of fungal or plant aquaporins (Sanchez-Romera et al., 2016). It is well acknowledged that 14-3-3 protein belongs to a group that consists of ubiquitous eukaryotic protein which plays significant role in regulation of various cellular functions viz., inhibition of disease or stress, growth control and acting as phosphoserine or phosphothreonine-binding proteins in plants (Huber et al., 2002; Camoni et al., 2018). Estrada et al. (2012) in his work have demonstrated that AMF has devised a strategy of tolerance by production of proteins in order to prevent misfolding or aggregation of proteins that were generated during abiotic stress condition. For that genes encoding specifically for a 14-3-3 protein and the luminal binding protein (BiP) have been described. In this perspective, the beneficial effects were envisaged when plants are associated with the arbuscular mycorrhizal (AM) which inhibits the harmful effects of drought in several plant species (Barzana et al., 2015; Ruiz-Lozano et al., 2016; Chitarra et al., 2016). The mechanisms by which AMF assists in avoiding drought related effects by several investigators includes enhanced leaf proline, mineral uptake, phenolics, increased turgor potential (Ψp), metabolites, glutathione, ascorbic acid, photosynthesis, leaf area index, growth parameters, plant biomass, fresh and dry weight of seeds (Mirshad and Puthur, 2016; Pal and Pandey, 2016; Jixiang et al., 2017; Pavithra and Yapa, 2018; Sara et al., 2018.) The AM fungi have shown its ability towards achieving sustainable alternatives for modern agricultural systems. Therefore, AMF plays crucial role which begins during nutrient deficit conditions, due to which plants signals via release of strigolactones (SLs) as exudation into nearby rhizosphere (Bonfante and Genre, 2015). So, SLs apparently plays significant role which leads to stimulation of AMF for initiating colonization process towards host plant roots (Besserer et al., 2006). So, towards ultimate successful colonization to occur Myc factors (lipochito-oligosaccharides and short chitin oligomers) are released in response to SLs by host plant (Bonfante and Genre, 2015). Moreover, there has been report of significant correlation between SLs and water stress condition which was showed in the work of Ha et al. (2014), where synthetic SL analogue GR24 was found to be more tolerant to drought like condition as compared to the plants where synthetic SL analogue GR24 were not treated. Also, it was found that plants were more susceptible towards drought like situation due to deficiency or response mutants for SLs. Thus, it suggests the very important role of AMF colonization with relation to SL signaling pathways towards achieving drought resistance for increase in crop production. Furthermore, the advent of modern scientific studies such as proteomics application has also provided with various insights like the presence of differential proteins that helps in modulation of sugar metabolism, cytoskeletal organization, cell wall rearrangement, sulfur-containing proteins and proteins related to plant drought stress responses. Hence, undertaking such kinds of studies like proteomic responses specific to AMF in relation drought stress will bring out further insights that are present in the process of mycorrhization. Also, it has prospective objective in discovering novel biological markers to be applied in breeding for new drought-resistant varieties in the near future (Bernardo et al., 2017).

Salinity stress

It is estimated that due to excessive salt over 800 million hectares of the land surfaces worldwide are affected (Rengasamy, 2006). Soils are getting converted into saline chiefly due to salt deposition by processes which may be natural or anthropogenic. Weathering of parent rocks, deposition from sea water and atmospheric deposition can be considered as natural. Also, several other processes comprises viz., lack of proper drainage systems, utilization of brackish groundwater for irrigation, nonstop irrigation, improper cultural methods in irrigated agriculture and management of water (Evelin et al., 2019). But, still there is rapid rise in case of soil salinity with each year is seeing continuous addition of 0.3–1.5 million ha of fertile land thereby leading to suppression and decreasing crop production by more than 20% (Porcel et al., 2012; Ahanger et al., 2017). And the global agricultural land that are cultivable could shrink by 50% due to salinity issue by the year 2050 (Ashraf and Akram, 2009). The higher level of salinity levels causes stress which may be hyperionic and hyperosmotic which results into decreased plants growth, low nutrient uptake and it finally leads to death of plants (Wang et al., 2018). Also, salt stress leads to several constraints that interfere with overall growth and development in plant by disturbances in several metabolic activities that ultimately lead to programmed cell death resulting into loss in crop production (Kumar et al., 2015). Therefore, there is challenge to undertake agriculture on such saline circumstances. Therefore, efforts being made by making use of AM fungi for justifying their beneficial role during adverse condition of salinity on the host plant (Chandrasekaran et al., 2014; Santander et al., 2019). In contrast, there are also reports available that indicates role of salinity that has discouraged AMF colonization (Hajiboland et al., 2010; Kumar et al., 2010). Nevertheless, there have been reports which suggests that AMF colonization can play significant role in imparting the level of salinity by growth regulators such as cytokinin concentration which showed increased photosynthate translocation and growth promotion achievement by alteration of polyamine pool (Talaat and Shawky, 2014; Kapoor et al., 2013). Also, Elhindi et al. (2017) reported significant improvement in the rate of photosynthesis, content of chlorophyll, increased water usage and other gas exchange traits in *Ocimum basilicum* L. under saline regimes. There have been improved accumulations of certain organic acids due to colonization of AMF in maize plants under saline condition and production of betaine, signifies indirect role played by AMF in the process of osmoregulation (Sheng et al., 2011). In *Allium sativum* plants when AMF-inoculation are done then there was improvement in growth parameters of fresh and dry biomass and leaf area index under saline conditions (Borde et al., 2010). In the study of Campanelli et al. (2013), it was observed that AM fungi inoculated *Medicago sativa* L. performed better on the parameters of growth and physiological activities as compared to nonAM control ones under saline conditions. In the another study undertaken by Hashem et al. (2018), it was reported that when *Cucumis sativus* plants was inoculated with AMF it showed upregulation in the synthesis of jasmonic acid, salicylic acid and several important inorganic nutrients when compared to nonAMF plants under saline conditions. Interestingly, a few studies have suggested enhanced uptake of nutrients such as phosphorus (P) being the mechanism for tolerance in salinity stress when colonized with AMF but other investigations have revealed reduced uptake of P by AMF-inoculated plants under salinity stress (Frosi et al., 2017; Li et al., 2019; Al-Karaki, 2000). Recent investigations have reported that AMF colonizations was especially beneficial with regard to several physiological parameters viz., biomass production, proline synthesis, N uptake, photosynthetic rate, stomatal conductance and leaf water relations under saline conditions (Ait-El-Mokhtar et al., 2019; Santander et al., 2019). Thus, better insights into AMF induced modulations for imparting tolerance by knowing exact mechanisms and the cross-talk for regulating plant performance can help improve crop productivity in future (Begum et al., 2019).

Heavy metal stress

At global level the heavy metal stress has been a major threat to existing terrestrial ecosystem. In the recent years, the various human activities which include sewage sludge amendments, mining, fossil fuels extraction and industrialization are contributing drastically towards becoming main source of heavy metal pollution that are affecting soil and productivity of crop plants (Prasad and de Oliveira Freitas, 2003; Shahid et al., 2015; Lenoir et al., 2016; Xiao et al., 2017). Therefore, the highly toxic heavy metals viz., arsenic (As), cadmium (Cd), lead (Pb), mercury (Hg), Al, Cr etc has the ability to contribute negatively towards crop production (Pierart et al., 2015). Lead (Pb) mainly as toxic compound have been entering the soil chiefly through lead-zinc mining (Yabe et al., 2018). And it estimated that almost 800,000 tons of Pb globally has been released into the environment which ultimate gets accumulated in the soil thereby interferes in the growth of plant (Yang et al., 2018). These toxic elements could cause abnormalities and disorders that reduce yield potential in the host plants (Amari et al., 2017). As a result of this, it can give rise to the production of reactive oxygen species (ROS) which may further deteriorate the redox homeostasis of a cell (Ibrahim et al., 2015). This imbalance in redox status is well acknowledged for its role in heavy metal toxicity in plants. So, for this constraint, phytoremediation seems to be economic and noninvasive elucidation that has been anticipated for remediation of soil due to heavy metal pollution (Mahar et al., 2016). And the role of phytoremediation has been gaining significant attention as it provides heavy metal resistance in the host plant (Ali et al., 2013; Dabrowska et al., 2017). Even though several other fungi are involved in phytoremediation but AM fungi also has significant role in alleviation of metal toxicity to the host plant (Joner and Leyval, 2001; Yang et al., 2016). In the perspective of AMF, several factors are involved (Pawlowska and Charvat, 2004). The mechanism by which host plant tolerates the heavy metal toxicity especially in presence of AMF has been reported to be mediated by direct or indirect influencing of plant biomass and plant photosynthesis or macronutrient acquisition respectively (Yang et al., 2016). Moreover, if legume tree are inoculated with AM fungal species then coplantation of leguminous herbs further provides phytoremediation outcome. Several other mechanisms may include absorption and restriction in certain fungal structures and AM fungal roots (Gonzalez-Chavez et al., 2004), selective barrier played by plasma membrane in plants and fungi, specific or nonspecific transporters activation, transportation through the hyphae, cytosol chelation, organic acids, amino acids and compound-specific chaperones (Cabral et al., 2015). The AMF possesses specific mechanism of retention which is helpful in decreasing the content of various elements in the leaves or shoot of host plant which basically caused by reduced translocation (Soares and Siqueira, 2008). Similarly, it was reported that extraradical mycelium of AM fungus possibly accumulates Cd 10–20 times more per unit of biomass when compared to the control nonmycorrhizal plants (Janouskova et al., 2006). Hence, the growth fungal mycelium provides special potential for accumulation of elements which results into metal tolerance potential (Christie et al., 2004). Furthermore, the several gene mediated studies have also shown significant regulation in terms of heavy metal tolerance like transcription factors *WRKY22, WRKY25, WRKY29* and *ZAT12* was involved in *Arabidopsis thaliana* plant where initially Cu and Cd led to cellular disturbances which resulted into ROS production. And this ROS production led to triggering mitogen-activated protein kinase (MAPK) cascades that eventually modulated antioxidative responses to metal stress (Opdenakker et al., 2012). Therefore, if more and more research in the areas of transcriptomics, ionomics, proteomics and metagenomics will be undertaken then novel genes with potential role in the field of biotechnology may be discovered which be beneficial for making phytoremediation achievable.

AMF mediated defense mechanisms

The AMF can induce tolerance in host plant incurred upon by various abiotic or biotic stresses via., alterations in root architecture, root exudates, improvement in nutritional status, induced systemic resistance, competition, activation of defense related mechanisms. Several investigations have looked at the significant role played by arbuscular mycorrhizal fungi (AMF) in nutrient uptake specifically P in reducing disease symptoms (Nouri et al., 2014). It was reported that simple root architecture when compared to complex root architecture suffered less due to association with AMF colonization which signifies that root architecture and its association with AMF are not independent in nature (Sikes et al., 2009). Alteration in root exudates as a mechanism by AMF in imparting bio-protection against *Fusarium oxysporum* in tomato plant has been reported by Hage-Ahmad et al. (2014), where chlorogenic acid as root exudates from AMF colonized plants inhibited the pathogen growth as compared to nonAMF plants. Recently the phenomenon of AMF competing for space in vine root against pathogen *Ilyonectria* was suggested as a mechanism for resistance (Holland et al., 2019). Quaglia et al. (2012) demonstrated that AMF colonization led to the activation of defense related mechanisms such as increased expression of defense-related genes and production of volatile compounds in the host plant. When host are plants are associated with AMF then it is known to inhibit pathogen via induced systemic resistance (ISR) (Jung et al., 2012). However, still adequate investigation is needed to know the exact mechanism in it.

Factors affecting AMF potential

Mycorrhizal symbioses in agro-ecosystems may be affected by several factors such as fertilization, soil reaction, moisture, toxicity, farming practices, soil tillage and plant protection measures (Sale et al., 2015). Generally, phosphorus fertilization decreases the efficacy of the energy investment for AMF to go into symbiotic associations (Johnson et al., 2015). The increase in availability of nutrient has been found to be negatively affecting AM fungal communities (Camenzind et al., 2018). Despite AMF being reportedly successfully enhancing inhibition of several diseases it does not always impart protection. Like in case of variability in dealing with plant pathogenic virus, instead it has regularly resulted into development of disease symptoms when in association with AMF (Shaul et al., 1999; Miozzi et al., 2011). Hence, AMF need to impart its protective properties from wide variety of pathogens under wide range of environmental factors.

Commercial potential of AMF

The various beneficial role of AMF especially as biofertilizer have increased their chance of development as an important commercial commodity. Moreover, increasing need to sustain our agriculture is as urgent as success of AMF inoculum as bio-fertilizer. As a result of this, the AMF associated agro-based products are considerably gaining in numbers during the past decades (Vosatka et al., 2008). Usually the commercially available inoculums of AMF are marketed as universally compatible for almost all variety of plants under diverse conditions without any scientific research and untrustworthy products are being marketed due to which the expected results are not often met (Tarbell and Koske, 2007; Faye et al., 2013). Often same AMF species gets separated due to different geographical location performs differently (Silvani et al., 2014). Also, the AMF being the obligate symbionts, it is often labored inclusive to get pure cultures which may not be economically feasible option especially for poor or developing countries. The viability of inoculum is also a concern that makes potential loss of desired AMF species during the course of subsequent culturing and possible contamination of stock cultures may further decrease its effectiveness. This makes production of AMF inocula in large scale a very difficult. The mass AMF inoculum may consists of soil based or sand/peat medium where it consist of spores of AMF, colonized root fragments and hyphae. In addition, lack of quality control, standardized method and AMF inoculum selection remains to be resolved for which even polyphasic approach have been undertaken to increase its effectiveness by microbiota was identified and characterized which were associated with AMF inoculum for commercial exploitation (Agnolucci et al., 2019). Even, Ijdo et al. (2011) have suggested use of high technological methods to produce AMF spores that were of pure quality which involves production based on hair root cultures. Therefore, concrete data is needed to establish the exact user-friendly role that AMF inoculums may play to gain confidence of suppliers and growers. Proper calibration of commercial mycorrhizal bio-fertilizer. Therefore, the lack of any kind of regulation on AMF inoculum production and distribution has allowed its market potential and growth at exponential rate (Owen et al., 2015; Vosatka et al., 2012). But, the existence of most AMF-based industries are present in small scales and the information regarding its dynamics in public domain is limited (Chen et al., 2018). Notwithstanding, the commercial AMF inocula has global presence and the Asian continent is presently being dominated mainly by India due to initiatives from its research institutes like The Energy and Resources Institute (TERI) and government that endorses mycorrhiza-based bio-stimulants which active since 1988 in India. Moreover, it has got Centre for Mycorrhizal Culture Collection (CMCC) since 1993 for conservation of mycorrhizal diversity and it has collected over 600 different isolates. With further growth in development of AM based biofertilizer at industrial level will make large scale application of AMF inocula in field possible in future agricultural practices (Salvioli and Bonfante, 2013; Schouteden et al., 2015). In Europe, initially the presence of AMF based companies ranged from a dozen (Vosatka et al., 2012), which grew to 75 companies by the year 2017 (Chen et al., 2018). And now it has significant presence in the various countries in the world. In Indian agricultural system, AMF alongwith various other microbes in several crops have been tested in greenhouse as well as fields and was subsequently adopted in farming for its property to increase yield which would be helpful in decreasing the use of agrochemical based products. Hence, the results are encouraging for commercial exploitation of AMF which exhibits significant characteristics of an ideal eco-friendly product.

Future challenges and strategies in AMF mediated disease management

Despite AMF inoculation offering several benefits to host plants during several unfavorable conditions, but the actual benefits has not been exploited. More essentially, the most important characteristic being associated with this type of association is that it assists improvement of crop productivity as one of the most promising practices under various constraints. Often, large scale production, application and variable results especially in fields in native condition has been posing serious challenge and it poses severe obstacle for its success among growers and commercial utilization. Not all combination of AMF inocula is found to be compatible with other beneficial microbes or plants. Hence, identification of different or suitable combinations with crop plant and other rhizospheric microbial population offers opening of new field of study for better perceptive of employing AMF. Further, bringing of appropriate planning through good legislature politically would be helpful to this scattered or unorganized agro-based industry which eventually will benefit its cultivators as well as economy of the country. Other challenge would be to convince the ultimate growers as industrial acceptance would be easier to get but not with the growers which generally are oriented towards low cost, synthetic chemicals and faster results over any other reason. Recent advancement in several scientific innovations in molecular technology have characterized almost whole mycorrhizal communities in rhizosphere (Berruti et al., 2013). Therefore, several other specific functions of genes that are mediated by AMF or quantitative trait loci (QTL) or other plant breeding programs could be helpful in further exploration. In this respect, the role of future research programs becomes quite inevitable to bring out more insights of precise mechanisms involved in this tri-party interaction. Globally, several institutes and organizations are currently undertaking to exhibit realistic application of AMF in agriculture system.

Conclusion

At present, the major challenge for contemporary agricultural system is to provide food security for growing population in a sustainable way. For that reason, AMF association which is ubiquitous symbiosis between plant and fungi has huge potential role in upregulation of nutritional status as well as in alleviation of plant stress under unfavorable conditions of abiotic and biotic stresses. Thus, it assists in overall improvement in crop production and growth. AMF can also be used singly or in combinations with other beneficial soil microbiota that supports plant growth and helps in improving fertility of soil. In this regards, the mycorrhizal company will play significant role along with scientific community to increase awareness and evidences among cultivators to make AMF inoculum successful for sustainability of our ecosystem. Therefore, for the current agricultural system, AMF associations may be viewed as ecosystem-friendly option for the promotion of crop productivity.

Acknowledgement

The authors are thankful and acknowledge the funding of University Grants Commission (UGC) New Delhi, India, for providing under UGC-PDF program.

References

Adeyemi, N.O., Atayese, M.O., Olubode, A.A., Akan, M.E., 2019. Effect of commercial arbuscular mycorrhizal fungi inoculant on growth and yield of soybean under controlled and natural field conditions. J. Exp. Res. 7 (2), 58–63. Available from: https://doi.org/10.1080/01904167.2019.1685101 (09 January 2020).

Agnolucci, M., Avio, L., Pepe, A., Turrini, A., Cristani, C., Bonini, P., et al., 2019. Bacteria associated with a commercial Mycorrhizal Inoculum: community composition and multifunctional activity as assessed by illumina sequencing and culture-dependent tools. Front. Plant. Sci. 9, 1–13. Available from: https://doi.org/10.3389/fpls.2018.01956 (09 January 2020).

Ahanger, M.A., Tittal, M., Mir, R.A., Agarwal, R.M., 2017. Alleviation of water and osmotic stress-induced changes in nitrogen metabolizing enzymes in *Triticum aestivum* L. cultivars by potassium. Protoplasma 254 (5), 1953–1963. Available from: https://doi.org/10.1007/s00709-017-1086-z (09 January 2020).

Ait-El-Mokhtar, M., Laouane, R.B., Anli, M., Boutasknit, A., Wahbi, S., Meddich, A., 2019. Use of mycorrhizal fungi in improving tolerance of the date palm (*Phoenix dactylifera* L.) seedlings to salt stress. Sci. Horticult. 253, 429–438. Available from: https://doi.org/10.1016/j.scienta.2019.04.066 (09 January 2020).

Akhtar, M.S., Siddiqui, Z.A., 2008. Biocontrol of a root-rot disease complex of chickpea by *Glomus intraradices*, *Rhizobium* sp. and *Pseudomonas straita*. Crop. Prot. 27 (3–5), 410–417. Available from: https://doi.org/10.1016/j.cropro.2007.07.009 (09 January 2020).

Alguacil, M.D., Torrecillas, E., Lozano, Z., Roldán, A., 2011. Evidence of differences between the communities of arbuscular mycorrhizal fungi colonizing galls and roots of *Prunus persica* infected by the root-knot nematode *Meloidogyne incognita*. Appl. Environ. Microbiol. 77 (11), 8656–8661. Available from: https://doi.org/10.1128/AEM.05577-11 (09 January 2020).

Ali, H., Khan, E., Sajad, M.A., 2013. Phytoremediation of heavy metals-concepts and applications. Chemosphere 91 (7), 869–881. Available from: https://doi.org/10.1016/j.chemosphere.2013.01.075 (09 January 2020).

Al-Karaki, G.N., 2000. Growth of mycorrhizal tomato and mineral acquisition under salt stress. Mycorrhiza 10, 51–54. Available from: https://doi.org/10.1007/s005720000055 (09 January 2020).

Amari, T., Ghnaya, T., Abdelly, C., 2017. Nickel, cadmium and lead phytotoxicity and potential of halophytic plants in heavy metal extraction. South. Afr. J. Bot. 111, 99–110. Available from: http://dx.doi.org/10.1016/j.sajb.2017.03.011 (09 January 2020).

Aseel, D.G., Rashad, Y.M., Hammad, S.M., 2019. Arbuscular mycorrhizal fungi trigger transcriptional expression of flavonoid and chlorogenic acid biosynthetic pathways genes in tomato against *Tomato Mosaic Virus*. Sci. Rep. 9, 1–10. Available from: https://doi.org/10.1038/s41598-019-46281-x (09 January 2020).

Ashraf, M., Akram, N.A., 2009. Improving salinity tolerance of plants through conventional breeding and genetic engineering: an analytical comparison. Biotechnol. Adv. 27 (6), 744–752. Available from: https://doi.org/10.1016/j.biotechadv.2009.05.026 (10 January 2020).

Barzana, G., Aroca, R., Ruiz-Lozano, J.M., 2015. Localized and non-localized effects of arbuscular mycorrhizal symbiosis on accumulation of osmolytes and aquaporins and on antioxidant systems in maize plants subjected to total or partial root drying. Plant. Cell Environ. 38 (8), 1613–1627. Available from: https://doi.org/10.1111/pce.12507 (10 January 2020).

Baslam, M., Garmendia, I., Goicoechea, N., 2011. Arbuscular mycorrhizal fungi (AMF) improved growth and nutritional quality of greehouse-grown lettuce. J. Agric. Food Chem. 59 (10), 5504–5515. Available from: https://doi.org/10.1021/jf200501c (10 January 2020).

Begum, N., Qin, C., Ahanger, M.A., Raza, S., Khan, M.I., Ashraf, M., et al., 2019. Role of arbuscular mycorrhizal fungi in plant growth regulation: implications in abiotic stress tolerance. Front. Plant. Sci. 10, 1068. Available from: https://doi.org/10.3389/fpls.2019.01068 (30 December 2019).

Bernard, G.C., Egnin, M., Bonsi, C., 2017. The impact of plant-parasitic nematodes on agriculture and methods of control, nematology-concepts, diagnosis and control. Intechopen 121–151. Available from: https://doi.org/10.5772/intechopen.68958 (07 December 2019).

Bernardo, L., Morcia, C., Carletti, P., Ghizzoni, R., Badeck, F.W., Rizza, F., et al., 2017. Proteomic insight into the mitigation of wheat root drought stress by arbuscular mycorrhizae. J. Proteom. 169, 21–32. Available from: https://doi.org/10.1016/j.jprot.2017.03.024 (02 December 2018).

Berruti, A., Borriello, R., Della Beffa, M.T., Scariot, V., Bianciotto, V., 2013. Application of non specific commercial AMF inocula results in poor mycorrhizationin *Camellia japonica* L. Symbiosis 61 (7), 263–276. Available from: http://doi.org/10.1007/s13199-013-0258-7 (02 December 2018).

Besserer, A., Puech-Pages, V., Kiefer, P., Gomez-Roldan, V., Jauneau, A., Roy, S., et al., 2006. Strigolactones stimulate arbuscular mycorrhizal fungi by activating mitochondria. Plos Biol. 4 (7), 1239–1247. Available from: https://doi.org/10.1371/journal.pbio.0040226 (03 December 2018).

Bona, E., Cantamessa, S., Massa, N., Manassero, P., Marsano, F., Copetta, A., et al., 2017. Arbuscular mycorrhizal fungi and plant growth-promoting pseudomonads improve yield, quality and nutritional value of tomato: a field study. Mycorrhiza 27 (1), 1–11. Available from: https://doi.org/10.1007/s00572-016-0727-y (02 December 2018).

Bonfante, P., Genre, A., 2015. Arbuscular mycorrhizal dialogues: do you speak 'plantish' or 'fungish'? Trends Plant. Sci. 20 (3), 150–154. Available from: http://dx.doi.org/10.1016/j.tplants.2014.12.002 (02 December 2018).

Borde, M., Dudhane, M., Jite, P.K., 2010. AM fungi influences the photosynthetic activity, growth and antioxidant enzymes in *Allium sativum* L. under salinity condition. Not. Sci. Biologicae 2 (4), 64–71. Available from: https://www.notulaebiologicae.ro/index.php/nsb/article/view/5434/8282 (23 Febraury 2011).

Cabral, L., Soares, C.R.F.S., Giachini, A.J., Siqueira, J.O., 2015. Arbuscular mycorrhizal fungi in phytoremediation of contaminated areas by trace elements: mechanisms and major benefits of their applications. World J. Microbiol. Biotechnol. 31 (11), 1655–1664. Available from: https://doi.org/10.1007/s11274-015-1918-y (02 December 2018).

Camenzind, T., Hammer, E.C., Lehmann, J., Solomon, D., Horn, S., Rillig, M.C., et al., 2018. Arbuscular mycorrhizal fungal and soil microbial communities in African Dark Earths. FEMS Microbiol. Ecol. 94 (4), 1–12. Available from: http://dx.doi.org/10.1093/femsec/fiy033 (02 December 2019).

Cameron, D., Neal, A., van Wees, S., Ton, J., 2013. Mycorrhiza-induced resistance: more than the sum of its parts? Trends Plant. Sci. 18 (10), 539–545. Available from: http://dx.doi.org/10.1016/j.tplants.2013.06.004 (20 April 2016).

Camoni, L., Visconti, S., Aducci, P., Marra, M., 2018. 14-3-3 Proteins in plant hormone signaling: doing several things at once. Front. Plant. Sci. 9, 1–8. Available from: https://doi.org/10.3389/fpls.2018.00297 (02 December 2019).

Campanelli, A., Ruta, C., De Mastro, G., Morone-Fortunato, I., 2013. The role of arbuscular mycorrhizal fungi in alleviating salt stress in *Medicago sativa* L. var. icon. Symbiosis 59 (2), 65–76. Available from: https://doi.org/10.1007/s13199-012-0191-1 (02 December 2018).

Cely, M.V.T., de Oliveira, A.G., de Freitas, V.F., de Luca, M.B., Barazetti, A.R., dos Santos, I. M.O., et al., 2016. Inoculant of arbuscular mycorrhizal fungi (*Rhizophagus clarus*) increase yield of soybean and cotton under field conditions. Front. Microbiol. 7, 1–9. Available from: https://doi.org/10.3389/fmicb.2016.00720 (09 January 2020).

Chandrasekaran, M., Boughattas, S., Hu, S.J., Oh, S.H., Sa, T.M., 2014. A meta-analysis of arbuscular mycorrhizal effects on plants grown under salt stress. Mycorrhiza 24 (8), 611–625PMID: 24770494. Available from: https://doi.org/10.1007/s00572-014-0582-7 (02 December 2018).

Chen, M., Arato, M., Borghi, L., Nouri, E., Reinhardt, D., 2018. Beneficial services of arbuscular mycorrhizal fungi - from ecology to application. Front. Plant. Sci. 9, 1–14. Available from: https://doi.org/10.3389/fpls.2018.01270 (02 December 2019).

Chitarra, W., Pagliarani, C., Maserti, B., Lumini, E., Siciliano, I., Cascone, P., et al., 2016. Insights on the impact of arbuscular mycorrhizal symbiosis on tomato tolerance to water stress. Plant. Physiol. 171 (2), 1009–1023. Available from: https://doi.org/10.1104/pp.16.00307 (02 December 2018).

Christie, P., Li, X., Chen, B., 2004. Arbuscular mycorrhiza can depress translocation of zinc to shoots of host plants in soils moderately polluted with zinc. Plant. Soil. 261 (1), 209–217. Available from: https://doi.org/10.1023/B:PLSO.0000035542.79345.1b (02 December 2018).

Dabrowska, G., Hrynkiewicz, K., Trejgell, A., Baum, C., 2017. The effect of plant growth-promoting rhizobacteria on the phytoextraction of Cd and Zn by *Brassica napus* L. Int. J. Phytoremediat. 19 (7), 597–604. Available from: https://doi.org/10.1080/15226514.2016.1244157 (02 December 2018).

De Ronde, D., Butterbach, P., Kormelink, R., 2014. Dominant resistance against plant viruses. Front. Plant. Sci. 5, 1–17. Available from: https://doi.org/10.3389/fpls.2014.00307 (02 December 2018).

Deja-Sikora, E., Mercy, L., Baum, C., Hrynkiewicz, K., 2019. The contribution of endomycorrhiza to the performance of potato *virus Y*-infected solanaceous plants: disease alleviation or exacerbation? Front. Microbiol. 10, 1–9. Available from: https://doi.org/10.3389/fmicb.2019.00516 (02 December 2019).

Diedhiou, P.M., Hallmann, J., Oerke, E.C., Dehne, H.W., 2003. Effects of arbuscular mycorrhizal fungi and non-pathogenic *Fusarium oxysporum* on *Meloidogyne incognita* infestation of tomato. Mycorrhiza 13 (4), 199–204. Available from: https://doi.org/10.1007/s00572-002-0215-4 (02 December 2018).

Doley, K., Dudhane, M., Borde, M., 2017. Biocontrol of *Sclerotium rolfsii* in groundnut by using microbial inoculants. Not. Sci. Biol. 9 (1), 124–130. Available from: https://doi.org/10.15835/nsb919992 (05 April 2017).

Doley, K., Dudhane, M., Borde, M., Jite, P.K., 2014. Effects of *Glomus fasciculatum* and *Trichoderma asperelloides* application on local groundnut (cv. Western-51) against pathogen *Sclerotium rolfsii*. Int. J. Phytopathol. 3 (2), 89–100. Available from: https://escijournals.net/index.php/phytopath/article/view/809/425 (02 December 2015).

Eke, P., Chatue, G.C., Wakam, L.N., Kouipou, R.M.T., Fokou, P.V.T. & Boyom, F.F. 2016, Mycorrhiza consortia suppress the fusarium root rot (*Fusarium solani* f. sp. *Phaseoli*) in common bean (*Phaseolus vulgaris* L.). Available from: https://doi.org/10.1016/j.biocontrol.2016.10.001 (11.08.2017).

Elhindi, K.M., El-Din, S.A., Elgorban, A.M., 2017. The impact of arbuscular mycorrhizal fungi in mitigating salt-induced adverse effects in sweet basil (*Ocimum basilicum* L.). Saudi J. Biol. Sci. 24, 170–179. Available from: http://dx.doi.org/10.1016/j.sjbs.2016.02.010 (02 December 2018).

Elliott, J., Deryng, D., Müller, C., Frieler, K., Konzmann, M., Gerten, D., et al., 2014. Constraints and potentials of future irrigation water availability on agricultural production under climate change. Proc. Natl Acad. Sci. U S Am. 111 (9), 3239–3244. Available from: https://doi.org/10.1073/pnas.1222474110 (09 January 2020).

Estrada, B., Barea, J.M., Aroca, R., Ruiz-Lozano, J., 2012. A native *Glomus irregularis* strain from a Mediterranean saline area exhibits salt tolerance and enhanced symbiotic efficiency with maize plants under salt stress conditions. Plant. Soil. 366 (1), 333–349. Available from: https://www.jstor.org/stable/42952386 (09 January 2020).

Evelin, H., Devi, T.S., Gupta, S., Kapoor, R., 2019. Mitigation of salinity stress in plants by arbuscular mycorrhizal symbiosis: current understanding and new challenges. Front. Plant. Sci. 10, 1–21. Available from: https://doi.org/10.3389/fpls.2019.00470 (09 January 2020).

Faye, A., Dalpé, Y., Ndung'u-Magiroi, K., Jefwa, J., Ndoye, I., Diouf, M., et al., 2013. Evaluation of commercial arbuscular mycorrhizal inoculants. Can. J. Plant. Sci. 93 (6), 1201–1208. Available from: https://doi.org/10.4141/CJPS2013-326 (09 January 2020).

Fiorilli, V., Vannini, C., Ortolani, F., Garcia-Seco, D., Chiapello, M., Novero, M., et al., 2018. Omics approaches revealed how arbuscular mycorrhizal symbiosis enhances yield and resistance to leaf pathogen in wheat. Sci. Rep. 8, 9625. Available from: https://doi.org/10.1038/s41598-018-27622-8 (09 January 2020).

Frank, B., 1885. Ueber die auf Wurzelsymbiose beruhende Ernahrung gewisser Baume durch unterirdische Pilze. Berichte Der Deutschen Botanischen Ges. 3 (4), 2128–2145. Available from: https://doi.org/10.1111/j.1438-8677.1885.tb04240.x (21 September 2018).

Frosi, G., Barros, V.A., Oliveira, M.T., Santos, M., Ramos, D.G., Maia, L.C., et al., 2017. Arbuscular mycorrhizal fungi and foliar phosphorus inorganic supply alleviate salt stress effects in physiological attributes, but only arbuscular mycorrhizal fungi increase biomass in woody species of a semiarid environment. Tree Physiol. 38 (1), 25–36. Available from: https://doi.org/10.1093/treephys/tpx105 (09 January 2020).

Gonzalez-Chavez, M.C., D'Haen, J., Vangronsveld, J., Dodd, J.C., 2004. Copper sorption and accumulation by the extraradical mycelium of different *Glomus* spp. (arbuscular mycorrhizal fungi) isolated from the same polluted soil. Plant. Soil. 240 (2), 287–297. Available from: https://doi.org/10.1023/A:1015794622592 (09 January 2020).

Gutjahr, C., Parniske, M., 2013. Cell and developmental biology of arbuscular mycorrhiza symbiosis. Annu. Rev. Cell Dev. Biol. 29, 593–617. Available from: https://doi:10.1146/annurev-cellbio-101512-122413.

Ha, C.V., Leyva-Gonzalez, M.A., Osakabe, Y., Tran, U.T., Nishiyama, R., Watanabe, Y., et al., 2014. Positive regulatory role of strigolactone in plant responses to drought and salt stress. Proc. Natl Acad. Sci. USA 111 (2), 851–856. Available from: www.pnas.org/cgi/doi/10.1073/pnas.1322135111 (09 January 2020).

Hage-Ahmad, K., Moyses, A., Voglgruber, A., Hadacek, F., Steinkellner, S., 2014. Alterations in root exudation of intercropped tomato mediated by the arbuscular mycorrhizal fungus *Glomus mosseae* and the soilborne pathogen *Fusarium oxysporum F.* sp. *lycopersici*. J. Phytopathology 161 (11–12), 763–773. Available from: https://doi.org/10.1111/jph.12130 (09 January 2020).

Hajiboland, R., Aliasgharzadeh, N., Laiegh, S.F., Poschenrieder, C., 2010. Colonization with arbuscular mycorrhizal fungi improves salinity tolerance of tomato (*Solanum lycopersicum* L.) plants. Plant. Soil. 331 (1–2), 313–327. Available from: https://doi.org/10.1007/s11104-009-0255-z (09 January 2020).

Hao, Z., Fayolle, L., van Tuinen, D., Chatagnier, O., Li, X., Gianinazzi, S., et al., 2012. Local and systemic mycorrhiza-induced protection against the ectoparasitic nematode *Xiphinema index* involves priming of defence gene responses in grapevine. J. Exp. Botany 63 (10), 3657–3672. Available from: https://doi.org/10.1093/jxb/ers046 (09 January 2020).

Hao, Z., van Tuinen, D., Fayolle, L., Chatagnier, O., Li, X., Chen, B., et al., 2018. Arbuscular mycorrhiza affects grapevine fanleaf virus transmission by the nematode vector *Xiphinema index*. Appl. Soil. Ecol. 129, 107–111. Available from: https://doi.org/10.1016/j.apsoil.2018.05.007 (09 January 2020).

Hasanuzzaman, M., Gill, S.S., Fujita, M., 2013. Physiological role of nitric oxide in plants grown under adverse environmental conditions. In: Tuteja, N., Gill, S.S. (Eds.), Plant Acclimation to Environmental Stress. Springer, New York, pp. 269–322. Available from: https://doi.org/10.1007/978-1-4614-5001-6_11 (09 January 2020).

Hashem, A., Alqarawi, A.A., Radhakrishnan, R., Al-Arjani, A.F., Aldehaish, H.A., Egamberdieva, D., et al., 2018. Arbuscular mycorrhizal fungi regulate the oxidative system, hormones and ionic equilibrium to trigger salt stress tolerance in *Cucumis sativus* L. Saudi J. Biol. Sci. 25 (6), 1102–1114. Available from: https://doi.org/10.1016/j.sjbs.2018.03.009 (09 January 2020).

Hashimoto, M., Neriya, Y., Yamaji, Y., Namba, S., 2016. Recessive resistance to plant viruses: potential resistance genes beyond translation initiation factors. Front. Microbiol. 7, 1–11. Available from: https://doi.org/10.3389/fmicb.2016.01695 (09 January 2020).

Holland, T., Bowen, P., Kokkoris, V., Urbez-Torres, J.R., Hart, M., 2019. Does inoculation with arbuscular mycorrhizal fungi reduce trunk disease in grapevine rootstocks? Horticulturae 5, 1–12. Available from: https://doi.org/10.3390/horticulturae5030061 (09 January 2020).

Huber, S.C., MacKintosh, C., Kaiser, W.M., 2002. Metabolic enzymes as targets for 14-3-3 proteins. Plant. Mol. Biol. 50 (6), 1053–1063. Available from: https://doi.org/10.1023/A:1021284002779 (09 January 2020).

Ibrahim, M.A., Khan, P.R., Hegazy, S.S., Hashim, E.A., Azamal, H., Ansari, M.K.A., et al., 2015. Improving the phytoextraction capacity of plants to scavenge heavy-metal infested sites. Environ. Rev. 23 (1), 44–65. Available from: https://doi.org/10.1139/er-2014-0043 (09 January 2020).

Ijdo, M., Cranenbrouck, S., Declerck, S., 2011. Methods for large-scale production of AM fungi: past, present, and future. Mycorrhiza 21, 1–16.

Janouskova, M., Pavlikova, D., Vosatka, M., 2006. Potential contribution of arbuscular mycorrhiza to cadmium immobilisation in soil. Chemosphere 65 (11), 1959–1965. Available from: https://doi.org/10.1016/j.chemosphere.2006.07.007 (09 January 2020).

Jixiang, L., Yingnan, W., Shengnan, S., Chunsheng, M., Xiufeng, Y., 2017. Effects of arbuscular mycorrhizal fungi on the growth, photosynthesis and photosynthetic pigments of *Leymus chinensis* seedlings under salt-alkali stress and nitrogen deposition. Sci. Total. Environ. 576, 234–241. Available from: https://doi.org/10.1016/j.scitotenv.2016.10.091 (09 January 2020).

Johnson, N.C., Wilson, G.W.T., Wilson, J.A., Miller, R.M., Bowker, M.A., 2015. Mycorrhizal phenotypes and the law of the minimum. N. Phytologist 205 (4), 1473–1484. Available from: https://doi.org/10.1111/nph.13172 (09 January 2020).

Joner, E., Leyval, C., 2001. Time-course of heavy metal uptake in maize and clover as affected by root density and different mycorrhizal inoculation regimes. Biol. Fertil. Soils 33 (5), 351–357. Available from: https://doi.org/10.1007/s003740000331 (09 January 2020).

Jung, S.C., Martinez-Medina, A., Lopez-Raez, J.A., Pozo, M.J., 2012. Mycorrhiza induced resistance and priming of plant defenses. J. Chem. Ecol. 38, 651–664. Available from: https://doi.org/10.1007/s10886-012-0134-6 (15 May 2026).

Kapoor, R., Evelin, H., Mathur, P., Giri, B., 2013. Arbuscular mycorrhiza: approaches for abiotic stress tolerance in crop plants for sustainable agriculture. In: Tuteja, N., Gill, S.S. (Eds.), Plant Acclimation to Environmental Stress. LLC: Springer Science + Business Media, pp. 359–401. Available from: https://doi.org/10.1007/978-1-4614-5001-6_14 (09 January 2020).

Kreuze, J.F., Valkonen, J.P.T., 2017. Utilization of engineered resistance to viruses in crops of the developing world, with emphasis on sub-Saharan Africa. Curr. Opin. Virol. 26, 90–97. Available from: https://doi.org/10.1016/j.coviro.2017.07.022 (09 January 2020).

Kumar, A., Dames, J.F., Gupta, A., Sharma, S., Gilbert, J.A., Ahmad, P., 2015. Current developments in arbuscular mycorrhizal fungi research and its role in salinity stress alleviation: a biotechnological perspective. Crit. Rev. Biotechnol. 35 (4), 1–14. Available from: https://doi.org/10.3109/07388551.2014.899964 (8 April 2014, 09 January 2020).

Kumar, A., Sharma, S., Mishra, S., 2010. Influence of arbuscular mycorrhizal (AM) fungi and salinity on seedling growth, solute accumulation, and mycorrhizal dependency of *Jatropha curcas* L. J. Plant. Growth Regul. 29 (3), 297–306. Available from: https://doi.org/10.1007/s00344-009-9136-1 (09 January 2020).

Lenoir, I., Fontaine, J., Lounès-Hadj Sahraoui, A., 2016. Arbuscular mycorrhizal fungal responses to abiotic stresses: a review. Phytochemistry 123, 4–15. Available from: https://doi.org/10.1016/j.phytochem.2016.01.002 (09 January 2020).

Li, J., Meng, B., Chai, H., Yang, X., Song, W., Li, S., et al., 2019. Arbuscular mycorrhizal fungi alleviate drought stress in C3 (*Leymus chinensis*) and C4 (*Hemarthria altissima*) grasses via altering antioxidant enzyme activities and photosynthesis. Front. Plant. Sci. 10, 1–2. Available from: https://doi.org/10.3389/fpls.2019.00499 (09 January 2020).

Li, X., Zhu, T., Peng, F., Chen, Q., Lin, S., Christie, P., et al., 2015. Inner Mongolian steppe arbuscular mycorrhizal fungal communities respond more strongly to water availability than to nitrogen fertilization. Environ. Microbiol. 17 (8), 3051–3068. Available from: https://doi.org/10.1111/1462-2920.12931 (09 January 2020).

Luginbuehl, L.H., Menard, G.N., Kurup, S., Erp, H.V., Radhakrishnan, G.V., Breakspear, A., et al., 2017. Fatty acids in arbuscular mycorrhizal fungi are synthesized by the host plant. Science 356 (6343), 1175–1178. Available from: https://doi.org/10.1126/science.ssn0081 (09 January 2020).

Maffei, G., Miozzi, L., Fiorilli, V., Novero, M., Lanfranco, L., Accotto, G.P., 2014. The arbuscular mycorrhizal symbiosis attenuates symptom severity and reduces virus concentration in tomato infected by *Tomato yellow leaf curl Sardinia virus* (TYLCSV). Mycorrhiza 24 (3), 179–186. Available from: https://doi.org/10.1007/s00572-013-0527-6 (09 January 2020).

Mahar, A., Wang, P., Ali, A., Awasthi, M.K., Lahori, A.H., Wang, Q., et al., 2016. Challenges and opportunities in the phytoremediation of heavy metals contaminated soils: a review. Ecotoxicol. Environ. Saf. 126, 111–121. Available from: https://dx.doi.org/10.1016/j.ecoenv.2015.12.023 (15 December 2019).

Miozzi, L., Vaira, A.M., Catoni, M., Fiorilli, V., Accotto, G.P., Lanfranco, L., 2019. Arbuscular mycorrhizal symbiosis: plant friend or foe in the fight against viruses? Front. Microbiol. 10, 1–9. Available from: https://doi.org/10.3389/fmicb.2019.01238 (09 January 2020).

Miozzi, L., Catoni, M., Fiorilli, V., Mullineaux, P.M., Accotto, G.P., Lanfranco, L., 2011. Arbuscular mycorrhizal symbiosis limits foliar transcriptional responses to viral infection and favors long-term virus accumulation. Mol. Plant. Microbe Interact. 24, 1562–1572. Available from: https://doi.org/10.1094/MPMI-05-11-0116 (09 January 2020).

Mirshad, P.P., Puthur, J.T., 2016. Arbuscular mycorrhizal association enhances drought tolerance potential of promising bioenergy grass *Saccharum arundinaceum*, Retz. Environ. Monit. Assess. 188 (425), 1–20. Available from: https://doi.org/10.1007/s10661-016-5428-7 (16 December 2019).

Nicol, J.M., Turner, S.J., Coyne, D.L., Nijs, L., Hockland, S., Maafi, Z.T., 2011. Current nematode threats to world agriculture. In: Jones, J., Gheysen, G., Fenoll, C. (Eds.), Genomics and Molecular Genetics of Plant-Nematode Interactions. Springer Netherlands, Dordrecht, pp. 21–43. Available from: https://doi.org/10.1007/978-94-007-0434-3_2 (09 January 2020).

Nouri, E., Breuillin-Sessoms, F., Feller, U., Reinhardt, D., 2014. Phosphorus and nitrogen regulate arbuscular mycorrhizal symbiosis in *Petunia hybrida*. PLoS One 9 (3), 1–14. Available from: https://doi.org/10.1371/journal.pone.0090841 (09 January 2020).

Nyaku, S.T., Affokpon, A., Danquah, A., Brentu, F.C., 2017. Harnessing useful rhizosphere microorganisms for nematode control. In: Shah, M.M., Mahamood, M. (Eds.), Nematology—Concepts, Diagnosis and Control. InTech, London, pp. 153–182.

Oldroyd, G.E.D., 2013. Speak, friend, and enter: signalling systems that promote beneficial symbiotic associations in plants. Nat. Rev. Microbiol. 11, 252–263. Available from: http://dx.doi.org/10.1038/nrmicro2990 (09 January 2020).

Opdenakker, K., Remans, T., Keunen, E., Vangronsveld, J., Cuypers, A., 2012. Exposure of *Arabidopsis thaliana* to Cd or Cu excess leads to oxidative stress mediated alterations in

MAP Kinase transcript levels. Environ. Exp. Bot. 83, 53–61. Available from: https://doi.org/10.1016/j.envexpbot.2012.04.003 (09 January 2020).

Owen, D., Williams, A.P., Griffith, G.W., Withers, P.J.A., 2015. Use of commercial bio-inoculants to increase agricultural production through improved phosphorus acquisition. Appl. Soil. Ecol. 86, 41–54. Available from: http://dx.doi.org/10.1016/j.apsoil.2014.09.012 (09 January 2020).

Pal, A., Pandey, S., 2016. Role of arbuscular mycorrhizal fungi on plant growth and reclamation of barren soil with wheat (*T. aestivum* L.) crop. Int. J. Soil. Sci. 12 (1), 25–31. Available from: https://doi.org/10.3923/ijss.2017.25.31 (09 January 2020).

Palukaitis, P., Yoon, J.Y., Choi, S.K., Carr, J.P., 2017. Manipulation of induced resistance to viruses. Curr. Opin. Virol. 26, 141–148. Available from: http://dx.doi.org/10.1016/j.coviro.2017.08.001 (09 January 2020).

Pavithra, D., Yapa, N., 2018. Arbuscular mycorrhizal fungi inoculation enhances drought stress tolerance of plants. Ground Water Sustain. Dev. 7, 490–494. Available from: https://doi.org/10.1016/j.gsd.2018.03.005 (09 January 2020).

Pawlowska, T.E., Charvat, I., 2004. Heavy-metal stress and developmental patterns of arbuscular mycorrhizal fungi. Appl. Environ. Microbiol. 70 (11), 6643–6649. Available from: https://doi.org/10.1128/AEM.70.11.6643-6649.2004 (09 January 2020).

Perry, R.N., Moens, M., 2011. Introduction to plant-parasitic nematodes; modes of parasitism. In: Jones, J.T., Gheysen, L., Fenoll, C. (Eds.), Genomics and Molecular Genetics of Plant-Nematode Interactions. Springer, Dordrecht, pp. 3–20. Available from: https://doi.org/10.1007/978-94-007-0434-3_1 (09 January 2020).

Pierart, A., Shahid, M., Séjalon-Delmas, N., Dumat, C., 2015. Antimony bioavailability: knowledge and research perspectives for sustainable agricultures. J. Hazard. Mater. 289, 219–234. Available from: http://dx.doi.org/10.1016/j.jhazmat.2015.02.011 (09 January 2020).

Porcel, R., Aroca, R., Ruiz-Lozano, J.M., 2012. Salinity stress alleviation using arbuscular mycorrhizal fungi. A review. Agron. Sustain. Dev. 32 (1), 181–200. Available from: https://doi.org/10.1007/s13593-011-0029-x (09 January 2020).

Pozo, M.J., Cordier, C., Dumas-Gaudot, E., Gianinazzi, S., Barea, J.M., Azcon-Aguilar, C., 2002. Localized versus systemic effect of arbuscular mycorrhizal fungi on defense responses to *Phytophthora* infection in tomato plants. J. Exp. Bot. 53, 525–534. Available from: https://doi.org/10.1016/j.pbi.2007.05.004 (09 January 2020).

Prasad, M., de Oliveira Freitas, H., 2003. Metal hyperaccumulation in plants: biodiversity prospecting for phytoremediation technology. Electron. J. Biotechnol. 6 (3), 285–321. Available from: https://doi.org/10.2225/vol6-issue3-fulltext-6 (05 December 2019).

Quaglia, M., Fabrizi, M., Zazzerini, A., Zadra, C., 2012. Role of pathogen-induced volatiles in the *Nicotiana tabacume, Golovinomyces cichoracearum* interaction. Plant. Physiol. Biochem. 52, 9–20. Available from: https://doi.org/10.3389/fpls.2017.01056 (09 January 2020).

Quiroga, G., Erice, G., Aroca, R., Chaumont, F., Ruiz-Lozano, J.M., 2017. Enhanced drought stress tolerance by the arbuscular mycorrhizal symbiosis in a drought-sensitive maize cultivar is related to a broader and differential regulation of host plant aquaporins than in a drought-tolerant cultivar. Front. Plant. Sci. 8, 1–15. Available from: https://doi.org/10.3389/fpls.2017.01056 (09 January 2020).

Rengasamy, P., 2006. World salinization with emphasis on australia. J. Exp. Bot. 57 (5), 1017–1023. Available from: https://doi.org/10.1093/jxb/erj108 (01 January 2020).

Roth-Bejerano, N., Navarro-Rodenas, A., Gutierrez, A., 2014. Types of mycorrhizal association. In: Kagan-Zur, V., Roth-Bejerano, N., Sitrit, Y., Morte, A. (Eds.), Desert Truffles. Springer-

Verlag, Berlin, Heidelberg, pp. 69–80. Available from: https://doi.org/10.1007/978-3-642-40096-4_5 (01 January 2020).

Rùa, M., Umbanhowar, J., Hu, S., Burkey, K., Mitchell, C., 2013. Elevated CO_2 spurs reciprocal positive effects between a plant virus and an arbuscular mycorrhizal fungus. N. Phytol. 199 (2), 541–549. Available from: https://doi.org/10.1111/nph.12273 (01 January 2020).

Ruiz-Lozano, J.M., Aroca, R., Zamarreno, A.M., Molina, S., Andreo-Jimenez, B., Porcel, R., et al., 2016. Arbuscular mycorrhizal symbiosis induces strigolactone biosynthesis under drought and improves drought tolerance in lettuce and tomato. Plant. Cell Environ. 39 (2), 441–452. Available from: https://doi.org/10.1111/pce.12631 (09 January 2020).

Sale, V., Aguilera, P., Laczko, E., Mader, P., Berner, A., Zihlmann, U., et al., 2015. Impact of conservation tillage and organic farming on the diversity of arbuscular mycorrhizal fungi. Soil. Biol. Biochem. 84, 38–52. Available from: http://doi.org/10.1016/j.soilbio.2015.02.005 (09 January 2020).

Salvioli, A., Bonfante, P., 2013. Systems biology and "omics" tools: a cooperation for next-generation mycorrhizal studies. Plant. Sci. 203–204, 107–114. Available from: https://doi.org/10.1016/j.plantsci.2013.01.001 (09 January 2020).

Sanchez-Romera, B., Ruiz-Lozano, J.M., Zamarreño, Á.M., García-Mina, J.M., Aroca, R., 2016. Arbuscular mycorrhizal symbiosis and methyl jasmonate avoid the inhibition of root hydraulic conductivity caused by drought. Mycorrhiza 26 (6), 111–122. Available from: https://doi.org/10.1007/s00572-015-0650-7 (09 January 2020).

Santander, C., Sanhueza, M., Olave, J., Borie, F., Valentine, C., Cornejo, P., 2019. Arbuscular mycorrhizal colonization promotes the tolerance to salt stress in lettuce plants through an efficient modification of ionic balance. J. Soil. Sci. Plant. Nutr. 19 (2), 321–331. Available from: https://doi.org/10.1007/s42729-019-00032-z (09 January 2020).

Sara, O., Ennajeh, M., Zrig, A., Gianinazzi, S., Khemira, H., 2018. Estimating the contribution of arbuscular mycorrhizal fungi to drought tolerance of potted olive trees (*Olea europaea*). Acta Physiol. Plant. 40, 1–81. Available from: https://doi.org/10.1007/s11738-018-2656-1 (09 January 2020).

Sawadogo, A., Thio, B., Kiemde, S., Drabo, I., Dabire, C., Ouedraogo, J., et al., 2009. Distribution and prevalence of parasitic nematodes of cowpea (Vigna unguiculata) in Burkina Faso. J. Nematol. 41 (2), 120–127 (09 January 2020).

Schouteden, N., De Waele, D., Panis, B., Vos, C.M., 2015. Arbuscular mycorrhizal fungi for the biocontrol of plant-parasitic nematodes: a review of the mechanisms involved. Front. Microbiol. 6, 1–12. Available from: https://doi.org/10.3389/fmicb.2015.01280 (09 January 2020).

Shah, S.J., Anjam, M.S., Mendy, B., Anwer, M.A., Habash, S.S., Lozano-Torres, J.L., et al., 2017. Damage-associated responses of the host contribute to defence against cyst nematodes but not root-knot nematodes. J. Exp. Bot. 68 (21–22), 5949–5960. Available from: https://doi.org/10.1093/jxb/erx374 (09 January 2020).

Shahid, M., Khalid, S., Abbas, G., Shahid, N., Nadeem, M., Sabir, M., et al., 2015. Heavy metal stress and crop productivity. In: Hakeem, K.R. (Ed.), Crop Production and Global Environmental Issues. Springer International Publishing, Switzerland, pp. 1–25. Available from: https://doi.org/10.1007/978-3-319-23162-41 (09 January 2020).

Shaul, O., Galili, S., Volpin, H., Ginzberg, I., Elad, Y., Chet, I., et al., 1999. Mycorrhiza-induced changes in disease severity and PR protein expression in tobacco leaves. Mol. Plant-Microbe Interact. 12 (11), 1000–1007. Available from: https://doi.org/10.1094/MPMI.1999.12.111000 (09 January 2020).

Sheng, M., Tang, M., Zhang, F., Huang, Y., 2011. Influence of arbuscular mycorrhiza on organic solutes in maize leaves under salt stress. Mycorrhiza 21, 423–430. Available from: https://doi.org/10.1007/s00572-010-0353-z (09 January 2020).

Sikes, B.A., Cottenie, K., Klironomos, J.N., 2009. Plant and fungal identity determines pathogen protection of plant roots by arbuscular mycorrhizas. J. Ecol. 97, 1274–1280. Available from: https://doi.org/10.1111/j.1365-2745.2009.01557.x (09 January 2020).

Silvani, V.A., Bidondo, L.F., Bompadre, M.J., Colombo, R.P., Pérgola, M., Bompadre, A., et al., 2014. Growth dynamics of geographically different arbuscular mycorrhizal fungal isolates belonging to the '*Rhizophagus* clade' under monoxenic conditions. Mycologia 106 (5), 963–975. Available from: https://doi.org/10.3852/13-118 (09 January 2020).

Smith, S.E., Read, D.J., 2008. Mycorrhizal Symbiosis, 3rd ed. Academic Press, London.

Soares, C.R.F.S., Siqueira, J.O., 2008. Mycorrhiza and phosphate protection of tropical grass species against heavy metal toxicity in multi-contamined soil. Biol. Fertil. Soils 44 (8), 833–841. Available from: https://doi.org/10.1007/s00374-007-0265-z (09 January 2020).

Song, Y., Chen, D., Lu, K., Sun, Z., Zeng, R., 2015. Enhanced tomato disease resistance primed by arbuscular mycorrhizal fungus. Front. Plant. Sci. 6, 1–13. Available from: https://doi.org/10.3389/fpls.2015.00786 (09 January 2020).

Song, Y.Y., Cao, M., Xie, L.J., Liang, X.T., Zeng, R.S., Su, Y.J., et al., 2011. Induction of DIMBOA accumulation and systemic defense responses as a mechanism of enhanced resistance of mycorrhizal corn (*Zea mays* L.) to sheath blight. Mycorrhiza 21 (8), 721–731. Available from: https://doi.org/10.1007/s00572-011-0380-4 (09 January 2020).

Strullu-Derrien, C., Selosse, M.-A., Kenrick, P., Martin, F.M., 2018. The origin and evolution of mycorrhizal symbioses: from palaeomycology to phylogenomics. N. Phytol. 220 (4), 1012–1030. Available from: http://doi.org/10.1111/nph.15076 (01 December 2019).

Talaat, N.B., Shawky, B.T., 2014. Protective effects of arbuscular mycorrhizal fungi on wheat (*T. aestivum* L.) plants exposed to salinity. Environ. Exp. Botany 98, 20–31. Available from: https://doi.org/10.1016/j.envexpbot.2013.10.005 (09 January 2020).

Tarbell, T.J., Koske, R.E., 2007. Evaluation of commercial arbuscular mycorrhizal inocula in a sand/peat medium. Mycorrhiza 18, 51–56. Available from: https://doi.org/10.1007/s00572-007-0152-3 (09 January 2020).

Tedersoo, L., Sánchez-Ramírez, S., Kõljalg, U., Bahram, M., Döring, M., Schigel, D., et al., 2018. High-level classification of the Fungi and a tool for evolutionary ecological analyses. Fungal Diversity 90 (1), 135–159. Available from: https://doi.org/10.1007/s13225-018-0401-0 (01 January 2020).

Teste, F.P., Kardol, P., Turner, B.L., Wardle, D.A., Zemunik, G., Renton, M., et al., 2019. Toward more robust plant-soil feedback research: comment. Ecology 100 (9), 1–5. Available from: http://doi.org/10.1002/ecy.2590 (01 January 2020, 09 January 2020).

Torres, N., Antolín, M.C., Goicoechea, N., 2018. Arbuscular mycorrhizal symbiosis as a promising resource for improving berry quality in grapevines under changing environments. Front. Plant. Sci. 9, 1–18. Available from: https://doi.org/10.3389/fpls.2018.00897 (17 June 2018).

van der Heijden, M.G.A., Martin, F.M., Selosse, M.A., Sanders, IR., 2015. Mycorrhizal ecology and evolution, the past, the present, and the future. N. Phytol. 205 (4), 1406–1423. Available from: https://doi.org/10.1111/nph.13288 (05 December 2019).

Vosatka, M., Albrechtova, J., Patten, R., 2008. "The international market development for mycorrhizal technology. In: Varma, A. (Ed.), Mycorrhiza. Springer, Berlin, Heidelberg, pp. 419–438. Available from: https://doi.org/10.1007/978-3-540-78826-3_21 (09 January 2020).

Vosatka, M., Látr, A., Gianinazzi, S., Albrechtová, J., 2012. Development of arbuscular mycorrhizal biotechnology and industry: current achievements and bottlenecks. Symbiosis 58, 29–37. Available from: https://doi.org/10.1007/s13199-012-0208-9 (09 January 2020).

Wang, Y., Wang, M., Li, Y., Wu, A., Huang, J., 2018. Effects of arbuscular mycorrhizal fungi on growth and nitrogen uptake of *Chrysanthemum morifolium* under salt stress. PLoS One 13 (4), 1–14. Available from: https://doi.org/10.1371/journal.pone.0196408 (09 January 2020).

Xiao, R., Wang, S., Li, R., Wang, J.J., Zhang, Z., 2017. Soil heavy metal contamination and health risks associated with artisanal gold mining in Tongguan, Shaanxi, China. Ecotoxicol. Environ. Saf. 141, 17–24. Available from: https://doi.org/10.1016/j.ecoenv.2017.03.002 (09 January 2020).

Yabe, J., Nakayama, S.M., Ikenaka, Y., Yohannes, Y.B., Bortey-Sam, N., Kabalo, A.N., et al., 2018. Lead and cadmium excretion in feces and urine of children from polluted townships near a lead-zinc mine in Kabwe, Zambia. Chemosphere 202, 48–55. Available from: https://doi.org/10.1016/j.chemosphere.2018.03.079 (05 December 2019, 09 January 2020).

Yang, Q., Li, Z., Lu, X., Duan, Q., Huang, L., Bi, J., 2018. A review of soil heavy metal pollution from industrial and agricultural regions in China: pollution and risk assessment. Sci. Total. Environ. 642, 690–700. Available from: https://doi.org/10.1016/j.scitotenv.2018.06.068 (05 December 2019).

Yang, Y., Liang, Y., Han, X., Chiu, T.Y., Ghosh, A., Chen, H., et al., 2016. The roles of arbuscular mycorrhizal fungi (AMF) in phytoremediation and tree-herb interactions in Pb contaminated soil. Sci. Rep. 6, 1–14. Available from: https://doi.org/10.1038/srep20469 (05 December 2019).

Zhang, F., Zou, Y.N., Wu, Q.S., 2018. Quantitative estimation of water uptake by mycorrhizal extraradical hyphae in citrus under drought stress. Sci. Horticult. 229, 132–136. Available from: https://doi.org/10.1016/j.scienta.2017.10.038 (05 December 2019).

Chapter 10

Endophytic fungal diversity of selected medicinal plants and their bio-potential applications

Yash Mishra, Lakshika Sharma, Mamta Dhiman and Madan Mohan Sharma*
Department of Biosciences, Manipal University Jaipur, Jaipur, India
**Corresponding author.* Email: drmadansharma@gmail.com

Diversity of fungal endophytes

The Plants of the World Online portal (POWO) (launched by Royal Botanic Gardens, Kew in 2017) aimed to enable users to access information on all the world's known seed-bearing plants by 2020. Approximately, all vascular plants studied were reported to harbor endophytic bacteria and/or fungi (Rodriguez et al., 2009; Hardoim et al., 2015). Owing to the vast phyto-diversity exit in India, our country offers huge scope for fungal diversity. Very less fungal diversity is explored and is reported to only 7% out of total of 1.5 million fungi on earth (Hawksworth, 2001, 2004; Strobel and Daisy, 2003; Kumar and Kaushik, 2013). Whatsoever, endophytic fungi recovered contribute to the diversity of microorganisms in natural environments (Nalini et al., 2014; Fernandes et al., 2015; Kumar et al., 2016; Santos Souza and dos Santos, 2017) and the contribution in the diversity is increasing gradually.

Fungal endophytes are supposed to be a treasure of unique natural compounds and produce various bioactive compounds that play major roles in their inherent surroundings (Tan and Zou, 2001; Strobel, 2003; Strobel et al., 2004; Zhang et al., 2006; Qadri et al., 2014; Tiwari, 2015; Qader et al., 2017). For the past almost 30 years, the terms endophyte and endophytic fungi have appeared frequently in the mycological literature (Arivudainambi et al., 2011). People in the Indian subcontinent encompass a long history of using medicinal plants to alleviate various diseases. Medicinal plants of Western Ghats of India (hot spots of global biodiversity) reported to have diverse community of endophytic fungi (Raviraja 2005; Krishnamurthy et al., 2008). However, the diversity of endophytic fungi and their seasonal

Fungi Bio-Prospects in Sustainable Agriculture, Environment and Nano-Technology.
DOI: https://doi.org/10.1016/B978-0-12-821394-0.00010-X

colonization pattern in medicinal plant species have been focused largely in the Malnad region, Western Ghats of Karnataka and Southern India (Naik et al., 2008). Besides, endophytic fungi occur in tropical forest plants across a rainfall gradient (Suryanarayanan et al., 2002) and in woody perennial trees of the Western Ghats in south India (Suryanarayanan et al., 2011). Subsequently, colonization rate of fungal endophytes in medicinal plants viz. *Adhathoda vasica, Ocimum sanctum, Viola odorata, Cannabis sativa* and *Withania somnifera* were very high in Himalayan region (Gautam, 2014). Moreover, five species of Memecylon such as *M. umbellatum, M. edule, M. talbotianum, M. malabaricum* and *M. wightii* have also been reported to harbor diverse group of fungal species viz. *Alternaria* (12.2%), *Cladosporium* (8.2%), *Fusarium* (7.3%), *Aspergillus* (12.2%), *Collectotrichum* (21.6%), *Phoma* (4.4%) and many more from Western Ghats. These fungal species are a source of chemical pool which possesses applications in agriculture and therapeutics (Bharathi and Prakash, 2018).

There is an inclination among the Indian researchers to select plants mainly based on accessibility and vicinity of host plants in the province of their work place for research viz. medicinal plants of Varanasi (Verma et al., 2007; Kharwar et al., 2008), *Tylophora indica* (Burm. f.) Merrill. from Delhi (Kumar et al., 2011), *Ocimum sanctum* Linn. from Amritsar (Bhagat et al., 2012), *Pongamia pinnata* L. commonly known as Karanja has been recognized as a major biodiesel producer plant from Sanganer region of Jaipur, Rajasthan for screening the endophytic fungi viz., *Alternaria alternata* reported from *Rhus mysorensis, Alternaria solani,* (40.85%) *Fusarium oxysporum* (15.95%), *Colletotrichum gloeosporioides* (14.39%), *Curvularia lunata* (9.72%), *Helminthosporium papulosum* (9.33%), *Aspergillus flavus* (6.61%), *Phomopsis viticola* (2.33%) and *Cladosporium cladosporioides* (0.77%) reported from *Pongamia pinnata* which accommodate a large number of fungal endophytes (Porwal et al., 2010; Tiwari and Chittora, 2013). Further, Momsia and Momsia (2013) reported *Alternaria alternata* (15.09% colonization frequency) from *Catharanthus roseus* (L.) G. Don from Kukas (Jaipur), Rajasthan, emphasizing frequency distribution of the fungus in this region. Nagda et al. (2017) documented endophytic fungi viz. Aspergillus sp. and Penicillium sp. from various explants of *Calotropis procera* (Aiton) W. T. Aiton from Udaipur, Rajasthan, India, *Aspergillus* sp. (22.22% CF) and *Penicillium* sp. (10.18% CF) from *Ensete ventricosum* (Chauhan et al., 2019), various species *Curvularia* from *Rauwolfia macrophylla* (Kaaniche et al., 2019) were documented. However, endophytic fungi provoke plant resistance resulting in the reduction of root feeding nematodes (Martinez-Medina et al., 2017) and herbivores (Cosme et al., 2016). Effects of endophytes are more under stress conditions, while under ambient environment, endophytes might have both positive (Newsham, 2011) or negative effects on plant performance (Mayerhofer et al., 2013; Kia et al., 2017). Consequently, adding collectively to their role in plant performance, many endophytes viz., *Fusarium* sp, *Alternaria* sp. etc. can survive and grow as saprophytes in soil which are primary decomposers of infected plant material (Peay et al., 2016; Song et al., 2017). According to Bennett and Cahill (2016), fungi continued their existence at a low rate for competing with plants growing at early stages as compared to the adult plants ensuing in higher plant-endophytic diversity. The most prominent production of metabolites by endophytic fungi with bioactive properties is the anti-cancer drug taxol, which was originally produced from the *Taxus brevifolia* (yew tree). The production of this anticancer compound from endophytic fungus was first reported by Stierle et al. (1995). They had produced first time taxol from *Taxomyces andreanae* isolated from pacific yew tree in in vitro conditions. Since then, researcher started exploring endophytic fungi from different species of Taxus for the production of taxol viz., *Pestalotiopsis microspora* isolated from *Taxus wallichiana* (Strobel et al., 1996), *Colletotrichum gloeosporioides* from *Taxus mairei* (Wang et al., 2008), *Fusarium solani, Diaporthe phaseolorum, Aspergillus versicolor, Cladosporium tenuissimum, Metarhizium anisopliae* etc. from *Taxus chinensis* (Liu et al., 2009), *Gliocladium* sp. and *Fusarium redolens* from *Taxus baccata* (Sreekanth et al., 2009; Garyali and Reddy, 2013), *Guignardia mangiferae* from *Taxus media* (Xiong et al., 2013). Additionally, many more antimicrobial compounds have also been identified from the endophytic fungi residing within plant tissues (Sturz et al., 2000; Arnold and Lutzoni, 2007; Polonio et al., 2015; Gouda et al., 2016). Therefore, the success story for the search of an anti-cancer drug i.e. taxol from species of *Taxus* had opened up new vistas for the exploration of other drugs from other fungi probably residing in many more plant species to cure dreaded diseases.

Hence, keeping in view the importance of plants for the production of drugs/ natural compounds from their fungal counter parts, following plants were selected for screening, isolation, purification of endophytic fungi for production of secondary metabolites, enzyme production, mycosynthesis mediated nanoparticles and their biological properties.

The plant

Tinospora cordifolia (Willd.) Miers ex Hook F & Thoms, *Adhatoda vasica* Nees. and *Murraya koenigii* (L.) Spreng. were investigated in author's laboratory at Manipal University Jaipur, India. These plants were poorly investigated for the search of endophytic fungi for their further applications.

Tinospora cordifolia (Willd.) Miers ex Hook F & Thoms

It is an important medicinal plant belongs to family Menispermaceae, commonly known as Giloy. The systematic position of this plant as per the system of classification given by Bentham and Hooker (1862) is as follows:

Systematic position

Division: Angiosperms; Class: Dicotyledonae; Sub-class: Polypetalae; Series: Thalamiflorae

Order: Ranales; Family: Menispermaceae; Genus: Tinospora; Species: cordifolia

Tinospora cordifolia is a hefty deciduous, broadly spreading climbing shrub through which branches emerge in coil form. Leaves of *T. cordifolia* are simple, alternate, glabrous exstipulate have long petiole approx. 12−15 cm. round, heart shaped, ovate lamina, 7−9 nerved. Stem of this plant is slightly succulent lengthy, filiform and climbing in nature. Aerial roots arise from the branches, which are characterized by tetra to penta-arch primary structure, removal of the exterior skin, the greenish color mucilaginous substance is observed. Bark is whitish creamy to greyish in color. Flowers are unisexual, yellow colored. Male flowers are in the form of bunch and female flowers acquire solitary inflorescence. Sepals are 6 (in 2 pair of 3 each). Petals are smaller than sepals and it is also 6, free and membranous. Flowering and fruiting occur during March to June, they are orange-reddish in color, fleshy, aggregate of 1−3 and ovoid on thick stalk (Kirtikar and Basu, 2012; Mittal et al., 2014). The plant contains a number of secondary metabolites like berberine, palmatine, tembetarine, tinosporin, tinosporide, tinosporaside, cordifolide, cordifol, tinocordiside and tinocordifolioside etc. (Devprakash et al., 2011; Pandey et al., 2012). Owing to the presence of these potent natural compounds, this plant has been in use for ameliorating various diseases viz. diabetes, cholesterol, fever, gonorrhea, syphilis, jaundice, chronic diarrhea, cancer, dysentery, bone fracture, pain, asthma, skin disease, poisonous insect, snake bite, eye disorders, etc. (Parthipan et al., 2011; Juyal and Ghildiyal, 2013; Rahul, 2013; Mittal et al., 2014; Thangaraj et al., 2017). The reported compounds also show various other activities i.e. antiperiodic, anti-spasmodic, anti-microbial, anti-osteoporotic, anti-inflammatory, antiarthritic, anti-allergic and anti-diabetic properties (Sharma et al., 2012) (Table 10.1). However, endophytic fungi such as *Fusarium culmorum* have been documented from this plant species through which taxol was extracted and used against human cancer cell lines and *Mycobacterium tuberculosis* H37Rv by Radiometric Bactec 460 assay (Sonaimuthu et al., 2010; Thangaraj et al., 2017). *Colletotrichum linicola*, *Fusarium oxysporum*, *Curvularia lunata* , *Chaetomium globosum* etc. showed cytotoxic and antimicrobial activities (Mishra et al., 2012), *Penicillium* sp. reported to be effective against *Bacillus subtilis* and *Staphylococcus aureus* and promote antibacterial action, *Phoma* sp. also reported to promote growth of maize (Desai et al., 2012; Kedar et al., 2014). Moreover, endophytic fungi *Fusarium solani* showed antimicrobial activity against the human pathogens such as *Staphylococcus aureus*, *Salmonella typhi*, etc. (Uzma and Chowdappa, 2017).

TABLE 10.1 List of natural compounds and their biological activities from the selected medicinal plants.

S. no.	Plant name	Secondary metabolites	Biological activity	Reference
1.	*Tinospora cordifolia*	Glycosides viz. Furanoid diterpene glucoside Tinocordiside Tinooordifolioside Cordioside Palmatosides	Antioxidant, Immunomodulatoy Anticancer	Saha and Ghosh (2012), Shanthi and Nelson (2013), Mittal et al. (2014), Kavya et al. (2015)
		Aliphatic compounds viz. Ootaoosanol Heptacosanol Nonacosan-15-one Dichlcromethane	Anti-inflammatory	Pandey et al. (2008)
		Steroids viz.hydroxy ecdysone Ecdysterone Giloinsterol	Antiosteoporotic Anti-inflammatory	Reddy and Reddy (2015)
		Alkaloids, terpenoids, glycosides, flavonoids, essential oils	Antimicrobial	Kumar et al. (2017)
2.	*Adhatoda vasica*	Alkaloids and Vasicine	Antimicrobial	Soni et al. (2008)
		Phenolic, flavonoid, and alkaloids	Antibacterial	Shukla et al. (2017)

(*Continued*)

TABLE 10.1 (Continued)

S. no.	Plant name	Secondary metabolites	Biological activity	Reference
		Vasicine acetate and 2-Acetyl benzylamine,	Antifeedant, larvicidal activity against moth *Plutella xylostella*	Paulraj et al. (2014)
		Vasicine	Anti-inflammatory and Antistress properties.	Keesara and Jat (2017)
		Crude extract	Inhibitory activity of HIV-Protease	Singh et al. (2010)
		Alkaloids, flavonoids, terpenoids, saponins, phenols and steroids.	Antioxidant and free radical scavenging potential	Bajpai et al. (2015)
		Crude extract	Antioxidants, Antimicrobial	Wankhede (2015)
3.	*Murraya koenigii*	Carbazole alkaloids	Antioxidant and radical-scavenging	Rao et al. (2007), Kavitha (2017)
		Flavonols	Antioxidant	Singh et al. (2011)
		Phenolic, quercetin, catechin, epicatechin, naringin and myricetin	Anticancer, Inhibiting the growth of breast cancer cell line (MDA-MB-231)	Ghasemzadeh et al. (2014)
		Terpenoid and alkaloids	Anticancer	Meena et al. (2017)
		Alkaloids, tannins, phenols, flavonoids	Antioxidant	Prasanth and Ramamurthy (2018)

Adhatoda vasica Nees

It is commonly known as Vasaka, is a well-known herb belongs to family Acanthaceae. The systematic position of this plant as per the system of classification given by Bentham and Hooker (1862) is as under:

Systematic position

Division: Angiosperms; Class: Dicotyledonae; Subclass: Gamopetalae; Series: Bicarpellatae

Order: Personnales; Family: Acanthaceae; Genus: Adhatoda; Species: vasica

Adhatoda vasica is a diminutive evergreen plant that grows in low availability of water. It is large shrub grows in crowded areas along waste land, road sides etc. Its leaves are simple, acute, shiny, lanceolate and broader approx. 10 – 15 cm. in length and 4–6 cm. in width. It is minutely puberulous when it is young and after maturity it becomes glabrous. It turns brownish-green when dried and contains bitter taste and its smell is similar to strong tea. Stem is soft and used to make charcoal for gun powder. Flowers have bulky eye-catching white petals. Flowers are dense pedunculate spikes approx 3–8 cm. streaked with purple on the lower side, peduncles approx. 4–8 cm. shorter than the leaves, glabrous, 4–8 nerved, reticulate veination, and its fruit is a small capsule with four seeds. Flowering and fruiting occurs during August to December. It is used in indigenous systems of medicine for its beneficial effects, particularly in bronchitis. It also exhibits antispasmodic, expectorant and blood purifying virtues (Hossain and Hoq, 2016). Besides, poultice of its leaves can be applied over fresh wounds for beneficial results. At the same time, it also cures in rheumatic joints, inflammatory swellings etc. (Jain, 1991). Warm decoction of its leaves is useful in treating scabies and other skin diseases as well (Kumar et al., 2010). While, its leaves, bark, fruits and flowers reported useful in the removal of intestinal parasites and more significantly leaf powder also showed anti-ulcer activity (Shrivastava et al., 2006). The plant contains various secondary metabolites like alkaloids viz., vasicine, oxyvascicine and vasicinone (Kumar et al., 2010); flavonoids viz., apigenin, astragalin, kaempferol, quercetin, vitexin (Maurya and Singh, 2010); glycosides, saponin, phenol, lignin, sterols and tannins (Chouhan et al., 2013). Moreover, having innumerable number of important and useful phytoconstituents for curing wide range of dreadful diseases, this plant becomes house of many important endophytic fungi used for treating cold, cough, chronic bronchitis and asthma (Claeson et al., 2000; Gangwar and Ghosh, 2014) (Table 10.1). Further, some endophytic fungi have been reported from *A. vasica* earlier were *Aspergillus oryzae*, *Emericella nidulans*, *Fusarium oxysporum*, *Gliocladium catenulatum*, *Penicillium chrysogenum* (Shukla and Mishra, 2012), *Colletotrichum dermatium*, *Fusarium avenaceum* (Manoharachary and Nagaraju, 2016). The chemical prospective of endophytic fungi has received consideration of pharmaceutical industries due to the recognition of precious chemical constituents as products of fungal biosynthesis (Kusari et al., 2012; Qadri et al., 2013).

Murraya koenigii (L.) Spreng

It is known as curry leaf plant belongs to the family Rutaceae. It is widely used as herb, spice, condiments and reported for treating various types of ailments in Indian traditional medicinal system. The systematic position of this plant as per the system of classification given by Bentham and Hooker (1862) is as follows:

Systematic position

Division: Angiosperms; Class: Dicotyledonae; Subclass: Polypetalae; Series: Disciflorae

Order: Geraniales; Family: Rutaceae; Genus: Murraya; Species: koenigii

Murraya koenigii is small spreading plant and attains height of about 2.5 m. The leaves of this shrub are exstipulate, bipinnately compound, reticulate venation, lanceolate 25–30 cm long, each bearing many leaflets. The main stem is dark green to brown with numerous dots on it and revealed the colorless wood beneath and its girth is 12–16 cm. Flowers of this plant is bisexual, funnel-shaped, scented odor, ebracteate, actinomorphic, pentamerous, hypogynous, calyx-5, lobed, corolla white, polypetalous, inferior with 5 petals, lanceolate and its fruits are round to oblong, fully ripe fruits black with a very shining surface. Flowering starts from the middle of April and ends in the middle of May followed by fruiting in July to August (Kirtikar and Basu, 1935; Parmar and Kaushal, 1982). Various parts of this plant have been widely used by tribal communities. The leaves of this plant have been used as tonic and cure stomachic, carminative, dysentery, vomiting, piles (Handral et al., 2012; Singh et al., 2014). This is also documented useful as analgesic, allays heat of the body, thirst, inflammation and itching. Knolker and Reddy (2008) reported that curry leaf has richest contributor of carbazole alkaloids which contain diverse pharmacological activities. However, other secondary metabolites viz., alkaloids i.e. mahanine, koenine, koenigine, koenidine, girinimbiol, girinimibine, koenimbine, O-methyl murrayamine A, O-methyl mahanine, isomahanine, bismahanine, bispyrayafoline and other phytoconstituents such as coumarin glycoside viz., scopotin, murrayanine, calcium, phosphorus, iron, thiamine, riboflavin, niacin, vitamin C, carotene and oxalic acid etc. have been reported (Savitha and Balamurugan, 2014) (Table 10.1). Moreover, endophytic fungi viz. *Curvularia lunata*, *Colletotrichum gloeosporioides*, *Aspergillus niger*, *Aspergillus flavus*, *Candida albicans*, *Phoma* hedericola and *Penicillium sublateritium*, etc. have been reported from this plant species earlier (Thalavaipandian et al., 2011; Nath et al., 2014; Kumar et al., 2015).

Endophytes in laboratory culture

Endophytes are valuable partner to the host plants. The ultimate aim of tissue culture is to develop axenic plants. Even after surface sterilization of the explants, autoclaving and UV treatment of nutrient medium for tissue culture, endophytic bacteria or fungi or actinomycetes start growing from tissues or from the cultured explant. These endophytes are generally considered as contaminants resulting complete loss of time, media and explants, which sometimes may be of some rare and endangered species of microbes, which need to be conserved by tissue culture techniques. Besides, endophytic species composition and plant genotype together under culture conditions are the key factors for attainment of plant tissue cultures with elevated renewal capability. Interaction between endophytes and specific secondary compounds leached from plant may be a major facet for browning and cell death (Pirttila et al., 2008). Some endophytes have been reported from cultures of roots and photosynthetic tissues of plant and the same could be developed in plants under controlled conditions i.e. tissue culture of plants (Sandberg et al., 2014; Mishra et al., 2015). Besides, influence of endophytic fungus *Fusarium* sp. isolated from symptomless weeds encompass significant role in cherry production at a speedy rate through tissue culture (Ilic et al., 2017). Further, Dubois et al. (2004) also reported the importance of endophytic fungus i.e. *Fusarium* sp. as an antagonist against pests *Cosmopolite sordidus* and a nematode *Radopholus similis*, which cause loss in banana production.

Gateway of fungi in host plant

Latent endophytes move from soil to the host plant probably by one of the various mechanisms such as wind action, fondness to soil particles, in water habitat, through agricultural tools and by vector-borne transmission including human, birds, mites, insects, etc., which differ in their ability to shift their range, intra continental range shifts, holobiome region, etc. (Armstrong et al., 1987; Berg et al., 2010; Morrien et al., 2010; Dostalek et al., 2016; Mitter et al., 2016). Although, underneath organisms have a restricted dispersal capacity. Fungal endophytes regarded as commensals rather than foes or symbiotic relationship between fungi and host under stress conditions influence plant performance, induce plant resistance resulting in reduction of root feeding nematodes (Newsham, 2011; Martinez-Medina et al., 2017). Diverse microbiome acquired from soils have an increased defence against specialized soil borne organisms together with disease causing microorganisms, which provide them with a competitive benefit to establish in plant communities (Malcolm et al., 2013; Geisen et al., 2017). Besides, colonization of endophytes depends on many factors including plant tissue type, genotype of plant, microbial taxon and strain type, and biotic & abiotic factors.

Histological studies

The detection of fungal endophytes on suitable nutrient media is time consuming. Hence, histological localization of endophytic fungi within plant tissues through microtomy would be valuable to notice fungal imprints in short period of time. According to Bayman (2007), endophytes residing in the plant tissues are asymptomatic and their detection in sterile form i.e. non-sporulating isolates could be difficult and a number of endophytes are not at all isolated so far. In order to understand the biology of endophytic fungi, there is a need to directly observe their hyphae/mycelium within the plant parts through anatomical observations. Such observations provide information about the mode of infection, the extent of colonization and the reaction of the plant against infection by the fungus. Many fungi develop small mycelium within plant tissues, which are difficult to observe directly through external observations. Therefore, direct assessments have been made to find out whether the fungal infectivity is intercellular or intracellular in plant anatomy. Hence, the presence of endophytic fungi within symptomless aerial parts of vascular plants has been documented (Zhang and Yao, 2015). These fungi have been isolated from healthy and symptomless tissues after proper surface sterilization with sodium hypochlorite/hydrogen peroxide or mercuric chloride as a surfactant (Tervet and Hollis, 1948; Norse, 1972; Oyebanji et al., 2009; Lakra et al., 2013; Sharma et al., 2015). These endophytes colonize interior of plants especially leaves, branches, stem and roots showing no apparent harm to the host. The interaction between endophytic fungi and host plants involves exchange of metabolic products between endophyte and host. Generally, endophytes grow intercellularly in their host and the site and extent of invasion by endophytic fungi reflects the mechanism of nutrient transfer to the fungal symbionts (Bacon and Siegel, 1988; White and Morgan-Jones, 1996). Following the histological techniques as a preliminary tool to identify fungal endophytes, several reports have been available which prove the presence of endophytic fungal community in various hosts including *Terminalia arjuna* (Tejesvi et al., 2005), *Spondias mombin* (Rodrigues and Samuels, 1999), *Eucalyptus globules* (Bettucci and Saravay, 1993), Nicotiana sp. (Spurr and Welty, 1975), *Azadirachta indica* (Rajagopal and Suryanarayanan, 2000; Mahesh et al., 2005; Verma et al., 2007), *Musa acuminata* (Pereira et al., 1999), *Aegle marmelos* (Gond et al., 2007), *Piper longum* (Verma et al., 2011), etc.

Further, dense colonization of endophytic fungi has been reported to locate the distribution of fungal endophytes stained with lactophenol cotton blue in *Catharanthus roseus* white variety within leaf and stem tissue (Lakra et al., 2013) and inside the healthy leaves and stem of *Tinospora cordifolia*, *Adhatoda vasica* (Mishra et al., 2015; Mishra et al., 2017), respectively. The efficacy of the lactophenol cotton blue stain have been used while observing widespread and disseminated colonization along the leaf within the parenchyma and adaxial epidermis, demonstrating close interaction among the endophytes in multiple structural and trophic subniches in the host (Sathe and Raghukumar, 1991; Bernardi-Wenzel et al., 2010; Mishra et al., 2015; Mishra et al., 2017). Subsequent to histological studies, stains for observing the mycelium of fungi from diverse plants have been reported viz., *Phoenix dactylifera* (Gomez-Vidal et al.,2006) in lactophenol cotton blue and toluidine blue; *Luehea divaricata* (Bernardi-Wenzel et al., 2010) in Trypan Blue, etc. However, other staining techniques have also been available to assist in differentiating hyphae from plant tissues for investigation of fungal endophyte assemblages through microscopic investigations (Hood and Shew, 1996) (Table 10.2).

TABLE 10.2 Various stains used to locate endophytic fungi in different plant species.

Stain	Host plant/tissue	Reference
Lactophenol cotton blue	*Uromyces phaseoli; Catharanthus roseus; Tinospora cordifolia; Adhatoda vasica*	Heath (1974), Lakra et al. (2013), Mishra et al. (2015), Mishra et al. (2017)
Trypan blue & Rose Bengal	Turf and forage grasses	Saha et al. (1988)
Safranin & Calcofluor	Corn Tissue	Bandoni (1979), Trese and Loschke (1990)
Safranin & fast green staining	*Nassella neesiana*	Flemmer et al. (2010)
Safranin with Astra blue	white rot fungi	Srebotnik and Messner (1994)
KOH-Aniline blue	*Silene alba, Triticum aestivum, Solanum melongena*	Hood and Shew (1996)
Pianese III B Stain	Cassava	Hagen et al. (2003), Jurus and Sundberg (1976), Krajian (1940), Wilcox (1964)
Aniline blue	Rye grass	Neill (1940)
Trypan blue & Sudan IV	*Atriplex canescens*	Barrow and Aaltonen (2001)
Rhodamine B/ Methyl green method	*Acer pseudoplatanus*	Pearce (1984)
Chlorazole Black E (CBE)	*Eucalyptus* (AM Fungi)	Brundrett et al. (1996)

Isolation and identification of endophytic fungi

To begin with plant selection initially, the fresh samples were collected in beaker half filled with water and rinsed gently under running tap water to remove dust and debris in the laboratory (Selvanathan et al., 2011). Subsequently, plant parts excised into small pieces and placed in sealed plastic bags after removing the excessive moisture. Every endeavor is made to store the materials at 4 °C until isolation of fungi was done. Afterwards, plant materials were treated under sterile conditions in laminar air flow cabinet and were cut into small pieces 1–2 cm^2 (leaf), 1 cm lengthy (stem) followed by surface sterilization with 1% sodium hypochlorite and 90% ethanol, respectively (Hallmann et al., 2006; Gautam et al., 2013). Before the treatment with alcohol, traces of sodium hypochlorite were removed with a rinse in sterile distilled water which eliminates epiphytic bacteria, yeasts and fast-growing Zygomycetes (Arnold et al., 2000; Davis et al., 2003; Pawłowska et al., 2014). Surface sterilization was carried out following the modified protocol (Qin et al., 2009; Sarangthem and Momota, 2012; Potshangbam et al., 2017). However, number of reports has been available which describe various surface sterilization methods including the disinfectant, concentration and duration of treatment for various host plant (Table 10.3). Light and electron microscopy reveal the accurate internal locations of endophytic fungi (Kuldau and Yates, 2000; Gyaneshwar et al., 2001). After direct confirmation of the presence of endophytic fungi through histological analysis, endophytic fungi have been isolated from different symptomless parts such as stem, leaves and roots of medicinal plants. The tissues were then cut into smaller pieces of 0.5–1 cm^2 and inoculated on different microbiological media such as water agar, potato dextrose agar etc. For the emergence of fungal endophytes, the agar plates were incubated at 27 ± 2 °C and kept inverted in the incubator. For about one week from the incubation, hyphal tips of fungi emerging out from the plant tissues, were picked and grown on potato dextrose agar medium for developing pure cultures. After the proper incubation of the PDA plates sealed with parafilm, seven day old cultures were preserved by placing mycelium in 15% glycerol in deep freezer till further use (Mishra et al., 2017; Kannan et al., 2017; Mahfooz et al., 2017). The fungi were identified on the basis of their morphological and cultural characteristics (Gilman, 2001). Colonization frequency of each fungal endophyte was calculated (Kumar and Hyde, 2004, Khiralla et al., 2015). The fungi were then identified based on their morphological, biochemical and molecular characterization.

The identification of isolated endophytic fungi has been done on the basis of their morphological (Paul et al., 2012; Zakaria et al., 2016; Suradkar and Hande, 2017), biochemical (Jalgaonwala and Mahajan, 2011; Sunitha et al., 2013; Shubha and Srinivas, 2017) and molecular (Guo, 2010; Tian et al., 2013; Gonzalez-Teuber et al., 2017) approach. Further, on the basis of

TABLE 10.3 Methods of surface sterilization.

DC (disinfectant, concentration)	Duration of treatment	Host	Reference
Ethanol 75–96%; NaOCl 2–4% available Cl	1 min; 3–5 min	Conifer needles	Carroll and Carroll (1978)
Ethanol 96%; NaOCl, 2% available Cl, (1:2 bleach); ethanol 96%,	1 min; 7 min; 30 s	Conifer twigs	Petrini and Muller (1979)
Ethanol 96%; NaOCl 2.5% available Cl; ethanol 96%,	30 s; 1–3 min; 30 s	*Rhododendron*, *Vaccinium* leaves Lichens, mosses, ferns *Arctostaphylos* leaves	Petrini (1985) Petrini (1986) Widler and Muller (1984)
Ethanol 96%; NaOCl 3% available Cl; ethanol 70%,	1 min; 1–3 min; 30 s	*Sequoia* leaves	Espinosa-Garcia and Langenheim (1990)
Ethanol 95%; NaOCl 20% available Cl; ethanol 95%,	1 min; 3 min;30 s	*Pteridium* rhizomes, rachis, pinnules	Petrini et al. (1992)
Formaldehyde 37–40%, NaOCl, 10% available Cl	5 min	*Festuca* leaves and culms, *Anemone*, *Crataegus*, *Glechoma*, *Potentilla*, *Salix*, *Sorbus*, *Teucrium*, *Vaccinium leaves*	Schulz et al. (1993)
70% ethanol, 3%, sodium hypochloride solution	3 min	*Theobroma cacao* L.	Rubini et al. (2005)
70% ethanol15%, Hydrogenperoxide solution	1 min	*Heterosmilax japonica*	Gao et al. (2005)
70% ethanol, 5% in sodium Hypochloride solution	1 min; 5 min	3-NPA-endophytic strain	Chomcheon et al. (2005)
70% ethanol, 5.3% Sodium hypochloride solution	1 min; 5 min	Tea plant	Agusta et al. (2005)
75% ethanol, 6%, 3 or 5 min sodium hypochloride solution	1 min; 3–5 min	Mangrove species	Raviraja et al. (2006)

morphological characteristics, various fungi have been microscopically analyzed as colony diameter spread at fast rate in *Fusarium solani* from *Catharanthus roseus* (Kumar et al., 2013); conidiophores begin from the basal region in *Aspergillus* sp. from slender phialides both macro and micro conidia appears in *Fusarium* sp., septate hyphae forming brown geniculate conidiophores in *Curvularia* sp. Colonies are brown, hyaline conidiogenous cells in *Phoma* sp. from *Ricinus communis* were reported (Sandhu et al., 2014). Colonies are usually rapid growing, bright-colored cottony mycelium in *Fusarium oxysporum* and pinkish-color, clamp hyphae in *Fomitopsis* sp. from *Bacopa monnieri* (Katoch et al., 2014). Besides, the endophytic fungi were also characterized based on their biochemical characteristics from medicinal plant i.e. *Centella asiatica* fungi *Penicillium* sp. produced cellulase enzyme, which appeared as continuous yellow patches around the fungal colony (Devi et al., 2012). Further, *Talaromyces emersonii* produces amylase enzyme, which can be authenticated due to the presence of transparent zone around the fungal colony, this fungus also produces cellulolytic enzyme (a sign of appearance of yellow area at peripheral part of fungal colony), pectinase enzyme due to the indication of the availability of clear zone around the fungal colony, laccase enzyme produced by the fungus, which converts fungal zone with bluish appearance from medicinally important plant *Calophyllum inophyllum* and *Catharanthus roseus*, etc. (Sunitha et al., 2013). Further, identification of fungus up to species level needs molecular characterization, which is being done by using 18S, 5.8S, and 28S rRNA technology. Makimura et al. (1994) used primers to amplify specific regions within the 18S subunit for medically important fungi. Baker et al. (2004) reported that Eurotiomycetes class belongs to fungi were confirmed to be more affluent than Dothideomycetes class using 18S rRNA-targeted probes. However, Chen et al. (2000) reported that rRNA genes are extremely conserved the Internal transcribed spacer (ITS) regions and thus have divergent and distinctive characteristics. Therefore, rRNA genes are excellent candidates and understanding of their detection may be dramatically increased by the use of PCR. The transcriptional unit comprises of 5.8S, 18S and 28S rRNA genetic constituent. In phylogenetic studies, 18S rRNA potentially make precious markers for distinctive features among samples at advanced taxonomical levels for screening diverse groups of fungi (Meyer et al., 2010; Tang et al., 2012). Moreover, molecular systematic site of fungi significantly based on the investigation of nuclear ribosomal RNA (rRNA) cistron, comprises of small (18S) and large (28S) subunits (Hebert and Gregory, 2005). Furthermore, molecular characterization using fungal specific 18S rRNA primers helps in making phylogenetic relationship and characterization up to species level. Production of taxol by the fungal strain Gliocladium sp. from *Taxus baccata*, *Aspergillus japonicus* from *Rhynchosia beddomei* were authenticated using 18S rRNA technology (Sreekanth et al., 2011; Dandu et al., 2014). Subsequently, molecular techniques such as 18S rRNA have been successfully used for identifying endophytic fungi in recent time from the roots of *Chenopodium quinoa* i.e. *Penicillium*, *Phoma* and *Fusarium* (Gonzalez-Teuber et al., 2017). The isolated and identified endophytic fungi encompass further activities for the production of biological compounds and for mycosynthesis of nanoparticles. Besides, characterization of endophytic fungi through molecular means showed that the roots of *Musa acuminata* harbor a diverse group of endophytic fungi such as *Curvularia lunata, Trichoderma atroviride, Calonectria gracilis, Rhizoctonia solani, Bionectria ochroleuca*, and *Stromatoneurospora phoenix* (Zakaria et al., 2016). *Colletotrichum gloeosporioides* and *Pestalotiopsis virgatula* were reported by 18S rRNA gene sequencing from the leaves of *Vitex negundo* L. (Ramesh and Rajendran, 2017). Similarly, through molecular identification and phylogenetic analysis, a diverse group of endophytic fungi i.e. *Lophodermium conigenum*, Leotiomycetes sp., *Septoria pini-thunbergii* and Polyporales sp. were reported from needle of coniferous trees viz. *Pinus densiflora* and *Pinus koraiensis*, respectively.

Metabolites from endophytic fungi

In the early 1990s, a novel taxol-producing endophytic fungus *Taxomyces andreanae* was isolated from *Taxus brevifolia* (Strobel et al., 1993). Similarly, the endophytic fungus *Pestalotiopsis microspora* was isolated from *Taxus wallichiana* which was reported to produce taxol (Strobel et al., 1996). Since then, the search for new drugs from microbial origin started. In this respect, Koehn and Carter (2005) and Newman and Cragg (2007) reported many secondary metabolites of microbial origin, which showed potent pharmaceutical applications against various diseases viz. cancer, anti-inflammatory indications etc. At present, there is a need to search new ecological niches for potential sources of natural bioactive agents for different pharmaceutical, agriculture and industrial applications. The sources should be renewable, eco-friendly and easily accessible (Liu et al., 2001). From 1983 to 1994, over 60% of all approved cancer drugs at the pre-new drug application stage were of natural origin. Faulkner (2000) reported 2500 new metabolites from a diverse marine organism from 1977 to 1987. However, 840 novel structures recognized in 1998 has now been increased to 1003 in 2010 (Blunt et al., 2010; Newman and Cragg, 2004). Newman and Cragg (2016) has also reported 1562 novel chemical entities from 1981 to 1914, out of them 21% derived naturally, 16% are from biological macromolecules, 10% comprises the natural imitate entities, 9% drugs from the plants, 6% represent vaccines and 4% are unaffected natural products. These chemical compounds which have been derived from natural sources during 2011–14 reported to treat harmful diseases/ pathogens including antitumor, antibacterial agents. Moreover, 78% of all newly approved antibacterial agents were also of natural origin (Hyde, 2001). However, efforts for natural-product research have lost popularity in many major drug companies and in some cases have been replaced entirely by combinatorial chemistry, which is the automated synthesis of structurally related small molecules (Bills et al., 2002; Beutler, 2009; Dias et al., 2012). Further, Micafungin, an anti-fungal metabolite reported from *Coleophoma empetri* (Frattarelli et al., 2004), Rosuvastatin from *Penicillium citrinum* and *Penicillium brevicompactum*, have been used for treating Dyslipidemias (Scott et al., 2004). Furthermore, mycophenolate an product of *Penicillium brevicompactum*, which has been used for preventing rejection of renal transplant (Curran and Keating, 2005). Besides, soil fungi have been the most studied fungi and typical soil genera such as *Acremonium, Aspergillus, Fusarium* and *Penicillium* have shown ability to synthesize a diverse range of bioactive compounds. As of now, more than 30% of isolated secondary metabolites are produced from various species of *Aspergillus* and *Penicillium* (Berdy, 2005; Bladt et al., 2013; Stierle and Stierle, 2015; Singh et al., 2017). Out of them, at least 200,000 natural metabolites with various bioactive properties (Berdy, 2005). Endophytic fungi Xylaria sp. isolated from *Azadirachta indica* with oxygenated guaiane sesquiterpenes and isopimarane diterpenes encompass a good impact against *Candida albicans* and *Hormodendrum compactumii.* Likewise, *Emericella qaudrilineata* isolated from Pteris pellucid with benzyl benzoate showed antibacterial activity against *Staphylococcus. aureus* (gram positive) and *Aeromonas hydrophila* (gram negative), respectively (Wu et al., 2014; Goutam et al., 2016). Besides, crude natural products have also been applied directly as low cost drugs. They also provided the basic chemical architecture for deriving semi-synthetic natural products (Suryanarayanan et al., 2009). The access to the human population of new diseases such as AIDS and severe acute respiratory syndrome requires the discovery and development of new drugs to combat them. Not only the diseases such as AIDS require drugs that target them specifically, but new therapies for treating additional infections which are a consequence of a weakened immune system. During the recent time, there is a copious challenge which needs to be clarified. Concerning to this, goals are to be fixed to explore endophytic fungi to discover microbial populations, which favor plant growth and make them fit in external environment. They have emerged as a boon and left good impact on plants, environment and also human beings in numerous conceivable behaviors and are also found to have some important roles in nutrient cycling, biodegradation and bioremediation etc.

Biological roles of fungal endophytes

Endophytes play vital roles in various aspects of life varying from its effects on host plants to the environment and human life. Effects of endophytes on photosynthesis have been reported. e.g. *Colletotrichum musae* in banana decreased photo-chemical capacity as compared to endophyte-free plants (Pinto et al., 2000). Environmentally, the endophyte may be metabolically aggressive by affecting host defense chemicals (Cabral et al., 1993; Schulz et al., 1999). However, natural products from endophytic microbes for modern medicine, agriculture and industry may be effective to treat newly developing diseases in human beings, plants and animals, which may perhaps facilitate the production of appropriate drugs from microbial machinery (Strobel and Daisy, 2003). According to Kumar and Hyde (2004), nearly 343 endophytic fungal isolates representing 60 taxa with 30 morphotypes were reported from different parts of the Chinese medicinal plant i.e. *Tripterygium wilfordii*. However, fungal endophytic assemblage comprises cosmopolitan species such as *Colletotrichum* sp., *Glomerella* sp., *Phomopsis* sp. and *Phyllosticta* sp., etc. Additionally, few endophytes isolated from twig's xylem and bark probably may not be isolated from the roots, leaves and flowers. Consequently, the species symphony and frequency of endophyte species were reliant on the type of tissue. The dominant fungi isolated from different host tissues expressed a fair degree of recurrence. Further, endophytes may enhance growth by producing phytohormones without any apparent facilitation of host nutrient uptake. Hence, endophytic fungi also enhance biomass by producing growth hormones (Petrini, 1991; Schulz and Boyle, 2005). However, *Oryza sativa* (L. cv. Dongjin) mitigate production of phytohormones and organic acids under long-lasting heat-stress in endophytic fungus *Paecilomyces formosus* LWL1 belong to ascomycetes (Waqas et al., 2015). Likewise, *Porostereum spadiceum* AGH786 belongs to basidiomycetes alleviating salt stress in soybean (*Glycine max* (L.) Merril), exhibit production of potential gibberellins and also promote plant growth, induce abiotic stress forbearance and avert stress damage (Hamayun et al., 2017). Furthermore, commensalism and mutualism necessitate a sophisticated steadiness between the defence responses of the plant and the nutrient demand of the endophyte (Kogel et al., 2006). Moreover, endophytes have lofty ability to produce various novel and known enzymes which are used in various biotechnological applications like environmental applications for degradation of enzymes, medical applications and biotransformations of organic compounds (Firakova et al., 2007; Pimentel et al., 2011). Consequently, capability of endophytic fungi is also to fabricate new and interesting bioactive secondary metabolites, which are of pharmaceutical, industrial and agricultural significance. The various natural products produced by these endophytic fungi possess unique structures and reveal bioactivities against various pathogens. Endophytic fungi encompass various activities for the synthesis of natural compounds to reveal a number of biocidal activities which are as follows:

Nutrient pedaling and phyto-stimulation

Nutrient pedaling is a vital process that occurs continuously to balance nutrients and make them available for every component of the ecosystem. The biomass after its degradation into simple units becomes available in the form of nutrients to the environment especially to the food chain of organism. The major important roles are in biodegradation of the litter of its host plants (Muller et al., 2001; Kumaresan and Suryanarayanan, 2002; Osono, 2003; Osono 2006; Korkama-Rajala et al., 2008; Fukasawa et al., 2009; Promputtha et al., 2010). Endophytes also play significant roles in the uptake of essential nutrients from soil necessary for plant growth. They elicit uptake of nitrogen and in giant fescue adaptation to phosphorous deficiency (Arachevaleta et al., 1989; Gupta and Chaturvedi, 2019; Malinowski et al., 2000). A novel strain of fungus *Cladosporium sphaerospermum* isolated from the roots of *Glycine max* (L.) Merril, showed charisma for higher amounts of bioactive GA3, GA4 and GA7, which induced maximum plant growth in both rice and soybean varieties (Hamayun et al., 2009). The applications of endophytes are well documented such as anchorage of plant in soil, absorption of water and ions, nutrient storage and plant vegetative growth etc. (Berg and Smalla, 2009). Additionally, biocontrol prospective and phytostimulation of *Penicillium citrinum* and *Aspergillus terreus* against the stem rot caused by *Sclerotium rolfsii*, *Trichoderma* sp. in *Phaseolus vulgaris* has been reported (Waqas et al., 2015; Pierre et al., 2016). Another important activity is bioremediation, which is a method of removal of contaminants and wastes from the atmosphere by the use of micro-organisms. It relies on the life processes of microbes to breakdown the waste material. Microbes especially bioremediate toxic pollutants including metals, hydrocarbons, radioactive waste, PCBs (polychlorinated biphenyl's) and PAHs (polyaromatic hydrocarbons) with the help of specially fungal endophytes (Krishnamurthy and Naik, 2017). Moreover, the phytostimulant fungus *Mucor* sp. MHR-7strain isolated from *Brassica campestris* L. reported to produce indole 3-acetic acid (IAA) and solubilize phosphate and has great bioremediation potential (Zahoor et al., 2017). Further, heavy metals are also released into the soil and water through various industries, agriculture, unorganized urban development etc. and laid negative effects on the growth of plants (Robinson et al., 2003; Rosa et al., 2004). Some of the plant such as *Alyssum bertoloni, Alnus firma, Brassica napus, Nicotiana tabacum, Thlapsi caerulescens, Thlapsi goesingense* and *Solanum nigrum* can sustain in the heavy metal contaminated soil because of the presence of these metal resistant fungal endophytes viz. *Microspaeropsis*, *Mucor*, *Phoma*, *Alternaria*, *Peyronellaea*, *Steganosporium* and *Aspergillus* respectively (Li et al., 2012).

Anticancer activity

Paclitaxel and some of its derivatives represent the first major group of anticancer agents produced by fungal endophytes. The mode of action of paclitaxel is to preclude tubulin molecules from depolymerization during the processes of cell division (Schiff and Horwitz, 1980). It is the world's first billion dollar anticancer drug which is used to treat a number of human tissue proliferating diseases. *Taxomyces andreanae* serve as an alternative for taxol production by fermentation (Wang et al., 2000). Cytotoxic quinone dimer, torreyanic acid is another important anticancer drug produced from *Pestalotiopsis microspora* isolated from *Torreya taxifolia* (Lee et al., 1996). Few reports are also available which showed that *Hypocrea lixii*, novel endophytic fungi produced anticancer agent cajanol, isolated from *Cajanus cajan* (Zhao et al., 2013). Mucor fragilis isolated from *Sinopodophyllum hexandrum* was reported to produce podophyllotoxin and kaempferol (Huang et al., 2014), guanacastane diterpenoids reported from the plant endophytic fungus Cercospora sp. obtained from *Fallopia japonica* (Feng et al., 2014). Chen et al. (2015) also reported that cytochalasins produced from *Trichoderma harzianum* isolated from leaves of *Cola nitida* exhibited anticancer activity against hepatocellular carcinoma cell line, breast cancer cell line, pancreatic cancer cell line, human myeloid leukemia cell line and lung cancer cell line. These cytochalasins have been isolated from the fungi of the genera *Aspergillus, Phomopsis, Penicillium, Zygosporium, Chaetomium, Phoma, Xylaria, Hypoxylon* and *Rhinocladiella* and revealed antimicrobial, antitumor, anti-HIV and herbicidal activities (Wagenaar et al., 2000; Xu et al., 2009; Zhang et al., 2014). A number of other plants have also been reviewed which exhibited anticancer properties owing to the presence of particular endophytic fungi (Tables 10.4 and 10.5).

Antimicrobial properties

Drug resistance is a serious problem worldwide, particularly in hospital environments, contributing to the resurgence of more complex infections and deleterious socioeconomic effects (Barriere, 2015; Martins et al., 2013; Tillotson and Zinner, 2017). This situation is exacerbated by increasing number of immunocompromised patients, continuous flow of travelers and molecular mechanisms of resistance acquisition (Blair et al., 2015; Millar et al., 2015; Hall and Mah, 2017; Chagas et al., 2017). The increasing number of drug resistance due to the erst while factors created interest amidst scientists to search new and effective alternative antimicrobial agents. Accordingly, different sources of natural compounds have been explored in order to exploit the chemical diversity exist in nature. In conjugation, plants can serve as reservoirs for diverse microorganisms as symbionts and classified depending on the locus where on they are found and perform their functions. Likewise, endophytic fungi have higher potential to produce a wide variety of bioactive molecules such as alkaloids, peptides, steroids, terpenoids, phenols, quinones and flavonoids. These chemical varieties of fungal metabolites are associated with the inhibition of a wide variety of pathogenic microorganism's viz. *Staphylococcus aureus, Enterococcus faecium, Streptococcus pneumonia*, etc. (Deshmukh et al., 2015; Dos Santos et al.,

TABLE 10.4 Biological roles of endophytic fungi isolated from various plant species.

Host	Endophytic fungi isolated	Biological activity	References
Acanthus ilicifolius	*Zalerion maritimum, Zygosporium masonii, Cylindrocarpon* sp.	Used as a nerve tonic, stimulant expectorant & treatment of leucorrhea	Ananda and Sridhar (2002)
Acrostichum aureum	*Aspergillus* sp., *Pestalotiopsis* sp.	Antimicrobial activity and enzyme production	Maria et al. (2005)
Adathoda vasica	*Phyllosticta* sp.	Anticough and analgesic activities	Naik et al. (2008)
Adhatoda zeylanica	*Phomopsis pittospori* Yeast1	Antitussive, analgesic, antiulcer, abortifacient	Raviraja et al. (2006)
Aegle marmelos	*Chaetomium globosum, Aspergillus, Alternaria, Fusarium* sp.	Dyspepsia, sinusitis and chronic onstipation	Gond et al. (2007)
Alstonia scholaris	*Curvularia* sp., *Aspergillus* sp., *Fusarium* sp., *Colletotrichum* sp.	In Ayurveda used in treating skin disorders, malarial fever, urticaria, chronic dysentery, diarrhea and snake bite, Antimicrobial activity	Mahapatra and Banerjee (2010)
Anogeissus latifolia	*Glomerella* sp., *Sporomiella* sp.	Anti-ulcer and antimicrobial activities	Naik et al. (2008)
Apodytes dimidata	*Fusarium solani*	Anti-inflammatory, antioxidant activities, Anticancerous metabolites	Shweta et al. (2010)
Arachish ypogaea	*Leptosphaerulina crassiasca*	Anti-inflammatory, aphrodisiac and decoagulant, antifertility activity	Suryanarayanan and Murali (2006)

(*Continued*)

TABLE 10.4 (Continued)

Host	Endophytic fungi isolated	Biological activity	References
Calamus thwaitesii	*Rattania setulifera*	Anti-inflammatory and antimicrobial activities	Prabhugaonkar and Bhat (2009)
Callicarpa tomentosa	*Alternaria longipes, Curvularia pallescens Nigrospora oryzae Pithomyces* sp.	Used as depurgative, carminative	Raviraja et al. (2006)
Calotropis gigantea	*Phyllosticta* sp.	Anti-allergic and anti-leprosy	Naik et al. (2008)
Camellia caduca	*Penicillium* sp.	Extracts used as an astringent, digestive, carmi-native and diuretic herbal medicine Lipase production	Bhagobaty and Joshi (2012)
Canavalia cathartica	*Chaetomium globosum*	Used in the treatment of skin diseases.	Seena and Sridhar (2004)
Canavalia maritima	*Chaetomium globosum*	Effective as blood purifier and in treatment of skin diseases.	Seena and Sridhar (2004)
Careya arborea	*Glomerella* sp., *Sporomiella* sp.	Antidiarrhoeal activity & cures leucoderma	Murali et al. (2007)
Carissa carandas L.	*Kumbhamaya, Gonatobotryum*	Anticancer, antidiabetic, antipyretic activities	Jacob and Bhat (2000)
Carissa carandus	*Phyllosticta* sp.	Antifungal activity	Naik et al. (2008)
Cassia alata	*Phomopsis* sp.	Antibacterial activity	Naik et al. (2008)
Cassia fistula	*Alternaria alternata, Cladosporium cladosporioides*	Anti-tussive, anti-inflammatory and hypoglycemic activity	Naik et al. (2008)

(Continued)

TABLE 10.4 (Continued)

Host	Endophytic fungi isolated	Biological activity	References
Citrus medica	*Glomerella* sp.	Effective in treating skin disorders	Naik et al. (2008)
Clerodendrum serratum (L) Moon	Sterile isolate1	Alleviate symptoms of diabetes, obesity and hypertension	Raviraja et al. (2006)
Cordia obliqua	*Colletotrichum gloeosporioides, Lasiodiplodia theobromae,*	Antidiarrhoeal & anti-inflammatory activity	Murali et al. (2007)
Cordia wallichi	*Sporomiella* sp., *Colletotrichum gloeosporioides*	Antioxidant & antimicrobial activity	Murali et al. (2007)
Crataeva magna	*Fusarium verticilloides, Verticillium* sp., *Nigrospora oryzae*	Cures infantile dirrahoea and applied externally in piles	Nalini et al. (2005)
Dalbergia lanceolaria	*Phyllosticta capitalensis, Colletotrichum gloeosporioides*	Cures dyspepsia, diarrhea and leprosy	Naik et al. (2008)
Datura metel	*Phomopsis* sp.	Antifungal activity	Naik et al. (2008)
Dillenia indica	*Hypocrea* sp.	Anti-inflammatory and cardiotonic activity, Antimicrobial activity	Gogoi et al. (2008a, 2008b)
Diospyros Montana	*Phomopsis* sp., *Sporomiella minima*	Free radical scavenging activity	Naik et al. (2008)
Elaeodendron glaucum	*Alternaria alternata, Phomopsis* sp.	Antidiabetic, anti-inflammatory, emetic	Naik et al. (2008)
Ervatomia coronaria	*Colletotrichum* sp.	Effective in treating eye disorders	Naik et al. (2008)
Erythroxylon monogynum	*Glomerella* sp., *Colletotrichum gloeosporioides*	Anti-plasmodial and aphrodisiac activity	Naik et al. (2008)

(*Continued*)

TABLE 10.4 (Continued)

Host	Endophytic fungi isolated	Biological activity	References
Ficus benghalensis	*Colletotrichum gloeosporioides, Phoma* sp., *Phomopsis* sp.	Used to treat diarrhea, diabetes, ulcers, Skin diseases, gonorrhea and leucorrhea	Suryanarayanan and Vijaykrishna (2001)
Givotia rottleriformis	*Alternaria alternata, Phomopsis* sp.	Used as purgative	Naik et al. (2008)
Gmelina arborea	*Colletotrichum* sp., *Phomopsis* sp.	Cures anemia, leprosy, ulcers and vaginal discharge	Murali et al. (2007)
Grewia tiliaefolia	*Colletotrichum* sp., *Phomopsis* sp.	Antioxidant and antibacterial activity	Murali et al. (2007)
Halophila ovalis	*Cladosporium* sp., *Phoma* sp., *Fusarium* sp.	Antibacterial properties	Devarajan and Suryanarayanan (2002)
Helicteres isora	*Phomopsis* sp., *Phyllosticta capitalensis*	Antidiabetic & hypolipidemic activity	Murali et al. (2007)
Hibiscus rosa sinensis	*Curvularia* sp.	Treats gastrointestinal disorder	Naik et al. (2008)
Holarrhena antidysenterica	*Myrothecium cinctum, Humicola fuscoatra*	Cures ailments like anemia, epilepsy, stomach pain and cholera.	Tejesvi et al. (2006)
Hopea ponga	*Kumbhamaya jalapriya*	Treatment of piles and snakebite in Mangalore	Nair (2002)
Ipomoea pos-cprae	Deuteromycetes, Ascomycetes mycelia sterile	Anti-inflammatory, anti-diarrheal, analgesic	Beena et al. (2000)
Ixora coccinea	*Phyllosticta* sp.	Treats dysentary	Naik et al. (2008)
Ixora nigricans	*Pestalotiopsis* sp., *Phomopsis* sp.	Treats diarrhea	Murali et al. (2007)
Jatropha curcus	*Leptosphaeria* sp.	Treats Eczema	Naik et al. (2008)

(Continued)

TABLE 10.4 (Continued)

Host	Endophytic fungi isolated	Biological activity	References
Justicia adathoda	*Curvalaria tubercularia*	Exhibits strong respiratory stimulant activity, moderate hypotensive activity and cardiac-depressant effect, Antibacterial activity	Sowparthani and Rajagopal (2011)
Kydia calycina	*Phoma* sp., *Phomopsis* sp.	Hepatoprotective activity	Murali et al. (2007)
Lagerstroemia microcarpa	*Colletotrichum* sp., *Phyllosticta capitalensis*	Treats gastro-intestinal diseases	Murali et al. (2007)
Lagerstroemia parviflora	*Pestalotiopsis* sp., *Phomopsis* sp.	Anti-inflammatory activity	Murali et al. (2007)
Lantana camara	*Colletotrichum* sp.	Cures blood clots	Naik et al. (2008)
Launea sarmentosa	*Acremonium* sp.	Used in treatment of urinary problems and painful joints	Beena et al. (2000)
Lobelia nicotinifolia	*Fusarium oxysporum*	Treats bronchitis and Asthma	Raviraja et al. (2006)
Madhuka neriifolia (Moon) H.J.	*Trichoderma koningi* Oud.	Rheumatism and asthma	Raviraja et al. (2006)
Melia azedarach	*Aspergillus* sp., *Nigrospora* sp.	Antihelmintic, antilithic, diuretic & deobstruent	Shekhawat et al. (2010)
Mimusops elengi	Sterile mycelium	Treats urinary tract infection, dirrahoea and dysentery, Ergoflavin metabolite with anticancer activity	Deshmukh et al. (2009)
Naregamia alata	*Pestalotia disseminate*	Treatment of asthma and	Raviraja et al. (2006)
Naringi crenulata	*Glomerella* sp.,	Bronchitis, colic disorders,	Murali et al. (2007)
Nerium indicum	*Sporomiella* sp., *Phyllosticta* sp.	Antitumor and antioxidant activities,	Naik et al. (2008)

(*Continued*)

TABLE 10.4 (Continued)

Host	Endophytic fungi isolated	Biological activity	References
Nyctanthes arbor-tristis	*Alternaria alternata,*	Immunostimulant, camptothecin with anticancerous activity Antiphytopathogenic	Gond et al. (2012)
Ocimum sanctum	*Cladosporium cladosporioides, Nigrospora oryzae, Colletotrichum dematium, Chaetomium globosum Alternaria* sp., *Phoma sojicola*	hepatoprotective, antileishmanial, antiviral and antifungal activities. Antihyper-lipidemic, cardioprotective effects and antibacterial activity, against *Mycobacterium smegmatis,* cytotoxic activity	Bhagat et al. (2012)
Ocimum sanctum	*Colletotrichum* sp., *Alternaria* sp.	Antihyperlipidemic, Biodiesel feedstock	Dey et al. (2011)
Oryzae sativa	*Fusarium oxysporum Penicillium chrysogenum, Chaetomium globosum*	Cardioprotective effects and antibacterial activity urinary retention, diarrhea, colic and general weakness	Naik et al. (2009)
Osbeckia chinensis	*Paecilomyces variabilis*	Used in treatment of watery diarrhea, Amaylase production dysentery and Excessive sputum production in coughing	Bhagobaty and Joshi (2012)
Osbeckia stellata	*Mortierella hyalina*	Treatment of cuts and wounds and as a remedy for toothaches, Cellulose, protease production	Bhagobaty and Joshi (2012)
Ougeinia oougeinensis	*Colletotrichum* sp., *Phoma* sp.	Anthelmintic, antidiabetic & hypolipidemic activity	Murali et al. (2007)

(*Continued*)

TABLE 10.4 (Continued)

Host	Endophytic fungi isolated	Biological activity	References
Polycarpaea corymbosa	*Fusarium* sp., *Chaetomium globosum*	Heals wounds	Beena et al. (2000)
Porteresia coarctata	*Penicillium chrysogenum*	Anti-inflammatory and Inhibitory activity against *Vibrio cholerae*	Devi et al. (2012)
Potentilla fulgens L.	*Penicillium verruculosum*	Anti-tumor, antioxidant, Antiulcerogenic antihyperlipidemic activities	Bhagobaty et al. (2010)
Premna tomentosa	*Phomopsis* sp., *Glomerella* sp.	Immunomodulatory & hepatoprotective potential	Murali et al. (2007)
Punica granatum	*Phyllosticta* sp.	Treats dysentary	Naik et al. (2008)
Rhizophora apiculata	*Sporormiella minima*, *Acremonium* sp.	Antiviral, antioxidant, antimicrobial activities	Suryanarayanan et al. (1998)
Rhizophora mucronata	*Ascotricha chartarum*, *Phoma herbarum*, *Cladosporium psoraleae*	Hepatoprotective and antioxidant properties	Ananda and Sridhar (2002)
Rhizophora mucronata	*Sporormiella minima*	Anti-diabetic and anti-hyperglycemic property	Suryanarayanan et al. (1998)
Salvadora oleoides	*Acremonium* sp., *Aspergillus* sp. JPY2,	Analgesic, Antioxidant activity	Dhankhar et al. (2012)
Sapindus detergens	*Aspergillus* sp. JPY1, *Penicillium chrysogenum*, *Phoma* sp. *Exserohilim* sp.	Anticonvulsant, antibacterial, antimycotic, cytotoxic activity Anti-inflammatory, anti-	Bhagat et al. (2012)
Schima khasiana	*Penicillium* sp.	microbial properties and mucolytic agent, Used as an anthelmintic,	Bhagobaty and Joshi (2012)

(*Continued*)

TABLE 10.4 (Continued)

Host	Endophytic fungi isolated	Biological activity	References
		Inhibitory activity against Mycobacterium smegmatis, cytotoxic activity Lipase and protease production and rubefacient	
Sesbania bispinosa	*Fusarium oxysporum*	Antibacterial activity	Anita et al. (2009)
Sinopodophyllum hexandrum	*Aspergillus niger* *Trametes hirusta*	Treats dysentery, Purgative, antirheumatic, Podophyllotoxin and Aryl tetralin lignans, production	Puri et al. (2006)
Sonneratia caseolaris *Stereospermum angustifolium*	*Lulworthia grandispora*, *Haplosporella acacia*, *Dactylaria purpurella* *Phomopsis* sp.	Anti-tussive, arrests hemorrhage, used as vermifuge. Antihyperglycemic and antidiabetic	Ananda and Sridhar (2002), Murali et al. (2007)
Strychnos potatorum	*Colletotrichum gloeosporioides*, *Glomerella* sp.	Aphrodisiac, tonic &	Murali et al. (2007)
Syzygium cumini	*Alternaria alternata* *Colletotrichum gloeosporioides*, *Pestalotiopsis* sp.	diuretic Antimicrobial & antioxidant activity	Murali et al. (2007)
Taxus baccata	*Gliocladium* sp. *Pestalotiopsis oxyanthi*	Treats asthma and bronchitis, Paclitaxel and 10-DAB-III with anticancer activity Treats asthma and bronchitis; Improved Taxol production	Sreekanth et al. (2009), Sonaimuthu et al. (2010)
Taxus celebica *Taxus Wallichiana*	*Fusarium solani* *Fusarium* sp.	Anti-tumor activity, Paclitaxel production with anticancer activity Anti-tumor activity, Paclitaxel production	Chakravarthi et al. (2008), Gogoi et al. (2008a, 2008b)

(Continued)

TABLE 10.4 (Continued)

Host	Endophytic fungi isolated	Biological activity	References
Tectona grandis	*Phomopsis* sp., *Alternaria* sp.	Antihyperglycemic, with anticancer activity	Murali et al. (2007)
Terminalia arjuna *Terminalia chebula*	*Botryodiplodia theobromae*, *Cochlonema* sp., *Myrothecium verrucaria* *Pestalotiopsis* sp.	Analgesic and anti-inflammatory properties Extracts used as rejuvenative, laxative anthelmintic, nervine, expectorant	Tejesvi et al. (2006)
Terminalia crenulata	*Phomopsis* sp.	Treats gonorrhea and leucorrhea	Murali et al. (2007)
Toddalia asiatica	*Colletotrichum* sp.	Treats gastrointestinal disorder	Naik et al. (2008)
Tylophora indica	*Alternaria tenuissima*, *Thielavia subthermophila*., *Colletotrichum truncatum*, and *Chaetomium* sp.	Anti-asthmatic and anticancer activities, Antiphytopathogenic activity	Kumar et al. (2010)
Vitex negundo L.	*Papulospora* sp., *Curvularia clavata* *Chaetomium globosum* *Phyllosticta* sp.	Antimicrobial, analgesic and anti-inflammatory activities, Antibacterial activity	Raviraja et al. (2006), Naik et al. (2008)
Zizyphus xylopyrus	*Phomopsis* sp., *Alternaria alternata*	Diabetes, diarrhea and digestive disorders	Murali et al. (2007)

2015; Carvalho et al., 2016; Gokhale et al., 2017). Additionally, antibiotic resistance marks in higher medical costs, longlasting hospital stays and augmented death and is mounting to hazardously high levels in all parts of the world (Pinheiro et al., 2004; Pinheiro et al., 2017). Consequently, antimicrobial activities have been reported through the production of bioactive compound Chaetoglobosin A from endophytic fungi viz. *Chaetomium globosum* from *Nymphaea nouchali*, *Fusarium* sp. from *Cinnamomum mercadoi*, which

TABLE 10.5 Biological activities of natural compounds extracted from endophytic fungi isolated form plant species.

Host Plant	Endophytic fungi	Chemical Compound	Biological Activity	Reference
Torreya taxifolia	*Pestalotiopsis microspora*	Torreyanic acid	Anticancer	Lee et al. (1996)
Catharanthus roseus	*Mycelia sterilia, Chaetomium globosum, Aspergillus, Alternaria* sp.	Vincristine	Anticancer; Treats diabetes, malaria, Hodgkin's disease and leukemia	Yang et al. (2004); Kharwar et al. (2008)
Rhizophora annamalayana	*Fusarium oxysporum*	Taxol	Anticancer	Elavarasi et al. (2012)
Tinospora cordifolia	*Fusarium culmorum* SVJM072; *Phyllosticta capitalensis, Penicillium* sp., *Colletotrichum* sp., *Acremonium* sp., *Guignardia* sp.	Taxol	Anticancer; Antimicrobial and antioxidant properties; *Improved Taxol production*	Sonaimuthu et al. (2010); Mishra et al. (2012)
Annovas quamosa	*Penicillium* sp.	Meleargine and Chrysogine	Anticancer/ Antibacterial	Yunianto et al. (2014)
Tripterygium wilfordii	*Rhino cladiella* sp.	22-oxa-(12)-cytochalasins	Anticancer	Wagenaar et al. (2000)
Podophyllum hexandrum	*Trametes hirsute, Fusarium solani*	Podophyllotoxin	Anticancer Agent; Analgesic & anti-tumor Activity, Antimicrobial activity	Puri et al. (2006); Nadeem et al. (2012)

(*Continued*)

TABLE 10.5 (Continued)

Host Plant	Endophytic fungi	Chemical Compound	Biological Activity	Reference
Ocimum basilicum	*Phyllosticta* sp.	Taxol	Anticancer	Gangadevi and Muthumary (2007)
Ocimum basilicum	*2 L-5*	Ergosterol, Cerevesterol	Antimicrobial	Haque et al. (2005)
Justicia gendarussa	*Colletotrichum gloeosporioides (strain JGC-9)*	Taxol	Anticancer	Gangadevi and Muthumary (2008)
Camptotheca acuminate	*Fusarium solani*	Camptothecin, 9-methoxycamptothecin, 10-hydroxycamptothecin	Anticancer Properties	Kusari et al. (2009)
Solanum nigrum	*Aspergillus flavus*	Solamargine	Anticancer	El-Hawary et al. (2016)
Nothapodytes Foetida	*Entrophospora infrequens*	Camptothecin	Anticancer, cardiotonic	Puri et al. (2005)
Terminalia morobensis	*Pestalotiopsis microspora*	1,3-Dihydro isobenzofurans	Antioxidant	Harper et al. (2003)
Torreya mairei	*Aspergillus clavatonanicus*	Clavatol	Antimicrobial	Leuchtman (2003)
Melia azedarach	*Penicillium janthinellum; Aspergillus sp., Nigrospora sp.*	Citrinin (Polyketide)	Antimicrobial; Antihelmintic, antilithic, diuretic & deobstruent	Marinho et al. (2005); Shekhawat et al. (2010)

Erythrina cristagalli	*Phomopsis* sp.	Isoflavonoids	Antimicrobial	Redko et al. (2006)
Plumeria acutifolia	*Phomopsis* sp.	Terpenoid	Antimicrobial	Nithya and Muthumary (2010)
Urospermum picroides *Kennedia nigriscans*	*Ampelomyces* sp. *Steptomyces* NRRL 30562	3-Omethylalaternin & altersolanol A Munumbicin A, B, C and D.	Antimicrobial	Miller et al. (1998)
Tripterigium wilfordii	*Crypto sporiopsis* sp.	Cryptocandin	Antimicrobial	Strobel et al. (1999)
Spondias mombin	*Guignardia* sp. *Phomopsis* sp. *Pestalotiopsis guepinii*	Phomopsidin & Phomopsichalasin	Antimicrobial Anticancer, Antimalarial	Kobayashi et al. (1995) Phongpaichit et al. (2007)
Garcinia sp.	*Phomopsis* sp.	Phomoxanthone A & B	Antimycobacterial Drug	Isaka et al. (2001)
Ananas ananassoides	*Muscodar crispans*	Volatile organic compounds/Propanoic acid, methyl ester,2-methyl butyl ester, Ethanol.	Antibacterial	Mitchell et al. (2010)
Aegiceras corniculatum	*Emericella* sp.	Aegicerascornicuiatum	Anti-viral	Zhang et al. (2011)
Scapania ciliate (Liverwort)	*Aspergillus sydowii*	Sydoxanthone A, B	Immunosuppresive activity	Song et al. (2013)

(*Continued*)

TABLE 10.5 (Continued)

Host Plant	Endophytic fungi	Chemical Compound	Biological Activity	Reference
Garcinia hombroniana	*Guignardia bidwelli* PSU-G11	Guignarenones (A-D)	Cytotoxic activity	Sommart et al. (2012)
Viscum album (Epiphytic parasite)	*Aspergillus flavus, Fusarium oxysporum, Fusarium moniliforme*	Lectin	Anticancer/Antioxidant	Sadananda et al. (2014)
Cinnamomum zeglanicum	*Muscodor albus*	1-Butanol, 3-methyl-, acetate	Antimicrobial	Strobel (2003)
Aphelandra tetragona	*Fusarium sambucinum Plectosporium tabacinum, Gliocladiumcibotii, Chaetosphaeria* sp.	*N*-(2-hydroxyphenyl) acetamide *N*-(2-hydroxyphenyl) malonamic acid.	Bioremedition	Zikmundova et al. (2002)
Terminalia morobensis	*Pestalotiopsismicrospora*	1,3-Dihydroisobenzofurans	Antioxidant activity	Harper et al. (2003)
Trachelospermum jasminoides	*Cephalosporium* sp. *IFB-E001*	"Graphislactone A"	Antioxidant Activity	Song et al. (2005)
Erythrina crista-galli	*Phomopsis* sp.	Isoflavonoids	Antimicrobial Activity	Redko et al. (2006)
Guazuma ulmifolia	*Muscodor albus E-6*	Caryophyllene, phenylethyl alcohol, 2-phenylethyl ester, bulnesene	Antibiotic Activity	Strobel et al. (2008)
Piptadenia adiantoides J. FMacbr	*Cochliobolus* sp. *(UFMGCB-555)*	Cochlioquinone A, isocochlioquinone A.	Anti-parasitical properties	Campos et al. (2009)

Azadirachta indica A. Juss	*Chloridium* sp.	Javanicin	Antibacterial Activity	Kharwar et al. (2009)
Eucryphia cordifolia	*Gliocladium roseum (NRRL 50072)*	2,6-Dimethyl, 3,3,5-trimethyl; cyclohexene, 4-methyl; decane, 3,3,6-trimethyl; and undecane, 4,4-dimethyl (Volatile hydrocarbons)	Bioremediation	Strobel et al. (2008)
Salvia officinalis	*Chaetomium* sp.	Cochliodinol, isocochliodinol	Cytotoxic Activity	Debbab et al. (2009)
Gastrodia Elata	*Armillaria mellea*	Sesquiterpene aryl esters	Antimicrobial Activity	Gao et al. (2009)
Plumeria acutifolia	*Phomopsis* sp.	Terpenoid	Antimicrobial Activity	Nithya and Muthumary (2010)
Costus speciosus	*Bipolaris sorokiniana*	Sesquiterpenes helminthosporal acid and helminthosporol	Antifungal	Qader et al. (2017)
Madhuca longifolia	*Colletotrichum* sp., *Alternaria* sp., *Pestalotiopsis* sp.	Crude extract	Antimicrobial/ Antibacterial Activity	Kuralarasi and Lingakumar (2018)

has been reported to show biocidal activities against *Bacillus subtilis*, *Staphylococcus aureus*, *Pseudomonas aeruginosa*, *Escherichia coli*, etc. (Dissanayake et al., 2016; Marcellano et al., 2017). Similarly, *Acremonium falciforme* isolated from *Litsea cubeba* reported highest zone of inhibition through biocidal activity against different test pathogens (Deka and Jha, 2018). Asides, a number of antimicrobial activities have been reported using the natural products obtained from endophytic fungi isolated from various host plants (Tables 10.4 and 10.5).

Antidiabetic activity

Diabetes is a major cause of blindness, kidney failure, heart attacks, stroke and lower limb amputation. The number of diabetic patients arose from 108 million in 1980 to 422 million in 2014 and it is expected that the number of diabetic patients will increase up to 9.9% of the adult population by the year of 2045 (www.statista.com/statistics/271464/Percentage of diabetes worldwide). Increase in numbers related to an aging population with diabetes: over the age of 65, 30% of the people are diabetic (123 million people out of 425 million people) and by 2045 figure will increase to 40% (253 million people out of 629 million people) added to the edition of the Atlas. To reduce avoidable mortality from diabetes, World Health Organization (WHO) global action plan for prevention and control of Noncommunicable diseases (2013–20) was came into force. The aim of this plan was to draw attention to the public health towards challenges of diabetes to generate momentum for national, regional and global action. Concurrently, several efforts have been made to cure diabetes using natural products produced by endophytic fungi apart from the existing treatments. Subsequently, a nonpeptidal fungal metabolite [L-783] was isolated from an endophytic fungus Pseudomasaris sp. collected from an African rainforest near Kinshasa in the Democratic Republic of the Congo. Besides, the nature has provided plentiful natural resources, which can be explored for their medicinal uses. The antidiabetic and hypolipidemic activity of the endophytic fungi isolated from *Salvadora oleoides* (Salvadoraceae) in glucose loaded, fasting and alloxan induced diabetic Wistar albino rats was reported (Dhankhar et al., 2013). They also investigated new antidiabetic drugs from fungal endophytes such as *Aspergillus* sp., *Phoma* sp. and some unidentified species which significantly reduced blood glucose level by glucose tolerance test. Moreover, αamylase inhibitor retards the liberation of glucose from dietary complex carbohydrates and delays the absorption of glucose to isolate and select α-amylase inhibitor-producing endophytic actinomycetes from the leaves and stem of *Leucas ciliata* and *Rauwolfia densiflora*, two of the well-known medicinal plants used in the treatment for diabetes (Akshatha et al., 2014). Likewise, the endophytic fungi *Penicillium* species reported from *Tabebuia argentea* produces octadecanoic acid, which has been reported to induce diabetic proteins and inactivate its activity (Murugan et al., 2017). Further, crude extract of endophytic fungi from *Cassia siamea*, *Syncephalastrum racemosum* from seaweed *Gracilaria corticata*, *Aspergillus* sp. from *Ficus religiosa* showed inhibition for α-amylase activity and indicate antidiabetic properties (Munim et al., 2013; Ushasri and Anusha, 2015; Tiwari et al., 2017). In addition to the elaborated activities, other endophytic fungi and their natural compounds were also reported to show the antidiabetic activities (Tables 10.4 and 10.5).

Immunosuppressive activity

Immunosuppression is a reduction of the activation of immune system. Some portion of the immune system itself has immunosuppressive effects on other parts of immune system. Immunosupression may also be induced with drugs in preparation of bone marrow to prevent the rejection of a transplant. An endophytic fungus *Fusarium subglutinans* isolated from *Tripterygium wilfordii* reported to produce subglutinol A and B, which act as the immunosuppressive agent. These drugs are used now days to prevent allograft rejection in transplant patients and in near future these could be used to treat autoimmune diseases such as rheumatoid arthritis and insulin dependent diabetes (Kumar et al., 2005; Padhi et al., 2013). Further, pseudomycin is an antifungal compound, which was very effective against human pathogen *Candida albicans* which cause Oropharyngeal candidiasis in human beings (Zhang et al., 2001). This is a peptide antibiotic containing unusual aminoacids like L-hydroxy aspartic acid, Lchlorothreonine and both D- and L-diaminobutyric acid (Castillo et al., 2003). In addition, ambuic acid is a cyclohexenone belongs to the family of pseudomycin isolated from *Pestalotiopsis microspora* and found effective against Gram-positive bacteria (Li and Strobel, 2001). Moreover, secondary metabolites viz., oxysporidinone, 3-hydroxyl-2-piperidinone, 4,6-anhydrooxysporidinone, fusarinolic acid, gibepyrone D, beauvercin, cerevisterol, fusaruside, 4-hydroxybenzoic acid, indole-3-acetic acid and gibepyrone D isolated from *Fusarium oxysporum*, subglutinol A and B isolated from *Fusarium subglutians* endophytic fungus residing within *Tripterygium wilfordii* reported to show immuosuppressive activity (Lee et al., 1995; Wang et al., 2011; Bogner et al., 2017).

Antiviral activity

The charming use of antibiotic products from endophytic fungi is the inhibition of viral growth. Two novel human cytomegalovirus protease inhibitors, cytonic acid A and B isolated from endophytic fungus *Cytonaema* sp. were elucidated by mass spectrometry and NMR methods are found to be effective against virus growth (Guo et al., 2000; Harper et al., 2001). Some metabolites from endophytic fungi of desert plants served as a viable source for identifying potent inhibitors of H1V1 replication (Wellensiek et al., 2013). Besides, Brefeldin A inhibited dengue viruses and other related viruses including Zika virus and Japanese encephalitis virus isolated from *Penicillium* sp. (FKI-7127 strain) (Raekiansyah et al., 2017). The increasing positive result of antiviral agents from fungal endophytes open new vistas amongst scientists to further research in this direction and reduce the uncurable viral diseases.

Besides, a variety of biological activities of isolated natural products from endophytic fungi isolated from different plants have been presented (Table 10.5).

Conclusion

Endophytism, in its action mechanism and ecological aspects is less understood which have arisen as one of the highly appreciated natural phenomenon recently in agriculture and forestry. Keeping in view, the increased awareness about the key roles of fungal endophytes in various domains and increasing scientific interest and government support, the diverse fungal endophytes has been coming into light. These micro entities efficiently impact on plant biology that accelerate quality production of various chemical compounds either irrespective of plant growth hormones or by stimulating their production, modifying secondary metabolite pathways, and gene expression inflection in the host plant. They can interact with host plant specifically and in coordinated manner as well. In this review, it is concluded that the plant–endophyte interaction is most efficient way to improve plant sustainability. Reported biological roles of endophytes suggests intimate association with plants, which helps increased crop produce, high secretion of metabolites and adaptability against biotic-abiotic stress.

The complex interaction of host plants with diverse group of microbial populations affects plant's ecological physiology. This association facilitates enhanced nutrient supply by providing sustainable flow of micro and macro elements in the host plants and also help developing phytohormones by the plant which are important for the plant growth & development. They facilitate the nutrient-cycle in the soil via releasing extracellular enzymes like phosphatase, amylase, lipase, cellulase, protease, urease, chitinase, etc. and thus mediate organic matter mineralization. These enzymes present in soil and their activities are key indicators of quality, value and strength of soil. This way, endophytic association are impactful in reducing use of chemical boosters used conventionally and thus benefitting the ecosystem and human health. In addition, they are reported to inhibit the pathogenic attacks by triggering defense hormones, even if the pathogenic elements are absent. Owing to the exceptional properties and their capability of growing fast, these microorganisms can be easily substantiated. Enzyme systems of these endophytes catalyze wide chemical reactions which are not reproducible in synthetic manner. They behave as potential drug producers and contribute to various processes leading to biological transformations. According to an estimated market understanding of Indian herbal commerce by National Medicinal Plant Board (New Delhi) India, it may grow up to Rs. 80–90 billion by 2020. Moreover, plants are the most abundant source of pharmaceuticals & nutraceuticals to the health industries. Association of host plants with endophytic fungi show significant importance in modifying the plant secondary metabolites, having nutritive and remedial properties. These interactions enhance phytoconstituents production more efficiently and ultimately contribute to economical turnover. Based on the wide literature survey, it can be clearly observed that plant-microbe interaction in natural environment are vital to their multifarious use in many domains. Although, endophytic interactions in natural environment are critical for plant development and phytoremediation processes, additional efforts are still prerequisite to completely understand the molecular mechanism behind the beneficial applications provided by endophytism.

References

Agusta, A., Maehara, S., Ohashi, K., Simanjuntak, P., Shibuya, H., 2005. Stereoselective oxidation at C-4 of flavans by the endophytic fungus *Diaporthe* sp. isolated from a tea plant. Chem Pharm Bull. 53 (12), 1565–1569.

Akshatha, V.J., Nalini, M.S., D'souza, C., Prakash, H.S., 2014. Streptomycete endophytes from anti-diabetic medicinal plants of the Western Ghats inhibit alpha-amylase and promote glucose uptake. Lett. Appl. Microbiol. 58 (5), 433–439.

Ananda, K., Sridhar, K.R., 2002. Diversity of endophytic fungi in the roots of mangrove species in the west coast of India. Can. J. Microbiol. 48 (10), 871–878.

Anita, D.D., Sridhar, K.R., Bhat, R., 2009. Diversity of fungi associated with mangrove legume Sesbania bispinosa (Jacq.) W. Wight (Fabaceae). Livest. Res. Rural. Dev 21 (5), 1–15.

Arachevaleta, M., Bacon, C.W., Hoveland, C.S., Radcliffe, D.E., 1989. Effect of the tall fescue endophyte on plant response to environmental stress. Agron. J. 81 (1), 83–90.

Arivudainambi, U.E., Anand, T.D., Shanmugaiah, V., Karunakaran, C., Rajendran, A., 2011. Novel bioactive metabolites producing endophytic fungus Colletotrichum gloeosporioides against multidrug-resistant Staphylococcus aureus. FEMS Immunol. Med. Microbiol. 61 (3), 340–345.

Armstrong, J.L., Knudsen, G.R., Seidler, R.J., 1987. Microcosm method to assess survival of recombinant bacteria associated with plants and herbivorous insects. Curr. Microbiol. 15 (4), 229–232.

Arnold, A.E., Lutzoni, F., 2007. Diversity and host range of foliar fungal endophytes: are tropical leaves biodiversity hotspots? Ecology 88 (3), 541–549.

Arnold, A.E., Maynard, Z., Gilbert, G.S., Coley, P.D., Kursar, T.A., 2000. Are tropical fungal endophytes hyperdiverse? Ecol. Lett. 3 (4), 267–274.

Bacon, C.W., Siegel, M.R., 1988. Endophyte parasitism of tall fescue. J. Prod. Agric. 1 (1), 45–55.

Bajpai, V.K., Agrawal, P., Bang, B.H., Park, Y.H., 2015. Phytochemical analysis, antioxidant and antilipid peroxidation effects of a medicinal plant, *Adhatoda vasica*. Front Life Sci. 8 (3), 305–312.

Baker, B.J., Lutz, M.A., Dawson, S.C., Bond, P.L., Banfield, J.F., 2004. Metabolically active eukaryotic communities in extremely acidic mine drainage. Appl. Environ. Microbiol. 70 (10), 6264–6271.

Bandoni, R.J., 1979. Safranin O as a rapid nuclear stain for fungi. *Mycologia*, 11 (4), 873.

Barriere, S.L., 2015. Clinical, economic and societal impact of antibiotic resistance. Expert. Opin. Pharmacotherapy 6 (2), 151–153.

Barrow, J., Aaltonen, R., 2001. Evaluation of the internal colonization of *Atriplex canescens* (Pursh) Nutt. roots by dark septate fungi and the influence of host physiological activity. Mycorrhiza 11 (4), 199–205.

Bayman, P., 2007. Fungal endophytes. In: Kubicek, C.P., Druzhinina, I.S. (Eds.), The Mycota Environmental and Microbial Relationships, 4. pp. 213–227.

Beena, K.R., Ananda, K., Sridhar, K.R., 2000. Fungal endophytes of three sand dune plant species of west coast of India. Sydowia 52 (1), 1–9.

Bennett, J.A., Cahill, J.F., 2016. Fungal effects on plant–plant interactions contribute to grassland plant abundances: evidence from the field. J. Ecol. 104 (3), 755–764.

Bentham, G., Hooker, J.D., 1862–1883. Genera Plantarum, 3.

Berdy, J., 2005. Bioactive microbial metabolites. J. antibiotics 58 (1), 1.

Berg, G., Smalla, K., 2009. Plant species and soil type cooperatively shape the structure and function of microbial communities in the rhizosphere. FEMS Microbiol. Ecol. 68 (1), 113.

Berg, M., Kiers, E., Driessen, G., Van Der Heijden, M., Kooi, B.W., Kuenen, F., et al., 2010. Adapt or disperse: understanding species persistence in a changing world. Glob. Change Biol. 16 (2), 587–598.

Bernardi-Wenzel, J., García, A., Celso Filho, J.R., Prioli, A.J., Pamphile, J.A., 2010. Evaluation of foliar fungal endophyte diversity and colonization of medicinal plant Luehea divaricata (Martius et Zuccarini). Biol. Res. 43 (4), 375–384.

Bettucci, L., Saravay, M., 1993. Endophytic fungi of Eucalyptus globulus: a preliminary study. Mycol. Res. 97 (6), 679–682.

Beutler, J.A., 2009. Natural products as a foundation for drug discovery. Curr. Protoc. pharmacology 46, 9–11.

Bhagat, J., Kaur, A., Sharma, M., Saxena, A.K., Chadha, B.S., 2012. Molecular and functional characterization of endophytic fungi from traditional medicinal plants. World J. Microbiol. Biotechnol. 28 (3), 963–971.

Bhagobaty, R.K., Joshi, S.R., 2012. Enzymatic activity of fungi endophytic on five medicinal plant species of the pristine sacred forests of Meghalaya, India. Biotechnology and bioprocess engineering 17 (1), 33–40.

Bhagobaty, R.K., Joshi, S.R., Kumar, R., 2010. Penicillium verruculosum RS7PF: a root fungal endophyte associated with an ethno-medicinal plant of the indigenous tribes of Eastern India. Afr. J. Microbiol. Res 4 (9), 766–770.

Bharathi, T.R., Prakash, H.S., 2018. Endophytic fungi isolation and identification from Memecylon species of Western Ghats, India. Int. J. Phytopharmacol. 8 (1), 06–11.

Bills, G., Dombrowski, A., Pelaez, F., Polishook, J., An, Z., 2002. Recent and future discoveries of pharmacologically active metabolites from tropical fungi. In: Watling, R., Ainsworth, J.C. A.M., Issac, S., Robinson, C.H. (Eds.), Tropical Mycology: Micromycetes, 2. CABI Publishing, New York, N.Y, pp. 165–194. , 2002.

Bladt, T.T., Frisvad, J.C., Knudsen, B., Larsen, T.O., 2013. Anticancer and antifungal compounds from Aspergillus, Penicillium and other filamentous fungi. Molecules 18 (9), 11338–11376.

Blair, J.M., Webber, M.A., Baylay, A.J., Ogbolu, D.O., Piddock, L.J., 2015. Molecular mechanisms of antibiotic resistance. Nat. Rev. Microbiol. 13 (1), 42.

Blunt, J.W., Copp, B.R., Munro, M.H.G., Northcote, P.T., Prinsep, M.R., 2010. Marine natural products. Nat. Prod. Rep. 27, 165–237.

Bogner, C.W., Kamdem, R.S., Sichtermann, G., Matthaus, C., Holscher, D., Popp, J., et al., 2017. Bioactive secondary metabolites with multiple activities from a fungal endophyte. Microb. Biotechnol. 10 (1), 175–188.

Brundrett, M., Bougher, N., Dell, B., Grove, T., Malajczuk, N., 1996. Working with Mycorrhiza For. Agric. ACIAR, Canberra.

Cabral, D., Stone, J.K., Carroll, G.C., 1993. The internal mycobiota of Juncus s: microscopic and cultural observations of infection patterns. Mycol. Res. 97 (3), 367–376.

Carvalho, J.M., Paixao, L., Dolabela, M.F., Marinho, S.B., Marinho, A., 2016. Phytosterols isolated from endophytic fungus Colletotrichum gloeosporioides (Melanconiaceae). Acta Amazon 46 (1), 69–72.

Campos, F.F., Rosa, L.H., Cota, B.B., Caligiorne, R.B., Rabello, A.L.T., Alves, T.M.A., Rosa, C.A., Zani, C.L., 2009. Correction: Leishmanicidal Metabolites from *Cochliobolus* sp., an Endophytic Fungus Isolated from *Piptadenia adiantoides* (Fabaceae). PLoS Neglect Trop. D 3 (1), 1–2.

Carroll, G.C., Carroll, F.E., 1978. Studies on the incidence of coniferous needle endophytes in the Pacific Northwest. Can. J. Bot. 56 (24), 3034–3043.

Castillo, U., Harper, J.K., Strobel, G.A., Sears, J., Alesi, K., Ford, E., et al., 2003. Kakadumycins, novel antibiotics from Streptomyces sp. NRRL 30566, an endophyte of Grevillea pteridifolia. FEMS Microbiol. Lett. 224 (2), 183–190.

Chagas, M.B., dos Santos, I., da Silva, L.C.N., dos Santos Correia, M.T., de Araújo, J.M., da Silva Cavalcanti, M., et al., 2017. Antimicrobial activity of cultivable endophytic fungi associated with Hancornia Speciosa Gomes Bark. open. microbiology J. 11, 179.

Chakravarthi, B.V.S.K., Das, P., Surendranath, K., Karande, A.A., Jayabaskaran, C., 2008. Production of paclitaxel by Fusarium solani isolated from Taxus celebica. J. Biosci. 33 (2), 259–267.

Chauhan, N.M., Gutama, A.D., Aysa, A., 2019. Endophytic fungal diversity isolated from different agro-ecosystem of Enset (Ensete ventericosum) in Gedeo zone, SNNPRS, Ethiopia. BMC Microbiol. 19, 172.

Chen, Y.C., Eisner, J.D., Kattar, M.M., Rassoulian-Barrett, S.L., LaFe, K., Yarfitz, S.L., et al., 2000. Identification of medically important yeasts using PCR-based detection of DNA sequence polymorphisms in the internal transcribed spacer 2 region of the rRNA genes. J. Clin. Microbiol. 38 (6), 2302–2310.

Chen, Z.M., Chen, H., Li, Y., Feng, T., Liu, J.K., 2015. Cytochalasins from cultures of endophytic fungus Phoma multirostrata EA-12. J. Antibiot. 68 (1), 23.

Chomcheon, P., Wiyakrutta, S., Sriubolmas, N., Ngamrojanavanich, N., Isarangkul, D., Kittakoop, P., 2005. 3-Nitropropionic acid (3-NPA), a potent antimycobacterial agent from endophytic fungi: is 3-NPA in some plants produced by endophytes? J. Nat. Prod. 68 (7), 1103–1105.

Chouhan, S., Bhatt, S., Dhyani, S., 2013. Preliminary phytochemical screening of secondary metabolites of Adhatoda vasica Nees flowers. Glob. J. Res. Med Plants Indigenous Med. 2 (5), 311–316.

Claeson, U., Malmfors, T., Wikman, G., Bruhn, J.G., 2000. Adhatoda vasica: a critical review of ethnopharmacological and toxicological data. J. Ethno pharmacology 72 (1-2), 1–20.

Cosme, M., Lu, J., Erb, M., Stout, M.J., Franken, Wurst, S., 2016. A fungal endophyte helps plants to tolerate root herbivory through changes in gibberellin and jasmonate signaling. N. Phytol. 211 (3), 1065–1076.

Curran, M., Keating, G.M., 2005. Mycophenolate sodium delayed release. Drugs 65 (6), 799–805.

Dandu, A., Reddy, N.V., Nagam, V., Bobbu, Tartte, V., 2014. Molecular identification of endophytic fungi isolated from Rhynchosia beddomei, an endemic medicinal plant of Tirumala hills. Res. J. Pharmaceut. Biol. Chem. Sci. 5 (3), 1280–1285.

Davis, E.C., Franklin, J.B., Shaw, A.J., Vilgalys, R., 2003. Endophytic Xylaria (Xylariaceae) among liverworts and angiosperms: phylogenetics, distribution, and symbiosis. Am. J. Bot. 90 (11), 1661–1667.

Debbab, A., Aly, H.A., Edrada-Ebel, R.A., Müller, W.E., Mosaddak, M., Hakiki, A., Ebel, R., Proksch, P., 2009. Bioactive secondary metabolites from the endophytic fungus *Chaetomium* sp. isolated from *Salvia officinalis* growing in Morocco. *Biotechnologie*. Agronomie, Societe et Environnement 13 (2), 229–234.

Deka, D., Jha, D.K., 2018. Antimicrobial activity of endophytic fungi from leaves and barks of Litsea cubeba Pers., a traditionally important medicinal plant of North East India. Jordan. J. Biol. Sci. 11 (1), 73–79.

Desai, S., Metrani, R., Vantamuri, S., Ginigeri, V., Phadke, K., Hungund, B., 2012. Phytochemical analysis, antimicrobial and antitumour screening of endophytes of Tinospora cordifolia. Methods 1 (4), 5–6.

Deshmukh, S.K., Mishra, D., Kulkarni-Almeida, A., Verekar, S., Sahoo, M.R., Periyasamy, G., et al., 2009. Anti-inflammatory and anticancer activity of ergoflavin isolated from an endophytic fungus. Chem. Biodivers. 6 (5), 784–789.

Deshmukh, S.K., Verekar, S.A., Bhave, S.V., 2015. Endophytic fungi: a reservoir of antibacterials. Front. microbiology 5, 715.

Devarajan, P.T., Suryanarayanan, T.S., 2002. Endophytic fungi associated with the tropical seagrass Halophila ovalis (Hydrocharitaceae). Indian J. Mar. Sci. 31, 73–74.

Devi, N.N., Prabakaran, J.J., Wahab, F., 2012. Phytochemical analysis and enzyme analysis of endophytic fungi from Centella asiatica. Asian Pac. J. Trop. Biomed. 2 (3), 1280–S1284.

Devprakash, S.K., Subburaju, T., Gurav, S., Singh, S., 2011. Tinospora cordifolia: a review on its ethnobotany, phytochemical & pharmacological profile. Asian J. Biochem. Pharm. Res. 4 (1), 291–302.

Dey, P., Banerjee, J., Maiti, M.K., 2011. Comparative lipid profiling of two endophytic fungal isolates–Colletotrichum sp. and Alternaria sp. having potential utilities as biodiesel feedstock. Bioresour. Technol. 102 (10), 5815–5823.

Dhankhar, S., Dhankhar, S., Kumar, S., Parkash Yadav, J., 2012. A novel and significant method for antioxidant activity utilizing microtitre-plate (resazurin reducing power assay). Curr. Chem. Biol. 6 (1), 70–76.

Dhankhar, S., Dhankhar, S., Parkash Yadav, J., 2013. Investigations towards new antidiabetic drugs from fungal endophytes associated with Salvadora oleoides Decne. Med. Chem. 9 (4), 624–632.

Dias, D.A., Urban, S., Roessner, U., 2012. A historical overview of natural products in drug discovery. Metabolites 2 (2), 303–336.

Dissanayake, R.K., Ratnaweera, B., Williams, D.E., Wijayarathne, C.D., Wijesundera, R.L., Andersen, R.J., et al., 2016. Antimicrobial activities of endophytic fungi of the Sri Lankan aquatic plant Nymphaea nouchali and chaetoglobosin A and C, produced by the endophytic fungus Chaetomium globosum. Mycology 7 (1), 1–8.

Dos Santos, I., Silva, L., Silva, M., Araujo, J., Cavalcanti, M., Lima, V., 2015. Antibacterial activity of endophytic fungi from leaves of Indigofera suffruticosa Miller (Fabaceae). Front. microbiology 6, 350.

Dostalek, T., Munzbergova, Z., Kladivova, A., Macel, M., 2016. Plant–soil feedback in native vs. invasive populations of a range expanding plant. Plant Soil 399 (1-2), 209–220.

Dubois, T., Gold, C.S., Coyne, D., Paparu, P., Mukwaba, E., Athman, S., et al., 2004. Merging biotechnology with biological control: Banana Musa tissue culture plants enhanced by endophytic fungi. Uganda J. Agric. Sci. 9 (1), 445–451.

Elavarasi, A., Rathna, G.S., Kalaiselvam, M., 2012. Taxol producing mangrove endophytic fungi *Fusarium oxysporum* from *Rhizophora annamalayana.* Asian Pac. J. Trop. Biomed. 2 (2), 1081–1085.

El-Hawary, S.S., Mohammed, R., Abou Zid, S.F., Bakeer, W., Ebel, R., Sayed, A.M., Rateb, M. E., 2016. Solamargine production by a fungal endophyte of *Solanum nigrum.* J. appl. Microbiol. 120 (4), 900–911.

Espinosa-Garcia, F.J., Langenheim, J.H., 1990. The endophytic fungal community in leaves of a coastal redwood population diversity and spatial patterns. New Phytol. 116 (1), 89–97.

Faulkner, D.J., 2000. Marine natural products. Nat. Prod. Rep. 17 (1), 7–55.

Feng, Y., Ren, F., Niu, S., Wang, L., Li, L., Liu, X., et al., 2014. Guanacastane diterpenoids from the plant endophytic fungus Cercospora sp. J. Nat. Prod. 77 (4), 873–881.

Fernandes, E.G., Pereira, O.L., da Silva, C.C., Bento, C.B., de Queiroz, M.V., 2015. Diversity of endophytic fungi in Glycine max. Microbiol. Res. 181, 84–92.

Firakova, S., Sturdíkova, M., Muckova, M., 2007. Bioactive secondary metabolites produced by microorganisms associated with plants. Biologia 62 (3), 251–257.

Flemmer, A.C., Anderson, F.E., Hansen, P.V., McLaren, D.A., 2010. Microscopic observations of a compatible host/pathogen interaction between a potential biocontrol agent (*Uromyces pencanus*) and its target weed (*Nassella neesiana*). Mycoscience 51 (5), 396–400.

Frattarelli, D.A., Reed, M.D., Giacoia, G., Aranda, J.V., 2004. Antifungals in systemic neonatal candidiasis. Drugs 64 (9), 949–968.

Fukasawa, Y., Osono, T., Takeda, H., 2009. Effects of attack of saprobic fungi on twig litter decomposition by endophytic fungi. Ecol. Res. 24 (5), 1067.

Gangadevi, V., Muthumary, J., 2007. Endophytic fungal diversity from young, mature and senescent leaves of *Ocimum basilicum* L. with special reference to Taxol production. *Indian.* J. Sci. Technol. 1 (1), 1–12.

Gangadevi, V., Muthumary, J., 2008. Isolation of *Colletotrichum gloeosporioides*, a novel endophytic taxol-producing fungus from the leaves of a medicinal plant. Justicia gendarussa. Mycologia Balcanica 5, 1–4.

Gangwar, A.K., Ghosh, A.K., 2014. Medicinal uses and pharmacological activity of Adhatoda vasica. Int. J. Herb. Med. 2 (1), 88–91.

Gao, X.X., Zhou, H., Xu, D.Y., Yu, C.H., Chen, Y.Q., Qu, L.H., 2005. High diversity of endophytic fungi from the pharmaceutical plant, *Heterosmilax japonica* Kunth revealed by cultivation-independent approach. FEMS Microbiol. Lett 249 (2), 255–266.

Gao, L.W., Li, W.Y., Zhao, Y.L., Wang, J.W., 2009. The cultivation, bioactive components and pharmacological effects of *Armillaria mellea.* Afr. J. Biotechnol 8 (25), 7383–7390.

Garyali, S., Reddy, M.S., 2013. Taxol production by an endophytic fungus, *Fusarium redolens*, isolated from Himalayan yew. J Microbiol Biotechnol 23 (10), 1372–1380.

Gautam, A.K., 2014. Diversity of fungal endophytes in some medicinal plants of Himachal Pradesh, India. Arch. Phytopathol. Plant Prot. 47 (5), 537–544.

Gautam, A.K., Kant, M., Thakur, Y., 2013. Isolation of endophytic fungi from Cannabis sativa and study their antifungal potential. Arch. Phytopathol. Plant Prot. 46 (6), 627–635.

Geisen, S., Kostenko, O., Cnossen, M.C., ten Hooven, F.C., Vres, B., van De Putten, W.H., 2017. Seed and root endophytic fungi in a range expanding and a related plant species. Front. microbiology 8, 1645.

Ghasemzadeh, A., Jaafar, H.Z., Rahmat, A. and Devarajan, T., Evaluation of bioactive compounds, pharmaceutical quality, and anticancer activity of curry leaf (*Murraya koenigii* L.). 2014, 1-8.

Gilman, J.C. (Ed.), 2001. A Manual of Soil fungi. 2nd Indian Edition Biotech Book Pvt. Ltd., India.

Gogoi, D.K., Boruah, D.H., Saikia, R., Bora, T.C., 2008a. Optimization of process parameters for improved production of bioactive metabolite by a novel endophytic fungus *Fusarium* sp. DF2 isolated from Taxus wallichiana of North East India. World J. Microbiol. Biotechnol. 24 (1), 79–87.

Gogoi, D.K., Mazumder, S., Saikia, R., Bora, T.C., 2008b. Impact of submerged culture conditions on growth and bioactive metabolite produced by endophyte Hypocrea spp. NSF-08 isolated from Dillenia indica Linn. in North-East India. J. Mycol. Med. 18 (1), 1–9.

Gokhale, M., Verma, M., Pandey, A., 2017. Isolation of endophytic fungi from different parts of *Oroxylum indicum* and ITS sequencing based identification. J. Mycopathol. Res. 55 (1), 73–79.

Gomez-Vidal, S., Lopez-Llorca, L.V., Jansson, H.B., Salinas, J., 2006. Endophytic colonization of date palm (Phoenix dactylifera L.) leaves by entomopathogenic fungi. Micron 37 (7), 624–632.

Gond, S.K., Mishra, A., Sharma, V.K., Verma, S.K., Kumar, J., Kharwar, R.N., et al., 2012. Diversity and antimicrobial activity of endophytic fungi isolated from Nycanthes arbor-tristis, a well known medicinal plant of India. Mycoscience 53 (2), 113–121.

Gond, S.K., Verma, V.C., Kumar, A., Kumar, V., Kharwar, R.N., 2007. Study of endophytic fungal community from different parts of Aegle marmelos Correae (Rutaceae) from Varanasi (India). World J. Microbiol. Biotechnol. 23 (10), 1371–1375.

Gonzalez-Teuber, M., Vilo, C., Bascunan-Godoy, L., 2017. Molecular characterization of endophytic fungi associated with the roots of Chenopodium quinoa inhabiting the Atacama Desert, Chile. Genomics Data 11, 109–112.

Gouda, S., Das, G., Sen, S.K., Shin, H.S., Patra, J.K., 2016. Endophytes: a treasure house of bioactive compounds of medicinal importance. Front. Microbiol. 7, 1538.

Goutam, J., Kharwar, R.N., Tiwari, V.K., Mishra, A., Singh, S., 2016. Isolation and identification of antibacterial compounds isolated from endophytic fungus Emericella qaudrilineata. Nat. Prod. Chem. Res. 4, 205.

Guo, B., Dai, J.R., Ng, S., Huang, Y., Leong, C., Ong, W., et al., 2000. Cytonic acids A and B: novel tridepside inhibitors of hCMV protease from the endophytic fungus Cytonaema species. J. Nat. Prod. 63 (5), 602–604.

Guo, L.D., 2010. Molecular diversity and identification of endophytic fungi. Molecular Identification of Fungi. Springer, Berlin, Heidelberg, pp. 277–296.

Gupta, S., Chaturvedi, P., 2019. Enhancing secondary metabolite production in medicinal plants using endophytic elicitors: a case study of Centella asiatica (Apiaceae) and Asiaticoside. Endophytes for a Growing World, p. 310.

Gyaneshwar, P., James, E.K., Mathan, N., Reddy, M., Reinhold-Hurek, B., Ladha, J.K., 2001. Endophytic colonization of rice by a diazotrophic strain of Serratia marcescens. J. Bacteriol. 183 (8), 2634–2645.

Hagen, J., Gasparotto, L., Moraes, V.H., Lieberei, R., 2003. Reaction of cassava leaves to *Microcyclus ulei*, causal agent of South American leaf blight of rubber tree. Fitopatol Bras 28 (5), 477–480.

Hall, C.W., Mah, T.F., 2017. Molecular mechanisms of biofilm-based antibiotic resistance and tolerance in pathogenic bacteria. FEMS Microbiol. Rev. 41 (3), 276–301.

Hallmann, J., Berg, G., Schulz, B., 2006. Isolation procedures for endophytic microorganisms. Springer Brelin Heidelberg, New York.

Hamayun, M., Hussain, A., Khan, S.A., Kim, H.Y., Khan, A.L., Waqas, M., et al., 2017. Gibberellins producing endophytic fungus Porostereum spadiceum AGH786 rescues growth of salt affected soybean. Front. microbiology 8, 686.

Hamayun, M., Khan, S.A., Ahmad, N., Tang, D.S., Kang, S.M., Na, C.I., et al., 2009. Cladosporium sphaerospermum as a new plant growth-promoting endophyte from the roots of Glycine max (L.) Merr. World J. Microbiol. Biotechnol. 25 (4), 627–632.

Handral, H.K., Pandith, A., Shruthi, S.D., 2012. A review on Murraya koenigii: multipotential medicinal plant. Asian J. Pharm. Clin. Res. 5 (4), 5–14.

Haque, M.A., Hossain, M.S., Rahman, M.Z., Rahman, M.R., Hossain, M.S., Mosihuzzaman, M., Nahar, N., Khan, S.I., 2005. Isolation of bioactive secondary metabolites from the endophytic fungus of *Ocimum basilicum*. J. Pharma. Sci 4 (2), 127–130.

Hardoim, R., Van Overbeek, L.S., Berg, G., Pirttilä, A.M., Compant, S., Campisano, A., et al., 2015. The hidden world within plants: ecological and evolutionary considerations for defining functioning of microbial endophytes. Microbiol. Mol. Biol. Rev. 79 (3), 293–320.

Harper, J.K., Arif, A.M., Ford, E.J., Strobel, G.A., Porco, J.A., Tomer, D.P., Oneill, K.L., Heider, E.M., Grant, D.M., 2003. Pestacin: a 1, 3-dihydro isobenzofuran from *Pestalotiopsis microspora* possessing antioxidant and antimycotic activities. Tetrahedron 59 (14), 2471–2476.

Harper, J.K., Mulgrew, A.E., Li, J.Y., Barich, D.H., Strobel, G.A., Grant, D.M., 2001. Characterization of stereochemistry and molecular conformation using solid-state NMR tensors. J. Am. Chem. Soc. 123 (40), 9837–9842.

Hawksworth, D.L., 2001. The magnitude of fungal diversity: the 1.5 million species estimate revisited. Mycol. Res. 105 (12), 1422–1432.

Hawksworth, D.L., 2004. Fungal diversity and its implications for genetic resource collections. Stud. mycology 50, 9–18.

Heath, M.C., 1974. Light and electron microscope studies of the interactions of host and non-host plants with cowpea rust—*Uromyces phaseoli* var. vignae. Physiol. Plant Pathol. 4 (4), 403–414.

Hebert, D., Gregory, T.R., 2005. The promise of DNA barcoding for taxonomy. Syst. Biol. 54 (5), 852–859.

Hood, M.E., Shew, H.D., 1996. Applications of KOH-aniline blue fluorescence in the study of plant-fungal interactions. Phytopathology 86 (7), 704–708.

Hossain, M.T., Hoq, M.O., 2016. Therapeutic use of Adhatoda vasica. Asian J. Med. Biol. Res. 2 (2), 156–163.

Huang, J.X., Zhang, J., Zhang, X.R., Zhang, K., Zhang, X., He, X.R., 2014. Mucor fragilis as a novel source of the key pharmaceutical agents podophyllotoxin and kaempferol. Pharm. Biol. 52 (10), 1237–1243.

Hyde, K.D., 2001. Increasing the likelihood of novel compound discovery from filamentous fungi. Bio-Exploit. Filamentous Fungi 77–91.

Ilic, J., Cosic, J., Vrandecic, K., Dugalic, K., Pranjic, A., Martin, J., 2017. Influence of endophytic fungi isolated from symptomless weeds on cherry plants. Mycosphere 8 (1), 18–30.

Isaka, M., Jaturapat, A., Rukseree, K., Danwisetkanjana, K., Tanticharoen, M., Thebtaranonth, Y., 2001. Phomoxanthones A and B, novel xanthone dimers from the endophytic fungus *Phomopsis* species. J. Nat. Pro 64 (8), 1015–1018.

Jacob, M., Bhat, D.J., 2000. Two new endophytic conidial fungi from India. Cryptogamie Mycol. 21 (2), 81–88.

Jain, S.K., 1991. Dictionary of Indian folk Medicine and Ethnobotany. Deep Publications.

Jalgaonwala, R.E., Mahajan, R.T., 2011. Evaluation of hydrolytic enzyme activities of endophytes from some indigenous medicinal plants. J. Agric. Technol. 7 (6), 1733–1741.

Jurus, A.M., Sundberg, W.J., 1976. Penetration of *Rhizopus oligosporus* into soybeans in tempeh. Appl Environ Microbiol 32 (2), 284–287.

Juyal, P., Ghildiyal, J.C., 2013. Medicinal phyto-diversity of Bhabar tract of Garhwal Himalaya. J. Med. Plants 1 (6), 43–57.

Kaaniche, F., Hamed, A., Abdel-Razek, A.S., Wibberg, D., Abdissa, N.E.L., Euch, I.Z., 2019. Bioactive secondary metabolites from new endophytic fungus Curvularia sp. isolated from Rauwolfia macrophylla. PLoS One 14 (6), 0217627.

Kannan, K., Thilakavathi, R., Kumar, M.D., Senthamarai, M., 2017. Biodiversity of edophytic fungi from Mukia maderespatana (L) M. Roem – a first report. J. Bacteriol. Mycol.: Open access. 4 (4), 00098.

Katoch, M., Singh, G., Sharma, S., Gupta, N., Sangwan, L., Saxena, A.K., 2014. Cytotoxic and antimicrobial activities of endophytic fungi isolated from Bacopa monnieri (L.) Pennell (Scrophulariaceae). BMC complementary alternative Med. 14 (1), 52.

Kavitha, M., 2017. Antibacterial and antioxidant activity of leaf organic extracts of local cultivars of *Murraya koenigii* (L.) Spreng from Tamilnadu. Biosci. Biotechnol. Res. Commun. 10 (3), 359–364.

Kavya, B., Kavya, N., Ramarao, V., Venkatateshwar, G., 2015. *Tinospora cordifolia* (Willd.) Miers: nutritional, ethanomedical and therapeutic utility. Int J Res Ayurveda Pharm 6 (2), 195–198.

Kedar, A., Rathod, D., Yadav, A., Agarkar, G., Rai, M., 2014. Endophytic Phoma sp. Isolated from medicinal plants promote the growth of Zea mays. Nusant. Biosci. 6 (2), 132–139.

Keesara, B.R., Jat, R.K., 2017. Isolation and characterization of Vasicine from *Adhatoda vasica* (Adusa). Int. J. Res. Dev. Pharm. Life Sci. 6, 2590–2596.

Kharwar, R.N., Verma, V.C., Kumar, A., Gond, S.K., Harper, J.K., Hess, W.M., Lobkovosky, E., Ma, C., Ren, Y., Strobel, G.A., 2009. Javanicin, an antibacterial naphthaquinone from an endophytic fungus of neem, *Chloridium* sp. Curr. Microbiol. 58 (3), 233–238.

Kharwar, R.N., Verma, V.C., Strobel, G., Ezra, D., 2008. The endophytic fungal complex of Catharanthus roseus (L.) G. Don. Curr. Sci. India 95 (2), 228–233.

Khiralla, A., Mohamed, I., Thomas, J., Mignard, B., Spina, R., Yagi, S., et al., 2015. A pilot study of antioxidant potential of endophytic fungi from some Sudanese medicinal plants. Asian Pac. J. Trop. Med. 8 (9), 701–704.

Kia, S.H., Glynou, K., Nau, T., Thines, M., Piepenbring, M., Macia-Vicente, J.G., 2017. Influence of phylogenetic conservatism and trait convergence on the interactions between fungal root endophytes and plants. ISME J. 11 (3), 777.

Kirtikar, K.R., Basu, B.D. (Eds.), 1935. Terminalia chebula. Indian Medicinal Plants 1, pp. 1020–1023.

second Edition Kirtikar, K.R., Basu, B.D. (Eds.), 2012. Indian Medicinal Plants, 1. Periodical Book Agency, Delhi, 77.

Knolker, H.J., Reddy, K.R., 2008. Biological and pharmacological activities of carbazole alkaloids. Alkaloids: Chem. Biol. 65, 181–193.

Kobayashi, H., Sunaga, R., Furihata, K., Morisaki, N., Iwasaki, S., 1995. Isolation and structures of an antifungal antibiotic, fusarielin A, and related compounds produced by a *Fusarium* sp. J. Antibiot. res 48 (1), 42–52.

Koehn, F.E., Carter, G.T., 2005. The evolving role of natural products in drug discovery. Nat. Rev. Drug. discovery 4 (3), 206.

Kogel, K.H., Franken, Hückelhoven, R., 2006. Endophyte or parasite – what decides? Curr. Opin. Plant Biol. 9 (4), 358–363.

Korkama-Rajala, T., Müller, M.M., Pennanen, T., 2008. Decomposition and fungi of needle litter from slow-and fast-growing Norway spruce (Piceaabies) clones. Microb. Ecol. 56 (1), 76.

Krajian, A.A., 1940. Histological technique. The C. V. Mosby Company, St Louis.

Krishnamurthy, Y.L., Naik, B.S., 2017. Endophytic fungi bioremediation. Endophytes: Crop Productivity and Protection. Springer, Cham, pp. 47–60.

Krishnamurthy, Y.L., Naik, S.B., Jayaram, S., 2008. Fungal communities in herbaceous medicinal plants from the Malnad region, Southern India. Microb. Environ. 23 (1), 24–28.

Kuldau, G.A., Yates, I.E., 2000. Evidence for Fusarium endophytes. Microb. endophytes 85.

Kumar, A., Abnave, Ahmad, A., 2013. Cultural, morphological and molecular characterization of vinca alkaloids producing endophytic fungus Fusarium solani isolated from Catharanthus roseus. Int. J. Bot. Res. 3 (2), 2277–4815.

Kumar, A., Shawl, M., Jain, R., Mathur, A., 2015. Antimicrobial potential of endophytes isolated and characterized from aerial and non aerial parts of Murraya koenigii L. Adv. Appl. Sci. Res. 6 (10), 21–29.

Kumar, D.S.S., Hyde, K.D., 2004. Biodiversity and tissue-recurrence of endophytic fungi in Tripterygium wilfordii. Fungal Divers. 17, 69–90.

Kumar, D.S.S., Lau, C.S., Wan, J.M., Yang, D., Hyde, K.D., 2005. Immunomodulatory compounds from Pestalotiopsis leucothes, an endophytic fungus from Tripterygium wilfordii. Life Sci. 78 (2), 147–156.

Kumar, D.V., Geethanjali, B., Avinash, K.O., Kumar, J.R., Chandrashekrappa, G.K., 2017. *Tinospora cordifolia*: The antimicrobial property of the leaves of *Amrutha balli*. J Bacteriol Mycol Open Access 5 (5), 363–371.

Kumar, K.S., Debjit, B., Pankaj, T., Rakesh, K., 2010. Indian traditional herbs Adhatoda vasica and its medicinal application. J. Chem. Pharm. Res. 2 (1), 240–245.

Kumar, S., Kaushik, N., 2013. Endophytic fungi isolated from oil-seed crop jatropha curcas produces oil and exhibit antifungal activity. PLoS One 8 (2), 56202.

Kumar, S., Kaushik, N., Edrada-Ebel, R., Ebel, R., Proksch, P., 2011. Isolation, characterization, and bioactivity of endophytic fungi of Tylophora indica. World J. Microbiol. Biotechnol. 27 (3), 571–577.

Kumar, S., Upadhyay, R., Aharwal, R., Sandhu, S.S., 2016. Antibacterial activity of some isolated endophytic fungi from Menthe viridis. Int. J. Appl. Biol. Pharm. Technol. 7 (1), 239–249.

Kumaresan, V., Suryanarayanan, T.S., 2002. Endophyte assemblages in young, mature and senescent leaves of Rhizophora apiculata: evidence for the role of endophytes in mangrove litter degradation. Fungal Divers. 9, 81–91.

Kuralarasi, R., Lingakumar, R.K.K., 2018. Isolation and antibacterial activity of endophytic fungi from *Madhuca longifolia* Bark. J. Med. plants 6 (1), 36–39.

Kusari, S., Hertweck, C., Spiteller, M., 2012. Chemical ecology of endophytic fungi: origins of secondary metabolites. Chem. Biol. 19 (7), 792–798.

Kusari, S., Zuhlke, S., Spiteller, M., 2009. An endophytic fungus from *Camptotheca acuminata* that produces camptothecin and analogues. J. Nat. Prod. 72 (1), 2–7.

Lakra, N.S., Koul, M., Chandra, R., Chandra, S., 2013. Histological investigations of healthy tissues of Catharanthus roseus to localize fungal endophytes. International. J. Pharm. Sci. Rev. Res. 20 (1), 205–209.

Lee, J., Lobkovsky, E., Pliam, N.B., Strobel, G.A., Clardy, J., 1995. Subglutinols A and B, immunosuppressive compounds from the endophytic fungus Fusarium subglutinans. J. Org. Chem. 60, 7076–7077.

Lee, J.C., Strobel, G.A., Lobkovsky, E., Clardy, J., 1996. Torreyanic acid: a selectively cytotoxic quinone dimer from the endophytic fungus Pestalotiopsis microspora. J. Org. Chem. 61 (10), 3232–3233.

Leuchtman, A., 2003. In, J.F White, J. Bacon, C.W, N.L Hywel–Jones, J.W Spatafora, eds. New York, NY,USA, Marcel –Dekker, 341-175.

Li, H.Y., Li, D.W., He, C.M., Zhou, Z.P., Mei, T., Xu, H.M., 2012. Diversity and heavy metal tolerance of endophytic fungi from six dominant plant species in a Pb–Zn mine wasteland in China. fungal ecology 5 (3), 309–315.

Li, H.Y., Wei, D.Q., Shen, M., Zhou, Z., 2012. Endophytes and their role in phytoremediation. Fungal Divers. 54 (1), 11–18.

Liu, C.H., Zou, W.X., Lu, H., Tan, R.X., 2001. Antifungal activity of Artemisia annua endophyte cultures against phytopathogenic fungi. J. Biotechnol. 88 (3), 277–282.

Liu, K., Ding, X., Deng, B., Chen, W., 2009. Isolation and characterization of endophytic taxol-producing fungi from Taxus chinensis. J. Ind. Microbiol. Biotechnol. 36 (9), 1171.

Mahapatra, S., Banerjee, D., 2010. Diversity and screening for antimicrobial activity of endophytic fungi from *Alstonia scholaris*. Acta Microbiol Imm H 57 (3), 215–223.

Mahesh, B., Tejesvi, M.V., Nalini, M.S., Prakash, H.S., Kini, K.R., Subbiah, V., et al., 2005. Endophytic mycoflora of inner bark of Azadirachta indica. Curr. Sci. 88 (2), 218–219.

Mahfooz, M., Dwedi, S., Bhatt, A., Raghuvanshi, S., Bhatt, M., Agrawal, K., 2017. Evaluation of antifungal and enzymatic potential of endophytic fungi isolated from Cupressus torulosa D. Don. Int. J. Curr. Microbiol. Appl. Sci. 6 (7), 4084–4100.

Makimura, K., Murayama, S.Y., Yamaguchi, H., 1994. Detection of a wide range of medically important fungi by the polymerase chain reaction. J. Med. Microbiol. 40 (5), 358–364.

Malcolm, G.M., Kuldau, G.A., Gugino, B.K., Jimenez-Gasco, M.D.M., 2013. Hidden host plant associations of soilborne fungal pathogens: an ecological perspective. Phytopathology 103 (6), 538–544.

Malinowski, D., Alloush, G.A., Belesky, D., 2000. Leaf endophyte Neotyphodium coenophialum modifies mineral uptake in tall fescue. Plant Soil 227 (1-2), 115–126.

Manoharachary, C., Nagaraju, D., 2016. Endophytic hyphomycetous fungi associated with some medicinal plants from Telangana state, India. Indian Phytopathol. 69 (2), 169–172.

Marcellano, J., Collanto, A.S., Fuentes, R.G., 2017. Antibacterial activity of endophytic fungi isolated from the Bark of Cinnamomum mercadoi. Pharmacogn. J. 9 (3), 405–409.

Maria, G.L., Sridhar, K.R., Raviraja, N.S., 2005. Antimicrobial and enzyme activity of mangrove endophytic fungi of southwest coast of India. J. Agric. Technol. 1, 67–80.

Marinho, A.M., Rodrigues-Filho, E., Moitinho, M., Santos, L.S., 2005. Biologically active polyketides produced by *Penicillium janthinellum* isolated as an endophytic fungus from fruits of *Melia azedarach*. J. Brazil. Chem. Soc. 16 (2), 280–283.

Martinez-Medina, A., Fernandez, I., Lok, G.B., Pozo, M.J., Pieterse, C.M., Van Wees, S., 2017. Shifting from priming of salicylic acid-to jasmonic acid regulated defences by Trichoderma protects tomato against the root knot nematode Meloidogyne incognita. N. Phytol. 213 (3), 1363–1377.

Martins, J.T., Bourbon, A.I., Pinheiro, A.C., Souza, B.W., Cerqueira, M.A., Vicente, A.A., 2013. Biocomposite films based on κ-carrageenan/locust bean gum blends and clays: physical and antimicrobial properties. Food Bioprocess. Technol. 6 (8), 2081–2092.

Maurya, S., Singh, D., 2010. Quantitative analysis of total phenolic content in Adhatoda vasica Nees extracts. Int. J. PharmTech Res. 2 (4), 2403–2406.

Mayerhofer, M.S., Kernaghan, G., Harper, K.A., 2013. The effects of fungal root endophytes on plant growth: a meta-analysis. Mycorrhiza 23 (2), 119–128.

Meena, S., Kumar, S.R., Dwivedi, V., Singh, A.K., Chanotiya, C.S., Akhtar, M.Q., Kumar, K., Shasany, A.K., Nagegowda, D.A., 2017. Transcriptomic insight into terpenoid and carbazole alkaloid biosynthesis, and functional characterization of two terpene synthases in curry tree (*Murraya koenigii*). Sci. Rep. 7, 44126.

Meyer, A., Todt, C., Mikkelsen, N.T., Lieb, B., 2010. Fast evolving 18S rRNA sequences from Solenogastres (Mollusca) resist standard PCR amplification and give new insights into mollusk substitution rate heterogeneity. BMC Evol. Biol. 10 (1), 70.

Millar, E.V., Chen, W.J., Schlett, C.D., Cui, T., Crawford, K.B., Lanier, J.B., et al., 2015. Frequent use of chlorhexidine-based body wash associated with a reduction in methicillin-resistant Staphylococcus aureus nasal colonization among military trainees. Antimicrob. agents chemotherapy 59 (2), 943–949.

Miller, C.M., Miller, R.V., Garton-Kenny, D., Redgrave, B., Sears, J., Condron, M.M., Teplow, D.B., Strobel, G.A., 1998. Ecomycins, unique antimycotics from *Pseudomonas viridiflava*. J. App. Microbiol. 84 (6), 937–944.

Mishra, A., Gond, S.K., Kumar, A., Sharma, V.K., Kharwar, R.N., Sieber, T.N., 2012. Season and tissue type affect fungal endophyte communities of the Indian medicinal plant Tinospora cordifolia more strongly than geographic location. Microb. Ecol. 64 (2), 388–398.

Mishra, Y., Batra, A., Sharma, M.M., 2017. Histological localization of fungal endophytes in healthy tissues of Adhatoda vasica Nees. Curr. Sci. 112 (10), 2112–2115.

Mishra, Y., Mittal, J., Singh, A., Batra, A., Sharma, M.M., 2015. In vivo and in vitro histological localization of endophytic fungi in Tinospora cordifolia (Willd.) Miers ex Hook F. & Thomas. J. Appl. Res. Med. Aromatic Plants 2 (1), 30–33.

Mitchell, A.M., Strobel, G.A., Moore, E., Robison, R., Sears, J., 2010. Volatile antimicrobials from *Muscodor crispans*, a novel endophytic fungus. Microbiology 156 (1), 270–277.

Mittal, J., Sharma, M.M., Batra, A., 2014. Tinospora cordifolia: a multipurpose medicinal plant. J. Med. Plants 2 (2), 32–47.

Mitter, B., Pfaffenbichler, N., Sessitsch, A., 2016. Plant–microbe partnerships in 2020. Microb. Biotechnol. 9 (5), 635–640.

Momsia, P., Momsia, T., 2013. Isolation, frequency distribution and diversity of novel fungal endophytes inhabiting leaves of Catharanthus roseus. Int. J. Life Sci. Biotechnol. Pharm. Res. 2 (4), 83–87.

Morrien, E., Engelkes, T., Macel, M., Meisner, A., Van der Putten, W.H., 2010. Climate change and invasion by intracontinental range-expanding exotic plants: the role of biotic interactions. Ann. botany 105 (6), 843–848.

Muller, M.M., Valjakka, R., Suokko, A., Hantula, J., 2001. Diversity of endophytic fungi of single Norway spruce needles and their role as pioneer decomposers. Mol. Ecol. 10 (7), 1801–1810.

Munim, A., Ramadhan, M.G., Soemiati, A., 2013. Screening of endophytic fungi from Cassia siamea Lamk leaves as a-glucosidase inhibitor. Int. Res. J. Pharm. 4 (5), 128–131.

Murali, T.S., Suryanarayanan, T.S., Venkatesan, G., 2007. Fungal endophyte communities in two tropical forests of southern India: diversity and host affiliation. Mycol. Prog. 6 (3), 191–199.

Murugan, K.K., Poojari, C.C., Ryavalad, C., Lakshmikantha, R.Y., Satwadi, R., Vittal, R.R., et al., 2017. Anti-diabetic activity of endophytic fungi, Penicillium species of Tabebuia argentea; in silico and experimental analysis. Phytochemistry 11 (2), 90–110.

Nadeem, M., Ram, M., Alam, P., Ahmad, M.M., Mohammad, A., Al-Qurainy, F., Khan, S., Abdin, M.Z., 2012. *Fusarium solani*, P1, a new endophytic podophyllotoxin producing fungus from roots of *Podophyllum hexandrum*. Afr. J. Microbiol. Res. 6 (10), 2493–2499.

Nagda, V., Gajbhiye, A., Kumar, D., 2017. Isolation and characterization of endophytic fungi from Caltropis procera for their antioxidant activity. Asian J. Pharm. Clin. Res. 10 (3), 254–258.

Naik, B.S., Shashikala, J., Krishnamurthy, Y.L., 2008. Diversity of fungal endophytes in shrubby medicinal plants of Malnad region, Western Ghats, Southern India. Fungal Ecol. 1 (2-3), 89–93.

Nair, S.K., 2002. *Studies on diversity, ecology and biology of aquatic fungi of some freshwater streams of Western Ghat forests in Goa state, India* (Doctoral dissertation, Goa University).

Naik, B.S., Shashikala, J., Krishnamurthy, Y.L., 2009. Study on the diversity of endophytic communities from rice (Oryza sativa L.) and their antagonistic activities in vitro. Microbiol. Res. 164 (3), 290–296.

Nalini, M.S., Mahesh, B., Tejesvi, M.V., Prakash, H.S., Subbaiah, V., Kini, K.R., et al., 2005. Fungal endophytes from the three-leaved caper Crataeva magna (Capparidaceae). Mycopathologia 159 (2), 245–249.

Nalini, M.S., Sunayana, N., Prakash, H.S., 2014. Endophytic fungal diversity in medicinal plants of Western Ghats, India. Int. J. Biodivers. 2014.

Nath, A., Pathak, J., Joshi, S.R., 2014. Bioactivity assessment of endophytic fungi associated with Centella asiatica and Murraya koengii. J. Appl. Biol. Biotechnol. 2 (05), 006–011.

Neill, J.C., 1940. The endophyte of Rye-Grass (*Lolium perenne*). N.Z.Jl Sci. Technol, Section A 21 (5), 280–291.

Newman, D.J., Cragg, G.M., 2004. Marine natural products and related compounds in clinical and advanced preclinical trials. J. Nat. Prod. 67 (8), 1216–1238.

Newman, D.J., Cragg, G.M., 2007. Natural products as sources of new drugs over the last 25 years. J. Nat. Prod. 70 (3), 461–477.

Newman, D.J., Cragg, G.M., 2016. Natural products as sources of new drugs from 1981 to 2014. J. Nat. Prod. 79 (3), 629–661.

Newsham, K.K., 2011. A meta-analysis of plant responses to dark septate root endophytes. N. Phytol. 190 (3), 783–793.

Nithya, K., Muthumary, J., 2010. Secondary metabolite from *Phomopsis* sp. isolated from *Plumeria acutifolia* Poiret. Recent res. sci. technol. 2 (4), 99–103.

Norse, D., 1972. Fungi isolated from surface-sterilized tobacco leaves. Trans. Br. Mycol. Soc. 58 (3), 515–518.

Osono, T., 2003. Effects of prior decomposition of beech leaf litter by phyllosphere fungi on substrate utilization by fungal decomposers. Mycoscience 44 (1), 0041–0045.

Osono, T., 2006. Role of phyllosphere fungi of forest trees in the development of decomposer fungal communities and decomposition processes of leaf litter. Can. J. Microbiol. 52 (8), 701–716.

Oyebanji, O.B., Nweke, O., Odebunmi, O., Galadima, N.B., Idris, M.S., Nnodi, U.N., et al., 2009. Simple, effective and economical explant-surface sterilization protocol for cowpea, rice and sorghum seeds. Afr. J. Biotechnol. 8 (20).

Padhi, L., Mohanta, Y.K., Panda, S.K., 2013. Endophytic fungi with great promises: a review. J. Adv. Pharm. Educ. Res. 3 (3), 152–170.

Pandey, M., Surendra, K., Chikara, S.K., Vyas, M.K., Sharma, R., Thakur, G.S., et al., 2012. Tinospora cordifolia: a climbing shrub in health care management. Int. J. Pharm. Biol. Sci. 3 (4), 612–628.

Pandey, M.M., Rastogi, S., Rawat, A.K., 2008. Indian herbal drug for general healthcare: an overview. Internet J. Alternat Med 6 (1), 3.

Parmar, C., Kaushal, M.K., 1982. Murraya koenigii. Wild Fruits. Kalyani Publishers, New Delhi, India, pp. 45–48.

Parthipan, M., Aravindhan, V., Rajendran, A., 2011. Medico-botanical study of Yercaud hills in the Eastern Ghats of Tamil Nadu, India. Anc. Sci. Life 30 (4), 104.

Paul, N.C., Deng, J.X., Shin, K.S., Yu, S.H., 2012. Molecular and morphological characterization of endophytic Heterobasidion araucariae from roots of Capsicum annuum L. in Korea. Mycobiology 40 (2), 85–90.

Paulraj, M., Shanmugam, N., Ignacimuthu, S., 2014. Antifeedant activity and toxicity of two alkaloids from *Adhatoda vasica* Nees leaves against diamond back moth *Plutella xylostella* (Linn.) (Lepidoptera: Plutellidae) larvae. Arch Phytopathology Plant Protect 47 (15), 1832–1840.

Pawłowska, J., Wilk, M., Śliwińska-Wyrzychowska, A., Mętrak, M., Wrzosek, M., 2014. The diversity of endophytic fungi in the above-ground tissue of two Lycopodium species in Poland. Symbiosis 63 (2), 87–97.

Pearce, R.B., 1984. Staining fungal hyphae in wood. Trans. Brit. Mycol. Soc. 82 (3), 564–567.

Peay, K.G., Kennedy, G., Talbot, J.M., 2016. Dimensions of biodiversity in the Earth mycobiome. Nat. Rev. Microbiol. 14 (7), 434.

Pereira, J.O., Vieira, M.C., Azevedo, J.L.D., 1999. Endophytic fungi from Musa acuminata and their reintroduction into axenic plants. World J. Microbiol. Biotechnol. 15 (1), 37–40.

Petrini, O., 1985. Wirtsspezifität endophytischer Pilze bei einheimischen Ericaceae. Bot. Helv 95 (2), 213–238.

Petrini, O., 1986. Taxonomy of endophytic fungi of aerial plant tissues. In: Fokkema, N.J., van den Heuvel, J. (Eds.), Microbiology of the phyllosphere. Cambridge University Press, Cambridge, UK, pp. 175–187.

Petrini, O., 1991. Fungal endophytes of tree leaves. Microbial Ecology of Leaves. Springer, New York, NY, pp. 179–197.

Petrini, O., Fisher, P.J., Petrini, L.E., 1992. Fungal endophytes of bracken (*Pteridium aquilinum*), with some reflections on their use in biological control. Sydowia 44, 282–293.

Petrini, O., Muller, E., 1979. Pilzliche Endophyten, am Beispiel von *Juniperus communis* L. Sydowia 32, 224–251.

Phongpaichit, S., Nikom, J., Rungjindamai, N., Sakayaroj, J., Hutadilok-Towatana, N., Rukachaisirikul, V., Kirtikara, K., 2007. Biological activities of extracts from endophytic fungi isolated from *Garcinia* plants. FEMS Immunol. Med. MIC. 51 (3), 517–525.

Pierre, E., Louise, N.W., Marie, T.K.R., Valere, T.F., Arc-en-ce, J.M., Fekam, B.F., 2016. Integrated assessment of phytostimulation and biocontrol potential of endophytic Trichoderma sp. against common bean (Phaseolus vulgaris L.) root rot fungi complex in centre region, Cameroon. Int. J. Pure Appl. Biosci. 4 (4), 50–68.

Pimentel, M.R., Molina, G., Dionisio, A., Maróstica Junior, M.R., Pastore, G.M., 2011. The use of endophytes to obtain bioactive compounds and their application in biotransformation process. Biotechnol. Res. Int. 576286. Available from: https://doi.org/10.4061/2011/576286.

Pinheiro, E.A., Pina, J.R., Feitosa, A.O., Carvalho, J.M., Borges, F.C., Marinho, S., et al., 2017. Bioprospecting of antimicrobial activity of extracts of endophytic fungi from Bauhinia guianensis. Rev. Argentina de. microbiologia 49 (1), 3–6.

Pinheiro, E.T., Gomes, B.F.A., Drucker, D.B., Zaia, A.A., Ferraz, C.C.R., Souza Filho, F.J., 2004. Antimicrobial susceptibility of Enterococcus faecalis isolated from canals of root filled teeth with periapical lesions. Int. Endod. J. 37 (11), 756–763.

Pinto, L.S.R.C., Azevedo, J.L., Pereira, J.O., Vieira, M.L.C., Labate, C.A., 2000. Symptomless infection of banana and maize by endophytic fungi impairs photosynthetic efficiency. N. Phytol. 147 (3), 609–615.

Pirttila, A.M., Podolich, O., Koskimaki, J.J., Hohtola, E., Hohtola, A., 2008. Role of origin and endophyte infection in browning of bud-derived tissue cultures of Scots pine (Pinus sylvestris L.). Plant Cell Tissue Organ. Cult. 95 (1), 47–55.

Polonio, J.C., Almeida, T.T., Garcia, A., Mariucci, G.E.G., Azevedo, J.L., Rhoden, S.A., et al., 2015. Biotechnological prospecting of foliar endophytic fungi of guaco (Mikania glomerata Spreng.) with antibacterial and antagonistic activity against phytopathogens. Genet. Mol. Res. 14 (3), 7297–7309.

Porwal, P., Shukla, K., Mishra, D.K., Mahajan, S.C., Tiwari, A., 2010. Comparative in vitro antioxidant activity of Pongamia pinnata Linn. Leaves extracts and isolated compound. Int. J. Pharm. Biol. Arch. 1 (1), 69–75.

Potshangbam, M., Devi, S.I., Sahoo, D., Strobel, G.A., 2017. Functional characterization of endophytic fungal community associated with Oryza sativa L. and Zea mays L. Front. microbiology 8, 325.

Prabhugaonkar, A., Bhat, D.J., 2009. *Rattania setulifera*, an undescribed endophytic hyphomycete on rattans from Western Ghats, India. Mycotaxon 108, 217–222.

Promputtha, I., Hyde, K.D., McKenzie, E.H., Peberdy, J.F., Lumyong, S., 2010. Can leaf degrading enzymes provide evidence that endophytic fungi becoming saprobes? Fungal Divers. 41 (1), 89–99.

Prasanth, A.S., Ramamurthy, V., 2018. Determination of secondary metabolites and antioxidant activity. Int. J. Pharm. Drug. Anal 6 (2), 279–285.

Puri, S.C., Nazir, A., Chawla, R., Arora, R., Riyaz-ul-Hasan, S., Amna, T., et al., 2006. The endophytic fungus Trametes hirsuta as a novel alternative source of podophyllotoxin and related aryl tetralin lignans. J. Biotechnol. 122 (4), 494–510.

Puri, S.C., Verma, V., Amna, T., Qazi, G.N., Spiteller, M., 2005. An Endophytic Fungus from *Nothapodytes foetida* that Produces Camptothecin. J. Nat. Prod. 68 (12), 1717–1719.

Qader, M.M., Kumar, N.S., Jayasinghe, L., Araya, H., Fujimoto, Y., 2017. Bioactive sesquiterpenes from an endophytic fungus Bipolaris sorokiniana isolated from a popular medicinal plant Costus speciosus. Mycology 8 (1), 17–20.

Qadri, M., Johri, S., Shah, B.A., Khajuria, A., Sidiq, T., Lattoo, S.K., et al., 2013. Identification and bioactive potential of endophytic fungi isolated from selected plants of the Western Himalayas. Springer Plus 2 (1), 8.

Qadri, M., Rajput, R., Abdin, M.Z., Vishwakarma, R.A., Riyaz-Ul-Hassan, S., 2014. Diversity, molecular phylogeny, and bioactive potential of fungal endophytes associated with the Himalayan blue pine (Pinus wallichiana). Microb. Ecol. 67 (4), 877–887.

Qin, S., Li, J., Chen, H.H., Zhao, G.Z., Zhu, W.Y., Jiang, C.L., et al., 2009. Isolation, diversity, and antimicrobial activity of rare actinobacteria from medicinal plants of tropical rain forests in Xishuangbanna, China. Appl. Environ. Microbiol. 75 (19), 6176–6186.

Raekiansyah, M., Mori, M., Nonaka, K., Agoh, M., Shiomi, K., Matsumoto, A., et al., 2017. Identification of novel antiviral of fungus-derived brefeldin A against dengue viruses. Trop. Med. health 45 (1), 32.

Rahul, J., 2013. An ethnobotanical study of medicinal plants in taindol village, District Jhansi, Region of Bundelkhand, Uttar Pradesh, India. J. Med. Plants Stud. 1 (5), 59–71.

Rajagopal, K., Suryanarayanan, T.S., 2000. Isolation of endophytic fungi from leaves of neem (Azadirachta indica A. Juss.). Curr. Sci. 1375–1378.

Ramesh, V., Rajendran, A., 2017. The molecular phylogeny and taxonomy of endophytic fungal species from the leaves of *Vitex negundo* L. Studies in Fungi 2 (1), 26–38.

Rao, L.J.M., Ramalakshmi, K., Borse, B.B., Raghavan, B., 2007. Antioxidant and radical-scavenging carbazole alkaloids from the oleoresin of curry leaf (*Murraya koenigii* Spreng.). Food Chem. 100 (2), 742–747.

Raviraja, N.S., 2005. Fungal endophytes in five medicinal plant species from Kudremukh Range, Western Ghats of India. J. Basic Microbiol. 45 (3), 230–235.

Raviraja, N.S., Maria, G.L., Sridhar, K.R., 2006. Antimicrobial evaluation of endophytic fungi inhabiting medicinal plants of the Western Ghats of India. Eng. Life Sci. 6 (5), 515–520.

Reddy, N.M., Reddy, R., 2015. *Tinospora cordifolia* chemical constituents and medicinal properties: a review. Sch. Acad. J. Pharm 4 (8), 364–369.

Redko, F., Clavin, M., Weber, D., Anke, T., Martino, V., 2006. Search for active metabolites of Erythrina crista-galli and its endophyte *Phomopsis* sp. Molecular Medicinal Chemistry 10, 24–26.

Robinson, B.H., Lombi, E., Zhao, F.J., McGrath, S., 2003. Uptake and distribution of nickel and other metals in the hyperaccumulator Berkheya coddii. N. Phytol. 158 (2), 279–285.

Rodrigues, K.F., Samuels, G.J., 1999. Fungal endophytes of Spondias mombin leaves in Brazil. J. Basic Microbiol.: Int. J. Biochem. Physiol. Genetics Morphol. Ecol. Microorg. 39 (2), 131–135.

Rodriguez, R.J., White Jr, J.F., Arnold, A.E., Redman, A.R.A., 2009. Fungal endophytes: diversity and functional roles. N. phytologist 182 (2), 314–330.

Rosa, G., Peralta-Videa, J.R., Montes, M., Parsons, J.G., Cano-Aguilera, I., Gardea-Torresdey, J.L., 2004. Cadmium uptake and translocation in tumbleweed (Salsola kali), a potential Cd-hyper accumulator desert plant species: ICP/OES and XAS studies. Chemosphere 55 (9), 1159–1168.

Rubini, M.R., Silva-Ribeiro, R.T., Pomella, A.W., Maki, C.S., Araújo, W.L., Dos Santos, D.R., Azevedo, J.L., 2005. Diversity of endophytic fungal community of cacao (*Theobroma cacao* L.) and biological control of *Crinipellis perniciosa*, causal agent of Witches' Broom Disease. Int J Biol Sci. 1 (1), 24.

Sadananda, T.S., Govindappa, M., Ramachandra, Y.L., 2014. In vitro antioxidant activity of lectin from different endophytic fungi of *Viscum album* L. Br. J. Pharmacol Res 4, 626–643.

Saha, S., Ghosh, S., 2012. *Tinospora cordifolia*: One plant, many roles. Anc sci life 31 (4), 151–159.

Saha, D.C., Jackson, M., Cicalese, J.M., 1988. A rapid staining method for detection of endophytic fungi in turf and forage grasses. Phytopathology 78, 237–239.

Sandberg, D.C., Battista, L.J., Arnold, A.E., 2014. Fungal endophytes of aquatic macrophytes: diverse host-generalists characterized by tissue preferences and geographic structure. Microb. Ecol. 67 (4), 735–747.

Sandhu, S.S., Kumar, S., Aharwal, R., 2014. Isolation and identification of endophytic fungi from Ricinus communis linn. and their antibacterial activity. Int. J. Res. Pharm. Chem. 4 (3), 611–618.

Santos Souza, B., dos Santos, T.T., 2017. Endophytic fungi in economically important plants: ecological aspects, diversity and potential biotechnological applications. J. Bioenerg Food Sci. 4 (2), 113–126.

Sarangthem, I.D., Momota, P., 2012. Isolation and characterization of endophytic microbiome from indigenous maize (Zea mays) variety of Manipur and its impact on Biological control. Int. J. Hum. Genet. Med. Biotechnol. Microbiol. Stud. 1, 11–17.

Sathe, V., Raghukumar, S., 1991. Fungi and their biomass in detritus of the seagrass Thalassia hemprichii (Ehrenberg) Ascherson. Botan. Marina 34 (4), 271–278.

Savitha, G., Balamurugan, S., 2014. Pharmacognostical and antibacterial evaluation of Murraya koenigii (L) Spreng. Int. Lett. Nat. Sci. 17, 44–50.

Schiff, B., Horwitz, S.B., 1980. Taxol stabilizes microtubules in mouse fibroblast cells. Proc. Natl. Acad. Sci. 77 (3), 1561–1565.

Schulz, B., Boyle, C., 2005. The endophytic continuum. Mycol. Res. 109 (6), 661–686.

Schulz, B., Rommert, A.K., Dammann, U., Aust, H.J., Strack, D., 1999. The endophyte-host interaction: a balanced antagonism? Mycol. Res. 103 (10), 1275–1283.

Schulz, B., Wanke, U., Draeger, S., Aust, H.J., 1993. Endophytes from herbaceous plants and shrubs: effectiveness of surface sterilization methods. Mycol. Res. 97 (12), 1447–1450.

Scott, L.J., Curran, M., Figgitt, D., 2004. Rosuvastatin. Am. J. Cardiovasc. Drugs 4 (2), 117–138.

Seena, S., Sridhar, K.R., 2004. Endophytic fungal diversity of 2 sand dune wild legumes from the southwest coast of India. Can. J. Microbiol. 50 (12), 1015–1021.

Selvanathan, S., Indrakumar, I., Johnpaul, M., 2011. Biodiversity of the endophytic fungi isolated from Calotropis gigantea (L.) R. Br. Recent Res. Sci. Technol. 3 (4), 94–100.

Shanthi, V., Nelson, R., 2013. Anitbacterial activity of *Tinospora cordifolia* (Willd) Hook. F. Thoms on urinary tract pathogens. Int J Curr Microbiol Appl Sci 2 (6), 190–194.

Sharma, M., Neerajarani, G., Kumar, A., Basak, D., Kumar, K.A., 2015. Studies on endophytes, phytochemicals, antioxidant and antimicrobial property of Vitex altissima. Int. J. Health Sci. Res. 5 (3), 173–182.

Sharma, U., Bala, M., Kumar, N., Singh, B., Munshi, R.K., Bhalerao, S., 2012. Immunomodulatory active compounds from Tinospora cordifolia. J. Ethnopharmacol. 141 (3), 918–926.

Shekhawat, K.K., Rao, D.V., Batra, A., 2010. Morphological study of endophytic fungi inhabiting leaves of Melia azedarach L. Int. J. Pharm. Sci. Rev. Res. 5 (3), 177–180.

Shrivastava, N., Srivastava, A., Banerjee, A., Nivsarkar, M., 2006. Anti-ulcer activity of Adhatoda vasica Nees. J. Herb. pharmacotherapy 6 (2), 43–49.

Shubha, J., Srinivas, C., 2017. Diversity and extracellular enzymes of endophytic fungi associated with Cymbidium aloifolium L. Afr. J. Biotechnol. 16 (48), 2248–2258.

Shukla, S., Ahirwal, L., Bajpai, V.K., Huh, Y.S., Han, Y.K., 2017. Growth Inhibitory Effects of *Adhatoda vasica* and Its Potential at Reducing Listeria monocytogenes in Chicken Meat. Front Microbiol 8.

Shukla, M., Mishra, M.K., 2012. Mycoflora associated with five commonly used medicinal plants of Karaikal (UT of Puducherry). Int. J. Sci. Res. Publ. 2 (1), 1–4.

Shweta, S., Zuehlke, S., Ramesha, B.T., Priti, V., Kumar, M., Ravikanth, G., et al., 2010. Endophytic fungal strains of Fusarium solani, from Apodytes dimidiata E. Mey. ex Arn (Icacinaceae) produce camptothecin, 10-hydroxycamptothecin and 9-methoxycamptothecin. Phytochemistry 71 (1), 117–122.

Singh, A.P., Wilson, T., Vorsa, V., Luthria, D., Freeman, M.R., Scott, R.M., Bilenker, D., Shah, S., Somasundaram, S., Vorsa, N., 2011. Corrigendum to "LC-MS-MS characterization of curry leaf flavonols and antioxidant activity". Food Chem. 127 (4), 80–85.

Singh, D., Rathod, V., Ninganagouda, S., Hiremath, J., Singh, A.K., Mathew, J., 2014. Optimization and characterization of silver nanoparticle by endophytic fungi Penicillium sp. isolated from Curcuma longa (turmeric) and application studies against MDR E. coli and S. aureus. Bioinorgan. Chem. Appl. 2014. Available from: http://dx.doi.org/10.1155/2014/408021.

Singh, D.K., Sharma, V.K., Kumar, J., Mishra, A., Verma, S.K., Sieber, T.N., Kharwar, R.N., 2017. Diversity of endophytic mycobiota of tropical tree *Tectona grandis* Linn. f.: Spatiotemporal and tissue type effects. Sci. Rep. 7 (1), 3745.

Singh, K.P., Upadhyay, B., Pra, R., Kumar, A., 2010. Screening of *Adhatoda vasica* Nees as a putative HIV-protease inhibitor. J. Phytol 2, 78–82.

Sommart, U., Rukachaisirikul, V., Trisuwan, K., Tadpetch, K., Phongpaichit, S., Preedanon, S., Sakayaroj, J., 2012. Tricycloalternarene derivatives from the endophytic fungus *Guignardia bidwellii* PSU-G11. Phytochem Lett 5 (1), 139–143.

Sonaimuthu, V., Krishnamoorthy, S., Johnpaul, M., 2010. Taxol producing endophytic fungus Fusarium culmorum SVJM072 from medicinal plant of Tinospora cordifolia – a first report. J. Biotechnol. 150, 425.

Song, X.Q., Zhang, X., Han, Q.J., Li, X.B., Li, G., Li, R.J., Jiao, Y., Zhou, J.C., Lou, H.X., 2013. Xanthone derivatives from *Aspergillus sydowii*, an endophytic fungus from the liverwort *Scapania ciliata* S. Lac and their immunosuppressive activities. Phytochem Lett 6 (3), 318–321.

Song, Y.C., Huang, W.Y., Sun, C., Wang, F.W., Tan, R.X., 2005. Characterization of graphislactone A as the antioxidant and free radicalscavenging substance from the culture of *Cephalosporium* sp. IFB-E001, an endophytic fungus in *Trachelospermum jasminoides*. Biol Pharma Bull 28 (3), 506–509.

Song, Z., Kennedy, G., Liew, F.J., Schilling, J.S., 2017. Fungal endophytes as priority colonizers initiating wood decomposition. Funct. Ecol. 31 (2), 407–418.

Soni, S., Anandjiwala, S., Patel, G., Rajani, M., 2008. Validation of different methods of preparation of *Adhatoda vasica* leaf juice by quantification of total alkaloids and vasicine. Indian J. Pharm. Sci. 70 (1), 36–42.

Sowparthani, K., Rajagopal, K., 2011. Screening of endophytic fungi from a medicinal plant *Justicia adathoda* L. for antibacterial activity. Int. J. Pharm. Technol. 3, 2992–3000.

Spurr Jr, H.W., Welty, R.E., 1975. Characterization of endophytic fungi in healthy leaves of Nicotiana sp. Phytopathology 65, 417–422.

Srebotnik, E., Messner, K., 1994. A simple method that uses differential staining and light microscopy to assess the selectivity of wood delignification by white rot fungi. Appl Environ Microbiol 60 (4), 1383–1386.

Sreekanth, D., Sushim, G.K., Syed, A., Khan, B.M., Ahmad, A., 2011. Molecular and morphological characterization of a taxol-producing endophytic fungus, Gliocladium sp., from Taxus baccata. Mycobiology 39 (3), 151–157.

Sreekanth, D., Syed, A., Sarkar, S., Sarkar, D., Santhakumari, B., Ahmad, A., et al., 2009. Production, purification and characterization of taxol and 10DAB III from a new endophytic fungus Gliocladium sp. isolated from the Indian yew tree, Taxus baccata. J. Microbiol. Biotechnol. 19 (11), 1342–1347.

Stierle, A., Strobel, G., Stierle, D., Grothaus, Bignami, G., 1995. The search for a taxol-producing microorganism among the endophytic fungi of the Pacific yew, Taxus brevifolia. J. Nat. Prod. 58 (9), 1315–1324.

Stierle, A.A., Stierle, D.B., 2015. Bioactive secondary metabolites produced by the fungal endophytes of conifers. Nat. product. Commun. 10 (10), 1671.

Strobel, G., Daisy, B., 2003. Bioprospecting for microbial endophytes and their natural products. Microbiol. Mol. Biol. Rev. 67 (4), 491–502.

Strobel, G., Daisy, B., Castillo, U., Harper, J., 2004. Natural products from endophytic microorganisms. J. Nat. Prod. 67 (2), 257–268.

Strobel, G., Stierle, A., Stierle, D., Hess, W.M., 1993. Taxomyces andreanae, a proposed new taxon for a bulbilliferous hyphomycete associated with Pacific yew (Taxus brevifolia). Mycotaxon 47, 71–80.

Strobel, G., Yang, X., Sears, J., Kramer, R., Sidhu, R.S., Hess, W.M., 1996. Taxol from Pestalotiopsis microspora, an endophytic fungus of Taxus wallachiana. Microbiology 142 (2), 435–440.

Strobel, G.A., 2001. Ambuic acid, a highly functionalized cyclohexenone with antifungal activity from Pestalotiopsis sp. and Monochaetia sp. Phytochemistry 56 (5), 463–468.

Strobel, G.A., 2003. Endophytes as sources of bioactive products. Microb. Infect. 5, 535–544.

Strobel, G.A., Dirkse, E., Sears, J., Markworth, C., 2001. Volatile antimicrobials from *Muscodor albus*, a novel endophytic fungus. Microbiology 147 (11), 2943–2950.

Strobel, G.A., Knighton, B., Kluck, K., Ren, Y., Livinghouse, T., Griffin, M., Spakowicz, D., Sears, J., 2008. The production of myco-diesel hydrocarbons and their derivatives by the endophytic fungus *Gliocladium roseum* (NRRL 50072). Microbiology 154 (11), 3319–3328.

Strobel, G.A., Miller, R.V., Martinez-Miller, C., Condron, M.M., Teplow, D.B., Hess, W.M., 1999. Cryptocandin, a potent antimycotic from the endophytic fungus *Cryptosporiopsis cf. quercina*. Microbiology 145 (8), 1919–1926.

Sturz, A.V., Christie, B.R., Nowak, J., 2000. Bacterial endophytes: potential role in developing sustainable systems of crop production. Crit. Rev. Plant Sci. 19 (1), 1–30.

Sunitha, V.H., Devi, D.N., Srinivas, C., 2013. Extracellular enzymatic activity of endophytic fungal strains isolated from medicinal plants. World J. Agric. Sci. 9 (1), 01–09.

Suradkar, K., Hande, D.V., 2017. Morphotaxonomy of endophytic fungi on Cissus Quadrangularis from Amravati (MS) India. J. Bacteriol. Mycol. (Open Access.) 5 (2), 00127.

Suryanarayanan, T.S., Vijaykrishna, D., 2001. Fungal endophytes of aerial roots of Ficus benghalensis. Fungal Divers. 8, 155–161.

Suryanarayanan, T.S., Kumaresan, V., Johnson, J.A., 1998. Foliar fungal endophytes from two species of the mangrove Rhizophora. Can. J. Microbiol. 44 (10), 1003–1006.

Suryanarayanan, T.S., Murali, T.S., 2006. Incidence of *Leptosphaerulina crassiasca* in symptomless leaves of peanut in southern India. J. Basic Microbiol. 46 (4), 305–309.

Suryanarayanan, T.S., Murali, T.S., Venkatesan, G., 2002. Occurrence and distribution of fungal endophytes in tropical forests across a rainfall gradient. Can. J. Bot. 80 (8), 818–826.

Suryanarayanan, T.S., Murali, T.S., Thirunavukkarasu, N., Rajulu, M.G., Venkatesan, G., Sukumar, R., 2011. Endophytic fungal communities in woody perennials of three tropical forest types of the Western Ghats, southern India. Biodivers. Conserv. 20 (5), 913–928.

Suryanarayanan, T.S., Thirunavukkarasu, N., Govindarajulu, M.B., Sasse, F., Jansen, R., Murali, T.S., 2009. Fungal endophytes and bioprospecting. Fungal Biol. Rev. 23 (1-2), 9–19.

Tan, R.X., Zou, W.X., 2001. Endophytes: a rich source of functional metabolites. Nat. Prod. Rep. 18, 448–459.

Tang, C.Q., Leasi, F., Obertegger, U., Kieneke, A., Barraclough, T.G., Fontaneto, D., 2012. The widely used small subunit 18S rDNA molecule greatly underestimates true diversity in biodiversity surveys of the meiofauna. Proc. Natl. Acad. Sci. 109 (40), 16208–16212.

Tejesvi, M.V., Mahesh, B., Nalini, M.S., Prakash, H.S., Kini, K.R., Subbiah, V., et al., 2005. Endophytic fungal assemblages from inner bark and twig of Terminalia arjuna W. & A. (Combretaceae). World J. Microbiol. Biotechnol. 21 (8-9), 1535–1540.

Tejesvi, M.V., Mahesh, B., Nalini, M.S., Prakash, H.S., Kini, K.R., Subbiah, V., Shetty, H.S., 2006. Fungal endophyte assemblages from ethnopharmaceutically important medicinal trees. Can. J. Microbiol. 52 (5), 427–435.

Tervet, I.W., Hollis, J., 1948. Bacteria in the storage organs of healthy plants. Phytopathology 38, 960–967.

Thalavaipandian, A., Ramesh, V., Bagyalakshmi, Muthuramkumar, S., Rajendran, A., 2011. Diversity of fungal endophytes in medicinal plants of Courtallam hills, Western Ghats, India. Mycosphere 2 (5), 575–582.

Thangaraj, A., Mythili, M., Fathima, M., 2017. Action of root-endophytic fungi of Tinospora cordifolia (Willd.) Miers against cancer cell lines and human pathogens. J. Pharm. Biomed. Sci. 7 (01).

Tian, J.J., Gao, X.X., Zhang, W.M., Wang, L., Qu, L.H., 2013. Molecular identification of endophytic fungi from Aquilaria sinensis and artificial agarwood induced by pinholes-infusion technique. Afr. J. Biotechnol. 12 (21).

Tillotson, G.S., Zinner, S.H., 2017. Burden of antimicrobial resistance in an era of decreasing susceptibility. Expert. Rev. Anti-Infective Ther. 15 (7), 663–676.

Tiwari, K., Chittora, M., 2013. Assessment of genetic diversity and distribution of endophytic fungal communities of Alternaria solani isolates associated with the dominant Karanja plants in Sanganer Region of Rajasthan. Springer Plus 2 (1), 313.

Tiwari, K., 2015. The future products: endophytic fungal metabolites. J. Biodivers. Bioprospect. Dev. 2 (145), 2376-0214.

Tiwari, P., Nathiya, B., Mahalingam, G., 2017. Antidiabetic activity of endophytic fungi isolated from Ficus religiosa. Asian J. Pharm. Clin. Res. 10, 59–61.

Trese, A.T., Loschke, D.C., 1990. High contrast resolution of the mycelia of pathogenic fungi in corn tissue after staining with calcofluor and destaining with cellulase. Phytopathology 80 (2), 196–200.

Ushasri, R., Anusha, R., 2015. In vitro anti-diabetic activity of ethanolic and acetone extracts of endophytic fungi Syncephalastrum racemosum isolated from the seaweed Gracilaria corticata by alpha-amylase inhibition assay method. Int. J. Curr. Microbiol. Appl. Sci. 4 (1), 254–259.

Uzma, F., Chowdappa, S., 2017. Antimicrobial and antioxidant potential of endophytic fungi isolated from ethnomedicinal plants of Western Ghats, Karnataka. J. Pure Appl. Microbiol. 11 (2), 1009–1026.

Verma, A.K., Kumar, M., Bussmann, R.W., 2007. Medicinal plants in an urban environment: the medicinal flora of Banares Hindu University, Varanasi, Uttar Pradesh. J. Ethnobiol. Ethnomed. 3 (1), 35.

Verma, V.C., Lobkovsky, E., Gange, A.C., Singh, S.K., Prakash, S., 2011. Piperine production by endophytic fungus Periconia sp. isolated from Piper longum L. J. antibiotics 64 (6), 427.

Wagenaar, M.M., Corwin, J., Strobel, G., Clardy, J., 2000. Three new cytochalasins produced by an endophytic fungus in the genus Rhinocladiella. J. Nat. Prod. 63 (12), 1692–1695.

Wang, J., Li, G., Lu, H., Zheng, Z., Huang, Y., Su, W., 2000. Taxol from Tubercularia sp. strain TF5, an endophytic fungus of Taxus mairei. FEMS Microbiol. Lett. 193 (2), 249–253.

Wang, Q.X., Li, S.F., Zhao, F., Dai, H.Q., Bao, L., Ding, R., et al., 2011. Chemical constituents from endophytic fungus Fusarium oxysporum. Fitoterapia 82 (5), 777–781.

Wang, Y.T., Lo, H.S., Wang, H., 2008. Endophytic fungi from Taxus mairei in Taiwan: first report of Colletotrichum gloeosporioides as an endophyte of Taxus mairei. Botan. Stud. 49 (1), 39–43.

Wankhede, T.B., 2015. Antioxidant and antimicrobial properties of *Adhatoda vasica* L. Nees. Sci 3 (2), 152–156.

Waqas, M., Khan, A.L., Hamayun, M., Shahzad, R., Kang, S.M., Kim, J.G., et al., 2015. Endophytic fungi promote plant growth and mitigate the adverse effects of stem rot: an example of Penicillium citrinum and Aspergillus terreus. J. plant. Interact. 10 (1), 280–287.

Wellensiek, B., Ramakrishnan, R., Bashyal, B., Eason, Y., Gunatilaka, A.L., Ahmad, N., 2013. Inhibition of HIV-1 replication by secondary metabolites from endophytic fungi of desert plants. Open. Virol. J. 7, 72.

White Jr, J.F., Morgan-Jones, G., 1996. Morphological and physiological adaptations of Balansiae and trends in the evolution of grass endophytes. In: Redlin, S.C., Carris, L.M. (Eds.), Endophytic Fungi in Woody Plants: Systematics, Ecology and Evolution. APS Press, St. Paul, MN, US, pp. 133–154.

Widler, B. and Muller EPilze., von, *Arctostaphylos.* 1984 Untersuchungen uber endophytische Pilze von Arctostaphylos uva-ursi (L) Sprengel (Ericaceae). *Bot. helv, 94*, 307-37.

Wilcox, H.E., 1964. Staining plant tissues with Chlorazol Black E and Pianese III-B. Stain. Technol. 39, 81–86.

Wu, S.H., He, J., Li, X.N., Huang, R., Song, F., Chen, Y.W., et al., 2014. Guaiane sesquiterpenes and isopimarane diterpenes from an endophytic fungus Xylaria sp. Phytochemistry 105, 197–204.

Xiong, Z.Q., Yang, Y.Y., Zhao, N., Wang, Y., 2013. Diversity of endophytic fungi and screening of fungal paclitaxel producer from Anglojap yew, Taxus x media. BMC Microbiol. 13 (1), 71.

Xu, S., Ge, H.M., Song, Y.C., Shen, Y., Ding, H., Tan, R.X., 2009. Cytotoxic cytochalasin metabolites of endophytic Endothia gyrosa. Chem. Biodivers. 6 (5), 739–745.

Yang, X., Zhang, L., Guo, B., Guo, S., 2004. Preliminary study of a vincristine producing endophytic fungus isolated from leaves of *Catharanthus roseus*. Chinese traditional and herbal drugs 35 (1), 79–81.

Yunianto, P., Rusman, Y., Saepudin, E., Suwarso, W.P., Sumaryono, W., 2014. Alkaloid (meleagrine and chrysogine) from endophytic fungi (*Penicillium* sp.) of *Annona squamosa* L. Pakistan J. Biol. Sci. 17 (5), 667–674.

Zahoor, M., Irshad, M., Rahman, H., Qasim, M., Afridi, S.G., Qadir, M., et al., 2017. Alleviation of heavy metal toxicity and phytostimulation of Brassica campestris L. by endophytic Mucor sp. MHR-7. Ecotoxicol. Environ. Saf. 142, 139–149.

Zakaria, L., Jamil, M.I.M., Anuar, I.S.M., 2016. Molecular characterisation of endophytic fungi from roots of wild banana (Musa acuminata). Trop. Life Sci. Res. 27 (1), 153.

Zhang, G., Sun, S., Zhu, T., Lin, Z., Gu, J., Li, D., Gu, Q., 2011. Antiviral isoindolone derivatives from an endophytic fungus *Emericella* sp. Associated with *Aegiceras corniculatum.* Phytochemistry 72 (11-12), 1436–1442.

Zhang, H.W., Song, Y.C., Tan, R.X., 2006. Biology and chemistry of endophytes. Nat. product. Rep. 23 (5), 753–771.

Zhang, Q., Xiao, J., Sun, Q.Q., Qin, J.C., Pescitelli, G., Gao, J.M., 2014. Characterization of cytochalasins from the endophytic Xylaria sp. and their biological functions. J. Agric. food Chem. 62 (45), 10962–10969.

Zhang, T., Yao, Y.F., 2015. Endophytic fungal communities associated with vascular plants in the high arctic zone are highly diverse and host-plant specific. PLoS One 10 (6), pe0130051.

Zhang, Y.Z., Sun, X., Zeckner, D.J., Sachs, R.K., Current, W.L., Gidda, J., et al., 2001. Syntheses and antifungal activities of novel 3-amido bearing pseudomycin analogues. Bioorganic Med. Chem. Lett. 11 (7), 903–907.

Zhao, J., Li, C., Wang, W., Zhao, C., Luo, M., Mu, F., et al., 2013. Hypocrea lixii, novel endophytic fungi producing anticancer agent cajanol, isolated from pigeon pea (Cajanus cajan [L.] Mills). J. Appl. Microbiol. 115 (1), 102–113.

Zikmundova, M., Drandarov, K., Hesse, M., Werner, C., 2002. Hydroxylated 2-amino-3H-phenoxazin-3-one derivatives as products of 2-hydroxy-1, 4-benzoxazin-3-one (HBOA) biotransformation by *Chaetosphaeria* sp., an endophytic fungus from *Aphelandra tetragona*. Zeitschrift für Naturforschung C 57 (7-8), 660–665.

Further reading

Bhagobaty, R.K., Joshi, S.R., 2011. Multi-loci molecular characterisation of endophytic fungi isolated from five medicinal plants of Meghalaya, India. Mycobiology 39 (2), 71–78.

Bhagobaty, R.K., Joshi, S.R., Kumar, R., 2010. Penicillium verruculosum RS7PF: a root fungal endophyte associated with an ethno medicinal plant of the indigenous tribes of Eastern India. Afr. J. Microbiol. Res. 4 (9), 766–770.

Blunt, J.W., Copp, B.R., Munro, M.H., Northcote, T., Prinsep, M.R., 2011. Marine natural products. Nat. product. Rep. 28 (2), 196–268.

Devarajan, T., Suryanarayanan, T.S., 2002. Endophytic fungi associated with the tropical seagrass Halophila ovalis (Hydrocharitaceae). Indian J. Mar. Sci. 31, 73–74.

Li, J.Y., Strobel, G.A., 2001. Jesterone and hydroxy-jesterone antioomycetcyclohexenenone epoxides from the endophytic fungus Pestalotiopsis jesteri. Phytochemistry 57, 261–265.

Li, J.Y., Harper, J.K., Grant, D.M., Tombe, B.O., Bashyal, B., Hess, W.M., et al., 2017. Diversity of endophytic mycobiota of tropical tree Tectona grandis Linn. f.: spatiotemporal and tissue type effects. Sci. Rep. 7 (1), 3745.

Mahapatra, S., Banerjee, D., 2010. Diversity and screening for antimicrobial activity of endophytic fungi from Alstonia scholaris. Acta Microbiol. Immunol. Hungarica 57 (3), 215–223.

Prabhugaonkar, A., Bhat, D.J., 2009. Rattania setulifera, an undescribed endophytic hyphomycete on rattans from Western Ghats, India. Mycotaxon 108, 217–222.

Rajendran, A., 2011. Diversity of fungal endophytes in medicinal plants of Courtallam hills, Western Ghats, India. Mycosphere 2 (5), 575–582.

Seena, S., Sridhar, K.R., 2004. Endophytic fungal diversity of 2 sand dune wild legumes from the southwest coast of India. Can. J. Microbiol. 50 (12), 1015–1021.

Souza, B.D.S., Santos, T.T.D., 2017. Endophytic fungi in economically important plants: ecological aspects, diversity and potential biotechnological applications. J. bioenergy food Sci. 4 (2), 113–126.

Sowparthani, K., Rajagopal, K., 2011. Screening of endophytic fungi from a medicinal plant Justicia adathoda L. for antibacterial activity. Int. J. Pharm. Technol. 3 (3), 2992–3000.

Suryanarayanan, T.S., Murali, T.S., 2006. Incidence of Leptosphaerulina crassiasca in symptomless leaves of peanut in southern India. J. basic. microbiology 46 (4), 305–309.

Chapter 11

Prospect of biofuel production by fungus

Arpitha Ramachandraiah[1,a], Juhi Kaushik[1,a], Lavanya Gowda[1,a], Premchand Subhash Chigadannavar[1,a], Ramya Krishnappa[1,a], Sandeep Venkatesh[1,a], Sheersha Sivadas[1,a], Sneha Judith[1,a], Unni Maya Ramesh[1,a], Zeba Quadri[1,a], Priyadarshini Dey[1,a,*], Mahendra Chinthala[2,a], Deepak Gola[3,a], Nitin Chauhan[4,a] and Randhir Kumar Bharti[5,a]

[1]*Department of Biotechnology, Ramaiah Institute of Technology, Bengaluru, India,* [2]*Department of Chemical Engineering, Ramaiah Institute of Technology, Bengaluru, India,* [3]*Department of Biotechnology, Noida Institute of Engineering and Technology, Noida, India,* [4]*Department of Microbiology, Shaheed Rajguru College of Applied Sciences for Women, University of Delhi, New Delhi, India,* [5]*University School of Environmental Management, Guru Gobind Singh Indraprastha University, New Delhi, India*

**Corresponding author.* Email: priyadey@msrit.edu

Introduction

Due to the limited nature of the fossil fuel reserves, there has been intensive development and improvement of fuels that are sustainable and renewable. Further, the usage of fossil fuels causes the release of greenhouse gases (GHG) viz. (95%) carbon dioxide (CO_2); (1%) methane (CH_4) and nitrous oxides (N_2O). The remaining 3% of GHG comes from the leaking hydro fluorocarbons (HFCs) from the air conditioners of the vehicles (Atabani et al., 2012). These emissions are consequently responsible for rise in global warming. Environmental friendly and renewable fuels are the only options for sustained supply of clean energy. Fuels such as bioethanol, biodiesel, biogas etc. are proving to be eco-friendly and economical alternative to the depleting fossil fuels of petrol, diesel and natural gas, respectively.

For production of bioethanol, feedstocks from first generation (corn, sugarcane, etc.) and second generation (agricultural wastes, forestry wastes, etc.) can be utilized. Second generation feedstocks contain lignocellulosic biomass that can be pretreated by fungal enzymes for the release of cellulose for

[a] These authors contributed equally.

Fungi Bio-Prospects in Sustainable Agriculture, Environment and Nano-Technology.
DOI: https://doi.org/10.1016/B978-0-12-821394-0.00011-1

fermentation process (Dashtban et al., 2009). To scale up the production of bioethanol from different feedstocks, the type of bioreactors required depend on the type of microbial strain, physicochemical parameters, etc.

Biodiesel can be prepared from various sources of lipids i.e., plants, animals or microbes, therefore it is an essential form of renewable energy. Studies have pointed that fungi are oleaginous when the growth medium is supplemented with glucose and sucrose. This fungal biomass can be utilized further for lipid extraction. This microbial lipid can be transformed later to fatty acid methyl esters (FAME) directly using cost effective methods (Bautista et al., 2012). In the process of biodiesel production, catalyst plays crucial role. There are two types catalysts i.e., homogeneous and heterogeneous; homogeneous catalysts are mostly sodium hydroxide (NaOH), sulphuric acid (H_2SO_4), sodium methoxide ($NaOCH_3$) and potassium hydroxide (KOH) which are difficult to separate from the transesterification products. Heterogeneous catalysts can be easily separated from the products and generally include, alumina based, silica-based, calcium oxide based, zirconium based, magnesium based, titanium oxide based, etc. (Refaat, 2011). For industrial scale biodiesel production by fungi, optimization of various parameters for efficient biodiesel production is the main challenge. Therefore, studies are limited at lab level biofuel production only.

Further, biogas has also emerged as an excellent form of renewable energy produced from the conversion of organic feedstock such as agricultural, animal, industrial and municipal wet wastes. Recent studies have also reported the potential of various saprotrophic fungi for the production of methane (Lenhart et al., 2012).

Latest studies have also highlighted the development of microbial fuel cells (MFCs), and fungal based MFC are most extensively studied, where the electron is transferred via cytochrome C. Studies showed that fungal species such as *Saccharomyces cerevisiae, Candida* sp., and filamentous fungi like *Rhizopus* sp., *Aspergillus* sp., and *Penicillium* sp. have the potential for power generation (Belniak and Mamińska, 2018).

Certain factors are considered while determining the type of feedstocks required for the production of the different biofuels that is based on its life cycle assessment. This assessment encompasses factors such as land availability, water availability and requirement, energy crops cultivation practices, energy input and output, release of greenhouse gases, requirement of pesticides, soil fertility, conservation of biodiversity, transport and storage of feedstocks and fuel, economic and socio benefit of the biofuels, contribution of the biofuels towards the air quality (Atabani et al., 2012).

Thus, fungus being easily available and its easy culturing/growth at various setups makes it an excellent tool of research in the bioenergy sector. The present chapter covers various important aspects of fungal assisted clean energy production, mechanism involved and its future scope in the light of existing fuel crisis.

Bioethanol production

Bioethanol is an alternative fuel produced from the fermentation of sugar and starchy substrates by anaerobic microorganisms of bacterial and fungal origin. Bioethanol is the most widespread alternative biofuel for gasoline. Renewable agricultural sources such as corn, molasses are utilized for the production of bioethanol. Besides bioethanol is less toxic when compared to other alcoholic fuels. Also, the byproducts e.g. acetic acid and acetaldehyde released on incomplete oxidation of bioethanol are less harmful. Further, bioethanol is blended with gasoline to make gasoline–ethanol mixtures viz. E15 (15% ethanol and 85% gasoline) and E85 (85% ethanol and 15% gasoline), respectively (Vohra et al., 2014).

Status of bioethanol production in the world and in India

According to the alternative fuel datacentre in 2018, USA was the largest producer of bioethanol (16,061 million gallons) from corn followed by Brazil with 7920 million gallons from sugarcane. Altogether US and Brazil account for 85% of the world's ethanol. While European Union and China produced 1430 million gallons and 1050 million gallons of bioethanol, respectively.

With respect to India, the Government has authorized the blending of ethanol in gasoline by 10% to reduce the burden of gasoline import. The Bureau of Indian Standards (BIS) has specified 20% ethanol blending with gasoline for usage in vehicles.

Feedstocks and biomass used for bioethanol production

The feedstocks used for the generation of bioethanol can be classified as first generation agricultural raw materials containing sugar (sugarcanes, sugar beet, molasses and fruits juices) and starch that are also edible. These substrates can be directly fermented by microorganisms for bioethanol production. Second generation include lignocellulosic materials from agricultural waste residues (rice husk, etc.) and forestry waste residues (tree trunks, wood pallets, etc.). Due to the poor digestion of lignin in lignocellulosic materials, some pretreatment methods are necessary to make the hemicellulose and cellulose carbohydrates available for fermentation (Vohra et al., 2014).

Fungus as microorganism for pretreatment of lignocellulosic substrates for bioethanol production

Pretreatment strategies of lignin are mainly of two types viz. physico chemical treatment and biological treatment. Physico-chemical treatments comprises of treatment of feedstocks with harsh and strong acid or base for hydrolysis process. In the long run, these physico chemical treatment makes the purification of ethanol difficult. This also adds on to the cost of the bioethanol production and leads to carbon footprint.

Thus, biological methods involving enzymes from microbes are an ideal choice for pretreatment of lignocellulosic wastes. Microbes such as fungi produce a plethora of these degrading enzymes (lignase and hemicellulase) in abundance. These enzymes are produced extracellularly which is harmless to humans. Thus, fungal enzymes are easy to cultivate, easy to be extracted from fungus and are an economical adjunct to the physicochemical method of pretreatment of lignocellulosic materials

The fungal strains utilized for the production bioethanol can be of lignolytic, hydrolytic, selective lignin degrading and non-selective lignin degrading types.

Several fungi principally from Basidiomycetes produce extracellular enzymes responsible for lignocellulose degradation. The Basidiomycetous fungi comprises of white rot, brown rot and litter-decomposing (Steffen et al., 2003). However only white-rot fungi and litter-decomposing fungi can mineralize lignin effectively due to their unique ligninolytic systems (Eriksson et al., 1990; Hatakka, 2005).

The white rot fungi produce various ligninolytic enzymes such as laccases, different peroxidases, including lignin peroxidase, manganese-dependent peroxidase and versatile peroxidase (Murcia et al., 2002). Few wood rotting fungi contain all three classes of lignin degrading enzymes while the others contain either of these enzymes (Reddy and Mathew, 2001). Additionally, these lignocellulosic substrate degrading enzymes also have the ability to degrade different xenobiotic compounds e.g. dyes (Pointing, 2001; Reddy, 1995; Wesenberg et al., 2003).

Conventionally, white rot fungi have been employed for biopulping, upgradation of forage and also for bioremediation of soil and wastewater (Anasonye et al., 2014; Sánchez, 2009; Wesenberg et al., 2003).

Lignin degradation by the white rot fungus occurs by oxidation process and include chemical oxidants such as hydroxyl radicals and singlet oxygen. The degradation of lignin by white rot fungus occurs by four mechanisms viz. oxidation, demethylation, side chain oxidation, and propyl side chain cleavage. However, brown rot fungi bypass the lignin barrier and degrade cellulose and hemicellulose (Eriksson et al., 1990). The following Table 11.1 depicts the variety of enzymes produced by different fungi for lignocellulosic degradation.

Bioreactors for bioethanol production by fungus

Bioethanol production from any feed stock/substrates involve three steps: pretreatment, enzymatic hydrolysis and fermentation. Bioethanol production by fermentation of sugar and starch containing substrate commercial level

TABLE 11.1 Various enzymes produced by different fungi for lignocellulosic degradation.

Enzyme	Fungi strain	Author
Peroxidases	*Phanerochaete chrysosporium, Ceriporia lacerata, Cyathus stercoreus,*	Saritha et al. (2013), Wan and Li (2012)
Laccases	*Pycnoporus cinnabarinus, Trametes hirsute, Pleurotus ostreatus*	Wan and Li (2012)
Lignin Peroxidase, Mn Peroxidase	*Phanerochaete chrysoporium*	Rüttimann-Johnson et al. (1993)
Manganese peroxidase, laccase	*Ceriporiopsis subvermispora*	Rajakumar et al. (1996)
Hydrolytic enzyme	*Pleurotus ostreatus*	Elisashvili et al. (2008), Sánchez (2009)
Cellulases, endoglucanase	*Pleurotus ostreatus*	Elisashvili et al. (2008), Garzillo et al. (1992), Okamoto et al. (2010), Velázquez-Cedeño et al. (2002)
Cellobiohydranase	*Ceriporiopsis subvermispora*	Dumonceaux et al. (2001)

requires scaling up of lab process to pilot and finally to industrial level. Optimization of various parameters for efficient bioethanol production is the main challenge in industrial scale. Production of bioethanol using fungal strain is limited to lab scale studies. Satari et al. (2016) investigated the production of ethanol via two Zygomycetes fungi, *Mucor indicus* and *Rhizopus oryzae*, using citrus waste free sugars as source of substrate in lab shake flasks and airlift bioreactor (3.5 L). The total sugar (glucose and fructose) in citrus waste was 25.1 g/L at room temperature. It was observed that during shake flask cultivation, both fungal strains utilized glucose first, while fructose remained intact in the media. Up to 69% ethanol production (theoretical value) was observed in *Mucor indicus* with 0.39 g/g biomass yield.

In another study (Alam et al., 2009), the bioethanol production was carried in stirred tank bioreactor utilizing multiple fungal strains i.e., *Trichoderma harzianum, Phanerochaete chrysosporium, Mucor hiemalis*, and *Saccharomyces cerevisiae* to maximize the production of bioethanol via palm oil mill effluent as the source of substrate. It was observed that maximum bioethanol production was obtained with the combination of

Trichoderma harzianum and *Saccharomyces cerevisiae.* The present combination produced approximately 31.6 g/L ethanol in 2 L stir tank bioreactor (working volume 1.5 L), under well optimized conditions. Various other studies have also indicated that parameters such as pH and temperature of bioreactor system are the main controlling factors for the production of bioethanol (Acharya et al., 2019; Priya et al., 2016).

In general, *Saccharomyces cerevisiae* efficiently convert the available substrate into ethanol at pH ranging from 3.5 to 6.0 and temperatures at 28–35 °C (Alam et al., 2009). As per Nigam (2000), *Saccharomyces cerevisiae* showed maximum production of bioethanol from food waste at pH 4.5 and temperature 30 °C in a continuous stirred-tank reactor of 1 L (working volume 0.4 L). In another study, Saha and Ueda (1983) obtained highest bioethanol production from glucoamylase-treated starch at 38 °C by *Saccharomyces cerevisiae.*

Jeon et al. (2007) configured a sequential series bioreactor system (4 bioreactors) for the production of ethanol, utilizing potato starch as substrate. The configured bioreactor contained four different chambers connected in series for different biochemical processes to produce ethanol using *Aspergillus niger* and *Saccharomyces cerevisiae. Aspergillus niger* was cultivated in the first bioreactor (I) system at 30 °C using broth media (15 g/L potato starch, 3 g/L $NaNO_3$ and 2.72 g/L KH_2PO_4). The glucose in this bioreactor is converted to ethanol via *Saccharomyces cerevisiae.* Up to 22 g/L/day ethanol was produced in this bioreactor configuration.

It can be conjectured that the bioethanol production using fungi in bioreactor system is largely dependent on the type of strain used, pH, temperature, etc. To scale up the process at commercial level, cost and energy analysis is important and further research is very much needed.

Biodiesel production

Biodiesel is defined as mono-alkyl esters of long chain fatty acids obtained from the reaction of vegetable oils or animal fats with alcohol in the presence of a catalyst. Among all the renewable alternative fuels, biodiesel is an attractive choice due to its biodegradability and negligible toxicity nature. Biodiesel fuel can substitute diesel in internal combustion engines with less adjustments. In terms of environmental emissions, complete combustion of biodiesel has zero emissions of sulfur, less carbon monoxide, smoke, hydrocarbon emissions and particulate matters. However, there is a slight net carbon dioxide (CO_2) release on combustion of biodiesel when considering the whole life cycle of this fuel that includes cultivation of feedstocks for production of oil and conversion to biodiesel. Also, engine modification is not required if biodiesel is blended up to 20% with diesel (B20) (Atabani et al., 2012).

Status of biodiesel production in the world and in India

According to Statista, the United States and Brazil are the leading producers of biodiesel in the world, totaling of 6.9 billion and 5.4 billion liters followed by Indonesia at 4.0 billion liters and Germany at 3.5 billion liters respectively in 2018.

In India, the Government is promoting biodiesel and usage of biodiesel blended with diesel is slowly making its way. The Indian railways are using different blends of biodiesel. The blends of Jatropha oil and diesel are used to run the diesel locomotives that operate from Thanjavur to Nagore, Tiruchirapalli to Lalgudi and Dindigul to Karur sections in Tamil Nadu, India.

Feedstocks and biomass used for biodiesel production

Feedstocks for production of biodiesel are the most significant factor for consideration. More than 350 oil-bearing crops which have been identified as potential sources for biodiesel production. The feedstocks are selected based on two criteria viz. low production costs and large production scale. Thus, it necessitates the selection of the cheapest feedstock for biodiesel production as 75% of the production cost of the biodiesel is from the feedstock alone.

First generation feedstock for biodiesel production are some of the edible oils such as, sunflower, safflower, soybeans, palm oil, coconut, peanut, rapeseed etc. It is estimated that about 95% of the world biodiesel is produced from edible oils such as rapeseed (84%), sunflower oil (13%), palm oil (1%), soybean oil and others (2%).

Due to an enormous demand and supply gap of edible oils (Ratledge and Wynn, 2002), second generation non edible oil resources for biodiesel production such as jatropha (*Jatropha curcas*), karanja (*Pongamia pinnata*), nagchampa (*Calophyllum inophyllum*), tobacco seed (*Nicotiana tabacum*), rubber seed tree (*Hevia brasiliensis*), neem (*Azadirachta indica*), mahua (*Madhuca indica*), jojoba (*Simmondsia chinensis*), etc. have been sought. Further, fats derived from animals such as beef tallow, poultry fat and pork lard, waste oils and grease are also sources for biodiesel production. However, biodiesel obtained from vegetable oils and animal fats have low performance in cold weather. Further, biodiesel from animal fats are difficult to convert to biodiesel transesterification process as it contains more of saturated fatty acids. Thus, other sources of oil such as crops (miscanthus, switchgrass, etc.) and microorganisms (algae and fungi) are being explored (Shahid and Jamal, 2011). Furthermore oil from microorganisms would be preferred as it can be produced in bioreactors and would do away with the plant based oil that competes with the land and water resources for feed crops.

Fungus as microorganism for oil production

Biodiesel from oils extracted from microorganism such as algae are known. However, literature regarding the oil extraction from fungi are scanty. Some of the second generation lipid-based biofuels are from oleaginous heterotrophic microorganisms such as yeasts and fungi which can be cultivated on various carbon sources viz sugars, glycerol, etc. (Huang et al., 2013). The lipid content which can be found in the mycelia pellets which may vary from less than 1% w/w to over 80% w/w (Ratledge and Wynn, 2002). Fungi can be a source for single cell oil (SCO) production. SCOs are composed of triacyl glycerides and their total fatty acid (FA) composition are similar to several plant oils (Bellou et al., 2016).

Around 64 of fungal species have been identified that can accumulate more than 25% by mass of lipids. Usually, lipid which is accumulated in fungi have a higher degree of unsaturation than those accumulated by yeasts. Thus, there are many notable studies regarding lipid accumulation by fungus for the production of polyunsaturated fatty acid (PUFAs) like linolenic acid, arachidonic acid, docosahexaenoic acid and eicosapentaenoic acid (Bellou et al., 2016; Ratledge and Wynn, 2002).

Particularly, the Zygomycetes is considered as important and the best producer of SCO which contains linolenic acid, a PUFA which is an important dietary and pharmaceutical component as well. *Mortierella isabelline*, an oleaginous Zygomycete has been studied which is capable of accumulating lipids around 84% w/w of dry cell weight (Papanikolaou and Aggelis, 2010). The cultivation of *Mortierella isabellina* can be carried out on several agro-industrial substrates that are carbon sources such as fructose, glucose molasses (Chatzifragkou et al., 2011), cheese whey (Vamvakaki et al., 2010), industrial glycerol, lignocellulosic sugars or hydrolysates (Fakas et al., 2009). During biodiesel production, glycerol which is produced as waste has been employed as substrate by various fungal species from the genus Zygomycetes. However, glucose has proved to be a better substrate concerning its conversion into single cell oil.

Chatzifragkou et al. (2010, 2011) studied the oil content in different fungi strains such as *Thamnidium elegans* and *Cunninghamella echinulata* which were cultivated in the flasks. The carbon sources used for this study were crude glycerol, blackstrap molasses and fructose. The oil content in the fungus was measured as 71.1%, 31.4% and 21.5% w/w, respectively. Similarly, Ruan et al. (2015) studied the oil content in *Mordellistena isabellina* which were cultivated in flask by using different carbon sources such as corn stover acid hydrolysate, corn stover alkali hydrolysate, giant reed hydrolysate, miscanthus hydrolysate and synthetic lignocellulosic hydrolysate with combination of inhibitors for which the observed oil content was 34.5%, 29.5%, 21.2%, 32.2% and 37.5% w/w, respectively. Zikou et al. (2013) studied the oil content in *Thamnidium elegans* by using flask and batch bioreactor as mode of cultivation. The oil content in *T. elegans* was 45.9%, with xylose as carbon source and the oil content of *T. elegans* when grown in batch bioreactor was 46.1% w/w with glucose as carbon source. Fakas et al. (2009) studied the oil content of *C. echinulata* which were cultured in flasks. Xylose and crude glycerol were used as carbon source for which the observed oil content was 25.6% and 53.6% w/w, respectively. Economou et al. (2011), Harde et al. (2016), Vamvakaki et al. (2010), Xing et al. (2012) studied the oil content in *M. isabellina* with corn fiber hydrolysate, lactose enriched cheese whey, rice hull hydrolysate and xylose as carbon sources. The oil content in the fungus obtained were 54.6%, 25.3%, 64.3% and 66.0% w/w, respectively. Koutb and Morsy (2011) in their study observed 80.0% w/w of oil content in *Epicoccum purpurascens* by using molasses as carbon source.

Production of biodiesel from fungal oil after transesterification process

The oil after extraction from the feedstock is not directly used as fuels due to its low volatility, high viscosity and polyunsaturated character. Hence the oil is converted to methyl ester by the process of transesterification. Transesterification (alcoholysis) is the reaction of an oil/fat with an alcohol to form esters and glycerol. The catalysts such as ferric sulfate, hydrochloric acid, sulfuric acid and organic sulfonic acid are usually used to enhance the reaction rate and also the yield (Singh and Singh, 2010).

The extraction of oil from the fungus requires advance enzymatic route (Carvalho et al., 2015). Some advanced process using ethanol which is coupled with microwave irradiation for oil extraction have been applied to reduce the steps which are required to obtain microbial oil. This green method of lipid extraction using ethanol as a solvent, lowers energy consumption, reduces unit operations and thus reducing the overall cost of biodiesel production. Thus, extraction and transesterification reactions can be performed simultaneously. This approach is also known as direct transesterification (Thliveros et al., 2014; Vicente et al., 2010).

Hence, more research is needed to upgrade the process of biodiesel from fungus.

Nano catalyzed biodiesel production

The lipids obtained from different fungi can be converted to fatty acid methyl esters via catalyzed or non-catalyzed esterification and transesterification reactions. The catalyzed reactions are of two types – homogeneous and heterogeneous. Homogeneous catalysts are mostly sulphuric acid (H_2SO_4), potassium hydroxide (KOH), sodium hydroxide (NaOH) and sodium methoxide ($NaOCH_3$) which are difficult to separate from the transesterification products. Heterogeneous catalyzed reactions are having several advantages over homogeneous catalyzed reactions, like easy separation of catalyst from the products and regeneration of the solid catalyst.

Heterogeneous catalysts are classified into heterogeneous acid and heterogeneous base catalysts. Base catalysts are mostly alkali and alkaline metal oxides. Most of the heterogeneous acid catalysts are based on inorganic mixed oxides like silica, alumina, zirconia-based oxides, heteropolyacids, mixed oxides, sulfated metal oxides, and sulfated carbon (Gardy et al., 2019). Solid acid catalysts could tolerate Free Fatty acids (FFA) and water content and hence offer more advantages over base catalysts. Also, the solid acid catalyst could effectively catalyze esterification and transesterification reactions simultaneously. However, the low reaction rates of acid catalysts made them less popular over base catalysts. The ideal features that are required for solid acid catalysts are more active sites, thermal stability, hydrophobicity, and high porous volume. Recently several nanomaterials and nanocomposites based on these metal oxides were reported for the esterification and transesterification reactions of FFA.

Alumina based catalysts

Amini et al. (2013) synthesized nano-alumina powders through the sol-gel method for the production of biodiesel with a conversion of about 95.8%. The synthesized nano-alumina was amorphous in nature with a high surface area of 312 m^2/g. It was regenerated by calcination at 700 °C for 4 h after each run. Bayat et al. (2018) incorporated alumina (Al_2O_3) on magnetic iron oxide (Fe_3O_4) nanoparticles at different ratios and synthesized a nanocomposite for the catalysis of transesterification of waste cooking oil. The oil to biodiesel conversion was observed to be 99.1% at 99.8 °C in 3 h of reaction time using Al_2O_3/Fe_3O_4 in the ratio of 1:2.

Silica-based catalysts

Lipases are biocatalysts used for the hydrolysis of triglycerides to FFA. Kanimozhi and Perinbam (2013) immobilized lipase onto amino-silane modified nanomagnetic Fe_2O_4 for the conversion of waste cooking oil to biodiesel. Kaur et al. (2018) used tungsten supported Ti/SiO_2 nanostructures for the conversion of waste cottonseed oil to biodiesel. With (5 wt.%) loading of this catalyst, the yield of biodiesel was 98% in 4 h at 65 °C. Tangy et al. (2016) doped strontium oxide (SrO) onto silica beads (SiO_2) in order to synthesize a novel, low-cost catalyst for the production of biodiesel from waste cooking oil. With 41.3% SrO on silica beads resulted in 99.4% conversion of oil to biodiesel within 10 s.

Calcium oxide-based catalysts

Among the metal oxides, calcium metal oxides were the most explored because of its availability, basic strength, and low cost. Calcium oxides easily dissociate into Ca^{2+} and O^{2-} ion pairs which are the reason for its basic strength in the FFA environment. Various combinations of calcium oxides were developed in the last decade for the conversion of fatty acids to fatty acid methyl esters. Zhao et al. (2013) examined the effect of particle size of CaO on its catalytic activity in the synthesis of biodiesel from canola oil. It was reported that 27.3 nm CaO had a surface area of about 89.3 m^2/g which yielded 100% biodiesel with (0.3 wt.%) nano CaO.

Zirconia based catalysts

Booramurthy et al. (2019) prepared biodiesel from the fat extracted from tannery waste using Ferric-Manganese doped sulfated Zirconia ($Fe-Mn-SO_4/ZrO_2$) nanocatalyst. The maximum yield of biodiesel was reported to be 96.6% at 65 °C at a (6 wt.%) catalyst load. Solid base catalyst based on zirconia was synthesized by Liu et al. (2017) for the production of biodiesel from soybean oils. Nano-magnetic $K/ZrO_2/\gamma\text{-}Fe_2O_3$ catalyst (5 wt.%) could produce biodiesel from soybean oil at 65 °C with a yield of 93.6% in 3 h reaction time.

Magnesium oxide-based catalysts

Vahid and Haghighi (2017) synthesized a nanocomposite $MgO/MgAl_2O_4$ which converted 95% of sunflower oil to biodiesel at 110 °C with a catalyst loading of (3 wt.%) in 3 h. Nanosized base catalyst Mg/Al in the ratio of 3:1 catalyzed the transesterification of jatropha oil to biodiesel with a yield of 95.2% in an ultrasonic reactor. Hydrotalcites (HTs), natural double layered hydroxide compounds, were used as a solid base catalyst. Mg–Al HTs enhanced the transesterification of triglycerides (Woodford et al., 2012).

Titanium dioxide-based catalysts

Madhuvilakku and Piraman (2013) used TiO_2-ZnO nano-catalyst for the transesterification of palm oil at 60 °C which yielded 92% of FAME with 200 mg catalyst loading. Ropero-Vega et al. (2010) studied the catalysis nature of SO_4/TiO_2 in the esterification of oleic/linoleic acids with ethanol. Because of its high acidic nature (Brønsted acidity), 82.2% conversion of FFA to biodiesel was observed at 80 °C with 2 wt.% loading.

Heteropolyacid based catalysts

A hybrid heteropolyacid based magnetic nanoparticle (SiO_2-MNP-HPA) showed 90.4% conversion of palmitic acid to FFA at mild reaction conditions (Duan et al., 2013).

Carbon-based catalysts

Guan et al. (2017) synthesized sulfonated multi-walled carbon nanotube (S-MWCNTs) solid acid catalyst which can convert 97.8% of triglyceride to biodiesel with 3.7 wt.% loading. Wu et al. (2019) synthesized sulfonated carbon nanofibers as solid acid catalysts for the transesterification of oleic acid. The sulfonated nanofiber apart from providing high surface area, it also had a large number of sulfonates, hydroxyl and carboxyl groups. Aryl sulfonic functionalized carbon nanofibers outperformed sulfonated carbon in the catalysis of triolein to biodiesel. Aryl sulfonic functionalization of carbon resulted in acidic carbon that could readily bind triglycerides on its surface (Stellwagen et al., 2013).

Thus, the nanometal oxides and their composites efficiently converted the triglycerides of different feedstocks when reacted with excess methanol to fatty acid methyl esters. The reported nanocatalyst could catalyze the transesterification of glycerides (lipids) extracted from different fungi cultures.

Biogas production

Biogas has emerged as a promising renewable and green energy, produced from the conversion of organic feedstocks (such as agricultural, animal, industrial and municipal wastes) to methane (a useful gaseous product) by microorganisms through anaerobic digestion. Biogas is majorly composed of 50–70% methane and 30–50% carbon dioxide (CO_2) (Kulkarni and Ghanegaonkar, 2019). Other components of biogas include hydrogen, nitrogen, oxygen, hydrogen sulfide, ammonia, volatile organic compounds (VOCs) and sulfur compounds (Papacz, 2011). The composition of these gases varies with the different biogas production plants (Rasi et al., 2007). The methane content in biogas can further be upgraded to natural gas quality of 75–99% methane content (Mittal et al., 2018). The gases i.e., methane, hydrogen and carbon monoxide can be combusted with oxygen. This release of energy allows biogas to be utilized as a fuel. The methane gas obtained can be introduced into a natural gas grid and can also be used as biofuel. It can further be transformed to heat energy and electricity. The solid residue obtained in the production of biogas is generally used as a source of fertilizer as it is rich in nitrogen and phosphorus (Rouches et al., 2016). Biogas production can be combined with the plans of improving sanitation and decrease in emission of greenhouse gases. Biogas has no odor and color which burns with a clear blue flame which is similar to that of liquefied petroleum gas (LPG) (Mittal et al., 2018). The additional advantage of biogas when compared to the liquid fuels of biodiesel and bioethanol, is its superior performance in terms of the energy recovered per hectare, its life cycle emissions and energy efficiency which is defined as ratio of energy produced by the biofuel to energy input in the conversion of biomass to biofuel (Börjesson and Mattiasson, 2008).

Status of biogas production in the world and in India

Global biogas generation has rapidly increased since 2000. From the year 2000–2014, the average growth of production was 11.2%. About 50% of the global biogas production occurs in Europe, 32% in Asia and 17% in America. Less than 2% of the production occurs in Africa and Oceania continents. However, China and USA are the leading biogas producers in the world as per World Bioenergy Association (WBA).

In the Indian scenario, according to the Ministry of New and Renewable Energy (MNRE) India, the estimated biogas production in India was about 20,757.0 lakh cubic meters in 2014–2015. Among the Indian states, Maharashtra lead the production with 3578.0 lakh cubic meters while Andhra Pradesh came next with 2165.0 lakh cubic meters followed by Karnataka with 1985.0 lakh cubic meters.

Feedstocks and biomass used for biogas production

Biomass resources that are suitable for biogas feedstocks are a variety of decomposing organic materials originating from forestry residues, agriculture residues that include animal manure and slurries, plants (crop) residue and energy crops (Wagner et al., 2018). Industrial feedstock includes animal by-products not meant for human consumption and by-products from biorefineries and the biofuels industry (Wagner et al., 2018). Municipal residues include the source-separated organic waste which is the house hold waste and sewage sludge. These decomposing organic feedstocks that otherwise contaminate the environment are a renewable biomass and comprise of simple to complex solid matters. These substrates usually have a high content of sugar, starch, proteins or fats, and a common feature is their ability to be easily decomposed through anaerobic decomposition (Seadi et al., 2013).

The first-generation fuel feedstocks from energy crops competes with the food crop farming. Thus, efforts are being channelized to convert the second-generation fuel feedstocks of crop residues and catch crops that includes sugarcane bagasse, wheat straw, corn stalks, rice straw etc. that are rich in lignocellulose (Wan and Li, 2012). Furthermore, lignocellulose accounts for the most abundant renewable biomass (Alvira et al., 2010). Lignocellulose comprises of 45% of cellulose as the major component, 30%

of hemicellulose, and 25% of lignin and this composition differs among various plants (Dahadha et al., 2017). Further, lignin is usually found in fusion with cellulose and hemicellulose, thus so-called lignocellulose. Lignin embeds cellulose and hemicellulose thereby making a physical barrier in the plant cell wall. Lignin is difficult to hydrolyze, so lignin breakdown is possible via co-metabolism. Furthermore, lignin cannot be digested under anaerobic conditions and protects cellulose and hemicellulose from enzymatic degradation. Thus, in plant biomass 40–60% of organic carbon remains undigested due to the recalcitrant chemical nature as well as the physical structure of this polymer. Since physical and chemical methods of pretreatment of lignocellulose result in production of inhibitory products. Hence the biological means of degradation of lignin that is economic, eco-friendly, requiring low energy input and minimum disposal costs as well as improves anaerobic degradation of lignocellulosic biomass are being sought after (Wagner et al., 2018).

Production of biogas by anaerobic digestion of organic feedstocks by microorganisms

Anaerobic decomposition is an action in which a complex mixture of symbiotic microorganisms such as bacteria, archaea, protozoa and fungi transforms complex organic matter under anaerobic condition into minerals forms such as biogas, nutrients and additional cell matter (Rouches et al., 2016). In nature this process occurs in environments like swamps, landfills, sewage digesters, marshes, ponds, hot springs, paddy fields, lakes, oceans and also in the intestinal tracts of humans and animals.

Aerobic decomposition of organic matter by microorganisms involves degradation of organic matter in the presence of oxygen as the electron acceptor to convert to CO_2, water, nitrates and sulfates. The lacunae in the aerobic decomposition is the incomplete degradation of some recalcitrant components in the organic matter that accounts to 70% of the chemical oxygen demand (COD). Another drawback is the accumulation of biomass sludge due to the active growth of the microorganism in the presence of oxygen (Christy et al., 2014).

On the other hand, anaerobic digestion results in stabilization of the organic compounds in wastewater solids, reduction in pathogens and odors and in lessening of total solids by converting the volatile solid fraction into biogas. This process also produces nutrients in available forms such as nitrogen and ammonia. The biogas production by anaerobic process has several advantages. The economic advantages include scope for diversification of the crop residues and also the production of manure due to the valorization of the digested sludge. From the agronomic point of view, there is transformation of the liquid slurry to fertilizer which are effectively assimilated by plants. This results in the reduction of odors and in insect elimination at the biogas storage pits. The environmental advantages are the recyclability and sustainability of the production of this gaseous fuel with the methane being considered as carbon neutral because there is no net increase in CO_2 in the atmosphere.

The biogas production by microorganisms is carried in four stages that include hydrolysis, acidogenesis, acetogenesis and methanogenesis. These four stages breakdown the complex insoluble substrates (polysaccharides, lipids, proteins) into smaller units by anaerobic microorganisms by production of hydrolytic enzymes such as xylanase, amylase, cellulase, protease, lipase etc. to the precursor compounds of propionate, butyric acid, valerate, acetic acid that are ultimately converted to methane, carbon dioxide, hydrogen sulfide, etc. (Cirne et al., 2007).

Fungus as microorganism for anaerobic digestion

Microorganisms from the bacterial and archaeal kingdoms carry out anaerobic digestion as they produce hordes of plant carbohydrate hydrolyzing enzymes. However, the potential role of anaerobic fungi for the production biogas has not been fully explored. Biological conversion of lignocellulosic residues is chiefly carried out by fungi. Since the cellulose, hemicellulose, and lignin are insoluble, this conversion process occurs extracellularly in syndication with the outer cell envelope. There are two types of fungal enzymes that are majorly known to degrade lignocellulose are the hydrolases that degrades polysaccharides. The other is the oxidative ligninolytic system that degrade lignin (Sánchez, 2009).

Both bacterial and fungi have the potential to digest cellulose (Dashtban et al., 2009). However, there are special group of fungi that possess enzymes to degrade lignin covered cellulose and their efficacy is 10–100-fold more than the bacterial enzymes. Further, due to the lower productivity of bacterial cellulases as compared to the fungal enzymes, only small amount of energy can be conserved. Also, cellulose degradation by bacteria is slower under anaerobic condition than under aerobic condition (Adney et al., 1991). The fungi possessing the cellulolytic arsenal of lignocellulose degradation belong to brown rot, white rot, and soft rot fungi. Soft and white fungi are able to degrade lignin and also to some extent cellulose, while brown rots mainly degrade cellulose. However, white rot fungi from the genera *Phanerochaete*, *Trametes*, *Ceripoioposis*, *Fusarium*, *Polyporus* and *Pleurotus* are a preferred choice for delignification as they have highly efficient enzymes system (Rodriguez et al., 2017; Rouches et al., 2016). Few white rot fungi can selectively attack lignin only. They require fewer inputs such as the delignification process can be carried out at room temperature and with solid state fermentation where usage of water can be minimized (Wan and Li, 2012). However, the uses of white rot fungi as delignifying agents are still at a nascent stage. The following Table 11.2 depicts few examples of fungi that aid in the production of biogas.

Microbial fuel cell

Microbial Fuel Cell (MFC) is a technical instrumentation that are ecological source of green bioelectricity produced from organic matter. An MFC is a system in which the microorganisms (fungi, bacteria, algae) convert chemical energy produced by the oxidation of organic compounds (in the wastewater) into electron carriers such as ATP that transfers the electrons to a terminal electron acceptor to generate electricity (Chaturvedi and Verma, 2016). MFC are an integration of interdisciplinary sciences such as microbiology, electrochemistry, environmental engineering and material science, these hybrid technologies offer sustainable solutions for wastewater treatment with the potential to generate electricity (Pontié et al., 2019).

Current scenario of MFC's in the world and in India

According to the report of International Energy Outlook (IEO) submitted by US energy information administration in 2019. The global market for MFC is forecasted to reach USD 19.5 million by 2026. In the last few decades extensive fundamental research on MFC has been carried out in many institutes around the world as these offer great promise in a wastewater treatment and electricity generation simultaneously. The developing nations with emerging economies have a greater influence on wastewater treatment and electrification in the industrial sector, as it demands the use of new technologies to address the primary need for drinking water and wastewater treatment. Further, energy requirements in developing Asia, south and central America is projected more than double between 1997 and 2020. The fastest-growing fuel consumed according to Organization for Economic Co-operation and Development (OECD) countries is the purchased electricity. The share of electricity of the total energy consumption is projected to increase from 2018 to 2050 from 56% to 62%. The Global market foretells that the Asia Pacific region, USA, UK and Canada will have substantial annual market rise by the creation of a manufacturing hub for cost-effective technologies, due to the availability of low-cost labor and raw materials. USA is the world leader in the development of fuel cells. United States Environmental Protection Agency (US EPA) report of 2008 states that US spends $23 billion annually on domestic wastewater treatment and for improvement of public-owned infrastructure which costs approximately $200 billion. About 3–4% of electrical energy load (110 Terawatt-hour/year) of US accounted for wastewater treatment, which is equivalent to the annual

TABLE 11.2 Fungi that aid in the production of biogas.

Fungi	Enzymes produced by fungi	Substrate used	Gases produced	References
Neocallimastigomycota	Cellulosome- glycosyl hydrolase	Municipal solid waste, animal waste and lignocellulose-rich substrates	Acetate, formate, CO_2, hydrogen	Dollhofer et al. (2015)
Piromyces	Cellulosome-isomerases	Municipal solid waste, animal waste and lignocellulose-rich substrates	Acetate, formate, CO_2, hydrogen	Dollhofer et al. (2015)
Orpinomyces	Cellulosome	Municipal solid waste, animal waste and lignocellulose-rich substrates	Acetate, formate, CO_2, hydrogen	Dollhofer et al. (2015)
Trichoderma reesei	Cellulases	Lignocellulosic substrates	Acetate, formate, CO_2, hydrogen	Muthangya et al. (2009)
Sporotrichum	Hydroxylase	Orange processing waste	Methane and CO_2	Srilatha et al. (1995)
Aspergillus	Hydrolase	Orange processing waste	Methane and CO_2	Srilatha et al. (1995)
Fusarium	Cellulases	Orange processing waste	Methane and CO_2	Srilatha et al. (1995)
Ceriporiopsis subvermispora	Laccases and manganese peroxidases	Albiza chips	Methane	Ge et al. (2015)

electricity usage of 9.6 million households (McCarty et al., 2011). The researchers from Bristol Bioenergy Center, University of West of England, United Kingdom have designed a miniaturized ceramic-based MFC for power generation from urine, and the team recently demonstrated it by charging a commercially available mobile phone first time using MFCs which is being fed with real neat urine (Gajda et al., 2018).

At the end of 2022 India will be the fifth largest electricity producer. On an average a total of 36,400 million litres per day (MLD) of wastewater generated by using numerous sectors, which has attracted the eye of the Indian scientific community that will serve multiple functions using MFCs. Indian researchers have developed and optimized fuel cell configurations over the past few years. Marine sludge collected from the Bay of Bengal, India was used as inoculum for construction of MFC (Patel et al., 2019; Yadav et al., 2012).

Mechanism of electricity production in MFC

MFC are bioelectrochemical system and are designed with two chambers, one with cathode and one with anode, divided by a membrane. In the anode chamber, the substrates such as acetate, glucose, sucrose, and fructose are digested (anaerobic digestion) by microbes through their metabolic activity that concurrently generates protons and electrons by means of specific substrates (e.g., cytochrome, pilli, filaments), or by direct transfer of electrons. The generated electrons reach the anode and are transferred via external electrical circuit to the cathode. Oxygen is utilized as the final electron acceptor in the cathode chamber. Oxygen is commonly used in cathode chamber due to its intrinsic sustainability and relatively high oxidation and reduction potential with consecutive/periodical feed of organic substrate. Oxygen combines with the protons generated at anode chamber to form water, thereby completing the circuit (Xu et al., 2016).

Microorganisms have the ability to utilize an array of organic compounds (carbohydrates, proteins, lipids) as carbon and energy sources for their metabolic activity. These organic compounds are electron contributors in krebs cycle reactions that form the energy currency adenosine triphosphate (ATP) prior to the cyclic chain reactions. Through the process of glycolysis, these complex lipids, proteins and carbohydrates convert to simple monomers that forms acetyl coenzyme A (CoA) that initiates the kreb cycle. The kreb cycle is completed with the electron conveyors, NADH and $FADH_2$. Through different intricate layer intermediates with the help of ATP synthase transmembrane protein adenosine diphosphate (ADP) is converted to ATP. In the anodic compartment, a microbial cell replaces a cathode as the terminal electron acceptor. Certain redox enzymes present in the cell membrane of the fungus like ferricyanide reductase or lactate dehydrogenase aid in the oxidation of organic compounds (Singh et al., 2019).

Mechanisms of electron transfer in MFC

Direct electron transfer

By direct cell attachment or through pili. Direct electron transfer necessitates a direct contact between the anode and the microorganisms. Pili are nanowires that connects the microorganism's membrane and the anode surface (Sayed and Abdelkareem, 2017).

Indirect electron transfer (exogenous or endogenous)

In this process, the redox active material which is called as mediator is responsible for the transfer of the electrons between microorganisms and the anode surface. This mediator is produced naturally by microorganisms or added externally. This type of mediator aided mechanism, transfers the electrons higher than that of direct electron transfer. Endogenous mediators have advantage over exogenous mediator such as they are carried by the microorganisms in the cost-effective way and have no toxic effect on the microorganisms. Exogenous mediators include methylene blue, yeast extract, bromocresol purple, cresol red, neutral red, methyl orange, thionine, etc. While endogenous mediators are flavin, quinones phenazines (Sayed and Abdelkareem, 2017).

Why fungus is superior to bacteria or algae?

Yeast is effectively utilized as a biological catalyst in MFC, that displays distinctive electron movements according to the species. *Kluyveromyces marxianus* is one of the most promising yeast strains as it could successfully utilize the natural complex organic materials with high power yield considerably under high working temperature conditions. In *Saccharomyces cerevisiae* and *Pichia anomala*, the electron movement happens through the surface-bound species; in *Candida melibiosica*, *Hansenula polymorpha* and *Arxula adeninivorans*, the transfer of electrons from yeast cells to the anode is both by the discharge of redox particles and by the movement of electron. The modification of biocatalysts in MFC essentially upgrades the electron transfer rate and also forcing living cells to modify their various catabolic pathways by channeling electrons from different energetic levels. This results in increase in the energy production, enhancing the transportation ability of electrons due to better networking ability of the fungal hyphae and generation of stable electricity that can drive the external electrochemical process. Therefore, yeast and fungus could be a preferred choice for waste water treatment and electricity generation than algae and bacteria (Sayed and Abdelkareem, 2017).

Fungus as anode and cathode

The first observation of electrical current being generated by microorganisms, *Escherichia coli* and *Saccharomyces cerevisiae* was done by Potter in 1911 (Włodarczyk and Włodarczyk, 2019). Due to the anodes' direct contact with the electron generating microbes, the anode is a crucial element of each MFC. The anode not only helps in electron conductivity, but also strongly impacts the adhesion of microorganisms, which regulates the biofilm formation needed for the electron transfer from the microbe to the electrode. Fungal based/yeast based MFC are most extensively studied where the electron is transferred via cytochrome C. The organisms used are *Saccharomyces cerevisiae*, *Candida* sp., *Galactomyces reessii* and filamentous fungi like *Rhizopus* sp., *Aspergillus* sp., and *Penicillium* sp. Among all the species, *Saccharomyces cerevisiae* produces the highest power density (Belniak and Mamińska, 2018).

Fungi/yeasts as biocatalyst in cathode

The fungal organisms make use of the substrates such as glucose (*S. cerevisiae*, *Shewanella oneidensis, Kluyveromyces marxianus*), yeast extract, peptone and fructose medium (*Candida melibiosica*) and dextrose and glucose (*Arxula adeninivorans*) and generated power of about 1500.0, 123.4, 850,000.0, 640.0 and 1030.0 mW/m^2, respectively. The organisms *S. cerevisiae*, *K. marxianus* and *A. adeninivorans* produces electricity under the influence of exogenous mediators like 2-hydroxy-1,4-naphthoquinone and TMPD (*N,N,N′,N′*-tetramethyl-*p*-phenylenediamine). The MFC constructed could be of double chambered and of continuous flow. Electron acceptors for these organisms could either be potassium permanganate or potassium ferricyanide. Separators used are nafion membrane and teflon. Cathode material are made up of graphite plate and carbon cloth (Belniak and Mamińska, 2018; Ganguli and Dunn, 2009; Haslett et al., 2011; Hubenova and Mitov, 2010; Kaneshiro et al., 2014; Lin et al., 2017).

Fungi/yeasts as biocatalyst in anode

The fungal organisms using substrates such as glucose (*Trametes versicolor, Rhizopus* sp., *Aspergillus* sp. and *Penicillium* sp.) have the potential to generate electricity of power of 320.0 mW/m^2, 317.3, 438.16 and 344.1 mW/m^3, respectively. The fungal strain *Galactomyces reessii* utilizes substrates such as sludge from the rubber industry and synthetic wastewater, producing electricity of 59.0 mW/m^2. The anolytes used were potassium ferricyanide, acetate and acid orange. Oxygen acts as an electron acceptor in the process. A double chambered MFC (H-type) was used, the separators used were salt bridge and nafion membrane. Anode material were made up of graphite plates and activated carbon fiber (Belniak and Mamińska, 2018; Chaijak et al., 2018; Morant et al., 2014; Wu et al., 2012).

Applications of MFC

Production of bio electricity

The primary goal of MFC is the microbial oxidation of natural starch substrates obtained from biomass generated from horticultural, agriculture etc. for the production of bioelectricity. MFC innovation can likewise be applied to bio-battery which can recharge machines and gadgets requiring little voltage (Bruno et al., 2018).

Biohydrogen production

An excellent potential of MFC can brought about by crossing the thermodynamic barrier for production of hydrogen at the cathode by the neutralization of electrons and protons (Tharali et al., 2016).

Waste water treatment containing organic matter and heavy metals

Wastewater effluent from different sources such as industrial effluents, dairy wastewater etc. are a prime hotspot for bioremediation of inorganic metals such as Cu, Zn, Cd, Sb, Pb, Mn and Fe. The essential three parameters on which the productivity of MFC works are power densities, coulombic efficiencies and chemical oxygen demand. Most power densities make use of unadulterated organic substrates, for example, acetic acid derivatives, glucose and sucrose by MFC innovation (Singh et al., 2019; Tharali et al., 2016).

Carbon sequestration

A photosynthetic MFC is capable to limit CO_2 emissions and release the O_2 into the environment. Liberated CO_2 gas from anode reach the cathodic chamber where microbes are found in dense populations scavenge CO_2, producing high power densities. MFC is a promising device which is carbon neutral and can produce zero emissions of CO_2 (Singh et al., 2019).

Biosensor

A biosensor is an analytical device used for the recognition of analytes by connecting an organic part (biological component) with a physicochemical detector. Researchers have structured a self-fueled MFC-based biosensor for detection of deleterious metal levels in wastewater (Singh et al., 2019).

Thus, overall MFCs proves to be a versatile technology for wastewater treatment and therefore requires further research and development.

Conclusions

Biofuels are considered to be the best economical alternative to many other commercial fuels. Researchers have identified various mechanisms and sources to develop clean eco-friendly fuel. According to recent developments, it has been observed that fungus is a potential candidate for the production of bio-fuel. The fungal enzymatic action to degrade lignocellulosic biomass into mixture of sugars is helpful to develop bioethanol. Production of bioethanol in bioreactors can be scaled to industrial level. Further, as many species of oleaginous fungi has also been recognized, therefore fungus can also act as a feedstock of biodiesel production. With advent of nanotechnology, the use of various nanocatalysts also proved to be essential in providing pace to the rate of transesterification process. Moreover, latest findings also indicated the role of fungus in biogas production and recently, through microbial fuel cell technology, various fungal species especially, *Saccharomyces cerevisiae* showed excellent results in electricity generation. Further, utilizing engineering tools, optimizing process parameters and designing innovative strategies, the near future would witness the power of fungi-based biofuels for a sustainable future.

Acknowledgement

Authors are grateful to the Department of Biotechnology and Department of Chemical Engineering, Ramaiah Institute of Technology, Bengaluru; Department of Biotechnology, Noida Institute of Engineering and Technology, Noida; University School of Environmental Management, GGSIPU, New Delhi and Department of Microbiology, Shaheed Rajguru College of Applied Sciences for Women, University of Delhi, New Delhi. One of the authors (Randhir K. Bharti) thank D.S. Kothari Post-Doctoral Fellowship, UGC, Govt. of India.

References

Acharya, B., Dutta, A., Basu, P., 2019. Ethanol production by syngas fermentation in a continuous stirred tank bioreactor using Clostridium ljungdahlii. Biofuels 10, 221–237. Available from: https://doi.org/10.1080/17597269.2017.1316143.

Adney, W.S., Rivard, C.J., Shiang, M., Himmel, M.E., 1991. Anaerobic digestion of lignocellulosic biomass and wastes. Appl. Biochem. Biotechnol. 30, 165–183. Available from: https://doi.org/10.1007/BF02921684.

Alam, Md.Z., Kabbashi, N.A., Hussin, S.N.I.S., 2009. Production of bioethanol by direct bioconversion of oil-palm industrial effluent in a stirred-tank bioreactor. J. Ind. Microbiol. Biotechnol. 36, 801. Available from: https://doi.org/10.1007/s10295-009-0554-7.

Alvira, P., Tomás-Pejó, E., Ballesteros, M., Negro, M.J., 2010. Pretreatment technologies for an efficient bioethanol production process based on enzymatic hydrolysis: a review. Bioresour.

Technol., Spec. Issue Lignocellulosic Bioethanol: Curr. Status Perspect. 101, 4851–4861. Available from: https://doi.org/10.1016/j.biortech.2009.11.093.

Amini, G., Najafpour, G.D., Rabiee, S.M., Ghoreyshi, A.A., 2013. Synthesis and characterization of amorphous nano-alumina powders with high surface area for biodiesel production. Chem. Eng. Technol. 36, 1708–1712. Available from: https://doi.org/10.1002/ceat.201300102.

Anasonye, F., Winquist, E., Kluczek-Turpeinen, B., Räsänen, M., Salonen, K., Steffen, K.T., et al., 2014. Fungal enzyme production and biodegradation of polychlorinated dibenzo-p-dioxins and dibenzofurans in contaminated sawmill soil. Chemosphere 110, 85–90. Available from: https://doi.org/10.1016/j.chemosphere.2014.03.079.

Atabani, A.E., Silitonga, A.S., Badruddin, I.A., Mahlia, T.M.I., Masjuki, H.H., Mekhilef, S., 2012. A comprehensive review on biodiesel as an alternative energy resource and its characteristics. Renew. Sustain. Energy Rev. 16, 2070–2093. Available from: https://doi.org/10.1016/j.rser.2012.01.003.

Bautista, L.F., Vicente, G., Garre, V., 2012. Biodiesel from microbial oil. In: Luque, R., Melero, J.A. (Eds.), Advances in Biodiesel Production: Processes and Technologies. pp. 179–203.

Bayat, A., Baghdadi, M., Bidhendi, G.N., 2018. Tailored magnetic nano-alumina as an efficient catalyst for transesterification of waste cooking oil: optimization of biodiesel production using response surface methodology. Energy Convers. Manag. 177, 395–405. Available from: https://doi.org/10.1016/j.enconman.2018.09.086.

Bellou, S., Triantaphyllidou, I.-E., Aggeli, D., Elazzazy, A.M., Baeshen, M.N., Aggelis, G., 2016. Microbial oils as food additives: recent approaches for improving microbial oil production and its polyunsaturated fatty acid content. Curr. Opin. Biotechnol., Food Biotechnol. Plant. Biotechnol. 37, 24–35. Available from: https://doi.org/10.1016/j.copbio.2015.09.005.

Belniak, A.S., Mamińska, R.T., 2018. Fungi-based microbial fuel cells. Energies 11, 2827. Available from: https://doi.org/10.3390/en11102827.

Booramurthy, V.K., Kasimani, R., Pandian, S., 2019. Biodiesel production from tannery waste using a nano catalyst (ferric-manganese doped sulphated zirconia). Energy Sources Part. Recov. Util. Environ. Eff. Available from: https://doi.org/10.1080/15567036.2019.1639849.

Börjesson, P., Mattiasson, B., 2008. Biogas as a resource-efficient vehicle fuel. Trends Biotechnol. 26, 7–13. Available from: https://doi.org/10.1016/j.tibtech.2007.09.007.

Bruno, L.B., Jothinathan, D., Rajkumar, M., 2018. Microbial fuel cells: fundamentals, types, significance and limitations. Microbial Fuel Cell Technology for Bioelectricity. Springer International Publishing, Cham, pp. 23–48. Available from: https://doi.org/10.1007/978-3-319-92904-0_2.

Carvalho, A.K.F., Rivaldi, J.D., Barbosa, J.C., de Castro, H.F., 2015. Biosynthesis, characterization and enzymatic transesterification of single cell oil of Mucor circinelloides – a sustainable pathway for biofuel production. Bioresour. Technol. 181, 47–53. Available from: https://doi.org/10.1016/j.biortech.2014.12.110.

Chaijak, P., Sukkasem, C., Lertworapreecha, M., Boonsawang, P., Wijasika, S., Sato, C., 2018. Enhancing electricity generation using a laccase-based microbial fuel cell with yeast Galactomyces reessii on the cathoden.d. J. Microbiol. Biotechnol. 28, 1360–1366.

Chaturvedi, V., Verma, P., 2016. Microbial fuel cell: a green approach for the utilization of waste for the generation of bioelectricity. Bioresour. Bioprocess. 3, 38. Available from: https://doi.org/10.1186/s40643-016-0116-6.

Chatzifragkou, A., Fakas, S., Panayotou, M.G., Komaitis, M., Aggelis, G., Papanikolaou, S., 2010. Commercial sugars as substrates for lipid accumulation in Cunninghamella echinulata and Mortierella isabellina fungi. Eur. J. Lipid Sci. Technol. 112, 1048–1057.

Chatzifragkou, A., Makri, A., Belka, A., Bellou, S., Mavrou, M., Mastoridou, M., et al., 2011. Biotechnological conversions of biodiesel derived waste glycerol by yeast and fungal species. Energy 36, 1097–1108. Available from: https://doi.org/10.1016/j.energy.2010.11.040.

Christy, P.M., Gopinath, L.R., Divya, D., 2014. A review on anaerobic decomposition and enhancement of biogas production through enzymes and microorganisms. Renew. Sustain. Energy Rev. 34, 167–173. Available from: https://doi.org/10.1016/j.rser.2014.03.010.

Cirne, D.G., Lehtomäki, A., Björnsson, L., Blackall, L.L., 2007. Hydrolysis and microbial community analyses in two-stage anaerobic digestion of energy crops. J. Appl. Microbiol. 103, 516–527. Available from: https://doi.org/10.1111/j.1365-2672.2006.03270.x.

Dahadha, S., Amin, Z., Bazyar Lakeh, A.A., Elbeshbishy, E., 2017. Evaluation of different pretreatment processes of lignocellulosic biomass for enhanced biomethane production. Energy Fuels 31, 10335–10347. Available from: https://doi.org/10.1021/acs.energyfuels.7b02045.

Dashtban, M., Schraft, H., Qin, W., 2009. Fungal bioconversion of lignocellulosic residues; opportunities & perspectives. Int. J. Biol. Sci. 5, 578–595.

Dollhofer, V., Podmirseg, S.M., Callaghan, T.M., et al., 2015. Anaerobic fungi and their potential for biogas production. Biogas Science and Technology, Advances in Biochemical Engineering/Biotechnology. Springer International Publishing, Cham, pp. 41–61. Available from: https://doi.org/10.1007/978-3-319-21993-6_2.

Duan, X., Liu, Y., Zhao, Q., Wang, X., Li, S., 2013. Water-tolerant heteropolyacid on magnetic nanoparticles as efficient catalysts for esterification of free fatty acid. RSC Adv. 3, 13748–13755. Available from: https://doi.org/10.1039/C3RA40219C.

Dumonceaux, T., Bartholomew, K., Valeanu, L., Charles, T., Archibald, F., 2001. Cellobiose dehydrogenase is essential for wood invasion and nonessential for kraft pulp delignification by Trametes versicolor. Enzyme Microb. Technol. 29, 478–489. Available from: https://doi.org/10.1016/S0141-0229(01)00407-0.

Economou, Ch.N., Aggelis, G., Pavlou, S., Vayenas, D.V., 2011. Single cell oil production from rice hulls hydrolysate. Bioresour. Technol. 102, 9737–9742. Available from: https://doi.org/10.1016/j.biortech.2011.08.025.

Elisashvili, V., Penninckx, M., Kachlishvili, E., Tsiklauri, N., Metreveli, E., Kharziani, T., et al., 2008. Lentinus edodes and Pleurotus species lignocellulolytic enzymes activity in submerged and solid-state fermentation of lignocellulosic wastes of different composition. Bioresour. Technol. 99, 457–462. Available from: https://doi.org/10.1016/j.biortech.2007.01.011.

Eriksson, K.-E.L., Blanchette, R., Ander, P., 1990. Microbial and enzymatic degradation of wood and wood components. Springer Series in Wood Science. Springer-Verlag, Berlin Heidelberg, p. 407. Available from: https://doi.org/10.1007/978-3-642-46687-8.

Fakas, S., Papanikolaou, S., Galiotou-Panayotou, M., Komaitis, M., Aggelis, G., 2009. Biochemistry and biotechnology of single cell oil. In: Pandey, A., Larroche, C. (Eds.), New Horizons in Biotechnology. Asiatech Publishers, pp. 38–60.

Gajda, I., Greenman, J., Ieropoulos, I.A., 2018. Recent advancements in real-world microbial fuel cell applications. Curr. Opin. Electrochem., Environ. Electrochem. Sol. Cell 11, 78–83. Available from: https://doi.org/10.1016/j.coelec.2018.09.006.

Ganguli, R., Dunn, B.S., 2009. Kinetics of anode reactions for a yeast-catalysed microbial fuel cell. Fuel Cell 9, 44–52. Available from: https://doi.org/10.1002/fuce.200800039.

Gardy, J., Rehan, M., Hassanpour, A., Lai, X., Nizami, A., 2019. Advances in nano-catalysts based biodiesel production from non-food feedstocks. J. Environ. Manage. 249.

Garzillo, A.M.V., Di Paolo, S., Burla, G., Buonocore, V., 1992. Differently-induced extracellular phenol oxidases from Pleurotus ostreatus. Phytochem. Int. J. Plant. Biochem. 31, 3685–3690. Available from: https://doi.org/10.1016/S0031-9422(00)97509-5.

Ge, X., Matsumoto, T., Keith, L., Li, Y., 2015. Fungal pretreatment of albizia chips for enhanced biogas production by solid-state anaerobic digestion. Chemistry 200–204.

Guan, Q., Li, Y., Chen, Y., Shi, Y., Gu, J., Li, B., et al., 2017. Sulfonated multi-walled carbon nanotubes for biodiesel production through triglycerides transesterification. RSC Adv. 7, 7250–7258. Available from: https://doi.org/10.1039/C6RA28067F.

Harde, S.M., Wang, Z., Horne, M., Zhu, J., Pan, X., 2016. Microbial lipid production from SPORL-pretreated Douglas fir by Mortierella isabellina. Fuel 175, 64–74. Available from: https://doi.org/10.1016/j.fuel.2016.02.023.

Haslett, N.D., Rawson, F.J., Barriëre, F., Kunze, G., Pasco, N., Gooneratne, R., et al., 2011. Characterisation of yeast microbial fuel cell with the yeast Arxula adeninivorans as the biocatalyst. Biosens. Bioelectron. 26, 3742–3747. Available from: https://doi.org/10.1016/j.bios.2011.02.011.

Hatakka, A., 2005. Biodegradation of lignin. Biopolymers Online. Wiley Online Library, pp. 129–180. Available from: https://doi.org/10.1002/3527600035.bpol1005.

Huang, C., Chen, X.-f., Xiong, L., Chen, X.-de., Ma, L., Chen, Y., 2013. Single cell oil production from low-cost substrates: the possibility and potential of its industrialization. Biotechnol. Adv. 31, 129–139. Available from: https://doi.org/10.1016/j.biotechadv.2012.08.010.

Hubenova, Y., Mitov, M., 2010. Potential application of Candida melibiosica in biofuel cells. Bioelectrochemistry 78, 57–61. Available from: https://doi.org/10.1016/j.bioelechem.2009.07.005.

Jeon, B.Y., Kim, S.J., Kim, D.H., Na, B.K., Park, D.H., Tran, H.T., et al., 2007. Development of a serial bioreactor system for direct ethanol production from starch using Aspergillus niger andSaccharomyces cerevisiae. Biotechnol. Bioprocess. Eng. 12, 566. Available from: https://doi.org/10.1007/BF02931356.

Kaneshiro, H., Takano, K., Takada, Y., Wakisaka, T., Tachibana, T., Azuma, M., 2014. A milliliter-scale yeast-based fuel cell with high performance. Biochem. Eng. J. 83, 90–96. Available from: https://doi.org/10.1016/j.bej.2013.12.011.

Kanimozhi, S., Perinbam, K., 2013. Biodesel production from Pseudomonas fluorescens Lp1 lipase immobilized on amino-silane modified super paramagnetic $Fe3O_4$ nanoparticles. J. Phys. Conf. Ser. 431, 1–7. Available from: https://doi.org/10.1088/1742-6596/431/1/012010.

Kaur, M., Malhotra, R., Ali, A., 2018. Tungsten supported Ti/SiO_2 nanoflowers as reusable heterogeneous catalyst for biodiesel production. Renew. Energy 116, 109–119. Available from: https://doi.org/10.1016/j.renene.2017.09.065.

Koutb, M., Morsy, F.M., 2011. A potent lipid producing isolate of Epicoccum purpurascens AUMC5615 and its promising use for biodiesel production. Biomass Bioenergy 35, 3182–3187. Available from: https://doi.org/10.1016/j.biombioe.2011.04.050.

Kulkarni, M.B., Ghanegaonkar, P.M., 2019. Methane enrichment of biogas produced from floral waste: a potential energy source for rural India. Energy Sources Part. Recovery Util. Environ. Eff. 41, 2757–2768. Available from: https://doi.org/10.1080/15567036.2019.1571126.

Lenhart, K., Bunge, M., Ratering, S., Neu, T.R., Schüttmann, I., Greule, M., et al., 2012. Evidence for methane production by saprotrophic fungi. Nat. Commun. 3, 1–8. Available from: https://doi.org/10.1038/ncomms2049.

Lin, T., Bai, X., Hu, Y., Li, B., Yuan, Y.-J., Song, H., et al., 2017. Synthetic Saccharomyces cerevisiae-Shewanella oneidensis consortium enables glucose-fed high-performance microbial fuel cell. AIChE J. 63, 1830–1838. Available from: https://doi.org/10.1002/aic.15611.

Liu, K., Wang, R., Yu, M., 2017. Biodiesel production from soybean oils by a novel nano-magnetic solid base catalyst (K/ZrO2/γ-Fe2O3). RSC Adv. 7, 51814–51821. Available from: https://doi.org/10.1039/C7RA10067A.

Madhuvilakku, R., Piraman, S., 2013. Biodiesel synthesis by TiO_2–ZnO mixed oxide nanocatalyst catalyzed palm oil transesterification process. Bioresour. Technol. 150, 55–59. Available from: https://doi.org/10.1016/j.biortech.2013.09.087.

McCarty, P.L., Bae, J., Kim, J., 2011. Domestic wastewater treatment as a net energy producer – can this be achieved? Environ. Sci. Technol. 45, 7100–7106. Available from: https://doi.org/10.1021/es2014264.

Mittal, S., Ahlgren, E.O., Shukla, P.R., 2018. Barriers to biogas dissemination in India: a review. Energy Policy 112, 361–370. Available from: https://doi.org/10.1016/j.enpol.2017.10.027.

Morant, K.V., da Silva, P.H., de Campos-Takaki, G.M., Hernández, C.E.L.R., 2014. Isolation and bioelectrochemical characterization of novel fungal sources with oxidasic activity applied in situ for the cathodic oxygen reduction in microbial fuel cells. Enzyme Microb. Technol. 66, 20–27. Available from: https://doi.org/10.1016/j.enzmictec.2014.07.007.

Murcia, M.A., Martínez-Tomé, M., Jiménez, A.M., Vera, A.M., Honrubia, M., Parras, P., 2002. Antioxidant activity of edible fungi (truffles and mushrooms): losses during industrial processing. J. Food Prot. 65, 1614–1622.

Muthangya, M., Mshandete, A.M., Kivaisi, A.K., 2009. Two-stage fungal pre-treatment for improved biogas production from sisal leaf decortication residues. Int. J. Mol. Sci. 10, 4805–4815. Available from: https://doi.org/10.3390/ijms10114805.

Nigam, J.N., 2000. Continuous ethanol production from pineapple cannery waste using immobilized yeast cells. J. Biotechnol. 80, 189–193. Available from: https://doi.org/10.1016/S0168-1656(00)00246-7.

Okamoto, K., Imashiro, K., Akizawa, Y., Onimura, A., Yoneda, M., Nitta, Y., et al., 2010. Production of ethanol by the white-rot basidiomycetes Peniophora cinerea and Trametes suaveolens. Biotechnol. Lett. 32, 909–913. Available from: https://doi.org/10.1007/s10529-010-0243-7.

Papacz, W., 2011. Biogas as vehicle fuel. J. KONES Powertrain Transp. 18.

Papanikolaou, S., Aggelis, G., 2010. Yarrowia lipolytica: a model microorganism used for the production of tailor-made lipids. Eur. J. Lipid Sci. Technol. 112, 639–654. Available from: https://doi.org/10.1002/ejlt.200900197.

Patel, R., Zaveri, P., Munshi, N.S., 2019. Microbial fuel cell, the Indian scenario: developments and scopes. Biofuels 10, 101–108. Available from: https://doi.org/10.1080/17597269.2017.1398953.

Pointing, S.B., 2001. Feasibility of bioremediation by white-rot fungi. Appl. Microbiol. Biotechnol. 57, 20–33. Available from: https://doi.org/10.1007/s002530100745.

Pontié, M., Jaspard, E., Friant, C., Kilani, J., Fix-Tailler, A., Innocent, C., et al., 2019. A sustainable fungal microbial fuel cell (FMFC) for the bioremediation of acetaminophen (APAP) and its main by-product (PAP) and energy production from biomass. Biocatal. Agric. Biotechnol. 22, 101376. Available from: https://doi.org/10.1016/j.bcab.2019.101376.

Priya, A., Dureja, P., Talukdar, P., Rathi, R., Lal, B., Sarma, P.M., 2016. Microbial production of 2,3-butanediol through a two-stage pH and agitation strategy in 150l bioreactor. Biochem. Eng. J. 105, 159–167. Available from: https://doi.org/10.1016/j.bej.2015.09.016.

Rajakumar, S., Gaskell, J., Cullen, D., Lobos, S., Karahanian, E., Vicuna, R., 1996. Lip-like genes in Phanerochaete sordida and Ceriporiopsis subvermispora, white rot fungi with no detectable lignin peroxidase activity. Appl. Environ. Microbiol. 62, 2660–2663.

Rasi, S., Veijanen, A., Rintala, J., 2007. Trace compounds of biogas from different biogas production plants. Energy 32, 1375–1380. Available from: https://doi.org/10.1016/j.energy.2006.10.018.

Ratledge, C., Wynn, J.P., 2002. The biochemistry and molecular biology of lipid accumulation in oleaginous microorganisms. Adv. Appl. Microbiol. 51, 1–51. Available from: https://doi.org/10.1016/s0065-2164(02)51000-5.

Reddy, C.A., 1995. The potential for white-rot fungi in the treatment of pollutants. Curr. Opin. Biotechnol. 6, 320–328. Available from: https://doi.org/10.1016/0958-1669(95)80054-9.

Reddy, C.A., Mathew, Z., 2001. Bioremediation potential of white rot fungi. In: Gadd, G. (Ed.), Fungi in Bioremediation. Cambridge University Press, p. 52.

Refaat, A.A., 2011. Biodiesel production using solid metal oxide catalysts. Int. J. Environ. Sci. Technol. 8, 203–221. Available from: https://doi.org/10.1007/BF03326210.

Rodriguez, C., Alaswad, A., Benyounis, K.Y., Olabi, A.G., 2017. Pretreatment techniques used in biogas production from grass. Renew. Sustain. Energy Rev. 68, 1193–1204. Available from: https://doi.org/10.1016/j.rser.2016.02.022.

Ropero-Vega, J.L., Aldana-Pérez, A., Gómez, R., Niño-Gómez, M.E., 2010. Sulfated titania [TiO2/SO42 –]: a very active solid acid catalyst for the esterification of free fatty acids with ethanol. Appl. Catal. Gen. 379, 24–29. Available from: https://doi.org/10.1016/j.apcata.2010.02.020.

Rouches, E., Gimbert, I.H., Steyer, J.P., Carrere, H., 2016. Improvement of anaerobic degradation by white-rot fungi pretreatment of lignocellulosic biomass: a review. Renew. Sustain. Energy Rev. 59, 179–198. Available from: https://doi.org/10.1016/j.rser.2015.12.317.

Ruan, Z., Hollinshead, W., Isaguirre, C., Tang, Y.J., Liao, W., Liu, Y., 2015. Effects of inhibitory compounds in lignocellulosic hydrolysates on Mortierella isabellina growth and carbon utilization. Bioresour. Technol. 183, 18–24. Available from: https://doi.org/10.1016/j.biortech.2015.02.026.

Rüttimann-Johnson, C., Salas, L., Vicuña, R., Kirk, T.K., 1993. Extracellular enzyme production and synthetic lignin mineralization by Ceriporiopsis subvermispora. Appl. Environ. Microbiol. 59, 1792–1797.

Saha, B.C., Ueda, S., 1983. Alcoholic fermentation of raw sweet potato by a nonconventional method using Endomycopsis fibuligera glucoamylase preparation. Biotechnol. Bioeng. 25, 1181–1186.

Sánchez, C., 2009. Lignocellulosic residues: biodegradation and bioconversion by fungi. Biotechnol. Adv. 27, 185–194. Available from: https://doi.org/10.1016/j.biotechadv.2008.11.001.

Saritha, M., Arora, A., Singh, S., Nain, L., 2013. Streptomyces griseorubens mediated delignification of paddy straw for improved enzymatic saccharification yields. Bioresour. Technol., Biorefineries 135, 12–17. Available from: https://doi.org/10.1016/j.biortech.2012.11.040.

Satari, B., Karimi, K., Taherzadeh, M.J., Zamani, A., 2016. Co-production of fungal biomass derived constituents and ethanol from citrus wastes free sugars without auxiliary nutrients in airlift bioreactor. Int. J. Mol. Sci. 17, 302. Available from: https://doi.org/10.3390/ijms17030302.

Sayed, E.T., Abdelkareem, M.A., 2017. Yeast as a biocatalyst in microbial fuel cell. Old. Yeasts – N. Quest. Available from: https://doi.org/10.5772/intechopen.70402.

Seadi, T.A., Rutz, D., Janssen, R., Drosg, B., 2013. Biomass resources for biogas production. In: Wellinger, A., Murphy, J.D., Baxter, D. (Eds.), The Biogas Handbook: Science, Production and Applications. Elsevier.

Shahid, E.M., Jamal, Y., 2011. Production of biodiesel: a technical review. Renew. Sustain. Energy Rev. 15, 4732–4745.

Singh, H.M., Pathak, A.K., Chopra, K., Tyagi, V.V., Anand, S., Kothari, R., 2019. Microbial fuel cells: a sustainable solution for bioelectricity generation and wastewater treatment. Biofuels 10, 11–31. Available from: https://doi.org/10.1080/17597269.2017.1413860.

Singh, S.P., Singh, D., 2010. Biodiesel production through the use of different sources and characterization of oils and their esters as the substitute of diesel: a review. Renew. Sustain. Energy Rev. 14, 200–216. Available from: https://doi.org/10.1016/j.rser.2009.07.017.

Srilatha, H.R., Nand, K., Babu, K.S., Madhukara, K., 1995. Fungal pretreatment of orange processing waste by solid-state fermentation for improved production of methane. Process. Biochem. 30, 327–331. Available from: https://doi.org/10.1016/0032-9592(95)87041-5.

Steffen, K.T., Hatakka, A., Hofrichter, M., 2003. Degradation of Benzo[a]pyrene by the Litter-Decomposing Basidiomycete Stropharia coronilla: role of Manganese Peroxidase. Appl. Environ. Microbiol. 69, 3957–3964. Available from: https://doi.org/10.1128/AEM.69.7.3957-3964.2003.

Stellwagen, D.R., van der Klis, F., van Es, D.S., de Jong, K.P., Bitter, J.H., 2013. Functionalized carbon nanofibers as solid-acid catalysts for transesterification. Chem. Sustain. 6, 1668–1672. Available from: https://doi.org/10.1002/cssc.201300372.

Tangy, A., Pulidindi, I.N., Gedanken, A., 2016. SiO_2 beads decorated with SrO nanoparticles for biodiesel production from waste cooking oil using microwave irradiation. Energy Fuels 30, 3151–3160. Available from: https://doi.org/10.1021/acs.energyfuels.6b00256.

Tharali, A.D., Sain, N., Osborne, W.J., 2016. Microbial fuel cells in bioelectricity production. Front. Life Sci. 9, 252–266. Available from: https://doi.org/10.1080/21553769.2016.1230787.

Thliveros, P., Kiran, E.U., Webb, C., 2014. Microbial biodiesel production by direct methanolysis of oleaginous biomass. Bioresour. Technol. 157, 181–187. Available from: https://doi.org/10.1016/j.biortech.2014.01.111.

Vahid, B.R., Haghighi, M., 2017. Biodiesel production from sunflower oil over $MgO/MgAl_2O_4$ nanocatalyst: effect of fuel type on catalyst nanostructure and performance. Energy Convers. Manag. 134, 290–300. Available from: https://doi.org/10.1016/j.enconman.2016.12.048.

Vamvakaki, A.-N., Kandarakis, I., Kaminarides, S., Komaitis, M., Papanikolaou, S., 2010. Cheese whey as a renewable substrate for microbial lipid and biomass production by Zygomycetes. Eng. Life Sci. 10, 348–360. Available from: https://doi.org/10.1002/elsc.201000063.

Velázquez-Cedeño, M.A., Mata, G., Savoie, J.-M., 2002. Waste-reducing cultivation of Pleurotus ostreatus and Pleurotus pulmonarius on coffee pulp: changes in the production of some lignocellulolytic enzymes. World J. Microbiol. Biotechnol. 18, 201–207. Available from: https://doi.org/10.1023/A:1014999616381.

Vicente, G., Bautista, L.F., Gutiérrez, F.J., Rodríguez, R., Martínez, V., Frómeta, R.A.R., et al., 2010. Direct transformation of fungal biomass from submerged cultures into biodiesel. Energy Fuels 24, 3173–3178. Available from: https://doi.org/10.1021/ef9015872.

Vohra, M., Manwar, J., Manmode, R., Padgilwar, S., Patil, S., 2014. Bioethanol production: feedstock and current technologies. J. Environ. Chem. Eng. 2, 573–584. Available from: https://doi.org/10.1016/j.jece.2013.10.013.

Wagner, A.O., Lackner, N., Mutschlechner, M., Prem, E.M., Markt, R., Illmer, P., 2018. Biological pretreatment strategies for second-generation lignocellulosic resources to enhance biogas production. Energies 11, 1797. Available from: https://doi.org/10.3390/en11071797.

Wan, C., Li, Y., 2012. Fungal pretreatment of lignocellulosic biomass. Biotechnol. Adv. 30, 1447–1457Special issue on ACB 2011. Available from: https://doi.org/10.1016/j.biotechadv.2012.03.003.

Wesenberg, D., Kyriakides, I., Agathos, S.N., 2003. White-rot fungi and their enzymes for the treatment of industrial dye effluents. Biotechnol. Adv., VI Int. Sympos. Environ. Biotechnol. 22, 161–187. Available from: https://doi.org/10.1016/j.biotechadv.2003.08.011.

Włodarczyk, P.P., Włodarczyk, B., 2019. Wastewater treatment and electricity production in a microbial fuel cell with Cu–B alloy as the cathode catalyst. Catalysts 9, 572. Available from: https://doi.org/10.3390/catal9070572.

Woodford, J.J., Dacquin, J.-P., Wilson, K., Lee, A.F., 2012. Better by design: nanoengineered macroporous hydrotalcites for enhanced catalytic biodiesel production. Energy Environ. Sci. 5, 6145–6150. Available from: https://doi.org/10.1039/C2EE02837A.

Wu, C., Liu, X.-W., Li, W.-W., Sheng, G.-P., Zang, G.-L., Cheng, Y.-Y., et al., 2012. A white-rot fungus is used as a biocathode to improve electricity production of a microbial fuel cell. Appl. Energy 98, 594–596. Available from: https://doi.org/10.1016/j.apenergy.2012.02.058.

Wu, Z.-Y., Yin, P., Ju, H.-X., Chen, Z.-Q., Li, C., Li, S.-C., et al., 2019. Natural nanofibrous cellulose-derived solid acid catalysts. Research 1–11. Available from: https://doi.org/10.34133/2019/6262719.

Xing, D., Wang, H., Pan, A., Wang, J., Xue, D., 2012. Assimilation of corn fiber hydrolysates and lipid accumulation by Mortierella isabellina. Biomass Bioenergy Biorefinery 39, 494–501. Available from: https://doi.org/10.1016/j.biombioe.2012.01.024.

Xu, L., Zhao, Y., Doherty, L., Hu, Y., Hao, X., 2016. The integrated processes for wastewater treatment based on the principle of microbial fuel cells: a review. Crit. Rev. Environ. Sci. Technol. 46, 60–91.

Yadav, A.K., Dash, P., Mohanty, A., Abbassi, R., Mishra, B.K., 2012. Performance assessment of innovative constructed wetland-microbial fuel cell for electricity production and dye removal. Ecol. Eng. 47, 126–131. Available from: https://doi.org/10.1016/j.ecoleng.2012.06.029.

Zhao, L., Qiu, Z., Stagg-Williams, S.M., 2013. Transesterification of canola oil catalyzed by nanopowder calcium oxide. Fuel Process. Technol. 114, 154–162. Available from: https://doi.org/10.1016/j.fuproc.2013.03.027.

Zikou, E., Chatzifragkou, A., Koutinas, A.A., Papanikolaou, S., 2013. Evaluating glucose and xylose as cosubstrates for lipid accumulation and γ-linolenic acid biosynthesis of Thamnidium elegans. J. Appl. Microbiol. 114, 1020–1032. Available from: https://doi.org/10.1111/jam.12116.

Chapter 12

Fungal endophytes and their applications as growth promoters and biological control agents

Vinay Kumar[1], Lata Jain[1], Pankaj Kaushal[1] and Ravindra Soni[2,*]
[1]ICAR-National Institute of Biotic Stress Management, Raipur, India, [2]Department of Agricultural Microbiology, College of Agriculture, Indira Gandhi Krishi Vishwavidyalaya, Raipur, India
**Corresponding author.* Email: rs31693@gmail.com

Introduction

Plants interact with microorganisms at all the tropic levels, adapting developmental, growth and defense responses within a complex network of community. Endophytes are ubiquitous and extremely diverse group of organisms which form associations with a various group of microbes all over the plant kingdom and offer indirect defense for plants against pathogens and herbivores (Hartley and Gange, 2009). An ecosystem exhibits the highest plant diversity and contains abundant populations of endophytes (Strobel and Daisy, 2003). Endophytes colonize internal plant tissues and organs such as roots, stems, leaves, flowers and fruits (Saikkonen et al., 2016). They are well adapted to their host plant, as they spend entire or part of their life inside colonized living tissues or organs without harming the hosts (Hardoim et al., 2015). Further, fungal endophytes have been widely reported as naturally occurring microbes in numerous host plants and they play an important role in providing defense of their host plants (Saikkonen et al., 1998; Suryanarayanan, 2013; Yan et al., 2019). In last few years, endophytes have been studied widely in several geographic and agro-climatic conditions and were found to be pervasive inside the plant tissues, organs and rich in the diversity at different levels (Li et al., 2012; Zhang et al., 2013; De Souza Sebastianes et al., 2013; Sandberg et al., 2014; Sridhar, 2019, Kumar et al., 2019, 2020). It has been well established that fungal endophytes can improve

Fungi Bio-Prospects in Sustainable Agriculture, Environment and Nano-Technology.
DOI: https://doi.org/10.1016/B978-0-12-821394-0.00012-3

growth and provide tolerance to various biotic and abiotic stresses including disease, heat, drought and salt stresses to their host plant (Ikram et al., 2018, Ismail et al., 2019). The plant growth promotion activity of endophytic fungi offers an added advantage for agro-ecosystems since they reduce application of chemical fertilizers in turn helps in the reducing the soil, water and environmental pollution (Khan et al., 2015).

Endophytes play a crucial role in maintaining plant health, as they provide protection to the host plant against various biotic and abiotic stresses and act as plant growth promoters and facilitate growth enhancement and yields (Lugtenberg et al., 2016; Lata et al., 2018; Khare et al., 2018). Fungal endophytes are a diverse, polyphyletic group of microbes which are broadly classified as the clavicipitaceous (C) and the non-clavicipitaceous (NC) based on taxonomy, host plant range, evolutionary relatedness and ecological functions (Rodriguez et al., 2009). Several vital functions of the host plant have been reported to be influenced in the presence of endophytes, Hence endophytes seems to be a viable option for improvement of quality and yield of plant products through microbial assisted breeding program and genetic engineering techniques.

Plant-endophyte interaction

Available reports suggest that interactions between host - endophytes are not fixed in either evolutionary time or geographically (Faeth and Bultman, 2002). Majority of the endophytes infect locally in the plant tissues and limited to a small region. Some fungal endophytes have shown tissue and organ specificity other may be restricted to the specific parts of host plant (Moricca et al., 2012; Vincent et al., 2016). Generally, fungal infections are of three types i.e. intercellular, intracellular or inter−intra cellular (Fig. 12.1) (Raja et al., 2016). As soon as the endophyte enters inside the plant system, a cross-talk of signal molecules is started between host and fungal endophytes (Khare et al., 2018). It was reported that plant hormone strigolactone play a vital in this cross-talk during the mutualistic interaction (Rozpądek et al., 2018). Further, reports revealed that nitric oxide, salicylic acid and hydrogen peroxide (H_2O_2) involved in salidroside and tyrosol accumulations induced by fungi (Cui et al., 2017).

Fungal endophytes and their plant host

Endophytic fungi are ubiquitous exist within the vascular plants, inside the healthy tissues of the plants, and are essential components of plant micro-ecosystems (Suryanarayanan, 2013; Jia et al., 2016). The diverse array of endophytes present within the host plants even plant tissue of the same plant namely root, stem and leaf can contain different group/species of endophytes

(Vega et al., 2008; Fürnkranz et al., 2012; Kumar et al., 2019). Endophytic fungi have been isolated from different plant tissues, including roots, stems, leaves, flowers and seeds of *Eucalyptus globulus* (Lupo et al., 2001) and roots and leaves of Lepanthes (Bayman et al., 1997), roots and leaves of healthy sugar beet (*Beta vulgaris*) (Shi et al., 2016). Xiang et al. (2016) isolated total of 208 fungal endophytes from different plant parts namely stems, leaves and flowers of twenty six medicinal plant species and majority of the isolates were belonged to *Acremonium, Alternaria, Phoma, Colletotrichum*, Mycelia sterilia and *Phomopsis*. A list of fungal endophytes recovered/ isolated from different crops is mentioned in Table 12.1. During the long period of co-existence and evolutionary processes, diverse associations have been established between endophytic fungi and their host plants. Colonization of endophytic fungi in the host plant is not an incidental opportunity because of the chemotaxis process. The source of colonization of fungal endophytes ranges from transmission via vegetative planting material and seeds to entry from the surrounding (rhizosphere and phyllosphere) environment (Lata et al., 2018). Fungal endophytes are found in plants grown in different regions namely temperate, tropical and forest regions (Zhang et al., 2006). Higher plants have been reported to host fungal endophytes in a symbiotic plant-fungus interaction. When endophytic fungi provide certain benefits to their hosts in the exchange for the nutrients derived from the host such interactions called as symbiotic relations (Bamisile et al., 2018).

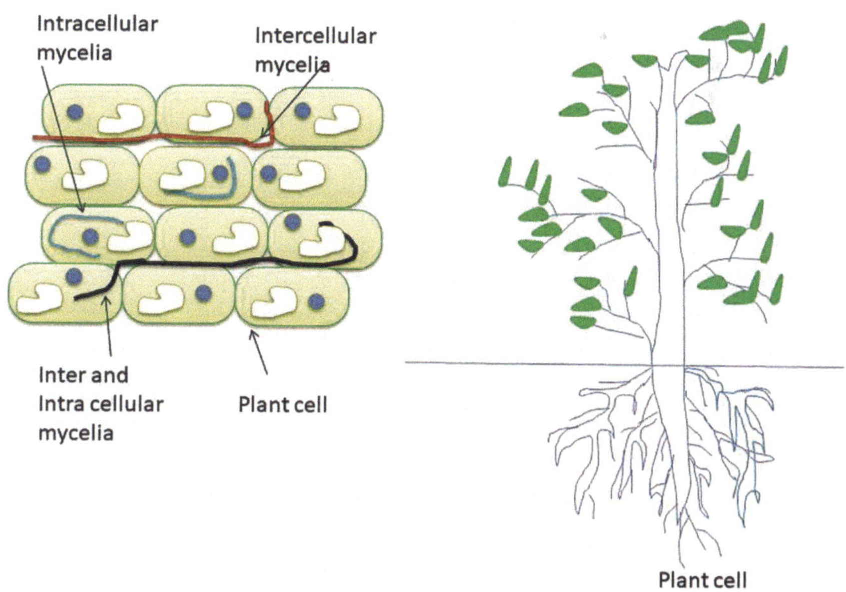

FIGURE 12.1 Types of infections during plant fungal interactions.

TABLE 12.1 List of fungal endophytes isolated different agricultural, horticultural and medicinal plants.

S. no.		Name of the fungal endophytes	Name of host plant/crop	Plant parts used	Reference
1.	Medicinal plants	*Cochliobolus lunatus, Cladosporium cladosporioides Cercophora caudate, Fusarium equiseti, Fusarium* sp., *Gibberella moniliformis, Gibberella moniliformis, Hypoxylon fragiforme* and *Nigrospora sphaerica*	Ashwagangha, *Withania somnifera*	Stems or twigs	Qadri et al. (2013)
2.		Six numbers of fungal endophytes	Ashwangandha, *Withania somnifera*	Leaf tissues	Kapoor et al. (2018)
3.		*Alternaria alternata Alternaria brassicae, Cladosporium cladosporioides, Diaporthe helianthi, Fusarium proliferatum, Lasiodiplodia theobromae* and *Glomerella acutata*	Sarpgandha, *Rauwolfia serpentina*	Stems or twigs	Qadri et al. (2013)
4.		*Ninety fungal endophytes isolated namely Chaetomium coarctatum Colletotrichum* sp., *Fusarium proliferatum Meyerozyma guilliermondii* and *Penicillium crustosum*	Indian Medicinal plant, *Ocimum sanctum*		Chowdhary and Kaushik (2015)
5.		Two numbers of fungal endophytes	Tulsi, *Ocimum basillicum*	Stem and internal stem tissues	Kapoor et al. (2018)
6.		*Chaetomium globosum, Diaporthe phaseolorum Thielavia subthermophila* and *Valsa sordid*	*Picrorhiza kurroa*	Stems or twigs	Qadri et al. (2013)

7.		*Alternaria alternata, Alternaria brassicae, Fusarium tricinctum, Fusarium flocciferum, Sordaria superba, Chaetomium* sp., *Paraphoma* sp. and *Gibberella avenacea,*	*Artemisia annua*	Stems or twigs	Qadri et al. (2013)
8.	Tree species	224 fungal endophytes isolated were represented by 29 genera of 15 orders of (93) Ascomycota (6) Basidiomycota and (1) Zygomycota percent	Podophyllum, *Dysosma versipellis*	Root, Rhizome, Stem and Leaves	Tan et al. (2018)
9.		*Bipolaris tetramera, Petriella* sp. *Penicillium expansum Trichophaea abundans and Ulocladium* sp.	*Pinus roxburgii*	Stems or twigs	Qadri et al. (2013)
10.		*Alternaria alternate, Penicillium frequentans Thielaviopsis basicola* and *Geotrichium albida*	*Pinus roxburghii*	Spike	Bhardwaj et al. (2015)
11.		Six numbers of fungal endophytes	Jamun, *Syzygium aromaticum*	Leaf, Stem and internal stem tissues	Kapoor et al. (2018)
12.		*Alternaria alternate and Aspergillus niger* were more dominated and *Chaetomium Aspergillus fumigatus globosum, Aspergillus japonicus, Aspergillus niger* strain were very rare species	Jamun, *Eugenia jambolana* (*Syzygium cumini*)	Leaf, Stem and Petiole	Yadav et al. (2016)
13.		*Aspergillus fumigatus, Penicillium oxalicum, Preussia* sp., *Peyronellaea eucalyptica, Peyronellaea sancta* and *Alternaria tenuissima.*	*Juniperus procera*	Twigs	Gherbawy and Elhariry (2014)
14.		*Apiosordaria otanii, Penicillium oxalicum, Daldinia fissa, Polyporus arcularius* and	*Abies pindrow*	Stems or twigs	Qadri et al. (2013)

(*Continued*)

TABLE 12.1 (Continued)

S. no.		Name of the fungal endophytes	Name of host plant/crop	Plant parts used	Reference
15.	Cereal crops	*Aspergillus, Acremonium, Arthrobotrys foliicola. Colletotrichum* spp., *Chaetomium* spp., *Curvularia* spp., *Fusarium, Penicillium, Gilmaniella, Phytophthora* spp., *Rhizopus* spp. and *Trichoderma* spp. and *Gibberella fujikuroi*	Rice/ Paddy *Oryza sativa*		Zakaria et al. (2010), Leewijit et al. (2016), Pili et al. (2016), Potshangbam et al. (2017)
16		*Alternaria* spp., *Cladosporium, Chaetomium* sp., *Cochliobolus* spp., *Darksidea* spp., *Embellisia* spp., *Periconia* spp. *Stemphylium*	Wheat Triticum		Bokati et al. (2016), Ofek-Lalzar et al. (2016)
17.		*Alternaria alternata, Acremonium, Aspergillus, Bipolaris zeicola, Botryodiplodia* sp., *Epicoccum nigrum, Fusarium, Cladosporium* sp., *Penicillium, Colletotrichum graminicola, Saccharomyces cerevisiae* and *Trichoderma koningii*	Maize *Zea mays*		Fisher et al. (1992), Orole and Adejumo (2011), Szilagyi-Zecchin et al. (2016)

Importance/functional significance of fungal endophytes

Endophytic microbes have recently drawn attention as a group of prospective plant growth promoters (biofertilizers) and controlling the biotic stresses caused by the pathogens (biocontrol agent). Endophytes assist in the improving plant health by deterring growth of pathogen and herbivory by producing different natural compounds namely flavonoids, alkaloids, isoflavonoids and diterpenes to increase the tolerance to various biotic and abiotic stresses. While others may facilitating plant growth through enhancing nutrient uptake, water use efficiency and curtailing of environmental stresses (Lata et al., 2018). Endophytes and their bioactive compounds/metabolites has been reported to improve and promote plant growth through production of growth hormones and phosphate solubilization, assimilable nitrogen to plants and reduce disease incidence and infestation from insects and mammals. Endophytes also assist in improving performance under extreme environmental stresses (Raps and Vidal, 1998; Rosenblueth and Martínez-Romero, 2006; Shahollari et al., 2007; Guo et al., 2008; Powell, 2009; Lata et al., 2018).

Role of fungal endophytes in plant growth promotion

Endophytes are known to enhance growth and uptake of nutrient of the host plant. These endophytes have ability to produce certain phyto-hormones and bioactive compounds (secondary metabolites, pharmaceutical drugs and enzymes) which find their applications in biotechnological and pharmaceutical industries (Joseph and Priya, 2011; Parthasarathi et al., 2012). They have been explored as possible microbial inoculants for increasing plant productivity (Hallmann et al., 1997). They may improve the plant's ability to tolerate various types of abiotic and biotic stresses, under the osmotic stress plants produces a broad array of responses at the molecular, cellular and whole plant level including increase in root growth, inhibition of shoot growth, adjustment in ion transport, changes in compatible solutes and carbon metabolism, etc. It has been reported that an endophytes produces a large number of bioactive metabolites in within a single plant or microbe that served as an exceptional source of drugs for curing various diseases and potential applications in medicine, agriculture, food and cosmetics industries (Jalgaonwala et al., 2011; Shukla et al., 2014; Godstime et al., 2014). List of fungal endophytes and bioactive compounds by these microbes are mentioned in Table 12.3.

Mode of action for growth promotion

Fungal endophytes may enhance the growth and fitness of the host plants by synthesizing or producing phytoharmones hormones namely, such as indole-3-acetic acid, cytokinins, indole-3-acetonitrile and salicylic acid (Jia et al.,

2016), some of them by activating the expression of a certain genes and enzymes involved in defense mechanism and production of bioactive compounds. For instance fungal endophyte (*Piriformospora indica*) stimulates the expression of the starch-degrading enzyme and nitrate reductase to increase the growth of tobacco roots (Chen and Guo, 2005; Sherameti et al., 2005).

Fungal endophytes can also promote growth by increasing uptake of nutrients (nitrogen, phosphorus, potassium and other essential elements) in the host plants (Bamisile et al., 2018), by obtaining nutritional components namely phosphorus, nitrogen and other useful molecules for plants (Hartley and Gange, 2009). Better understanding and knowledge of these functions would assist the user in optimum use of inorganic and organic fertilizers to ensure judicious and ecofriendly applications (Bamisile et al., 2018). Endophytic isolates also used as biofertilizers as they serve as plant growth promoting agent which facilitate uptake of nutrient not only through root system of the plants, but also through the transfer of the insect-derived nitrogen to plants. Fungal endophyte (*Metarhizium robertsii*) infects and kills the soil borne insects and from the dead insects it produces mycelia and forms an endophytic association with the plant roots, therefore increasing translocation of nitrogen (Behie et al., 2012). Some endophytes have been reported to enhance plant height, weight and other growth parameters of the host plant. The height, fresh weight of shoots and roots of *Vicia faba* plants has been improved by artificial inoculation of *Beauveria bassiana, B. brongniartii* and *M. brunneum* (Jaber and Enkerli, 2017).

Fungal endophytes as biocontrol agents

The application of endophytic microorganism as biocontrol agents is a smart choice for plant diseases management (Sapak et al., 2008). It was reported that some fungal endophyte play a key part in improving plant health and have been proven as an important resource of biocontrol agent to reduce agricultural pests including both insects and microbial pathogens (Branine et al., 2019). Several endophytic microorganisms consist of mutualistic symbionts that can be used as potential biocontrol agent against some severe phytopathogens. The potential of endophytic fungi during symbiont/host interaction has been revealed in number of studies (Rabiey et al., 2019; Latz et al., 2018). The best examples of fungal biocontrol agents are *Gliocladium virens* and *Trichoderma harzianum*. They have been reported as potential biocontrol agents by several groups (Romão-Dumaresq et al., 2016; Anjum et al., 2019). A list of fungal endophytes isolated from different crops and used as biocontrol agent against phytopathogens/ insects and their probable mode of action is mentioned in Table 12.2.

TABLE 12.2 A list of fungal endophytes isolated from different crops and used as biocontrol agent of phytopathogens/insects.

S. no.	Name of the fungal endophytes	Host plant	Biotic stresses	Mode of action/mechanism of biocontrol	References
1.	*Cylindrocarpon* sp, *Choiromyces aboriginum, Stachybotrys elegans*	*Phragmites australis*	Fungal pathogen	By producing cell wall-degrading enzymes	Cao et al. (2009)
2.	*Myxormia* sp.	*Angelica sinensis*	*Fusarium solani* and *F. oxysporum*	By producing certain chemicals toxic to pathogen	Yang et al. (2012)
3.	*Phomopsis cassiae*	*Cassia spectabilis*	*Cadosporium sphaerospermum*, and *C. cladosporioides*	Produce cadinane sesquiterpenoids toxic to pathogens	Silva et al. (2006)
4.	*Gilmaniella* sp. *AL12.*	*Atractylodes lancea*	Fungal pathogen	Production of jasmonic acid and induction of defense responses	Ren and Dai (2012)
5.	*Chaetomium globosum L18*	*Curcuma wenyujin*	Fungal pathogen	Produce some chemicals toxic to pathogens	Wang et al. (2012)
6.	*Trichothecium roseum*	*Maytenus hookeri*	Fungal pathogen	By producing Trichothecin chemical toxic to pathogen	Zhang et al. (2010)
7.	*Chaetomium sp, Phoma* sp.	*Triticum aestivum*	*Puccinia recondite*	Activation of plant defense response	Dingle and McGee (2003)
8.	*Cryptosporiopsis* cf. *quercina*	*Triptergyium wilfordii*	*Pyricularia oryzae*	Cryptocandin and cryptocin production which is toxic to pathogens	Strobel et al. (1999)

(*Continued*)

TABLE 12.2 (Continued)

S. no.	Name of the fungal endophytes	Host plant	Biotic stresses	Mode of action/mechanism of biocontrol	References
9.	*Acremonium blochii, Aspergillus fumigatus, C. destructans, Dactylaria* sp., *Fusarium equiseti, Phoma herbarum*	*Hordeum vulgare*	*Gaeumannomyces graminis* var. *Tritici*	Improve the competence for space inhibiting the colonization of pathogens	Maciá-Vicente et al. (2008)
10.	*Epichloë festucae*	*Festuca rubra* ssp. *rubra.*	*Sclerotinia homoeocarpa*	Secretion of anti-fungal protein	Tian et al. (2017)
11.	*Penicillium simplicissimum*	Cotton Roots	*Verticillium dahliae*	Increased transcript levels for PAL, PPO, POD, which leads to the increase of cotton defense reactions	Yuan et al. (2017)
12.	*Gliocladium catenulatum*	*Theobroma cacao*	*Crinipellis perniciosa*	Combination of antibiosis and mycoparasitism	Rubini et al. (2005)
13.	*Rhizopycnis vagum*	*Zingiber officinale*	*Rhizoctonia solani, Colletotrichum acutatum, Fusarium oxysporum, Sclerotium rolfsii*	Broad spectrum plant probiotic agent	Anisha et al. (2018)
14.	*Epicoccum nigrum*	Sugarcane	*Fusarium verticillioides, Ceratocystis paradoxa,*	Production of secondary metabolites and compounds that inhibit the growth of fungal pathogen	Fávaro et al. (2012)

15.	*Trichoderma viride*	*Spilanthes paniculata*	*Alternaria solani, Fusarium solani, Colletotrichum capsici, Pythium aphanidermatum*	By producing lytic enzymes which increases antagonism	Talapatra et al. (2017)
16.	*Trichoderma asperellum*	Lettuce leaves	*Corynespora cassiicola; Curvularia aeri,*	Induction of defense responses	Baiyee et al. (2019)
17.	*Chaetomium cochliodes, Cladosporium cladosporioides, Trichoderma viride*	*Cirsium arvense*	Foliar feeding insects	Produce some chemicals toxic to pathogens	Gange et al. (2012)
18.	*Chaetomium* Ch1001	*Cucumis sativus*	Root-knot nematode *Meloidogyne incognita*	Produced abscisic acid affecting motility of the second stage juveniles of insects	Yan et al. (2011)
19.	*Beauveria bassiana, Lecanicillium dimorphum*	*Phoenix dactylifera*	Date palm pests	Modulate the expression of cell division-related proteins in host	Gómez-Vidal et al. (2009)
20.	*Leucocoprinus gongylophorus*	*Cordia alliodora*	Insect	Produces chemical antagonistic to ants' fungal symbiont	Bittleston et al. (2011)

TABLE 12.3 List of bioactive compound/secondary metabolites of different importance produced by the fungal endophytes.

S. no.	Natural compounds/ secondary metabolites	Properties/ application/uses	Name of the endophytic fungi	Name of the host plants	Reference
1.	Camptothecin	Antitumor	*Entrophospora; Fusarium solani*	*Nothapodytes foetida; Apodytes dimidiata*	Rehman et al. (2008), Shweta et al. (2010)
2.	Paclitaxel	Antitumor	Taxomyces andreanae; Aspergillus niger; Phyllosticta dioscoreae; Fusarium solani	Taxus brevifolia; Taxus cuspidate; Hibiscus rosasinensis; Taxus chinensis	Stierle et al. (1995), Kim and Ford (1999), Kumaran et al. (2009), Deng et al. (2009)
3.	Podophyllotoxin	Antitumor	*Monilia species; Penicillium implicatum; Fusarium oxysporum*	*Dysosma veitchii; Diphylleia sinensis; Juniperus recurva*	Yang et al. (2003), Zeng et al. (2004), Kour et al. (2008)
4.	Taxol and Taxane	Anticancer	*Taxomyces andreanae*	*Pacific yew*	Stierle et al. (1993)
5.	Chlorogenic acid	Antitumor and Antimicrobial	*Sordariomycete* species	*Eucommia ulmoides*	Chen et al. (2010)
6.	Diosgenin	Antitumor, anti-inflammatory	*Cephalosporium* species	*Paris polyphylla*	Cao et al. (2007)
7.	Piperine	Anticancer, anti-inflammatory and Antimicrobial,	*Colletotrichum gloeosporioides*	*Piper nigrum*	Chithra et al. (2014)
8.	Quercetin	Anti- inflammatory	*Aspergillus oryzae* and *Aspergillus nidulans*	*Ginkgo biloba*	Qiu et al. (2010)

9.	Borneol	Anti-inflammatory, antioxidant	*Cochliobolus nisikadoi*	*Cinnamomum camphora*	Chen et al. (2011)
10.	Hypericin	Anti-depressant	*Chaetomium globosum*	*Hypericum perforatum*	Kusari et al. (2008)
11.	Huperzine A	Anticholinesterase	Penicillium chrysogenum	Lycopodium serratum	Zhou et al. (2009)
12.	Ginkgolide B	Anti-inflammatory and antiallergic	*Fusarium oxysporum*	*Ginkgo biloba*	Cui et al. (2012)
13.	Cajaninstilbene acid	Antioxidant and hypoglycemic	*Fusarium oxysporum, F. solani, F. proliferatum and Neonectria macrodidym*	*Cajanus cajan*	Zhao et al. (2012)
14.	Graphislactone A	Antioxidant Compounds	*Cephalosporium* sp.	*Trachelospermum jasminoides*	Song et al. (2005)
15.	Griseofulvin	Antimicrobial	*Xylaria* sps.	*Abies holophylla*	Park et al. (2005)
16.	Sipeimine	Antibechic and anti-ulcer	*Cephalosporium corda*	*Fritillaria ussuriensis*	Yin and Chen (2008)
17.	Spoxazomicins	antitrypanosomal alkaloids	*Streptosporangium oxazolinicum*		Inahashi et al. (2011)
18.	Vinblastine	Antitumor	*Alternaria* sp, *Fusarium oxysporum*	*Catharanthus roseus*	Zhang et al. (1998)
19.	Vincristine	anticancer drug	*Fusarium oxysporum*	*Catharanthus roseus*	Lingqi et al. (2000)
20.	Forskolin alkaloid	Glaucoma, antitumor, anti-HIV, cardiovascular protective Antibacterial	*Rhizoctonia bataticola*	*Coleus forskohlii*	Mir et al. (2015)

Mechanism for biocontrol activities

Fungal endophytes has been reported as a major source for production of antimicrobial compounds namely alkaloids, flavanoids, steroids, peptides and phenol which possses numerous applications in the agricultural industry and medical sciences (Strobel and Daisy, 2003). Numerous publications have been reported that fungal endophyte produce compounds with antifungal activity (Kusari and Spiteller, 2012; Campos et al., 2008). Furthermore, bioactive metabolites synthesized by endophytes can be used directly or indirectly as bio-control agents against plant diseases management (Park et al., 2017; Saad et al., 2019). Bioactive compounds synthesized by the fungal endophytes *viz.*, *Acremonium* spp. and *Fusarium* spp. exhibited potential antimicrobial property (Powthong et al., 2012), *Pestalotiopsis* species isolated from leaf tissues of *Pinus caneriensis* exhibited antimicrobial activities against broad range of bacteria (Bagyalakshmi et al., 2012).

Fungal endophytes were reported to act against plant pathogens by triggering their host resistance via two known mechanisms, systemic acquired resistance (SAR) and induced systemic resistance (ISR) (Kavroulakis et al., 2007; Latz et al., 2018). Fungi like *Trichoderma, Aureobasidium, Fusarium* and *Penicillium* antagonize plant pathogens by inducing SAR (Bae et al., 2011). Generally, various mechanisms of antagonistic actions of a bioagent are described as mycoparasitism, synthesis of lytic enzymes and/or antibiotics, initiation of plant defense and competition for nutrients and ecological niches (De Silva et al., 2019). In addition, competition for nutrients and/or space and induction of plant defense also responsible for antagonism (Busby et al., 2016). A leaf endophyte *Colletotrichum tropicale* triggers the expression of several host defense-related genes in *Theobroma cacao* (Mejía et al., 2014; Busby et al., 2016).

Synthesis of bioactive compounds

Microorganisms are the rich and important source of biologically active compounds and able to produce >20 000 compounds, which are influencing the survival and performance of other organisms (Demain and Sanchez, 2009). Especially fungal endophytes have capability of producing plant related metabolites and their analogs with beneficial value. They could produce various classes of phytochemicals, secondary metabolites of plant origin, including the well-known compounds such as paclitaxel (Stierle et al., 1993). It is worth mentioning here that secondary metabolites are biologically active compounds that are very useful source of anticancer, antioxidant, antidiabetic, immunosuppressive, antifungal, antioomycete, antibacterial, insecticidal, nematicidal, and antiviral agents (Gunatilaka, 2006). Moreover, endophytes produce metabolites those are involved in mechanisms of signaling between host and partner fungi, defense, and genetic regulation of the establishment of symbiosis (Sun et al., 2014; Hardoim et al., 2015).

Furthermore, these compounds can provoke the production of some novel biologically active secondary metabolites (Zhang et al., 2006; Firáková et al., 2007; Rodriguez et al., 2009) that can be exploited and useful for human beings as important medicinal resources.

One of the best application is to exploit the benefits of fungal endophytes that can uphold the accumulation of secondary metabolites originally made by plants (Jia et al., 2016). The most of the bioactive produces endophytic fungi, excluding those produced by their host plant, are very vital to enhance the adaptability of both fungal endophytic partners and their host plants, such as the tolerances to biotic and abiotic stresses.

Conclusion

Exploring the plants microbiome and its diversity is of great significance to identify new and effective microorganisms which can be used as biological control agent (BCAs) and production of bioactive substances/compounds. Endophytic microbes hold an enormous potential to be deployed for product development to practical applications and can be used as plant bio-fertilizer and bio-control agents. Significant progress in research has been made in past few years on isolation and characterization of novel endophytic microbes and their bioactive metabolites, it is high time to explore endophytes from extreme and diverse habitats and identify their natural products like antibiotics which can be effective against multi-drug resistant (MDR) bacteria. Molecular studies are required for understanding of host-endophyte interactions in a more comprehensive manner and development of endophytes for production of bioenergy crops. Unraveling the microbiome associated with plants offer genetic variability to plants thus opening up new horizons for breeding of next-generation crops for developing high yielding, pest and disease resistant plants and resilient to climatic change.

References

Anisha, C., Jishma, P., Bilzamol, V.S., Radhakrishnan, E.K., 2018. Effect of ginger endophyte Rhizopycnis vagum on rhizome bud formation and protection from phytopathogens. Biocatal. Agric. Biotechnol. 14, 116–119.

Anjum, R., Afzal, M., Baber, R., Khan, M.A.J., Kanwal, W., Sajid, W., et al., 2019. Endophytes: as potential biocontrol agent—review and future prospects. J. Agric. Sci. 11, 113.

Bae, H., Roberts, D.P., Lim, H.-S., Strem, M., Park, S.-C., Ryu, C.-M., et al., 2011. Endophytic Trichoderma isolates from tropical environments delay disease and induce resistance against Phytophthora capsici in hot pepper using multiple mechanisms. Mol. Plant–Microbe Interact. 24 (3), 336–351.

Bagyalakshmi, T.A., Ramesh, V., Arivudainambi, U.S.E., Rajendran, A., 2012. A novel endophytic fungus Pestalotiopsis sp. Inhibiting Pinus canariensis with antibacterial and antifungal potential. Int J Adv Life Sci. 1, 1–7.

Baiyee, B., Ito, S., Sunpapao, A., 2019. Trichoderma asperellum T1 mediated antifungal activity and induced defense response against leaf spot fungi in lettuce (Lactuca sa- tiva L.). Physiol. Mol. Plant Pathol. 106, 96–101.

Bamisile, B.S., Dash, C.K., Akutse, K.S., Keppanan, R., Wang, L., 2018. Fungal endophytes: beyond herbivore management. Front. Microbiol. 9, 1–11.

Bayman, P., Lebron, L.L., Tremblay, R.L., 1997. Variation in endophytic fungi from roots and leaves of Lepanthes (Orchidaceae). N. Phytol. 135, 143–149.

Bhardwaj, A., Sharma, D., Jadon, N., Agrawal, P.K., 2015. Antimicrobial and phytochemical screening of endophytic fungi isolated from spikes of *Pinus roxburghii*. Arch. Clin. Microbiol. 6 (3), 1–9.

Behie, S.W., Zelisko, P.M., Bidochka, M.J., 2012. Endophytic insect- parasitic fungi translocate nitrogen directly from insects to plants. Science 336, 1576–1577.

Bittleston, L.S., Brockmann, F., Wcislo, W., Van Bael, S.A., 2011. Endophytic fungi reduce leaf-cutting ant damage to seedlings. Biol. Lett. 7, 30–32.

Bokati, D., Herrera, J., Poudel, R., 2016. Soil influences colonization of root-associated fungal endophyte communities of maize, wheat, and their progenitors. J. Mycol. 9. ID 8062073.

Branine, M., Bazzicalupo, A., Branco, S., 2019. Biology and applications of endophytic insect-pathogenic fungi. PLoS Pathog. 15 (7), e1007831. Available from: https://doi.org/10.1371/journal.ppat.1007831.

Busby, P.E., Ridout, M., Newcombe, G., 2016. Fungal endophytes: modifiers of plant disease. Plant Mol. Biol. 90, 645–655.

Campos, F., Rosa, L., Cota, B., Caligiorne, R., Rabello, A., Alves, T., et al., 2008. Leishmanicidal metabolites from Cochliobolus sp., an endophytic fungus isolated from Piptadenia adiantoides (Fabaceae). PLoS Negl. Trop. Dis. 2, e348.

Cao, R., Liu, X., Gao, K., Mendgen, K., Kang, Z., Gao, J., et al., 2009. Mycoparasitism of endophytic fungi isolated from reed on soilborne phytopathogenic fungi and production of cell wall-degrading enzymes in vitro. Curr. Microbiol. 59, 584–592.

Cao, X., Li, J., Zhou, L., 2007. Determination of diosgenin content of the endophytic fungi from *Paris polyphylla* var. yunnanensis by using an optimum ELISA. Nat. Prod. Res. Dev. 19, 1020–1023.

Chen, B., Wang, M., Hu, Y., Lin, Z., Yu, R., Huang, L., 2011. Preliminary study on promoting effects of endophytic fungi to growth of *Rehmannia glutinosa*. Chin. J. Chin. Mater. Med. 36, 1137–1140.

Chen, X., Sang, X., Li, S., Zhang, S., Bai, L., 2010. Studies on a chlorogenic acid-producing endophytic fungi isolated from *Eucommia ulmoides* Oliver. J. Ind. Microbiol. Biotechnol. 37, 447–454.

Chen, X.M., Guo, S.X., 2005. Effects of four species of endophytic fungi on the growth and polysaccharide and alkaloid contents of Dendrobium nobile. Chin. J. Chin. Mater. Med. 30, 253–257.

Chithra, S., Jasim, B., Sachidanandan, P., Jyothis, M., Radhakrishnan, E.K., 2014. Piperine production by endophytic fungus *Colletotrichum gloeosporioides* isolated from *Piper nigrum*. Phytomedicine 21, 534–540.

Chowdhary, K., Kaushik, N., 2015. Fungal endophyte diversity and bioactivity in the indian medicinal plant *Ocimum sanctum* Linn. PLoS One 10 (11), e0141444.

Cui, J., Wang, Y., Jiao, J., et al., 2017. Fungal endophyte-induced salidroside and tyrosol biosynthesis combined with signal cross-talk and the mechanism of enzyme gene expression in Rhodiola crenulata. Sci. Rep. 7, 12540. Available from: https://doi.org/10.1038/s41598-017-12895-2.

Cui, Y., Yi, D., Bai, X., Sun, B., Zhao, Y., Zhang, Y., 2012. Ginkgolide B produced endophytic fungus (*Fusarium oxysporum*) isolated from *Ginkgo biloba*. Fitoterapia 83, 913–920.

Demain, A.L., Sanchez, S., 2009. Microbial drug discovery: 80 years of progress. J. Antibiot. (Tokyo) 62 (1), 5–16. Available from: https://doi.org/10.1038/ja.2008.16.

De Silva, N.I., Brooks, S., Lumyong, S., Hyde, K.D., 2019. Use of endophytes as biocontrol agents. Fungal Biol. Rev. 33, 133–148.

De Souza Sebastianes, F.L., Romao-Dumaresq, A.S., Lacava, P.T., Harakava, R., Azevedo, J.L., de Melo, I.S., et al., 2013. Species diversity of culturable endophytic fungi from Brazilian mangrove forests. Curr. Genet. 59 (3), 153–166.

Deng, B.W., Liu, K.H., Chen, W.Q., Ding, X.W., Xie, X.C., 2009. *Fusarium solani,* Tax-3, a new endophytic taxol-producing fungus from *Taxus chinensis*. World J. Microb. Biotechnol. 25, 139–143.

Dingle, J., McGee, P.A., 2003. Some endophytic fungi reduce the density of pustules of *Puccinia recondita f. sp. tritici* in wheat. Mycol. Res. 107, 310–316.

Faeth, S.H., Bultman, T.L., 2002. Endophytic fungi and interaction among host plants, herbivores, and natural enemies. In: Tscharntke, T., Hawkins, B.A. (Eds.), Multitritrophic Level Interactions. Cambridge University Press, Cambridge, pp. 89–123.

Fávaro, L.C.D.L., Sebastianes, F.L.D.S., Araújo, W.L., 2012. Epicoccum nigrum P16, a sugarcane endophyte, produces antifungal compounds and induces root growth. PLoS One 7 (6), e36826.

Firáková, S., Šturdíková, M., Múčková, M., 2007. Bioactive secondary metabolites produced by microorganisms associated with plants. Biologia 62, 251–257.

Fisher, P.J., Petrini, O., Lappin Scott, H.M., 1992. The distribution of some fungal and bacterial endophytes in maize (*Zea mays* L). N. Phytol. 122, 299–305.

Fürnkranz, M., Lukesch, B., Müller, H., Huss, H., Grube, M., Berg, G., 2012. Microbial diversity inside pumpkins: microhabitat-specific communities display a high antagonistic potential against phytopathogens. Microb. Ecol. 63, 418–428.

Gange, A.C., Eschen, R., Wearn, J.A., Thawer, A., Sutton, B.C., 2012. Differential effects of foliar endophytic fungi on insect herbivores attacking a herbaceous plant. Oecologia 168, 1023–1031.

Gherbawy, Y.A., Elhariry, H.M., 2014. Endophytic fungi associated with high-altitude Juniperus trees and their antimicrobial activities. Plant Biosyst. Available from: https://doi.org/10.1080/11263504.2014.984011.

Godstime, O.C., Enwa, F.O., Augustina, J.O., Christopher, E.O., 2014. Mechanisms of antimicrobial actions of phytochemicals against enteric pathogens-a review. J. Pharm. Chem. Biol. Sci. 2, 77–85.

Gómez-Vidal, S., mez-Vidal, S., Salinas, J., Tena, M., Lopez-Llorca, L.V., 2009. Proteomic analysis of date palm (*Phoenix dactylifera* L.) responses to endophytic colonization by entomopathogenic fungi. Electrophoresis 30, 2996–3005.

Gunatilaka, A.A.L., 2006. Natural products from plant-associated micro- organisms: distribution, structural diversity, bioactivity, and implications of their occurrence. J. Nat. Prod. 69, 509–526.

Guo, B., Wang, Y., Sun, X., Tang, K., 2008. Bioactive natural products from endophytes: a review. Appl. Biochem. Microbiol. 44, 136–142.

Hallmann, J., Quadt-Hallmann, A., Mahaffee, W.F., Kloepper, J.W., 1997. Bacterial endophytes in agricultural crops. Can. J. Microbiol. 43, 895–914. Available from: https://doi.org/10.1139/m97-131.

Hardoim, P.R., Overbeek, L.S., Van, Berg, G., Pirttilä, M., Compant, S., et al., 2015. The hidden world within plants ecological and evolutionary considerations for defining functioning of microbial endophytes. Microbiol. Mol. Biol. Rev. 79, 293–320. Available from: https://doi.org/10.1128/MMBR.00050-14.

Hartley, S.E., Gange, A.C., 2009. Impacts of plant symbiotic fungi on insect herbivores: mutualism in a multitrophic context. Annu. Rev. Entomol. 54, 323–342.

Ikram, M., Ali, N., Jan, G., Jan, F.G., Rahman, I.U., Iqbal, A., et al., 2018. IAA producing fungal endophyte Penicillium roqueforti Thom., enhances stress tolerance and nutrients uptake in wheat plants grown on heavy metal contaminated soils. PLoS One 13 (11), e0208150. Available from: https://doi.org/10.1371/journal.pone.0208150.

Inahashi, Y., Iwatsuki, M., Ishiyama, A., Namatame, M., Nishihara- Tsukashima, A., Matsumoto, A., et al., 2011. Spoxazomicins A-C, novel antitrypanosomal alkaloids produced by an endophytic actinomycete, *Streptosporangium oxazolinicum* K07-0460(T). J. Antibiot. 64, 303–307.

Ismail, M., Hamayun, A., Hussain, S., Afzal Khan, A., Iqbal, I.-J., 2019. Lee Aspergillus flavus promoted the growth of soybean and sunflower seedlings at elevated temperature. BioMed. Res. Int. 2019, 1–13.

Jaber, L.R., Enkerli, J., 2017. Fungal entomopathogens as endophytes: can they promote plant growth? Biocontrol Sci.Technol. 27, 28–41.

Jalgaonwala, R.E., Mohite, B.V., Mahajan, R.T., 2011. Natural products from plant associated endophytic fungi. J. Microbiol. Biotechnol. Res. 1, 21–32.

Jia, M., Chen, L., Xin, H.-L., Zheng, C.-J., Rahman, K., Han, T., et al., 2016. A friendly relationship between endophytic fungi and medicinal plants: a systematic review. Front. Microbiol. 7.

Joseph, B., Priya, R.M., 2011. Bioactive compounds from endophytes and their potential in pharmaceutical effect: a review. Am. J. Biochem. Mol. Bio. 1, 291–309.

Kapoor, N., Rajput, P., Mushtaque, M.A., Gambhir, L., 2018. Bio-prospecting fungal endophytes of high altitude medicinal plants for commercially imperative enzymes. Biosci. Biotech. Res. Commun. 1 (3), 370–375.

Kavroulakis, N.S., Zervakis, G.I., Ehaliotis, C., Haralampidis, K., Papadopoulou, K.K., 2007. Role of ethylene in the protection of tomato plants against soil-borne fungal pathogens conferred by an endophytic Fusarium solani strain. J. Exp. Bot. 58, 3853–3864.

Khan, A.L., Hussain, J., Al-Harrasi, A., Al-Rawahi, A., Lee, I.-J., 2015. Endophytic fungi: resource for gibberellins and crop abiotic stress resistance. Crit. Rev. Biotechnol. 35 (1), 62–74.

Khare, E., Mishra, J., Arora, N.K., 2018. Multifaceted interactions between endophytes and plant: developments and prospects. Front. Microbiol. 9, 2732. Available from: https://doi.org/10.3389/fmicb.2018.02732.

Kim, S.U., Ford, E., 1999. Screening of taxol-producing endophytic fungi from *Ginkgo biloba* and *Taxus cuspidata* in Korea. Nat. Prod. Organ. Chem. 4, 97–99.

Kour, A., Shawl, A.S., Rehman, S., Sultan, P., Qazi, P.H., Suden, P., et al., 2008. Isolation and identification of an endophytic strain of *Fusarium oxysporum* producing podophyllotoxin from *Juniperus recurva*. World J. Microbiol. Biotech. 24, 115–1121.

Kumar, V., Jain, L., Soni, R., Kaushal, P., Goel, R., 2020. Bioprospecting of endophytic microbes from higher altitude plants: recent advances and their biotechnological applications. In: Goel, R., et al., (Eds.), Microbiological Advancements for Higher Altitude Agroecosystems & Sustainability, Rhizosphere Biology. Springer. Available from: https://doi.org/10.1007/978-981-15-1902-4_18.

Kumar, V., Soni, R., Jain, L., Dash, B., Goel, R., 2019. Endophytic fungi: recent advances in identification and explorations. In: Singh, B.P. (Ed.), Advances in Endophytic Fungal

Research, Fungal Biology. Springer Nature, Switzerland AG. Available from: https://doi.org/10.1007/978-3-030-03589-1_13.

Kumaran, R.S., Muthumary, J., Kim, E.K., Hur, B.K., 2009. Production of taxol from *Phyllosticta dioscoreae*, a leaf spot fungus isolated from *Hibiscus rosasinensis*. Biotechnol. Bioprocess. Eng. 14, 76–83.

Kusari, S., Lamshöft, M., Zühlke, S., Spiteller, M., 2008. An endophytic fungus from *Hypericum perforatum* that produces hypericin. J. Nat. Prod. 71, 159–162.

Kusari, S., Spiteller, M., 2012. Metabolomics of endophytic fungi producing associated plant secondary metabolites: progress, challenges and opportunities. In: Roessner, U. (Ed.), Metabolomics. In Tech, Rijeka, Croatia, pp. 241–266.

Lata, R., Chowdhury, S., Gond, S.K., White Jr, J.F., 2018. Induction of abiotic stress tolerance in plants by endophytic microbes. Lett. Appl. Microbiol. Available from: https://doi.org/10.1111/lam.12855.

Latz, M.A.C., Jensen, B., Collinge, D.B., Jørgensen, H.J.L., 2018. Endophytic fungi as biocontrol agents: elucidating mechanisms in disease suppression. Plant. Ecol. Divers. 11 (5–6), 555–567.

Leewijit, T., Pongnak, W., Soytong, K., Poeaim, S., 2016. Isolation of Soil and Endophytic Fungi from Rice (Oryza sativa L.). Int. J. Agri. Tech. 12 (7.2), 2191–2202.

Li, H.Y., Shen, M., Zhou, Z.P., Li, T., Wei, Y.L., Lin, L.B., 2012. Diversity and cold adaptation of endophytic fungi from five dominant plant species collected from the Baima Snow Mountain, Southwest China. Fungal Divers. 54, 79–86.

Lingqi, Z., Bo, G., Haiyan, L., Songrong, Z., Hua, S., Su, G., et al., 2000. Preliminary study on the isolation of endophytic fungus of *Catharanthus roseus* and its fermentation to produce products of therapeutic value. Zhong Cao Yao 31, 805–807.

Lugtenberg, B.J.J., Caradus, J.R., Johnson, L.J., 2016. Fungal endophytes for sustainable crop production. FEMS Microbiol. Ecol. 92, fiw194. Available from: https://doi.org/10.1093/femsec/fiw19.

Lupo, S., Tiscornia, S., Bettucci, L., 2001. Endophytic fungi from flowers, capsules and seeds of *Eucalyptus globulus*. Rev. Iberoam. Microl. 18, 38e41.

Maciá-Vicente, J., Jansson, H., Mendgen, K., Lopez-Llorca, L.V., 2008. Colonization of barley roots by endophytic fungi and their reduction of take-all caused by *Gaeumannomyces graminis* var. *tritici*. Can. J. Microbiol. 54, 600–609.

Mejía, L.C., Herre, E.A., Sparks, J.P., Winter, K., Garcia, M.N., Van Bael, S.A., et al., 2014. Pervasive effects of a dominant foliar endophytic fungus on host genetic and phenotypic expression in a tropical tree. Front. Microbiol. 5, 479.

Mir, R.A., Kaushik, P.S., Chowdery, R.A., Anuradha, M., 2015. Elicitation of forskolin in cultures of Rhizactonia bataticola a phytochemical synthesizing endophytic fungi. Int. J. Pharm. Pharm Sci. 7 (10), 185–189.

Moricca, S., Ginetti, B., Ragazzi, A., 2012. Species- and organ-specificity in endophytes colonizing healthy and declining Mediterranean oaks. Phytopathol. Medit. 51, 587–598.

Ofek-Lalzar, M., YGur, S., Ben-Moshe, O., Sharon, E., Kosman, E., Mochli, A., et al., 2016. Diversity of fungal endophytes in recent and ancient wheat ancestors *Triticum dicoccoides* and *Aegilops sharonensis*. FEMS Microbiol. Ecol. 92, 10.

Orole, O.O., Adejumo, T.O., 2011. Bacterial and fungal endophytes associated with grains and roots of maize. J. Ecol. Nat. Env. 3, 298–303.

Park, J.H., Choi, G.J., Lee, H.B., et al., 2005. Griseofulvin from *Xylaria sp* . strain F0010, an endophytic fungus of *Abies holophylla* and its antifungal activity against plant pathogenic fungi. J. Microbiol. Biotechnol. 15 (1), 112–117.

Park, Y., Kim, Y., Mishra, R.C., et al., 2017. Fungal endophytes inhabiting mountain-cultivated ginseng (Panax ginseng Meyer): diversity and biocontrol activity against ginseng pathogens. Sci. Rep. 7, 16221. Available from: https://doi.org/10.1038/s41598-017-16181-z.

Parthasarathi, S., Sathya, S., Bupesh, G., Samy, D.R., Mohan, M.R., Selva, G.K., et al., 2012. Isolation and characterization of antimicrobial compound from marine Streptomyces hygroscopicus BDUS 49. World J. Fish Mar. Sci. 4, 268–277.

Pili, N.N., de Carvalho França, S., Kyndt, T., Makumba, B.A., Skilton, R., Höfte, M., et al., 2016. Analysis of fungal endophytes associated with rice roots from irrigated and upland ecosystems in Kenya. Plant Soil 405 (1–2), 371–380.

Potshangbam, M., Indira Devi, S., Sahoo, D., Strobel, G.A., 2017. Functional characterization of endophytic fungal community associated with *Oryza sativa* L. and *Zea mays* L. Front. Microbiol. 8, 1–15.

Powell, R.G., 2009. Plant seeds as sources of potential industrial chemicals, pharmaceuticals, and pest control agents. J. Nat. Prod. 72, 516–523.

Powthong, P., Jantrapanukorn, B., Thongmee, A., Suntornthiticharoen, P., 2012. Evaluation of endophytic fungi extract for their antimicrobial activity from Sesbania grandiflora (L.) Pers. Pharm. Biomed. Res. 3 (2), 132–136.

Qadri, M., Johri, S., Shah, B.A., Khajuria, A., Sidiq, T., Lattoo, S.K., et al., 2013. Identification and bioactive potential of endophytic fungi isolated from selected plants of the Western Himalayas. Springer Plus 2, 8.

Qiu, M., Xie, R., Shi, Y., Zhang, H., Chen, H., 2010. Isolation and identification of two flavonoid-producing endophytic fungi from *Ginkgo biloba* L. Ann. Microbiol. 60, 143–150.

Rabiey, M., Hailey, L.E., Roy, S.R., et al., 2019. Endophytes vs tree pathogens and pests: can they be used as biological control agents to improve tree health? Eur. J. Plant Pathol. 155, 711.

Raja, S., Subhashini, P., Thangaradjou, T., 2016. Differential methods of localisation of fungal endophytes in the seagrasses. Mycology 7 (3), 112–123. Available from: https://doi.org/10.1080/21501203.2016.1218966.

Raps, A., Vidal, S., 1998. Indirect effects of an unspecialized endophytic fungus on specialized plant–herbivorous insect interactions. Oecologia 114, 541–547.

Rehman, S., Shawl, A., Verma, V., Kour, A., Athar, M., Andrabi, R., et al., 2008. An endophytic N*eurospora sp* . from *Nothapodytes foetida* producing camptothecin. Prikl. Biokhim. Mikrobiol. 44, 225–231.

Ren, C.G., Dai, C.C., 2012. Jasmonic acid is involved in the signaling pathway for fungal endophyte-induced volatile oil accumulation of *Atractylodes lancea* plantlets. BMC Plant. Biol. 12, 128.

Rodriguez, R.J., White Jr., J.F., Arnold, A.E., Redman, R.S., 2009. Fungal endophytes: diversity and functional roles. N. Phytol. 182, 314–330.

Romão-Dumaresq, A.S., Dourado, M.N., Fávaro, L.C.D.L., Mendes, R., Ferreira, A., Araújo, W. L., 2016. Diversity of cultivated fungi associated with conventional and transgenic sugarcane and the interaction between endophytic Trichoderma virens and the host plant. PLoS One 11 (7), e0158974.

Rosenblueth, M., Martínez-Romero, E., 2006. Bacterial endophytes and their interactions with hosts. Mol. Plant. Microbe Interact. 19, 827–837.

Rozpądek, P., Domka, A.M., Nosek, M., Ważny, R., Jędrzejczyk, R.J., Wiciarz, M., et al., 2018. The role of Strigolactone in the cross-talk between Arabidopsis thaliana and the endophytic fungus Mucor sp. Front. Microbiol. 9, 441. Available from: https://doi.org/10.3389/fmicb.2018.00441.

Rubini, M.R., Silva-Ribeiro, R.T., Pomella, A.W.V., Maki, C.S., Araújo, W.L., dos Santos, D. R., et al., 2005. Diversity of endophytic fungal community of cacao (Theobroma cacao L.) and biological control of Crinipellis perniciosa, causal agent of Witches' Broom Disease. Int. J. Biol. Sci. 1, 24–33.

Saad, M.M.G., Ghareeb, R.Y., Saeed, A.A., 2019. The potential of endophytic fungi as bio-control agents against the cotton leafworm, Spodoptera littoralis (Boisd.) (Lepidoptera: Noctuidae). Egypt. J. Biol. Pest. Control. 29, 7. Available from: https://doi.org/10.1186/s41938-019-0108-x.

Saikkonen, K., Faeth, S.H., Helander, M., Sullivan, T., 1998. Fungal endophytes: a continuum of interactions with host plants. Annu. Rev. Ecol. Syst. 29, 319–343.

Saikkonen, K., Young, C.A., Helander, M., Schardl, C.L., 2016. Endophytic Epichloë species and their grass hosts: from evolution to applications. Plant Mol. Biol. 90, 665–675.

Sandberg, D.C., Battista, L.J., Arnold, A.E., 2014. Fungal endophytes of aquatic macrophytes: diverse host-generalists characterized by tissue preferences and geographic structure. Microb. Ecol. 67 (4), 735–747.

Sapak, Z., Meon, S., Ahmad, Z.A.M., 2008. Effect of endophytic bacteria on growth and suppression of Ganoderma infection in Oil Palm. Int. J. Agric. Biol. 10, 127–132.

Shahollari, B., Vadassery, J., Ajit, V., Oelmuller, R., 2007. A leucine-rich repeat protein is required for growth promotion and enhanced seed production mediated by the endophytic fungus Piriformospora indica in Arabidopsis thaliana. Plant. J. 50, 1–13.

Sherameti, I., Shahollari, B., Venus, Y., Altschmied, L., Varma, A., Oelmüller, R., 2005. The endophytic fungus Piriformospora indica stimulates the expression of nitrate reductase and the starch-degrading enzyme glucan- water dikinase in tobacco and Arabidopsis roots through a homeodomain transcription factor that binds to a conserved motif in their promoters. J. Biol. Chem. 280, 26241–26247.

Shi, Y.W., Li, C., Yang, H.M., Zhang, T., Gao, Y., Zeng, J., et al., 2016. Endophytic fungal diversity and space-time dynamics in sugar beet. Eur. J. Soil Biol. 77, 77–85.

Shukla, S.T., Habbu, P.V., Kulkarni, V.H., Jagadish, K.S., Pandey, A.R., Sutariya, V.N., 2014. Endophytic microbes: a novel source for biologically/pharmacologically active secondary metabolites. Asian J. Pharmacol. Toxicol. 2, 1–16.

Shweta, S., Zuehlke, S., Ramesha, B.T., Priti, V., Kumar, P.M., Ravikanth, G., et al., 2010. Endophytic fungal strains of *Fusarium solani*, from *Apodytes dimidiata* E. Mey. ex Arn (Lcacinaceae) produce camptothecin, 10-hydroxycamptothecin and 9-methoxycamptothecin. Phytochemistry 71, 117–122.

Silva, G.H., Teles, H.L., Zanardi, L.M., Young, M.C.M., Eberlin, M.N., Hadad, R., et al., 2006. Cadinane sesquiterpenoids of *Phomopsis cassiae*, an endophytic fungus associated with *Cassia spectabilis* (*Leguminosae*). Phytochemistry 67, 1964–1969.

Song, Y.C., Huang, W.Y., Sun, C., Wang, F.W., Tan, R.X., 2005. Characterization of graphislactone A as the antioxidant and free radical-scavenging substance from the culture of Cephalosporium sp. IFB-E001, an endophytic fungus in *Trachelospermum jasminoides*. Biol. Pharm. Bull. 28 (3), 506–509.

Sridhar, K.R., 2019. Diversity, ecology, and significance of fungal endophytes. In: Jha, S. (Ed.), Endophytes and Secondary Metabolites. Reference Series in Phytochemistry. Springer, Cham.

Stierle, A., Strobel, G., Stierle, D., 1993. Taxol and taxane production by *Taxomyces andreanae*, an endophytic fungus of Pacific yew. Science 260, 214–216.

Stierle, A., Strobel, G., Stierle, D., Grothaus, P., Bignami, G., 1995. The search for a taxol-producing microorganism among the endophytic fungi of the Pacific yew, *Taxus brevifolia*. J. Nat. Prod. 58, 1315–1324.

Strobel, G., Daisy, B., 2003. Bioprospecting for microbial endophytes and their natural products. Microbiol. Mol. Biol. Rev. 67 (4), 491–502.

Strobel, G.A., Miller, R.V., Martinez-Miller, C., Condron, M.M., Teplow, D.B., Hess, W.M., 1999. Cryptocandin, a potent antimycotic from the endophytic fungus *Cryptosporiopsis cf. quercina*. Microbiology 145, 1919–1926.

Sun, J., Zhang, Q., Zhou, J., Wei, Q., 2014. Illumina amplicon sequencing of 16S rRNA tag reveals bacterial community development in the rhizosphere of apple nurseries at a replant disease site and a new planting site. PLoS One 9.

Suryanarayanan, T.S., 2013. Endophyte research: going beyond isolation and metabolite documentation. Fungal Eco. 6, 561–568.

Szilagyi-Zecchin, V.J., Adamoski, D., Gomes, R.R., Hungria, M., Ikeda, A.C., Kava-Cordeiro, V., et al., 2016. Composition of endophytic fungal community associated with leaves of maize cultivated in south Brazilian field. Acta Microbiol. Immunol. Hung. 63, 449–466.

Talapatra, K., Das, A.R., Saha, A.K., Das, P., 2017. In vitro antagonistic activity of a root endophytic fungus towards plant pathogenic fungi. J. Appl. Biol. Biotechnol. 5, 68–71.

Tan, X., Zhou, Y., Zhou, X., Xia, X., Wei, Y., He, L., et al., 2018. Diversity and bioactive potential of culturable fungal endophytes of *Dysosma versipellis*; a rare medicinal plant endemic to China. Sci. Rep. 8, 5929.

Tian, Z., Wang, R., Ambrose, K.V., Clarke, B.B., Belanger, F.C., 2017. The Epichloë festucae antifungal protein has activity against the plant pathogen Sclerotinia homoeocarpa, the causal agent of dollar spot disease. Sci. Rep. 7 (1), 5643.

Vega, F.E., Posada, F., Aime, M.C., Pava-Ripoll, M., Infante, F., Rehner, S.A., 2008. Entomopathogenic fungal endophytes. Biol. Control. 46, 72–82.

Vincent, J.B., Weblen, G.D., May, G., 2016. Host associations and beta diversity of fungal endophyte communities in New Guinea rainforest trees. Mol. Ecol. 25, 825–841.

Wang, Y., Xu, L., Ren, W., Zhao, D., Zhu, Y., Wu, X., 2012. Bioactive metabolites from *Chaetomium globosum* L18, an endophytic fungus in the medicinal plant *Curcuma wenyujin*. Phytomedicine 19, 364–368.

Xiang, L., Gong, S., Yang, L., Hao, J., Xue, M.F., Zeng, F.S., et al., 2016. Biocontrol potential of endophytic fungi in medicinal plants from Wuhan Botanical Garden in China. Biol. Control. 94, 47–55.

Yadav, M., Yadav, A., Kumar, S., Yadav, J.P., 2016. Spatial and seasonal influences on culturable endophytic mycobiota associated with different tissues of *Eugenia jambolana* Lam. and their antibacterial activity against MDR strains. BMC Microbiol. 16, 44.

Yan, L., Zhu, J., Zhao, X., Shi, J., 2019. Beneficial effects of endophytic fungi colonization on plants. Appl. Microbiol. Biotech. 103, 3327–3340.

Yan, X.N., Sikora, R.A., Zheng, J.W., 2011. Potential use of cucumber (*Cucumis sativus* L.) endophytic fungi as seed treatment agents against root- knot nematode *Meloidogyne incognita*. J. Zhejiang Univ. Sci. B 12, 219–225.

Yang, N.Y., Jiang, S., Shang, E.X., Tang, Y.P., Duan, J.A., 2012. A new phenylpentanamine alkaloid produced by an endophyte *Bacillus subtilis* isolated from *Angelica sinensis*. J. Chem. Res. 36, 647.

Yang, X., Guo, S., Zhang, L., Shao, H., 2003. Select of producing podophyllotoxin endophytic fungi from podophyllin plant. Nat. Prod. Res. Dev. 5, 419–422.

Yin, H., Chen, J.L., 2008. Sipeimine-producing endophytic fungus isolated from Fritillaria ussuriensis. Z Naturforsch C J Biosci. 63 (11-12), 789–793.

Yuan, Y., Feng, H., Wang, L., Li, Z., Shi, Y., Zhao, L., et al., 2017. Potential of endophytic fungi isolated from cotton roots for biological control against Verticillium wilt disease. PLoS One 1–12.

Zakaria, L., Yaakop, A.S., Salleh, B., Zakaria, M., 2010. Endophytic fungi from paddy. Trop. Life Sci. Res. 21 (1), 101–107.

Zeng, S., Shao, H., Zhang, L., 2004. An endophytic fungus producing a substance analogous to podophyllotoxin isolated from *Diphylleia sinensis*. J. Microbiol. 24, 1–2.

Zhang, T., Zhang, Y.Q., Liu, H.Y., Wei, Y.Z., Li, H.L., Su, J., et al., 2013. Diversity and cold adaptation of culturable endophytic fungi from bryophytes in the Fildes Region, King George Island, maritime Antarctica. FEMS Microbiol. Lett. 341 (1), 52–61.

Zhang, L.Q., Guo, B., Li, H.Y., 1998. Isolation of an fungus producting Vinbrastine. J. Yunnan. Univ. (Nat. Sci.) 20, 214–215.

Zhang, H.W., Song, Y.C., Tan, R.X., 2006. Biology and chemistry of endophytes. Nat. Prod. Rep. 23, 753–771.

Zhang, X., Li, G., Ma, J., Zeng, Y., Ma, W., Zhao, P., 2010. Endophytic fungus *Trichothecium roseum* LZ93 antagonizing pathogenic fungi *in vitro* and its secondary metabolites. J. Microbiol. 48, 784–790.

Zhao, J., Fu, Y., Luo, M., Zu, Y., Wang, W., Zhao, C., et al., 2012. Endophytic fungi from pigeon pea [*Cajanus cajan* (L.) Millsp.] produce antioxidant cajaninstilbene acid. J. Agric. Food Chem. 60, 4314–4319.

Zhou, S.L., Yang, F., Lan, S.L., Xu, N., Hong, Y.H., 2009. Huperzine A producing conditions from endophytic fungus in SHB *Huperzia Serrata*. J. Microbiol. 3, 32–36.

Chapter 13

Rhizosphere fungi and their plant association: Current and future prospects

Deep Chandra Suyal[1], Pramod Prasad[2], Balram Sahu[3], Ravindra Soni[3,*] and Reeta Goel[4]

[1]*Department of Microbiology, Akal College of Basic Sciences, Eternal University, Sirmour, Himachal Pradesh, India,* [2]*Regional Station, ICAR-Indian Institute of Wheat and Barley Research, Flowerdale, Shimla, India,* [3]*Department of Agricultural Microbiology, College of Agriculture, Indira Gandhi Krishi Vishwavidyalaya, Raipur, India,* [4]*Department of Microbiology, College of Basic Sciences & Humanities, GBPUAT, Pantnagar, India*

**Corresponding author.* Email: rs31693@gmail.com

Introduction

Earth is richly populated in plant and much different type of microorganisms grows in close association with them. While the microbial activity to determined plant growth may be the obvious, microorganisms are also beneficial to plants. Over time the difference interaction between the microorganisms (bacteria, fungi or algae) and plant have been identified and several of the activity is indicated.

The microorganisms in the rhizosphere play important role for the growth of plant and also ecological health of their host plant (Tomer et al., 2016, 2017; Kumar et al., 2018). In native status of soil the microorganisms as fungi are organotrophic fungi, better also known as saprophytic fungi present in the rhizosphere. The rhizosphere saprophytic fungi appear to combined community both yeasts and filamentous fungi that is representative of the all major phyla (Ascomycota and Basidiomycota) and some sub-phyla (Mucoromycotina) (Hilber-Bodmer et al., 2017; Koricha et al., 2019).

Plant–microbe communication

Plant root exert a strong chemical composition in the form of rhizodeposition and provide the suitable ecological niches for microbial growth (Bais et al.,

Fungi Bio-Prospects in Sustainable Agriculture, Environment and Nano-Technology.
DOI: https://doi.org/10.1016/B978-0-12-821394-0.00013-5

2008; Goel et al., 2018; Jeyakumar et al., 2020). These chemicals are commonly known as secondary metabolites due to their reported less importance in plant growth and survival as compare to primary metabolites, with a broad range of chemical structure. These plant secondary metabolites have several biological importance for example they are poisonous to insects and phytopathogen or sometime they act as signaling molecules, in the rhizosphere (Turlings and Ton, 2006; Bais et al., 2008). Plant uses these secondary metabolites to protect itself against the detrimental organisms and attract beneficial microorganisms. The attraction of beneficial microbes to plant root help for the mutualistic relationship with the help of signaling molecules called the strigolactones (Massalha et al., 2017). These are specially secreted by the plant roots in low concentration and also induce the growth of both obligate biotrophes microbes. This opportunity provides the sophisticated symbiotic relationship between the host plant and microorganisms (Goel et al., 2017; Dash et al., 2019).

In many plant compound of different secondary metabolites are present which have germination stimulants activity. Most common secondary metabolites are sorgoleone, isoflavanone, sesquiterpene lactone and strigolactones (Matusova et al., 2005, Majdi et al., 2016; Massalha et al., 2017). The presence of strigolactones plays the essential role in plant biology (Awad et al., 2006; Akiyama and Hayashi, 2006). The mycorrhizal symbiosis depends on a suitable and constant molecular combination between the two partners, of which the first important steps was host recognition. During the co-evaluation of the two partners, strigolactones facilitate for the process of host recognition.

Rhizosphere fungi and their potential bio-prospects in agriculture/plants

Non-mycorrhizal fungal endophytes studies have multi-fold attention owing to their basic and applied importance (Sridhar, 2019). Rhizosphere consist highly diverse microbial population which exhibited strong connection between plants and soil. Further, Morgan et al. (2005) reported that fungal abundance is about 20 times more in this zone than in the rhizoplane. Fungal population inhabiting in this zone have more potential of spreading through the soil than rhizobacteria (Castro et al., 2009). There are so many bioactive substances secreted by plant's root into the rhizosphere soil to perform numerous functions. Some of the major fungal genera belongs to various rhizosphere are summarized in Table 13.1.

Bio-control

Wheat (*Triticum* spp.), the utmost staple cereal produced in the worldwide after maize and paddy, is the leading source of protein and calories in vegetarian diet. The future demand of wheat production to feed the global

TABLE 13.1 **Major fungal genera of some important crop rhizosphere.**

Crop/ rhizosphere	Major fungal genera	References
Wheat plants	*Mortierella polycephala, E. javanicum, Symbiotaphrina buchneri, Symbiotaphrina buchneri, Neocosmospora vasinfecta, Neocosmospora vasinfecta, Paecilomyces tenuipes, Verticillium dahlia, Sordaria fimicola, Chaetomium elatum, Exophiala mansonii, Cladosporium cladosporioides, Pleospora rudis, Alternaria alternate, Pleospora herbarum, Mycosphaerella mycopappi, Blumeria graminis, Geosmithia lavendula, Eupenicillium javanicum, Spizellomyces acuminatus*	Smit et al. (1999)
	Fusarium, Gibberella, Haematonectria, Ilyonectria, Acremonium strictum, Clonostachys, Chaetomium, Gliocladium, Trichoderma spp., *Alternaria, Cladosporium, Gaeumannomyces graminis, Glomerella graminicola, Oculimacula yallundae, Phoma* spp., *Pythium, Rhizoctonia*	Lenc et al. (2015)
Rice	*Boudiera acanthospora, Chytridium polysiphoniae, Nowakowskiella hemisphaerospora, Taeniolella typhoides, Fusarium verticillioides, Fungal sp. SLZ62, Galactomyces geotrichum, Issatchenkia occidentalis*	Asiloglu and Murase (2016, 2017)
Cotton	*Mortierella, Gliomastix, Engyodontium, Rhodosporidium, Trichoderma, Guehomyces, Fusarium, Penicillium, Alternaria,* and *Preussia*	Qiao et al. (2019)
Pea	*Olpidium virulentus, Penicillium janthinellum, usarium* spp. *F. graminearum, Clonostachys rosea* f. *catenulate, Periconia macrospinosa, F. merismoides*	Niu et al. (2018)
	Fusarium oxysporum, Exophiala salmonis, Leohumicola minima, Epicoccum nigrum, Neonectria radicicola, Glomus mosseae, Bionectria ochroleuca, Dendryphion nanum, Fusarium culmorum, Verticillium dahlia, Trichocladium asperum, Bionectria ochroleuca, Cryptococcus aerius, Mortierella sp., *Phoma eupyrena, Mortierella elongate, Fusarium merismoides, Exophiala salmonis*	Xu et al. (2012)
Chickpea	*Acremonium, Alternaria, Bionectria, Chaetomium, Cryptococcus, Cladosporium,*	Bazghaleh et al. (2015)

(*Continued*)

TABLE 13.1 (Continued)

Crop/ rhizosphere	Major fungal genera	References
	Fusarium, Microdochium, Mortierella, Paecilomyces, Pyrenophora, Rhizoctonia, Rhodotorula, and *Trichoderma*	
Mulberry	*Mortierella, Schizothecium, Pseudombrophila, Tetracladium, Fusarium, Peziza, Humicola*	Zhang et al. (2019)
Alfalfa	*Mortierella, Schizothecium, Tausonia, Bolbitius, Tetracladium, Humicola, Nectria*	Zhang et al. (2019)

population is estimated to increase by 60% by 2050 which is projected to surpass nine billion people by 2050 (Figueroa et al., 2018). However, declining cultivable land, climate change posed threats and different abiotic and biotic stresses are few of the most important limiting factors in achieving the desired wheat production for the future. Biotic stresses including rusts, smuts, bunts, leaf blight, powdery mildew etc., which cause huge losses to the quality and quantity of the produce, are the most significant constraint to wheat production worldwide. Recent emergence of Ug99 and other devastating races of *Puccinia graminis tritici* in Ug99 lineage and their spread to East-Africa, Middle-East, South-East Asia together with wheat blast disease caused by *Magnaporthe oryzae* are some of the serious concerns (Prasad et al., 2016; Singh et al., 2015; Bhardwaj et al., 2014).

Wheat is prone to several soil borne diseases including Alternaria leaf blight (*Alternaria triticina*), spot blotch (*Cochliobolus sativus*), Anthracnose (*Colletotrichum graminicola, teleomorph, Glomerella graminicola*), Ascochyta Leaf Spot (*Ascochyta tritici*), *Aureobasidium* Decay (*Microdochium bolleyi*), *Cephalosporium stripe* (Hymenula cerealis), common Root Rots (*C. sativus, Fusarium graminearum* and *F. culmorum* etc.), Crown/Foot Rot, Seedling Blight, Dry land Root Rot (*F. pseudograminearum, F. graminearum, F. avenaceum and F. culmorum*), Downy Mildew/ Crazy Top (*Sclerophthora macrospora*), ergot (*Claviceps purpurea, Sphecelia segetum*), false eyespot (*Gibellina cerealis*), Phoma Rot (*Phoma* spp.), Pythium root rot (*Pythium aphanidermatum, Pythium arrhenomanes, Pythium graminicola*), Rhizoctonia root rot (*Rhizoctonia solani*), Sclerotium Wilt/Southern Blight (*Sclerotium rolfsi*), Take-all (*Gaeumannomyces graminis* var. *tritici*), Loose smut (*Ustilago tritici*) etc. (Sharma, 2012). Controlling soil-borne diseases in wheat is a highly challenging task due to the weird features of soil-borne plant pathogens such as (i) long-lived and persistent nature of these pathogens as they mostly produce microscopic structures that resist drying and weathering (ii) they produce tiny propagules which enable them to easily move around (iii) majority of these pathogens are capable of infecting a broad number of crops (iv) they have multiple attack strategies i.e. through soil, seed, and vegetatively propagated material. There are limited number of commercially available fungicides that are effective against soil-borne pathogens in wheat unlike other high-value crops. These diseases are being managed using different conventional approaches like chemicals, biocontrol, resistance, and cultural practices. Fungicide application through seed/soil treatment have had limited disease controlling potential for diseases such as take-all disease. The availability of host resistance to soil-borne necrotrophs is very limited in the cultivated wheat cultivars, which requires continuous work in this area. Host resistance has been found effective against *Cephalosporium stripe*, tan spot, *Septoria tritici* blotch, and *Stagonospora nodorum* and have been utilized to some extent (Bockus, 1998). The cultural control practices such as tillage that destroy residue have been found effective against the diseases such as tan spot.

Furthermore, there is growing demand over the use of biological agents as natural commercial products to manage soil-borne pathogens and to reduce the use of hazardous pesticides (Cook, 1993). Several researches have been demonstrated the suppression of soil-borne wheat diseases after supplementing soils with fungal or bacterial antagonists (Daguerre et al., 2017). Management of soil-borne plant diseases using different biological agents such as bacteria, fungi, nematodes etc. is principally a complex and multifaceted phenomenon that occur in dynamic environments within the rhizosphere. Different mechanisms involved in disease suppression by biological agents include the production of antimicrobial metabolites such as antibiotics, hydrolytic enzymes, hydrogen cyanide, siderophores, volatile extracellular metabolites, etc., parasitism, competition for nutrients, promotion of plant growth and, finally, induced resistance within the plants (Fig. 13.1) (Moeinzadeh et al., 2010). Rhizosphere fungi such as *Trichoderma harzianum, T. virens, Gliocladium virens, Chaetomium globosum, Coniothyrium minitans, Verticillium biguttatum, Pythium oligandrum, Sporidesmium sclerotivorum, Limonomyces roseipellis,* etc. are some of most promising biocontrol agents (Daguerre et al., 2017). These fungi have long been known to interact with different plant pathogens, through antibiosis, mycoparasitism, competition, etc. Majority of these fungi modify the quality and richness of rhizosphere microflora and alter overall microbial activity of the rhizosphere in favor of plants. They are also known to have strong positive effects on the plant's health. Despite antibiosis, mycoparasitism and competition; induced systemic resistance has also been documented and proved as one of the important disease management tactics of biological control agents (Harman, 2004).

Further, strong antagonistic effects of *Chaetomium* sp., *Idriella bolleyi* and *Gliocladium roseum*, isolated from wheat rhizosphere, on seedborne

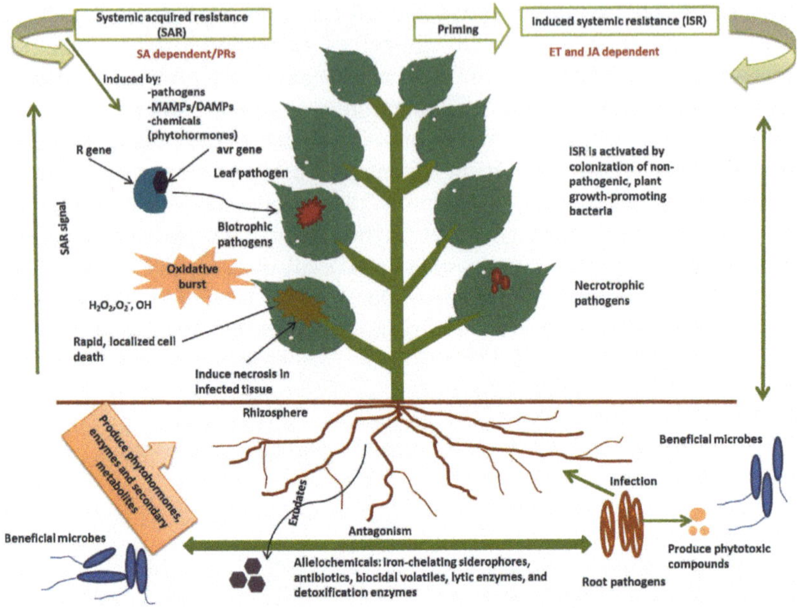

FIGURE 13.1 Mechanism of biological control, acquired systemic resistance and induced systemic resistance in plants https://doi.org/10.1016/j.plantsci.2017.11.012 (Ab Rahman et al., 2018).

wheat pathogens *F. culmorum* and *B. sorokiniana* have also been reported by Knudsen et al. (1995). Despite minimizing the disease severity these fungal bioagents also improved growth and yield components in wheat. Three biocontrol agents *T. hamatum, C. globosum* and *Fusarium* sp., isolated from wheat rhizosphere are also reported to significantly reduce the average tan spot (*D. tritici-repentis*) disease severity in wheat under greenhouse conditions (Larran et al., 2016). Biocontrol of *P. tritici-repentis* using the fungus *Limonomyces roseipellis* has been demonstrated through the suppression of fruiting body formation of *P. tritici repentis* (Pfender et al., 1990; Pfender et al., 1993; Bockus, 1998). Although the exact mode of action is still poorly understood, it appears that *L. roseipellis* is a mycoparasite and attacks the mycelium of *P. tritici-repentis*, preventing pathogen reproduction. Nonpathogenic *Phialophora* spp., originally isolated from wheat rhizosphere, was tested under field conditions for its ability to suppress take-all disease in wheat. Its introduction into wheat rhizosphere as a source of competition or for induced host plant resistance resulted in complete control of the disease (Mathre et al., 1998).

Additionally, Aggarwal et al. (2004) reported the role of antibiosis in the biological control of spot blotch (*C. sativus*) of wheat by *C. globosum*. Other reports showed that *Chaetomium* spp. inhibited the *Pyrinophora tritici-*

repentis (Istifadah et al., 2006; Istifadah and McGee, 2006) and *Drechslera sorokiniana* (wheat spot blotch) growth in in vitro and in planta assays (Biswas et al., 2000). The interaction between *C. globosum* and *D. sorokiniana* produced inhibition zone, indicating biocontrol mechanism of *C. globosum* through antibiosis (Biswas et al., 2000). Fungal population in the medicinal and aromatic plants has been studied broadly. It was investigated that 37 fungal species present in the roots surrounding area of *Abutilon indicum, Argemone maxicana, Aloevera, Amaranthus polygamus* and *Achyranthus aspera*. Majority of these are found to cultivate in wild and northern plains of India. These species were *Fusarium* sp., *Verticillium* sp., *Acaulospora scrobiculata* and *Glomus aggregatum* (Dal-Bello et al., 2002).

Trichoderma*: a promising biocontrol agent*

Trichoderma spp., one of the most studied rhizosphere fungi, have been reported to compete with plant pathogens for nutrients, secrete antifungal compounds, parasitize fungal pathogens, and induce systemic resistance in the host plant (Whipps and Lumsden, 2001). The effectiveness of four *T. harzianum* strains and one *T. koningii* strain in reducing the incidence and severity of the wheat leaf blotch was studied under field conditions (Perelló et al., 2009). The results indicated that *Trichoderma* isolates had strong antagonistic effect on the pathogen which reduced leaf blotch score significantly especially when applied at early stage of the pathogen infection. Biological control potential of *T. harzianum, T. aureoviride, T. koningii* was evaluated against tan spot disease in wheat under in vitro and greenhouse conditions. The results demonstrated that the isolates of *Trichoderma* spp. significantly inhibited the mycelial growth of *D. tritici-repentis* in vitro and reduced tan spot severity on wheat plants under greenhouse conditions (Perello et al., 2003). Mónaco et al. (2004) evaluated the effect of *T. harzianum and T. koningii* strains on the growth of *Bipolaris sorokiniana* and *Alternaria alternata* and determined their effect on wheat black point disease incidence, infection percentage and wheat seedling emergence in a greenhouse assay. All the *Trichoderma* isolates inhibited the mycelial growth of both *B. sorokiniana* and *A. alternata* subsequent to plasmolysis and vacuolization of hyphae of both the pathogens in the presence of the antagonists. However, the effect of antagonists on disease incidence, infection percentage etc. was non-significant under greenhouse conditions (Mónaco et al., 2004). *Trichoderma* spp. isolated from the wheat rhizosphere were evaluated for biocontrol of root rot, food rot, crown rot, stem rot and head blight of wheat caused by *F. graminearum* under in vitro and in the greenhouse conditions. Some of the *Trichoderma* isolates resulted in up to 79% mycelial growth inhibition of *F. graminearum* in vitro and six isolates exhibited a pronounced suppressive effect under greenhouse conditions (Dal-Bello et al., 2002). Similar control of *Fusarium* spp. in cotton, wheat and muskmelon by

Trichoderma isolates was also reported by Sivan and Chet (1986). Likewise, *Trichoderma* spp. showed inhibitions of pathogens mycelial growth, conidial and hyphal plasmolysis (Perelló et al., 2006) and the reduction in wheat tan spot and leaf blotch (*Mycosphaerella graminicola*) severity under field conditions (Perelló and Dal Bello, 2011; Perelló et al., 2006).

Growth improvement and diversity

Rhizosphere fungal diversity of some aromatic rice varieties (Basmati 370, Jiri, Ambemohar Tambda, Ambemohar 157, Indrayani, and Raibhog) were studied in past. From these rhizosphere *Aspergillus niger, A. flavus, A. fumigatus, A. terreus, Alternaria tenuous, Curvularia lunata, T. viride, Rhizopus oryzae, T. lignorum, R. nigricans, C. inaequalis, Mucor racemosus, Nigrospora oryzae, F.oxyporum* and *Phoma* spp. were predominant (Shaikh et al., 2013). Shaikh et al. (2013) Several other plant growth promoting fungal genera were reported in a review compiled by Ding et al. (2019). Furthermore, inoculation *Trichoderma harzianum* along with *A. niger* significantly increase in growth of chickpea (Yadav et al., 2011). Similarly, *Aspergillus* sp. and *Penicillium* sp. were main P solubilizer in common bean, fava bean, cabbage, tomato, and sugarcane rhizosphere (Elias et al., 2016). Some more species of *Aspergillus* like *A. candidus, A. fumigatus, A. niger, A. parasiticus, A. rugulosus, A. terreus*, were also reported as P solubilizer in mung bean (Kalayu, 2019).

It total of 71 isolates of rhizosphere from *Santalum album* was obtained. These all belongs to 11 species of fungi. Out of these, ten (10) species were the member of class Hyphomycetes namely *A. fumigatus, A. niger, A. funiculosus, A. flavus, A. restrictus, A. terricola, F. oxyporum, A. flavipes, A. terreus, Penicillium* spp. and one species i.e. Mycelia sterilia belongs to Basidomycetes (Thombre et al., 2016). Further, rhizosphere of *Ocimum sanctum* and *Centella asiatica* showed 16–17 species of fungi (Sagar et al., 2009). Similarly, rhizospheres of 14 tree peony (*Paeonia suffruticosa*) varieties were colonized by AM fungi (Shi et al., 2013). Fungal isolates *P. pinophilum, A. niger* and *A. fumigatus*, with plant growth promoting properties were obtained from in the different plants rhizosphere (Wahid and Mehana, 2000).

Advance tools and techniques to study rhizospheric fungi

FAME and PLFA analysis

For many years, microbial lipids have been routinely used for their own identification. The two most common methods which are being used for this purpose are fatty acid methyl ester (FAME) analysis and phospholipid fatty acid (PLFA) analysis. FAME is rapid but indiscriminate while, PLFA is precise but time consuming. These methods involve the analysis of the microbial fatty acids and identify them on the basis of signature molecules. Signature fatty acids are known to make a relatively constant proportion within a cell and can be differentiated among major taxonomic groups of the microorganisms (Frostegard and Baath, 1996; Siles et al., 2018). Therefore, any variation in the fatty acid profile of the microorganisms represents the change in the microbial community structure. This technique is precise, high throughput, cost effective with higher resolution capacity (Nelson et al., 2010). The limitation of this technique is that, cellular fatty acid composition depends on growth conditions, media and temperature used to grow the organism, thus, may leads to the misinterpretation.

Random amplification of polymorphic DNA (RAPD)

It is a PCR based fingerprinting technique. RAPD markers are the DNA fragments produced from the random amplification of the genomic DNA using single primer of arbitrary nucleotide sequence. After purifying the genomic DNA, PCR amplification can be done by using randomly designed primers (Clerc et al., 1998). By selecting the primers and amplification conditions judiciously, all such pairs of sequences represented in the genome result in a set of fragments that is characteristic of the species or strain from which the DNA was prepared. These fragments are resolved by gel electrophoresis. The band pattern generated in the analysis represents genome characterization of a respective microbial strain. Recently, Hassan et al. (2019) have used this technique for rapid identification of the *Trichoderma* sp.

Amplified fragment length polymorphism (AFLP)

It is a variation of RAPD technique, and able to detect polymorphic restriction sites without prior sequence knowledge using PCR amplification. Restriction enzyme (RE) digested genomic DNA can be used as a template for PCR amplification. The primers contain the recognition sites of the RE as well as additional 'arbitrary' nucleotides that extend beyond the restriction sites (Blears et al., 1998; Bertani et al., 2019). The fixed portion gives the primer stability and the random portion allows it to detect many loci. The amplified fragments are separated and visualized on denaturing polyacrylamide gels. This multiple-locus fingerprinting technique is highly sensitive and robust and has been evaluated for genotypic characterization of the fungi (Kathuria et al., 2015). Furthermore, it has higher reproducibility, resolution, and sensitivity compared to other techniques.

Repetitive sequence-based PCR (Rep-PCR)

Microbial genomes possess several low-copy-number repeated sequences *viz.* rRNA operons, tRNA genes, insertion elements, etc. These sequences contribute to the evolution of the genome and function through DNA rearrangements. It also helps in creating the genetic fingerprints. Therefore, Rep-PCR fingerprinting is considered a well established technique for microbial diversity analysis and identification (Shin et al., 2012; Masanto et al., 2019). This method is based on PCR-mediated amplification of DNA fragments located between specific interspersed repeated sequences in microbial genomes. It has high resolution, but results may vary due to the PCR biasness.

Multiple locus variable number tandem repeat analysis (MLVA)

It is a molecular technique which explores the natural variation in the number of tandem repeats found in the multiple loci of the microorganism. This method is extensively used for molecular typing of the microorganisms (Johansson et al., 2006). In this technique, variable number tandem repeats (VNTR) loci are subjected to PCR amplification followed by amplicon sequencing. The amplicon size is used to assess the number of repeated units in each locus (Singh et al., 2019). Thus, total numbers of repeats of the VNTR loci are combined and used to prepare the MLVA profile; which can be compared for the fungal identification. This technique has high resolution and accuracy but often imperfect repeats containing mutations are encountered which affect the reproducibility of the results.

Single strand conformation polymorphism (SSCP)

It is a conformational difference in single-stranded nucleotide sequences of identical length (Schwieger and Tebbe, 1998). These nucleotide sequences with different confirmation can be separated by gel electrophoresis technique. Moreover, the gel patterns thus observed can be used for the fungal identification and characterization. The change in single nucleotide in the amplified region is sufficient to produce the distinct PCR-SSCP patterns. This technique is a rapid and convenient for the mutational analysis and allelic variance (Martynov et al., 2019). Problem in reproducibility is the major limitation for this SSCP technique.

Denaturing gradient gel electrophoresis (DGGE) & temporal gradient gel electrophoresis (TGGE)

These two techniques are the well-known techniques for the microbial ecology analysis and involve both PCR as well as polyacrylamide gel electrophoresis (Rajwar et al., 2018; Joshi et al., 2019; Rawat et al., 2019). The metagenomic DNA is amplified using GC clamp containing primers and allowed to separate on a polyacrylamide gel. The denaturation of the amplicons is achieved by urea and formamide in DGGE while, temperature in case of TGGE. The amplicons gets denatured and separated on the basis of their nucleotide sequences. Thus, a profile can be generated which can be compared further for assessing the microbial diversity within a respective sample (Kumar et al., 2014). These techniques are fast and labor-intensive.

However, primer selection, electrophoresis conditions and PCR reactions require optimization to achieve reproducibility.

Pulsed-field gel electrophoresis (PFGE)

This technique can be used for the separation of DNA fragments under the influence of an electric field by changing their directions periodically on the gel matrix. It is a powerful genetic fingerprinting technique to construct a genome amp of the microorganisms (Basim and Basim, 2001; Kwon et al., 2019). Microbial DNA can be restricted digested using RE and allowed to separate through gel electrophoresis. However, the direction of the electric field is changed continuously to get a discrete band pattern. These patterns are then compared and matched with the available databases for the identification of the microorganisms.

Fluorescent in situ hybridization (FISH)

This technique is extensively used for the microbial identification under culture-dependent as well culture-independent approaches. It can detect the complementary DNA sequence within a chromosome with the help of a fluorescence tag (Amann et al., 2001; Witchley et al., 2019). For this purpose, fluorescent probes are designed which are actually the complementary sequence of the desired DNA fragment. The probe: DNA binding is detected by fluorescent microscopy technique. This technique is highly reproducible with good resolution ability. Cellular permeability, sensitivity, target site specificity, and accessibility are the major concerns with this technique.

DNA microarray

A DNA microarray can be defined as the collection of small DNA spots on a solid surface. It can be identified as a DNA chip, biochip or gene chip (Wang et al., 2002). Initially, this technique was used for expressional analysis of the genes, however, now days it is frequently used for microbial identification and characterization. In microarray, randomly fragmented microbial genomes are allowed to hybridize to with the microbial genome spotted on a solid surface (DNA chip) (Ye et al., 2001; Nilsson et al., 2019). Resulting hybridization profiles are then analyzed and compared. This genetic fingerprinting has shown high reproducibility, accuracy and resolution. However, it is considered laborious to perform.

Mass spectrometry (MS)

It offers high-throughput, robust and sensitive way of microbial identification and characterization. Fungal proteins can be extracted, purified and allowed to mass spectrometric analysis for their detailed characterization (Demirev and Fenselau, 2008; Welker, 2011). In the field of microbial proteomics, MS can be used for both gel-based as well as gel-less approaches. In recent years, 2-Dimensional-Gel electrophoresis (2D-GE) coupled with MS having ionization with Matrix-assisted laser desorption/ionization (MALDI)-Time of Flight (TOF) have shown its potential under gel-based proteomic approach (Soni et al., 2015; Suyal et al., 2014, 2017). While, under gel-less approach Liquid Chromatography analysis coupled with MS (LC-MS) is in great demand (Suyal et al., 2018, 2020). Both the approaches produce a profile of the microbial proteins which is then used to compare and characterize the respective strains.

Bromodeoxyuridine (BrdU) method

This method can be used to identify a metabolically active population within a niche (Sebastián and Gasol, 2019). In this technique, BrdU (a labeled nucleotide) is added to the system and microbes are allowed to grow (Yin et al., 2000). Metabolically active individuals will incorporate BrdU into their nucleic acid and thus identified by using a label. This strategy is widely used in the bioremediation, especially for the isolation of the fungi which can use xenobiotics, heavy metals and other compounds.

Stable isotope probing (SIP)

It is also an enrichment method in which ^{13}C-labeled substrate is provided to the microorganisms. Metabolically active microorganisms incorporate ^{13}C in their DNA and thus, make it denser than normal DNA. Density gradient centrifugation can be used to separate both the DNA which can be analyzed further with the help of the specific primers (Achouak and Haichar, 2019). Therefore, SIP offers broad opportunity to study microbial communities and can be expanded further to stable isotopes of nitrogen and/or phosphorus (Buckley et al., 2007).

Conclusion

Exploring the microbiome from plant rhizopshere is of great importance to identify new and effective fungal bioinoculants which can be used as for plant growth improvement and production of valuable compounds. Rhizospheric fungi hold an enormous potential to be deployed for agricultural and industrial applications. Significant progress in research has been made in last decade on isolation and characterization of novel fungi from rhizopshere and their bioactive metabolites. It is high time to explore new fungal group from plants of extreme and diverse habitats using advanced molecular tools and approaches like metagenomics and approaches and identify their potential to reduce the use of harmful chemical in agriculture.

References

Ab Rahman, S.F., Singh, E., Pieterse, C.M., Schenk, P.M., 2018. Emerging microbial biocontrol strategies for plant pathogens. Plant Sci. 267, 102–111.

Achouak, W., Haichar, F.Z., 2019. Stable isotope probing of microbiota structure and function in the plant rhizosphere. In: Dumont, M., Hernández García, M. (Eds.), Stable Isotope Probing. Methods in Molecular Biology, vol 2046. Humana, New York, NY.

Aggarwal, R., Tiwari, A.K., Srivastava, K.D., Singh, D.V., 2004. Role of antibiosis in the biological control of spot blotch (Cochliobolus sativus) of wheat by Chaetomium globosum. Mycopathologia 15, 369–377.

Akiyama, K., Hayashi, H., 2006. Strigolactones: chemical signals for fungal symbionts and parasitic weeds in plant roots. Ann. Bot. (Lond.) 97, 925–931.

Amann, R., Fuchs, B.M., Behrens, S., 2001. The identification of microorganisms by fluorescence in situ hybridisation. Curr. Opin. Biotechnol. 12, 231–236.

Asiloglu, R., Murase, J., 2016. Active community structure of microeukaryotes in a rice (Oryza sativa L.) rhizosphere revealed by RNA-based PCR-DGGE. Soil Sci. Plant. Nutr. 62, 440–446.

Asiloglu, R., Murase, J., 2017. Microhabitat segregation of heterotrophic protists in the rice (Oryza sativa L.) rhizosphere. Rhizosphere 4, 82–88.

Awad, A.A., et al., 2006. Characterisation of strigolactones, germination stimulants for the root parasitic Striga and Orobanche, produced by maize, millet and sorghum. Plant Growth Regul. 48, 221–227.

Bais, H.P., Broeckling, C.D., Vivanco, J.M., 2008. Root exudates modulate plant–microbe interactions in the rhizosphere. In: Karlovsky, P. (Ed.), Secondary Metabolites in Soil Ecology. Soil Biology 14. Springer-Verlag, Berlin Heidelberg.

Basim, E., Basim, H., 2001. Pulsed-field gel electrophoresis (PFGE) technique and its use in molecular biology. Turk. J. Biol. 25, 405–418.

Bazghaleh, N., Hamel, C., Gan, Y., Tar'an, B., Knight, J.D., 2015. Genotype specific variation in the structure of root fungal communities is related to chickpea plant productivity. Appl. Environ. Microbiol. 81, 2368–2377.

Bertani, G., Savo Sardaro, M.L., Neviani, E., Lazzi, C., 2019. AFLP protocol comparison for microbial diversity fingerprinting. J. Appl. Genet. 60, 217.

Bhardwaj, S.C., Prashar, M., Prasad, P., 2014. Ug99 – future challenges. In: Goyal, A., Manoharachary, C. (Eds.), Future Challenges in Crop Protection Against Fungal Pathogens, Fungal Biology. Springer Science and Business Media, New York, pp. 231–248.

Biswas, S.K., Srivastav, K.D., Aggarwal, R., Dureja, P., Singh, D.V., 2000. Antagonism of *Chaetomium globosum* to *Drechslera sorokiniana*, the spot blotch pathogen of wheat. Indian Phytopath. 53 (4), 436–440.

Blears, M.J., DeGrandis, S.A., Lee, H., Trevros, J.T., 1998. Amplified fragment length polymorphism (AFLP): a review of the procedure and its applications. J. Indust Microbiol. Biotechnol. 21, 99–114.

Bockus, W.W., 1998. Control strategies for stubble-borne pathogens of wheat. Can. J. Plant. Pathol. 20 (4), 371–375. 1.

Buckley, D.H., Huangyutitham, V., Hsu, S.F., Nelson, T.A., 2007. Stable isotope probing with 15N achieved by disentangling the effects of genome G + C content and isotope enrichment on DNA density. Appl. Environ. Microbiol. 73, 3189–3195.

Castro, O.R., Contreras-Cornejo, H.A., Macías Rodríguez, L., López-Bucio, J., 2009. The role of microbial signals in plant growth and development. Plant Signal. Behav. 4 (8), 701–712.

Clerc, A., Manceau, C., Nesme, X., 1998. Comparison of randomly amplified polymorphic DNA with amplified fragment length polymorphism to assess genetic diversity and genetic relatedness within genospecies III of Pseudomonas syringae. Appl. Environ. Microbiol. 64 (4), 1180.

Cook, R.J., 1993. Making greater use of introduced microorganisms for biological control of plant pathogens. Annu. Rev. Phytopathol. 31, 53–80.

Daguerre, Y., Edel-Hermann, V., Steinberg, C., 2017. Fungal genes and metabolites associated with the biocontrol of soil-borne plant pathogenic fungi. Fungal Metabol. 33–104.

Dal-Bello, G.M., Monaco, C.I., Simon, M.R., 2002. Biological control of seedling blight of wheat caused by Fusarium graminearum with beneficial rhizosphere microorganisms. World J. Microbiol. Biotechnol. 18 (7), 627–636.

Dash, B., Soni, R., Kumar, V., Suyal, D.C., Dash, D., Goel, R., 2019. Mycorrhizosphere: microbial interactions for sustainable agricultural production. In: Varma, A., Choudhary, D. (Eds.), Mycorrhizosphere and Pedogenesis. Springer Nature, Singapore, pp. 321–338.

Demirev, P.A., Fenselau, C., 2008. Mass spectrometry for rapid characterization of microorganisms. Annu. Rev. Anal. Chem. 1, 3.1–3.23.

Ding, L.J., Cui, H.L., Nie, S.A., Long, X.E., et al., 2019. Microbiomes inhabiting rice roots and rhizosphere. FEMS Microbiol. Ecol. 95, fiz040. 2019.

Elias, F., Woyessa, D., Muleta, D., 2016. Phosphate solubilization potential of rhizosphere fungi isolated from plants in Jimma Zone, Southwest Ethiopia. Int. J. Microbiol. 2016, 5472601.

Figueroa, M., Hammond-Kosack, K.E., Solomon, P.S., 2018. A review of wheat diseases—a field perspective. Mol. Plant Pathol. 19, 1523–1536.

Frostegard, A., Baath, E., 1996. The use of phospholipid fatty acid analysis to estimate bacterial and fungal biomass in soil. Biol. Fert. Soils 22, 59.

Goel, R., Kumar, V., Suyal, D.C., Dash, B., Kumar, P., Soni, R., 2017. Root-associated bacteria: rhizoplane and endosphere. In: Singh, D., Singh, H., Prabha, R. (Eds.), Plant-Microbe Interactions in Agro-Ecological Perspectives. Springer Nature, Singapore, pp. 161–176.

Goel, R., Suyal, D.C., Kumar, V., Jain, L., Soni, R., 2018. Stress-tolerant beneficial microbes for sustainable agricultural production. In: Panpatte, et al., (Eds.), Microorganisms for Green Revolution, Microorganisms for Sustainability, vol. 7. Springer Nature, Singapore, pp. 141–159.

Harman, G.E., 2004. Overview of new insights into mechanisms and uses of Trichoderma based products. Phytopathology 94 (6), S138.

Hassan, M.M., Farid, M.A., Gaber, A., 2019. Rapid identification of Trichoderma koningiopsis and Trichoderma longibrachiatum using sequence-characterized amplified region markers. Egypt. J. Biol. Pest Control. 29, 13. Available from: https://doi.org/10.1186/s41938-019-0113-0.

Hilber-Bodmer, M., Schmid, M., Ahrens, C.H., Freimoser, F.M., 2017. Competition assays and physiological experiments of soil and phyllosphere yeasts identify Candida subhashii as a novel antagonist of filamentous fungi. BMC Microbiol. 17 (1).

Istifadah, N., McGee, P., 2006. Endophytic Chaetomium globosum reduces development of tan spot in wheat caused by Pyrenophora tritici-repentis. Aust. Plant Pathol. 35, 411–418.

Istifadah, N., Saleeba, J., McGee, P., 2006. Isolates of endophytic *Chaetomium* spp. inhibit the fungal pathogen Pyrenophora tritici-repentis in vitro. Can. J. Bot. 84, 1148–1155.

Johansson, A., Koskiniemi, S., Gottfridsson, P., Wistrom, J., Monsen, T., 2006. Multiple-locus variable-number tandem repeat analysis for typing of Staphylococcus epidermidis. J. Clin. Microbiol. 44 (1), 260.

Jeyakumar, S.P., Dash, B., Singh, A.K., Suyal, D.C., Soni, R., 2020. Nutrient cycling at higher altitudes. In: Goel, R., Soni, R., Suyal, D.C. (Eds.), Microbiological Advancements for Higher Altitude Agro-Ecosystems & Sustainability. Springer Nature, Singapore, pp. 293–305.

Joshi, D., Chandra, R., Suyal, D.C., Kumar, S., Goel, R., 2019. Impact of bioinoculants Pseudomonas jesenii MP1 and Rhodococcus qingshengii S10107 on Cicer arietinum yield and soil nitrogen status. Pedosphere 29 (3), 388–399.

Kalayu, G., 2019. Phosphate solubilizing microorganisms: promising approach as biofertilizers. Int. J. Agron. 7. Article ID 4917256.

Kathuria, S., Sharma, C., Singh, P.K., Agarwal, P., Agarwal, K., Hagen, F., et al., 2015. Molecular epidemiology and in-vitro antifungal susceptibility of *Aspergillus terreus* species complex isolates in Delhi, India: evidence of genetic diversity by amplified fragment length polymorphism and microsatellite typing. PLoS One 10 (3), e0118997.

Knudsen, I.M., Hockenhull, J., Jensen, D.F., 1995. Biocontrol of seedling diseases of barley and wheat caused by Fusarium culmorum and Bipolaris sorokiniana: effects of selected fungal antagonists on growth and yield components. Plant Pathol. 44 (3), 467–477. Jun.

Koricha, A.D., Han, D.Y., Bacha, K., Bai, F.Y., 2019. Occurrence and molecular identification of wild yeasts from Jimma Zone, South West Ethiopia. Microorganisms 7, 633.

Kumar, S., Suyal, D.C., Dhauni, N., Bhoriyal, M., Goel, R., 2014. Relative plant growth promoting potential of Himalayan psychrotolerant Pseudomonas jesenii Strain MP1 against Native Cicer arietinum L., Vigna mungo (L.) Hepper; Vigna radiata (L.) Wilczek., Cajanus cajan (L.) Millsp. and Eleusine coracana (L.). Gaertn. Afr. J. Microbiol. 8 (50), 3931–3943.

Kumar, S., Suyal, D.C., Bhoriyal, M., Goel, R., 2018. Plant growth promoting potential of psychrotolerant Dyadobacter sp. for pulses and finger millet and impact of inoculation on soil chemical properties and diazotrophic abundance. J. Plant Nut 41 (8), 1035–1046.

Kwon, Y.J., Shin, J.H., Byun, S.A., Choi, M.J., Won, E.J., Lee, D., et al., 2019. Candida auris clinical isolates from South Korea: identification, antifungal susceptibility, and genotyping. J. Clin. Microbiol. 57 (4). Available from: https://doi.org/10.1128/JCM.01624-18.

Larran, S., Simon, M.R., Moreno, M.V., Siurana, M.S., Perelló, A., 2016. Endophytes from wheat as biocontrol agents against tan spot disease. Biol. Control. 1 (92), 17–23.

Lenc, L., Kwasna, H., Sadowski, C., Grabowski, A., 2015. Microbiota in wheat roots, rhizosphere and soil in crops grown in organic and other production systems. J. Phytopathol. 163, 245–263.

Majdi, M., Ashengroph, M., Abdollahi, M.R., 2016. Sesquiterpene lactone engineering in microbial and plant platforms: parthenolide and artemisinin as case studies. Appl. Microbiol. Biotechnol. 100, 1041.

Martynov, V., Chizhik, V., Sokolova, E., Kuznetsova, M., Khavkin, E., 2019. Polymorphism of avirulence genes in potato late blight pathogen Phytophthora infestans as characterized by SSCP analysis. Agric. Gene 13, 100093.

Masanto, H.A., Wibowo, A., Subandiyah, S., Shimizu, M., Suga, H., Kageyama, K., 2019. Genetic diversity of Phytophthora palmivora isolates from Indonesia and Japan using rep-PCR and microsatellite markers. J. Gen. Plant Pathol. 85, 367.

Massalha, H., et al., 2017. Live imaging of root–bacteria interactions in a microfluidics setup. Proc. Natl Acad. Sci. USA. 114, 4549–4554.

Mathre, D.E., Johnson, R.H., Grey, W.E., 1998. Biological control of take-all disease of wheat caused by Gaeumannomyces graminis var. tritici under field conditions using a Phialophora sp. Biocontrol Sci. Technol. 8, 449–457.

Matusova, R., Rani, K., Verstappen, F.W., Franssen, M.C., Beale, M.H., Bouwmeester, H.J., 2005. The strigolactone germination stimulants of the plant-parasitic Striga and Orobanche spp. are derived from the carotenoid pathway. Plant Physiol. 139 (2), 920–934.

Moeinzadeh, A., Sharif-Zadeh, F., Ahmadzadeh, M., et al., 2010. Biopriming of sunflower (*Helianthus annuus* L.) seed with *Pseudomonas fluorescens* for improvement of seed invigoration and seedling growth. Aust. J. Crop. Sci. 4, 564.

Mónaco, C., Sisterna, M., Perelló, A., Dal Bello, G., 2004. Preliminary studies on biological control of the blackpoint complex of wheat in Argentina. World J. Microbiol. Biotechnol. 20 (3), 285–290.

Morgan, J.A.W., Bending, G.D., White, P.J., 2005. Biological costs and benefits to plant–microbe interactions in the rhizosphere. J. Exp. Bot. 56, 1729–1739.

Nelson, K.Y., Razban, B., Mc Martin, D.W., Cullimore, D.R., Takaya, O., Patrick, D.K., 2010. A rapid methodology using fatty acid methyl esters to profile bacterial community structures in microbial fuel cells. Bioelectrochemistry 78 (1), 80–86.

Nilsson, R.H., Anslan, S., Bahram, M., Wurzbacher, C., Baldrian, P., Tedersoo, L., 2019. Mycobiome diversity: high-throughput sequencing and identification of fungi. Nat. Rev. Microbiol. 17, 95–109. Available from: https://doi.org/10.1038/s41579-018-0116-y.

Niu, Y., Bainard, L.D., May, W.E., Hossain, Z., Hamel, C., Gan, Y., 2018. Intensified pulse rotations buildup pea rhizosphere pathogens in cereal and pulse based cropping systems. Front. Microbiol. 9, 1909.

Perelló, A.E., Moreno, M.V., Mónaco, C., Simón, M.R., Cordo, C., 2009. Biological control of Septoria tritici blotch on wheat by Trichoderma spp. under field conditions in Argentina. Biocontrol. 54 (1), 113–122. 1.

Perelló, A., Dal Bello, G., 2011. Suppression of tan spot and plant growth promotion of wheat by synthetic and biologic inducers in field conditions. Ann. Appl. Biol. 158, 267–274.

Perelló, A., Mónaco, C., Moreno, M.V., Cordo, C., Simón, M.R., 2006. The effect of Trichoderma harzianum and T. koningii on the control of tan spot (*Pyrenophora tritici-repentis*) and leaf blotch (*Mycosphaerella graminicola*) of wheat under field conditions in Argentina. Biocontrol Sci. Technol. 16, 803–813.

Perello, A., Monaco, C., Simon, M.R., Sisterna, M., Dal Bello, G.S., 2003. Biocontrol efficacy of Trichoderma isolates for tan spot of wheat in Argentina. Crop. Prot. 22 (9), 1099–1106. 1.

Pfender, W.F., Sharma, U., Zhang, W., 1990. Effect of water potential on microbial antagonism to *Pyrenophora tritici-repentis* in wheat residue. Mycol. Res. 95, 308–314.

Pfender, W.F., Zhang, W., Nus, A., 1993. Biological control to reduce inoculum of the tan spot pathogen *Pyrenophora tritici-repentis* in surface-borne residues of wheat fields. Phytopathology 83, 371–375.

Prasad, P., Bhardwaj, S.C., Khan, H., Gangwar, O.P., Kumar, S., Singh, S.B., 2016. Ug99: saga, reality and status. Curr. Sci. 110 (9), 1614–1616.

Qiao, Q., Zhang, J., Ma, C., Wang, F., Chen, Y., Zhang, C., et al., 2019. Characterization and variation of the rhizosphere fungal community structure of cultivated tetraploid cotton. PLoS One 14 (10), e0207903.

Rajwar, J., Chandra, R., Suyal, D.C., Tomer, S., Kumar, S., Goel, R., 2018. Comparative phosphate solubilizing efficiency of psychrotolerant *Pseudomonas jesenii* MP1 and *Acinetobacter* sp. ST02 against chickpea for sustainable hill agriculture. Biologia 73 (8), 793–802.

Rawat, N., Sharma, M., Suyal, D.C., Singh, D.K., Joshi, D., Singh, P., et al., 2019. Psyhcrotolerant bio-inoculants and their co-inoculation to improve *Cicer arietinum* growth and soil nutrient status for sustainable mountain agriculture. J. Soil Sci. Plant Nutr. 19 (3), 639–647.

Sagar, A., Kumari, R., 2009. Fungal associates of Centella asiatica and Ocimum sanctum. J. Pure Appl. Microbiol. 3, 243–248.

Schwieger, F., Tebbe, C.C., 1998. A new approach to utilize PCR-Single-Strand Conformation Polymorphism for 16S rRNA-based microbial community analysis. Appl. Environ. Microbiol. 64, 4870–4876.

Sebastián, M., Gasol, J.M., 2019. Visualization is crucial for understanding microbial processes in the ocean. Philos. Trans. R. Soc. B 374 (1786), 20190083.

Sharma, I., 2012. Disease Resistance in Wheat, vol. 1. Cabi.

Shi, J., Chen, F., Luo, Y.Q., Wang, Z., Xie, B.Y., 2013. First isolation of pine wood nematode from Pinus tabuliformis forests in China. Forest Path 43, 59–66.

Shin, H.D., Kim, D.U., Seong, C.N., Song, H.G., Jong-Ok, K., 2012. Genetic and phenotypic diversity of carbofuran-degrading bacteria isolated from agricultural soils. J. Microbiol. Biotechnol. 22 (4), 448–456.

Siles, J.A., Öhlinger, B., Cajthaml, T., et al., 2018. Characterization of soil bacterial, archaeal and fungal communities inhabiting archaeological human-impacted layers at Monte Iato settlement (Sicily, Italy). Sci. Rep. 8, 1903.

Singh, M., Malik, M.A., Singh, D.K., Doimari, S., Bhavna, Sharma, R., 2019. Multilocus variable number tandem repeat analysis (MLVA)-typing of Brucella abortus isolates of India reveals limited genetic diversity. Trop. Anim. Health Prod. (2019).

Singh, R.P., Hodson, D.P., Jin, Y., et al., 2015. Emergence and spread of new races of wheat stem rust fungus: continued threat to food security and prospects of genetic control. Phytopathology 105, 872–874.

Sivan, A., Chet, I., 1986. Biological control of Fusarium spp. in cotton, wheat and muskmelon by Trichoderma harzianum. J. Phytopathol. 116 (1), 39–47.

Smit, E., Leeflang, P., Glandorf, B., van Elsas, J.D., Wernars, K., 1999. Analysis of fungal diversity in the wheat rhizosphere by sequencing of cloned PCR-amplified genes encoding 18S rRNA and temperature gradient gel electrophoresis. Appl. Environ. Microbiol. 65 (6), 2614–2621.

Soni, R., Suyal, D.C., Agrawal, K., Yadav, A., Souche, Y., Goel, R., 2015. Differential proteomic analysis of Himalayan psychrotolerant diazotroph Pseudomonas palleroniana N26 Strain under low temperature diazotrophic conditions. CryoLetters 36 (2), 74–82.

Sridhar, K.R., 2019. Diversity, ecology, and significance of fungal endophytes. In: Jha, S. (Ed.), Endophytes and Secondary Metabolites. Reference Series in Phytochemistry. Springer, Cham.

Suyal, D.C., Joshi, D., Kumar, S., Soni, R., Goel, R., 2020. Differential protein profiling of soil diazotroph Rhodococcus qingshengii S10107 towards low-temperature and nitrogen deficiency. Sci. Rep. 9, 20378.

Suyal, D.C., Kumar, S., Yadav, A., Shouche, Y., Goel, R., 2017. Cold stress and nitrogen deficiency affected protein expression of psychrotrophic Dyadobacter psychrophilus B2 and Pseudomonas jessenii MP1. Front. Microbiol. 8 (430), 1–6.

Suyal, D.C., Kumar, S., Joshi, D., Soni, R., Goel, R., 2018. Quantitative proteomics of psychotrophic diazotroph in response to nitrogen deficiency and cold stress. J. Proteomics 187, 235–242.

Suyal, D.C., Yadav, A., Shouche, Y., Goel, R., 2014. Differential proteomics in response to low temperature diazotrophy of Himalayan psychrophilic nitrogen fixing Pseudomonas migulae S10724 strain. Curr. Microbiol. 68 (4), 543–550.

Thombre, S.S., Kalamkar, S.S., Shaikh, M.N., Torawane, S.D., Mokat, D.N., 2016. Studies on rhizosphere fungi and allelopathic potential of Santalum album L. Biosci. Dis 2, 158–161.

Tomer, S., Suyal, D.C., Goel, R., 2016. Biofertilizers: a timely approach for sustainable agriculture. In: Choudhary, D.K., Varma, A., Tuteja, N. (Eds.), Plant-Microbe Interaction: An Approach to Sustainable Agriculture. Springer Nature, Singapore, pp. 375–395.

Tomer, S., Suyal, D.C., Rajwar, J., Yadav, A., Shouche, Y., Goel, R., 2017. Isolation and characterization of Phosphate solubilizing bacteria from Western Indian Himalayan soils. 3Biotech 7 (2), 1–8.

Turlings, T.C.J., Ton, J., 2006. Exploiting scents of distress: the prospect of manipulating herbivore-induced plant odours to enhance the control of agricultural pests. Curr. Opin. Plant Biol. 9, 421–427.

Wang, R.F., Beggs, M.L., Robertson, L.H., Cerniglia, C.E., 2002. Design and evaluation of oligonucleotide-microarray method for the detection of human intestinal bacteria in fecal samples. FEMS Microbiol. Lett. 213, 175–182.

Wahid, O.A., Mehana, T.A., 2000. Impact of phosphate-solubilizing fungi on the yield and phosphorus uptake by wheat and faba bean plants. Microbiol. Res. 155, 221–227.

Welker, M., 2011. Proteomics for routine identification of microorganisms. Proteomics 11, 3143–3153.

Whipps, J.M., Lumsden, R.D., 2001. Commercial use of fungi as plant disease biological control agents: status and prospects. In: Butt, T.M., Jackson, C., Magan, N. (Eds.), Fungi as Biocontrol Agents: Progress, Problems and Potential. CABI Publishing, Wallingford, pp. 9–22.

Witchley, J.N., Penumetcha, P.M., Noble, S.M., 2019. Visualization of Candida albicans in the murine gastrointestinal tract using fluorescent in situ hybridization. J. Vis. Exp. (153), .

Xu, L., Ravnskov, S., Larsen, J., Nilsson, R.H., Nicolaisen, M., 2012. Soil fungal community structure along a soil health gradient in pea fields examined using deep amplicon sequencing. Soil Biol. Biochem. 46, 26–32.

Yadav, J., Verma, J.P., Tiwari, K.N., 2011. Plant growth promoting activities of fungi and their effect on chickpea plant growth. Asian J. Biol. Sci. 4, 291–299.

Ye, R.W., Wang, T., Bedzyk, L., Croker, K.M., 2001. Applications of DNA microarrays in microbial systems. J. Microbiol. Methods 47, 257–272.

Yin, B., Crowley, D., Sparovek, G., DeMelo, W.J., Borneman, J., 2000. Bacterial functional redundancy along a soil reclamation gradient. Appl. Environ. Microbiol. 66, 4361–4365.

Zhang, M., Wang, N., Zhang, J., Hu, Y., Cai, D., Guo, J., et al., 2019. Soil physicochemical properties and the rhizosphere soil fungal community in a mulberry (Morus alba L.)/Alfalfa (Medicago sativa L.) intercropping system. Forests 10, 167.

Chapter 14

Mycotoxins associated food safety concerns of agricultural crops, prevention and control

A. Sankaranarayanan[1,*], N. Amaresan[1], Abhishek Sharma[1] and Ashraf Y.Z. Khalifa[2,3]

[1]*C.G. Bhakta Institute of Biotechnology, Uka Tarsadia University, Surat, India,* [2]*Biological Sciences Department, College of Science, King Faisal University, Al-Ahsa, Kingdom of Saudi Arabia,* [3]*Botany and Microbiology Department, Faculty of Science, University of Beni-Suef, Beni-Suef, Egypt*

**Corresponding author.* Email: drsankarkamal@gmail.com

Introduction

In a recent years with high concern about food safety and quality towards the high concern about health aspects of consumer. Comparing to other microbes, fungi affect plant crops and leads to reduction of yield, diminishing the quality and ultimately leads to the loss of economy (Adejumo and Adejoro, 2014; Richard, 2007). The secondary metabolites produced by fungi, called as mycotoxins present in almost all food substances especially the crops infected by fungi and during the storage of crop products. The mycotoxin which responsible for mycotoxicoses, a low molecular weight secondary metabolite substances, which may contaminate the raw material by the way contaminate the food, feed and cause severe health damages to humans including infants, domestic animals and live stocks (Ashiq, 2015; Jeswal and Kumar, 2015). Various International agencies like Food and Agriculture Organization (FAO), United Nations Environment Programme (UNEP), World Health Organization (WHO), European Spice Association (ESA), American Spice Trade Association (ASTA), International Organization for Standardization (ISO) (Esmaillzadeh et al., 2013) European commission Scientific Committee on Food (SCF, 1998), Joint FAO/WHO Expert Committee on Food Additives (JECFA, 2001) are much concerned

Fungi Bio-Prospects in Sustainable Agriculture, Environment and Nano-Technology.
DOI: https://doi.org/10.1016/B978-0-12-821394-0.00014-7

about the mycotoxin, its level in various food commodities and associated health concerns.

Mycotoxins generally comprises and responsible for diversified problems including reduce the value of agricultural products, loss of yield, human health concerns, diminish the productivity of animal (Fig. 14.1). It seems difficult to quantify the health impacts due to mycotoxins especially through agricultural based products in human beings. It leads to acute toxicoses and chronic health impacts. At this juncture, a red alert report on the presence of mycotoxins in unaltered and altered (metabolized form) in breast milk (Hassan et al., 2006a; Biasucci et al., 2011) and the recent report also revealed the seriousness about this issue to human communities (Fakhri et al., 2019; Coppa et al., 2019) and its presence in cereal based infant food too (Coppa et al., 2019). Among the mycotoxins, aflatoxins cause serious human health impacts and loss of human lives in developed (Robens and Cardwell, 2003), developing and under developed countries (Daniel et al., 2011).

The association between fungi and plants were revealed by mycologist more than a century ago in corn (Butler, 1902) and neurological disorder was first reported in horse (Graham, 1912) and swine (McNutt et al., 1928) due to moldy feed. Presence of mycotoxin in maize (Ranum et al., 2014), grapes (Guo et al., 2019), spices (Ismaiel and Papenbrock, 2015), chickpea (Ramirez et al., 2018), barley (Oliveira et al., 2012), sorghum (Sreenivasa et al., 2013), oat (Fredlund et al., 2013), cabbages, carrot, onion (Lugauskas et al., 2004), tomato (Sarubbi et al., 2016), apple (Castoria et al., 2005), peanut (Adjou et al., 2012), soybean (Al-Seeni, 2012), almonds (Rodrigues et al., 2013), Chestnut (Yassin et al., 2013) and rice (Ok et al., 2014).

More than 300 mycotoxins are reported, however certain mycotoxins are highly important regarding food safety, economic aspects and attract more research attentions (Iqbal et al., 2014). However among them, five mycotoxins are considered to be the most important in agriculture, animal and human health aspects. They are Aflatoxin produced by *Aspergillus* species,

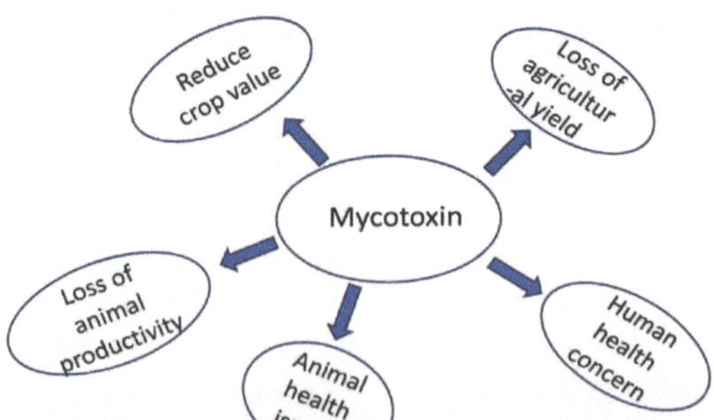

FIGURE 14.1 Mycotoxins general impacts on human, animal and agricultural crops.

Ochratoxin A produced by *Aspergillus* and *Penicillium* species, deoxynivalenol, zearalenone and fumonisins produced by *Fusarium* species (Iqbal et al., 2014). The broad list of mycotoxins is presented in Table 14.1.

Various detection procedures are followed for the detection of mycotoxins and reported including HPLC using UV fluorescent detectors (Herzallah, 2009), LC-MS/MS (Jestoi et al., 2009), immunochromatographic strip test (Shim et al., 2009) and GC–MS method (Cunha et al., 2009).

Control and prevention approaches are used to come out from the toxigencity of toxigenic fungi. Organic acids (formic and propionic acids) (Quintela et al., 2012), oxidizing agents (hydrogen peroxide and ozone) (Agriopoulou et al., 2016), bases (calcium hydroxide) (Venter, 2014), potassium and sodium bicarbonate (Amezqueta et al., 2008), filtering agents and adsorption methods (Avantaggiato et al., 2004; Liu et al., 2011) and biocontrol agents (Fuchs et al., 2008).

Mycotoxins and agriculture

Mycotoxins association with plants has been reported more than a century ago (Butler, 1902; Sheldon, 1904; Graham, 1912). It indicated the strong association between fungi and plant crops a long decade ago. Despite, until Turkey X disease, the toxigenicity of fungi was not seriously provoked by mycologists (CAST, 2003) As per Food and Agriculture Organization (FAO & WHO report, 2011) report, 25% of World's agricultural products are bearing the mycotoxins. In another observation by Bhat and Miller (1991) merely 25–50% of agricultural commodities in global scenario intrusion with mycotoxins. Huffman et al. (2010) reported the aflatoxin, ochratoxin A, zearalenone, trichothecenes, fusarium and patulin are considered as agriculturally important myctoxins and represent in various agricultural products and thus causing various diseases to humans.

Due to saprophytic fungi, mycotoxins are present in the field and subsequently present in the crop and after harvesting may be due to endophytic fungi may responsible for its presence in the products (Streit et al., 2012). Mycotoxins occurred in crops throughout the Word. Due to its presence in agricultural associated crops and their products, responsible for health hazards to animals and human beings. Though research initiatives, good agricultural practices and best storage guidelines, despite myctoxins as an universal problem (Eskola et al., 2019).

During the pre harvest time (in the field itself) or post harvesting (transportation and storage) and further processing period, the fungi can invade, colonize and produce mycotoxins (Bhat et al., 2010). The fungi invaded through seed called "field fungi" (*Cladosporium* sp., *Fusarium* sp., and *Alternaria* sp.) and fungi proliferated during storage and cause damage to the crop yield called "storage fungi" (*Aspergillus* sp. and *Penicillium* sp.). (Ismaiel and Papenbrock, 2015). Comparing to "field fungi", "storage fungi"

TABLE 14.1 Name of mycotoxin, organisms involved in the mycotoxins production, its diversity in foods and agricultural crops and their hazardous effects to human and animals.

S. No.	Name of the mycotoxins	Organism involved	Agriculture products/food	Health hazards in human and animals	Reference
1.	Ochratoxin A 'Ochratoxin α'	*Aspergillus carbonaris* *Aspergillus* sp. *Penicillium* sp.	Cereals, wine, beer, cocoa, coffee, spices, meat	Nephrotoxicity; immunotoxicity; Inhibition of protein synthesis; Urinary tract tumors; Cytotoxicity in human embryonic kidney cell lines	Aish et al. (2004), Creppy et al. (1983a, 1983b), Schulz et al. (2018), Gong et al. (2019), Al-Sheikh (2015)
2.	Aflatoxin B1 Aflatoxin B2 Aflatoxin G1 Aflatoxin G2 Aflatoxin M1 Aflatoxin M2	*Aspergillus* sp. *A. flavus,* *A. parasiticus,* *A. nomius,* *A. tamarii*	Corn, sorghum Cocoa	Liver cancer; Aflatoxicosis; Imunosuppression;	Blankson and Mill-Robertson (2016), Goto et al. (1996), Copetti et al. (2011, 2012), Williams et al. (2004),
3.	Fumonisins	*Fusarium* sp *F. verticillioides* *F. proliferatum*		Interfere folate and other protein binding in cell membrane; Equine leucoencephalomalacia; Esophageal cancer; Induce carcinogenicity in rat; Cytotoxicity in mammalian cell lines	IARC (2002), Krska et al. (2007), Weidenbörner (2001), Chang et al. (2015), Sundaramoorthy et al. (2012)
4.	Patulin	*Penicillium* sp. *Aspergillus* sp. and *Eupenicillium* sp.	Apple, bear, barley	Genotoxicity, immunotoxicity, neurotoxicity	Yang et al. (2018), Ruiyu et al. (2012), Glaser & Stopper (2012),
5.	Alternaria toxin	*Alternaria alternata,* *A. arborescens,* *A. radicina,* *A. brassicola*	Apple, tomato, citrus fruit and olives	Esophageal cancer;	Meena and Samal (2019), Gammbacorta et al. (2019)

6.	Ergot alkaloids	*Claviceps purpurea; Periglandula ipomoeae*	Cereals, rye, barley, pearl millet	St. Anthony's fire,	Schardl (2015)
7.	Trichothecenes	*Fusarium* sp. *Trichoderma Trichothecium* sp.	Corn, wheat and barley	Inhibition of DNA, RNA and protein synthesis; immunosuppression; Gastro intestinal poisoning	Escrivá et al. (2015), McCormick et al. (2011)
8.	Beauvericin	*Fusarium oxysporoum, F. proliferatum*	Corn	Cholesterol acyltransferase inhibitor in rat liver; cell death in human tumor cell lines	Kokkonen et al. (2010), Sørensen et al. (2008)
9.	Enniatin A Enniatin A1 Enniatin B Enniatin B1 Enniatin B4	*Fusarium avenaceum* *F. orthoceras* var. *enniatinum* *F. fujikuroi*	Cereals	Hepatotoxicity	Serrano et al. (2015), Juan-García et al. (2015), Liuzzi et al. (2017)
10.	Citrinin	*Penicillium citrinum; Aspergillus* sp.	Cereal, barley, pepper, rice, corn	Cytotoxicity in human embryonic kidney cell lines; Nephropathy in pigs	Gong et al. (2019), Rocha et al. (2014), Omotayo et al. (2019)
11.	Citreoviridin	*P. citreonigrum*	Rice	Cardiac beriberi	Kushiro (2015)
12.	Zearalenone	*Fusarium graminearum; F. culmorum; F. equiseti*	Rice, maize and peanuts	Hyperesterogenic syndrome in pigs;	Sangare-Tigori et al. (2006), Reddy et al. (2010)

grow profusely with less moisture condition and produce toxin in the stored grains (Santin, 2005). However according to Miller (1995), a different perception regarding mycogenic fungi (Fig. 14.2) access into plant system and damage the yields of the plants.

Maize, used both in human and animal food (Baffes et al., 2017; Ranum et al., 2014). Reports revealed the mycotoxin contamination during the pre and post harvesting process of maize (Geary et al., 2016). Aflatoxins, fumonisins, zearalenone, deoxynivalenol and ochratoxins are the predominant mycotoxins frequently reported in maize crops (Udomkun et al., 2017). *Aspergillus* sp. *Fusarium* sp. and *Penicillium* sp. are the fungal strains are reported in maize and maize products (Wu et al., 2014). Patulin, one among the fungal toxin produced by *Penicillium expansum* recorded in apple, pear, peach, grape and apricot fruits (Omurtag, 2008) especially the juices of apple contains high level of patulin (Marin et al., 2006).

Soil, soil debris and soils of cereal crops reported with *Fusarium* species. Various species of *Fusarium* which include *F. graminearum*, *F. sporotrichioides*, *F. poae*, *F. culmorum*, *F. langsethiae* and *F. paranaense* were caused various diseases in cereal crops. The report of various species of *Fusarium* in corn (Becker et al., 2014), wheat (Cui et al., 2013), barley (Bělakova et al., 2014), sorghum (Divakara et al., 2014) caused various diseases and damaged the yield of the crop through its toxin. Fusarium also reported in various vegetables like soyabean (Costa et al., 2016), chilli (Sundaramoorthy et al., 2012) and tomato (Loganathan et al., 2014). Different diseases like head blight (Zhang et al., 2013), root rot disease in soy bean (Chang et al., 2015), stem rot and wilt diseases (Qiu et al., 2012) in

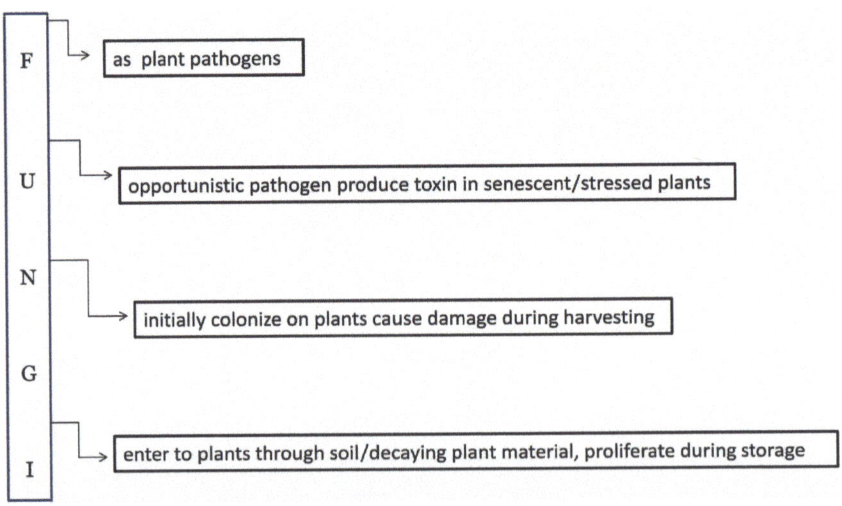

FIGURE 14.2 Different views of mycogenic fungi colonizing on the plants.

cucumber caused by *Fusarium* species and responsible for necrosis, chlorosis, growth inhibition, wilting seed germination inhibition in affected plants (Nelson et al., 1994; McLean, 1996).

Rice, wheat, corn, barley, sorghum has been severely affected by *Aspergillus* sp. Cherry, coffee, cocoa also been affected by *Aspergillus* sp. (Toth et al., 2012; Ruadrew et al., 2013; Kange et al., 2015). AFB1, AFB2, AFG1, AFG2 are the four major aflatoxins produced by various species of *Aspergillus* species (Pitt, 2000). Most of the aflatoxins produced by *A. flavus*, *A. parasiticus*. The phyto toxicity of aflatoxins were reported in mung bean (Sinha and Kumari, 1990), sesame (Samuel and Valentine, 2014) and cowpea (Adekunle and Bassir, 1997). Aflatoxin in general affected the yield, leaf and root developments, leaf mass development in various crops and especially in maize leaves the grana formation in chloroplast was reported two decades ago (Slowatizky et al., 1969).

Patulin, another important mycotoxin produced by *Penicillium*, *Aspergillus*, *Byssochlamys* species. This toxin predominantly present in apple, grape, pears and peaches (Yang et al., 2014; Moake et al., 2005). As per the report, improper handling of post harvest management especially fruits responsible for patulin toxin. During harvest, the organism entered into the injured fruit, then colonize on the whole fruit and later spreaded in the stored fruits by *Penicillium expansum*; (McCallum et al., 2002). The intake of this toxin cause nausea, vomiting and hemorrhage in human beings. Further chronic damage to kidney, suppression to immunity, carcinogenic and genotoxicity also reported (Liu et al., 2003) (Table 14.1).

Ochrotoxin in rice was reported as 341 μg/kg, maize 201 μg/kg in Nigeria and China respectively (Makun et al., 2011; Tan, 2006). Enniatin and beauvericin are the mycotoxins synthesized by the enzyme enniatin synthetase that contain peptides cause damages in cereal crops and produced by *Fusarium* sp. (Liuzzi et al., 2017) and based on Polymerase Chain Reaction (PCR) analysis, *esyn* 1 gene is responsible behind the production of enniatin (Kulik et al., 2007).

Mycotoxin and human health

Mycotoxin present in the agricultural field during pre harvest and post harvest period then during the processing time, storage duration and severely affect the quality of the food (Sforza et al., 2006) and human health (Leung et al., 2006). This influences the agribusiness in developing and under developed countries (Jelinek et al., 1989). Mycotoxins enter into the biotic system through direct and indirect route. In the direct, the fungi inhibit the food grains and the toxin occur during the processing, storage and preparing the food and cause diseases to the biotic organism, but in indirect the fungi damaging the food grains, but before processing the toxigenic fungi removed but the toxins are exist with the food (Frisvad and Samson, 1991). The infection caused through mycotoxin either directly consumed by the mycotoxin contaminated food and the food products derived from mycotoxin contaminated feed materials. Among the 300 known mycotoxins, vast majority of mycotoxins are withstand heat and a partial destruction may be occur when mycotoxigenic food was boiled, fried even pasteurized (Freire and da Rocha, 2016). Hence it is a need in this hour to seriously concern about the impact of mycotoxins on human health.

Mycotoxicoses, a disease caused by mycotoxin which produced by several fungi. Aflatoxin, one among the important fungal secondary metabolite which produced by *Aspergillus flavus*, *A. parasiticus* and *A. nomius* (Pitt, 2000). Aflatoxin, is highly carcinogenic and in 1960 is has been linked to liver cancer in human (Stoloff, 1977; Freire and da Rocha, 2016). Consumption of this toxin which leads to hypo albuminemia especially among the children (Ramjee et al., 1992). Due to various and serious pathogenicity of aflatoxin, the International agency for research on cancer (IARC, 1993) listed Aflatoxin B1 under group 1 of carcinogens under the red alert category. Next to Aflatoxin B1, B2, G1 and G2 are also considered as carcinogens (Bennett and Klich, 2003). Aflatoxins are responsible for diminishing the humoral and cell mediated immunities and causing the diseases (Bbosa et al., 2013). Ochratoxins reported in various plant products such as rice, maize, cereals, coffee, beans and pulses and it causes nephropathy in humans, further it may be suspected behind Tunisian nephropathy and Balken endemic nephropathy (BEN) (Hassan et al., 2006b; Malir et al., 2012). Teratogenic effect was reported in this mycotoxin on fetus developed in the womb by crossing the placenta, cause dis function in the central nervous system and damage the brain (Soleas et al., 2001). Citrinin, a toxin present in ethnic fermented food especially miso, sake and soy sauce. In association with ochratoxin, responsible for nephrotoxic mycotoxin. Reports revealed its responsible on embryocidal fetotoxic (Freire and da Rocha, 2016). Humans are frequency exposed to this toxin, may be due to ochratoxin, a common food mycotoxin and citrinin produced by same fungi. Ergotism caused by ergot alkaloids caused edema of the legs followed by amputation of legs and in extreme case it may leads to death based on the research report published four decades ago (Lorenz, 1979) and also responsible for gastro intestinal disorders and effects on central nervous system (Freire and da Rocha, 2016). Comparing to other toxins due to its heat labile nature, toxin may be destroyed during the processing of food (Peraica et al., 1999). Fumonisin caused cancer in esophagus in human (Shephard, 2011). Due to hydo soluble nature of fumonisin, more dangerous to human health and unable to detect from human samples.

As per International Agency for Research on Cancer (IARC), patulin toxin classified under group 3 and not cause carcinogenicity to humans (IARC report, 1986). This toxin produced immunotoxicity issues especially to reduce the cytokine secretion IFN-γ and IL-4 by human macrophages

(Wichmann et al., 2002). It reduced the production of interlukin 2 and 5 with the concentration of 500 ng patulin /ml in cell lines (Marin et al., 1996). The toxins produced by *Alternaria* caused high levels of human esophageal cancer in China (Liu et al., 1992). *Alternaria* toxin, a high concern for public health recommended by European food safety authority (EFSA report, 2011). Due to the infection cereal grains by *Fusarium* trichothecenes, it poses danger to young children through infant food and adult human population also, the toxin was detected in urine (Papageorgiou et al., 2018).

Based on the diversified threats poses to human population, it is a time to serious concentration on the miscellaneous fungal toxins.

Mycotoxin prevention and control in crops

Fungal toxin may occur both in field and storage stages (Gnonlonfin et al., 2013). Reports revealed the contribution of *Fusarium* and *Penicillium* on field and played a vital role by *Aspergillus* sp. and *Penicillium* sp. in storage stage in the production of mycotoxins (Pitt, 2000). Factors which played a role in the production of mycotoxin may be classified into two: (i) Controllable factors. (ii) Non controllable factors. Controllable factors which include good practices in the field, proper harvest, reduce mechanical damages, resistant crop varieties, efficient drying, good storage practices, cleaning, segregation and sort out infected crop yields. Concern with the above parameters may helpful to diminish the mycotoxin impacts on crop yields and human populations. The non controllable factors contain mainly the environmental parameters like temperature, moisture, humidity and other feasible environmental parameters which support the production of mycotoxins. These parameters are unable to modify / change and it is mainly based on the environment. Decontamination (Gullino et al., 2017) and

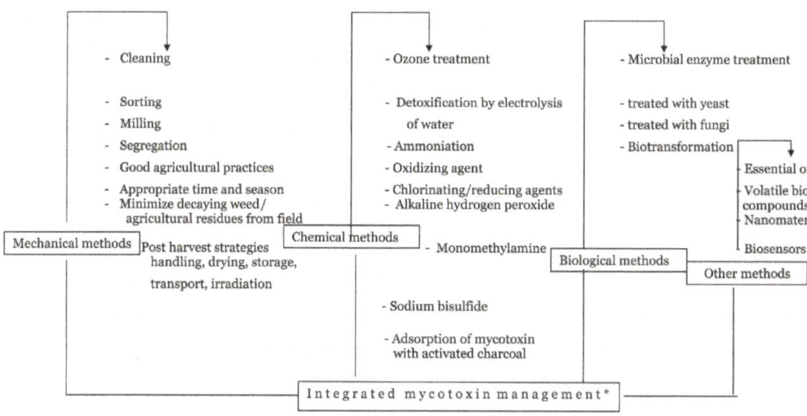

FIGURE 14.3 Integrated mycotoxin management strategies to prevent and control the mycotoxins from agricultural products (*Luo et al., 2018; Shi et al., 2018; Omotayo et al., 2019).

detoxification (Shi et al., 2018) may reduce or minimize the level of mycotoxins in agricultural products. It is advisable to prevent, reduce, destroy and inactivate the toxins from the existing mycotoxin contaminated agricultural products. Because it is impossible to completely eliminate the mycotoxins from the food chain (Shi et al., 2018). Hence an integrated mycotoxin management, is a need in this hour to prevent and control the mycotoxin in crop yields and their impact on human health (Fig. 14.3).

Conclusion

Mycotoxins as an inevitable part of agricultural products and its impact on human health leads upto mortality. It is unavoidable in food industry not only in under developing countries, throughout the World. The efficient control of mycotoxins. This review envisaged the relationship between mycotoxins and agriculture, listed out the important toxins and its pathogenicity on animals and human beings. An outline about various health hazards by consumption of mycotoxins on human being. Various integrated mycotoxin management on the prevention and control of mycotoxins by physical, chemical, biological and other miscellaneous approaches.

References

Adejumo, T.O., Adejoro, D.O., 2014. Incidence of aflatoxins, fumonisins, trichothecenes and ochratoxins in Nigerian foods and possible intervention strategies. Food Sci. Qual. Manag. 31, 127–146.

Adekunle, A.A., Bassir, O., 1997. The effects of afatoxinB1 and G1 on the germination and leaf colour of cowpea (*Vigna sinensis*). Mycopathologia et Mycologia Applicata. Klower Academic Publisher, Ibadan, Nigeria, pp. 1–12.

Adjou, E.S., Yehouenou, B., Sossou, C.M., Soumanou, M.M., de Souza, C.A., 2012. Occurrence of mycotoxins and associated mycoflora in peanut cake product (kulikuli) marketed in Benin. African J. Biotech. 11, 14354–14360.

Agriopoulou, S., Koliadima, A., Karaiskakis, G., Kapolos, J., 2016. Kinetic study of aflatoxins' degradation in the presence of ozone. Food Control. 61, 221–226.

Aish, J.L., Rippon, E.H., Barlow, T., Hattersley, H.J., 2004. Ochratoxin A. In: Magan, N., Olsen, M. (Eds.), Mycotoxins in Food: Detection and Control. CRC Press, Taylor and Francis, UK, p. 471.

Al-Seeni, M.N., 2012. Natural occurrence of heavy metal, fungi and mycotoxins in soybean meal samples used in animal feeding in Saudi Arabia. Afr. J. Biotechnol. 11, 9288–9294.

Al-Sheikh, H.M., 2015. LAMP-PCR detection of ochratoxigenic Aspergillus species collected from peanut kernel. Genet. Mol. Res. 14, 634–644. Available from: https://doi.org/10.4238/2015.January.30.5.

Amezqueta, S., Gonzalez-Penas, E., Lizarraga, T., Murillo-Arbizu, M., Lopez De Cerain, A., 2008. A simple chemical method reduces OTA in contaminated cocoa shells. J. Food Prot. 71, 1422–1426.

Ashiq, S., 2015. Natural occurrence of mycotoxins in food and feed: Pakistan perspective. Compr. Rev. Food Sci. Food Saf. 14, 159–175. Available from: https://doi.org/10.1111/crf3.2015.14.issue-2.

Avantaggiato, G., Havenaar, R., Visconti, A., 2004. Evaluation of the intestinal absorption of deoxynivalenol and nivalenol by an in vitro gastrointestinal model, and the binding efficacy of activated carbon and other adsorbent materials. Food Chem. Toxicol. 42 (5), 817–824.

Baffes, J., Kshirsagar, V., Mitchell, D., 2017. What drives local food prices? Evidence from the tanzanian maize market. World Bank. Economic Rev., 1 e25. Available from: https://watermark.silverchair.com.

Bbosa, G.S., Kitya, D., Lubega, A., Ogwal-Okeng, J., Anokbonggo, W.W., Kyegombe, D.B., 2013. Review of the biological and health effects of aflatoxins on body organs and body systems. In: Aflatoxins-Recent Advances and Future Prospects, InTech.

Becker, E.-M., Herrfurth, C., Irmisch, S., Kollner, T.G., Feussner, I., Karlovsky, P., et al., 2014. Infection of corn ears by *Fusarium* spp. induces the emission of volatile sesquiterpenes. J. Agric. Food Chem. 62, 5226–5236.

Bělakova, S., Benešova, K., Časlavsky, J., Svoboda, Z., Mikulikova, R., 2014. The occurrence of the selected fusarium mycotoxins in Czech malting barley. Food Control. 37, 93–98.

Bennett, J.W., Klich, M., 2003. Mycotoxins. Clin. Microbiol. Rev. 16 (3), 497–516.

Bhat, R., Rai, R.V., Karim, A.A., 2010. Mycotoxins in food and feed: Present status and future concerns. Compr. Rev. Food Sci. Food Saf. 9 (1). Available from: https://doi.org/10.1111/j.1541-4337.2009.00094.x.

Bhat, R.V., Miller, J.D., 1991. Mycotoxins and food supply. Food Nutr. Agriculture 1, 27–31.

Biasucci, G., Calabrese, G., Giuseppe, R., Carrara, G., Colombo, F., Mandelli, B., et al., 2011. The presence of ochratoxin A in cord serum and in human milk and its correspondence with maternal dietary habits. Eur. J. Nutr. 50, 211–218.

Blankson, G.K., Mill-Robertson, F.C., 2016. Aflatoxin contamination and exposure in processed cereal-based complementary foods for infants and young children in greater Accra, Ghana. Food Control. 64, 212–217.

Butler, T., 1902. Notes on a feeding experiment to produce leucoencephalitis in a horse with positive results. Am. Vet. Rev. 26, 748–751.

CAST, 2003. Mycotoxins: Risks in Plant, Animal, Human Systems, Task Force Report No. 38, Ames.

Castoria, R., Morena, V., Caputo, L., Panfili, G., De Curtis, F., De Cicco, V., 2005. Effect of the biocontrol yeast *Rhodotorula glutinis* strain LS11 on patulin accumulation in stored apples. Phytopathology 95, 1271–1278.

Chang, K.F., Hwang, S.F., Conner, R.L., Ahmed, H.U., Zhou, Q., Turnbull, G.D., et al., 2015. First report of *Fusarium proliferatum* causing root rot in soybean (*Glycine max* L.) in Canada. Crop. Prot. 67, 52–58.

Copetti, M.V., Iamanaka, B.T., Pereira, J.L., Fungaro, M.H., Taniwaki, M.H., 2011. Aflatoxigenic fungi and aflatoxin in cocoa. Int. J. Food Microbiol. 148, 141–144.

Copetti, M.V., Iamanaka, B.T., Pereira, J.L., Lemes, D.P., Nakano, F., Taniwaki, M.H., 2012. Determination of aflatoxins in by-products of industrial processing of cocoa beans. Food Addit. Contam.: Part. A 29, 972–978.

Coppa, C.F.S.C., Khaneghah, A.M., Alvito, P., Assuncao, R., Martins, C., Es, I., et al., 2019. The occurrence of mycotoxins in breast milk, fruit products and cereal based infant formula: a review. Trends Food Sci. Technol. 92, 81–93.

Costa, S.S., Matos, K.S., Tessmann, D.J., Seixas, C.D.S., Pfenning, L.H., 2016. *Fusarium paranaense* sp. nov., a member of the *Fusarium solani* species complex causes root rot on soybean in Brazil. Fungal Biol. 120, 51–60.

Creppy, E.E., Kern, D., Steyn, P.S., Vleggaar, R., Roschenthaler, R., Dirheimer, G., 1983b. Comparative study of the effect of ochratoxin A analogues on yeast aminoacyl-tRNA synthetases and on the growth and protein synthesis of hepatoma cells. Toxicol. Lett. 19, 217–224.

Creppy, E.E., Stormer, F.C., Kern, D., Roschenthaler, R., Dirheimer, G., 1983a. Effects of ochratoxin A metabolites on yeast phenylalanyl-tRNA synthetase and on the growth and *in vivo* protein synthesis of hepatoma cells. Chem. Biol. Interact. 47, 239–247.

Cui, L., Selvaraj, J.N., Xing, F., Zhao, Y., Zhou, L., Liu, Y., 2013. A minor survey of deoxynivalenol in Fusarium infected wheat from Yangtze–Huaihe river basin region in China. Food Control. 30, 469–473.

Cunha, S.C., Faria, M.A., Fernandes, J.O., 2009. Determination of patulin in apple and quince products by GC–MS using 13C5–7 patulin as internal standard. Food Chem. 115, 352–359.

Daniel, J.H., Lewis, L.W., Redwood, Y.A., Kieszak, S., Breiman, R.F., Flanders, W.D., et al., 2011. Comprehensive assessment of maize aflatoxin levels in eastern Kenya, 2005–2007. Environ. Health Perspect. 119 (12), 1794–1799.

Divakara, S.T., Santosh, P., Aiyaz, M., Venkata Ramana, M., Hariprasad, P., Nayaka, S.C., et al., 2014. Molecular identification and characterization of *Fusarium* spp. associated with sorghum seeds. J. Sci. Food Agriculture 94, 1132–1139.

EFSA report, 2011. EFSA Panel on Contaminants in the Food Chain (CONTAM) Scientific opinion on the risks for animal and public health related to the presence of Alternaria toxins in feed and food. EFSA J. 9, 2407.

Escrivá, L., Font, G., Manyes, L., 2015. In vivo toxicity studies of *Fusarium* mycotoxins in the last decade: a review. Food Chem. Toxicol. 78, 185–206.

Esmaillzadeh, A., Keshteli, A.H., Hajishafiee, M., Feizi, A., Feinle-Bisset, C., Adibi, P., 2013. Consumption of spicy foods and the prevalence of irritable bowel syndrome. World J. Gastroenterol. 19 (38), 6465–6471. Available from: https://doi.org/10.3748/wjg.v19.i38.6465.

Fakhri, Y., Rahmani, J., Oliveira, C.A.F., Franco, L.T., Corassin, C.H., Saba, S., et al., 2019. Aflatoxin M1 in human breast milk: a global systematic review, meta-analysis, and risk assessment study (Monte Carlo simulation). Trends Food Sci. Technol. 88, 333–342.

Food and Agriculture Organization of the United Nations, World Health Organization, 2011. FAO/WHO Guide for Application of Risk Analysis Principles and Procedures During Food Safety Emergencies. Food and Agriculture Organization of the United Nations, Rome.

Fredlund, E., Gidlund, A., Sulyok, M., Borjesson, T., Krska, R., Olsen, M., et al., 2013. Deoxynivalenol and other selected Fusarium toxins in Swedish oats — occurrence and correlation to specific Fusarium species. Int. J. Food Microbiol. 167, 276–283.

Freire, F.D.C.O., da Rocha, M.E.B., 2016. Impact of mycotoxins on human health. In: Merillion, J.M., Ramawat, K.G. (Eds.), Fungal Metabolites. Springer International publishing, Switzerland.

Frisvad, J.C., Samson, R.A., 1991. Filamentous fungi in foods and feeds: ecology, spoilage and mycotoxin production. In: Arora, D.K., Mukerjii, K.G., Marth, E.H. (Eds.), Handbook of Applied Mycology: Food and Feeds. Marcel Dekker, New York, pp. 31–68.

Fuchs, S., Sontag, G., Stidl, R., Ehrlich, V., Kundi, M., Knasmüller, S., 2008. Detoxification of patulin and OTA, two abundant mycotoxins, by lactic acid bacteria. Food Chem. Toxicol. 46, 1398–1407.

Gammbacorta, L., Darra, N.E., Fakhoury, R., Logrieco, A.F., Solfrizzo, M., 2019. Incidence and levels of *Alternaria* mycotoxins in spices and herbs produced worldwide and commercialized in Lebanon. Food Control. Available from: https://doi.org/10.1016/j.foodcont.2019.106724.

Geary, P.A., Chen, G., Kimanya, M.E., Shirima, Oplatowska-Stachowiak, E., Gong, C.T., 2016. Determination of multi-mycotoxin occurrence in maize based porridges from selected regions of Tanzania by liquid chromatography tandem mass spectrometry (LC-MS/MS), a longitudinal study. Food Control. 68, 337e343. Available from: https://doi.org/10.1016/j.foodcont.2016.04.018.

Glaser, N., Stopper, H., 2012. Patulin: mechanism of genotoxicity. Food Chem. Toxicol. 50, 1796–1801.

Gnonlonfin, G.J.B., Hell, K., Adjovi, Y., Fandohan, P., Koudande, D.O., Mensah, G.A., et al., 2013. A review on aflatoxin contamination and its implications in the developing world: a sub-Saharan African perspective. Crit. Rev. Food Sci. Nutr. 53, 349–365.

Gong, L., Zhu, H., Li, T., Ming, G., Duan, X., Wang, J., et al., 2019. Molecular signatures of cytotoxic effects in human embryonic kidney 293 cells treated with single and mixture of ochratoxin A and citrinin. Food Chem. Toxicol. 123, 374–384.

Goto, T., Wicklow, D.T., Ito, Y., 1996. Aflatoxin and cyclopiazonic acid production by a sclerotium-producing *Aspergillus tamarii* strain. Appl. Env. Microbiol. 62, 4036–4038.

Graham, R., 1912. Forage Poison in Horses, Cattle and Mules: So-Called Cerebro-Spinal Meningitis and Commonly Termed "Staggers" Kentucky Agricultural Experiment Station, pp. 369–383. Bull. No. 17.

Gullino, M.L., Stack, J.P., Fletcher, J., Mumford, J.D., 2017. Practical Tools for Plant and Food Biosecurity: Results From a European Network of Excellence. Springer International Publishing AG, Cham, Switzerland.

Guo, W., Fan, K., Nie, D., Meng, J., Huang, Q., Yang, J., et al., 2019. Development of a QuEChERS-based UHPLC-MS/MS method for simultaneous determination of six alternaria toxins in grapes. Toxins (Basel) 11, 1–16. 2019.

Hassan, A.M., Sheasha, H.A., Abdel Fatah, M.F., Ibrahim, A.Z., Gaber, O.A., 2006a. Does aflatoxin as an environmental mycotoxin adversely affect the renal and hepatic functions of Egyptian lactating mothers and their infants? A preliminary report. Int. Urol. Nephrol. 38, 339–342.

Hassan, A.M., Sheashaa, H.A., Fattah, M.F.A., Ibrahim, A.Z., Gaber, O.A., Sobh, M.A., 2006b. Study of ochratoxin A as an environmental risk that causes renal injury in breast-fed Egyptian infants. Pediatr. Nephrol. 21, 102–105.

Herzallah, S.M., 2009. Determination of aflatoxins in eggs, milk, meat and meat products using HPLC fluorescent and UV detectors. Food Chem. 114, 1141–1146.

Huffman, J., Gerber, R., Du, L., 2010. Recent advancements in the biosynthetic mechanisms for polyketide erived mycotoxins. Biopolymers 93 (9), 764–776.

IARC (International Agency for Research on Cancer), 2002. Some Traditional Herbal Medicines, Some Mycotoxins, Naphthalene and Styrene, 82. International Agency for Research on Cancer [IARC Monographs on the Evaluation of Carcinogenic Risks to Humans], Lyon, France.

IARC, 1986. Monographs on the Evaluation of the Carcinogenic Risk of Chemicals to Humans. Some Naturally Occurring and Synthetic Food Components, Furocoumarins and Ultraviolet Radiation, Lyon, France, pp. 83–98, IARC.

International Agency for Research on Cancer (IARC), 1993. Ochratoxin A. Monographs on the Evaluation of Carcinogenic Risks to Humans, Some Naturally Occurring Substances: Food Items and Constituents, Heterocyclic Aromatic Amines and Mycotoxins, vol. 56. International Agency for Research on Cancer, Lyon.

Iqbal, S.Z., Rabbani, T., Asi, M.R., Jinap, S., 2014. Assessment of aflatoxins, ochratoxin A and zearalenone in breakfast cereals. Food Chem. 157, 257–262.

Ismaiel, A.A., Papenbrock, J., 2015. Mycotoxins: producing fungi and mechanisms of phytotoxicity. Agriculture 5, 492–537.

Jelinek, C.F., Pohland, A.E., Wood, G.E., 1989. Worldwide occurrence of mycotoxins in foods and feeds, an update. J. Assoc. Off. Anal. Chem. 72, 223–230.

Jestoi, M., Rokka, M., Jarvenpaa, E., Peltonen, K., 2009. Determination of *Fusarium* mycotoxins beauvericin and enniatins (A, A1, B, B1) in eggs of laying hens using liquid chromatography–tandem mass spectrometry (LC–MS/MS). Food Chem. 115, 1120–1127.

Jeswal, P., Kumar, D., 2015. Mycobiota and natural incidence of aflatoxins, ochratoxin A, and citrinin in Indian spices confirmed by LC-MS/MS. Int. J. Microbiol. 242486. Available from: https://doi.org/10.1155/2015/242486.

Joint FAO/WHO Expert Committee on Food Additives (JECFA), 2001. Safety evaluation of certain mycotoxins in food. Food and Agriculture Organisation of the United Nations (FAO) food and nutrition paper 74, World Health Organisation (WHO) food additive series, 47, 281–415.

Juan-García, A., Ruiz, M.-J., Font, G., Manyes, L., 2015. Enniatin A1, enniatin B1 and beauvericin on HepG2: evaluation of toxic effects. Food Chem. Toxicol. 84, 188–196.

Kange, A.M., Cheruiyot, E.K., Ogendo, J.O., Arama, P.F., 2015. Effect of sorghum (*Sorghum bicolor* L. Moench) grain conditions on occurrence of mycotoxin-producing fungi. Agric. Food Security 4, 1.

Kokkonen, M., Ojala, L., Parikka, P., Jestoi, M., 2010. Mycotoxin production of selected *Fusarium* species at different culture conditions. Int. J. Food Microbiol. 143, 17–25.

Krska, R., Welzig, E., Boudra, H., 2007. Analysis of *Fusarium* toxins in feed. Anim. Feed. Sci. Technol. 137, 241–264.

Kulik, T., Pszczolkowska, A., Fordonski, G., Olszewski, J., 2007. PCR approach based on the esyn1 gene for the detection of potential enniatin-producing *Fusarium* species. Int. J. Food Microbiol. 116 (3), 319–324. 30.

Kushiro, M., 2015. Historical review of researches on yellow rice and mycotoxigenic fungi adherentto rice in Japan. Japanese Soc. Mycotoxicol. (JSM) Mycotoxins 65 (1), 19–23.

Leung, M.C.K., Diaz-Llano, G., Smith, T.K., 2006. Mycotoxins in pet food: a review on worldwide prevalence and preventative strategies. J. Agric. Food Chem. 54, 9623–9635.

Liu, B.H., Yu, F.Y., Wu, T.S., Li, S.Y., Su, M.C., Wang, M.C., et al., 2003. Evaluation of genotoxic risk and oxidative DNA damage in mammalian cells exposed to mycotoxins, patulin and citrinin. Toxicol. Appl. Pharmacol. 191, 255–263.

Liu, G.T., Qian, Y.Z., Zhang, P., Dong, W.H., Qi, Y.M., Guo, H.T., 1992. Etiological role of *Alternaria alternata* in human esophageal cancer. Chin. Med. J. (Engl.) 105 (5), 394–400.

Liu, Y.L., Meng, G.Q., Wang, H.R., Zhu, H.L., Hou, Y.Q., Wang, W.J., et al., 2011. Effect of three mycotoxin adsorbents on growth performance: nutrient retention and meat quality in broilers fed on mould-contaminated feed. Br. Poult. Sci. 52, 255–263.

Liuzzi, V.C., Mirabelli, V., Cimmarusti, M.T., Haidukowski, M., Leslie, J.F., Logrieco, A.F., et al., 2017. Enniatin and Beauvericin biosynthesis in *Fusarium* species: Production profiles and structural determinant prediction. Toxins . Available from: https://doi.org/10.3390/toxins9020045.

Loganathan, M., Garg, R., Venkataravanappa, V., Saha, S., Rai, A., 2014. Plant growth promoting rhizobacteria (PGPR) induces resistance against *Fusarium* wilt and improves lycopene content and texture in tomato. Afr. J. Microbiol. Res. 8, 1105–1111.

Lorenz, K., 1979. Ergot on cereal grains. Crit. Rev. Food Sci. Nutr. 11, 311–354.

Lugauskas, A., Repeckiene, J., Novosinskas, H., 2004. Micromycetes, producers of toxins, detected on stored vegetables. Ann. Agric. Environ. Medicine: AAEM 12, 253–260.

Eskola, M., Kos, G., Elliott, C.T., Hajšlová, J., Mayar, S., Krska, R., 2019. Worldwide contamination of food-crops with mycotoxins: validity of the widely cited 'FAO estimate' of 25%. Crit. Rev. Food Sci. Nutr. Available from: https://doi.org/10.1080/10408398.2019.1658570.

Makun, H.A., Dutton, M.F., Njobeh, P.B., Mwanza, M., Kabiru, A.Y., 2011. Natural multi-occurrence of mycotoxins in rice from Niger State, Nigeria. Mycotoxin Res. 27, 97–104.

Malir, F., Ostry, V., Pfohl-Leszkowicz, A., Roubal, T., 2012. Ochratoxin A exposure biomarkers in the Czech Republic and comparison with foreign countries. Biomarkers 17, 577–589.

Marin, M.L., Murtha, J., Dong, W., Pestka, J.J., 1996. Effects of mycotoxins on cytokine production and proliferation in EL-4 thymoma cells. J. Toxicol. Env. Health 48 (4), 379–396.

Marin, S., Morales, H., Hasan, H., Ramos, A.J., Sanchis, V., 2006. Patulin distribution in fuji and golden apples contaminated with *Penicillium expansum*. Food Addit. Contam. 23 (12), 1316–1322.

McCallum, J.L., Tsao, R., Zhou, T., 2002. Factors affecting patulin production by *Penicillium expansum*. J. Food Prot. 65 (12), 1937–1942.

McCormick, S.P., Stanley, A.M., Stover, N.A., Alexander, N.J., 2011. Trichothecenes: from simple to complex mycotoxins. Toxins 3, 802–814.

McLean, M., 1996. The phytotoxicity of *Fusarium* metabolites: an update since 1989. Mycopathologia 133, 163–179.

McNutt, S., Purwin, P., Murray, C., 1928. Vulvovaginitis in swine. J. Am. Vet. Med. Assoc. 73, 484.

Meena, S., Samal, M., 2019. *Alternaria* host specific (HSTs) toxins: An overview of chemical characterization, target sites, regulation and their toxic effects. Toxicol. Rep. 6, 745–758.

Miller, J.D., 1995. Fungi and mycotoxins in grain: implications for stored product research. J. Stored Prod. Res. 31, 1–16.

Moake, M.M., Padilla-Zakour, O.I., Worobo, R.W., 2005. Comprehensive review of patulin control methods in foods. Compr. Rev. Food Sci. Food Saf. 1, 8–21.

Nelson, E.P., Dignani, M.C., Anaissie, E.J., 1994. Taxonomy, biology, and clinical aspects *of Fusarium* species. Clin. Microbiol. Rev. 7, 479–504.

Ok, H.E., Kim, D.M., Kim, D., Chung, S.H., Chung, M.-S., Park, K.H., et al., 2014. Mycobiota and natural occurrence of aflatoxin, deoxynivalenol, nivalenol and zearalenone in rice freshly harvested in South Korea. Food Control. 37, 284–291.

Oliveira, P.M., Mauch, A., Jacob, F., Waters, D.M., Arendt, E.K., 2012. Fundamental study on the influence of *Fusarium* infection on quality and ultrastructure of barley malt. Int. J. Food Microbiol. 156, 32–43.

Omotayo, O.P., Omotayo, A.O., Mwanza, M., Babalola, O.O., 2019. Prevalence of mycotoxins and their consequences on human health. Toxicol. Res. 35 (1), 1–7.

Omurtag, G.Z., 2008. Fumonisins, trichothecenes and zearalenone in cereals. Int. J. Mol. Sci. 9 (11), 2062–2090.

Papageorgiou, M., Wells, L., Williams, C., White, K., De Santis, B., Liu, Y., et al., 2018. Assessment of urinary deoxynivalenol biomarkers in UK children and adolescents. Toxins 10, 50.

Peraica, M., Radica, B., Lucivca, A., Pavlovic, M., 1999. Toxic effects of mycotoxins in humans. Bull. World Health Organ. 9, 754–766.

Pitt, J.I., 2000. Toxigenic fungi and mycotoxins. Br. Med. Bull. 56, 184–192.

Qiu, M., Zhang, R., Xue, C., Zhang, S., Li, S., Zhang, N., et al., 2012. Application of bio-organic fertilizer can control *Fusarium* wilt of cucumber plants by regulating microbial community of rhizosphere soil. Biol. Fertil. Soils 48, 807–816.

Quintela, S., Villarán, M.C., López de Armentia, I., Elejalde, E., 2012. Ochratoxin A removal in red wine by several oenological fining agents: bentonite, egg albumin, allergen free adsorbents, chitin and chitosan. Food Addit. Contam.: Part. A 29 (7), 1168–1174.

Ramirez, M.L., Cendoya, E., Nichea, M.J., Zachetti, V.G.L., Chulze, S.N., 2018. Impact of toxigenic fungi and mycotoxins in chickpea: a review. Curr. Opin. Food Sci. 23, 32–37.

Ramjee, G., Berjak, P., Adhikari, M., Dutton, M.F., 1992. Aflatoxins and kwashiorkor in Durban, South Africa. Ann. Trop. Paediatr. 12, 241–247.

Ranum, P., Pe∼na-Rosas, J.P., Garcia-Casal, M.N., 2014. Global maize production, utilization and consumption. Ann. N. Y. Acad. Sci. 1312 (1), 105–112. Available from: https://doi.org/10.1111/nyas.12396.

Reddy, K., Salleh, B., Saad, B., Abbas, H., Abel, C., Shier, W., 2010. An overview of mycotoxin contamination in foods and its implications for human health. Toxin Rev. 29, 3–26.

Richard, J.L., 2007. Some major mycotoxins and their mycotoxicoses—an overview. Int. J. Food Microbiol. 119, 3–10. Available from: https://doi.org/10.1016/j.ijfoodmicro.2007.07.019.

Robens, J., Cardwell, K., 2003. The costs of mycotoxin management to the USA: management of aflatoxins in the United States. J. Toxicol. Toxin Rev. 22 (2–3), 139–152.

Rocha, M.E.B., Freire, F.C.O., Maia, F.E.F., Guedes, M.I.F., Rondina, D., 2014. Mycotoxins and their effects on human and animal health. Food Control. 36 (2014), 159–165.

Rodrigues, P., Venancio, A., Lima, N., 2013. Incidence and diversity of the fungal genera *Aspergillus* and *Penicillium* in Portuguese almonds and chestnuts. Eur. J. Plant. Pathol. 137, 197–209.

Ruadrew, S., Craft, J., Aidoo, K., 2013. Occurrence of toxigenic *Aspergillus* spp. and aflatoxins in selected food commodities of Asian origin sourced in the West of Scotland. Food Chem. Toxicol. 55, 653–658.

Ruiyu, Z., Juan, C., Qingqiang, Y., Ping, W., Wei, C., Xiao, W., et al., 2012. Inhibiting *Aspergillus flavus* growth and degrading aflatoxin B1 by combined beneficial microbes. Afr. J. Biotechnol. 11, 12903–12909.

Samuel, A.T., Valentine, I.T., 2014. Effect of total aflatoxin on the growth characteristics and chlorophyll level of sesame (*Sesamum indicum* L.). N. Y. Sci. J. 7, 8–13.

Sangare-Tigori, B., Moukha, S., Kouadio, H.J., Betbeder, A.M., Dano, D.S., Creppy, E.E., 2006. Co-ccurrence of aflatoxin B1, fumonisin B1, ochratoxin A and zearalenone in cereals and peanuts from Cote d'Ivoire. Food Addit. Contam. 23, 1000–1007.

Santin, E., 2005. Mould growth and mycotoxin production. In: Diaz, D.E. (Ed.), The Mycotoxin Blue Book. Nottingham University press, Nottingham, UK, p. 349.

Sarubbi, F., Formisano, G., Auriemma, G., Arrichiello, A., Palomba, R., 2016. Patulin in homogenized fruit's and tomato products. Food Control. 59, 420–423.

Schardl, C.L., 2015. Introduction to the toxins. Special issue on Ergot alkaloids. Toxins 7 (10), 4232–4237.

Schulz, M.C., Schumann, L., Rottkord, U., Humpf, H.-U., Gekle, M., Schwerdt, G., 2018. Synergistic action of the nephrotoxic mycotoxins ochratoxin A and citrinin at nanomolar concentrations in human proximal tubule-derived cells. Toxicol. Lett. 291, 149–157.

Scientific Committee on Food (SCF), 1998. Opinion of the Scientific Committee on Food on Ochratoxin A. http://europa.eu.int/comm/food/fs/sc/scy/out14_en.html.

Serrano, A.B., Capriotti, A.L., Cavaliere, C., Piovesana, S., Samperi, R., Ventura, S., et al., 2015. Development of a rapid LC-MS/MS method for the determination of emerging *Fusarium* mycotoxins enniatins and beauvericin in human biological fluids. Toxins 7, 3554–3571.

Sforza, S., Dall'asta, C., Marchelli, R., 2006. Recent advances in mycotoxin determination in food and feed by hyphenated chromatographic techniques/mass spectrometry. Mass. Spectr. Rev. 25, 54–76.

Sheldon, J., 1904. A corn mold (*Fusarium moniliforme* n. sp.). Annu. Rep. Neb. Agric. Experim. Stn. 17, 23–32.

Shephard, G., 2011. *Fusarium* mycotoxins and human health. Plant. Breed. Seed Sci. 64, 113–121.

Shi, H., Li, S., Bai, Y., Prates, L.L., Lei, Y., Yu, P., 2018. Mycotoxin contamination of food and feed in China: occurrence, detection techniques, toxicological effects and advances in mitigation technologies. Food Control. Available from: https://doi.org/10.1016/j.foodcont.2018.03.036.

Shim, W.B., Dzantiev, B.B., Eremin, S.A., Chung, D.H., 2009. One-step simultaneous immunochromatographic strip test for multianalysis of ochratoxin A and zearalenone. J. Microbiol. Biotechnol. 19, 83–92.

Sinha, K.K., Kumari, P., 1990. Some physiological abnormalities induced by aflatoxin B1 in mung seeds (*Vigna radiata* var. Pusa Baishakhi). Mycopathologica 110, 77–79.

Slowatizky, I., Mayer, A.M., Plojakoff-Mayber, A., 1969. The effect of aflatoxin on greening of etiolated leaves. Isr. J. Bot. 18, 31–36.

Soleas, G.J., Yan, J., Goldberg, D.M., 2001. Assay of ochratoxin A in wine and beer by high-pressure liquid chromatography photodiode array and gas chromatography mass selective detection. J. Agric. Food Chem. 49, 2733–2740.

Sørensen, J.L., Nielsen, K.F., Rasmussen, P.H., Thrane, U., 2008. Development of a LC MS/MS Method for the analysis of Enniatins and Beauvericin in whole Fresh and Ensiled Maize. J. Agric. Food Chem. 56, 10439–10443.

Sreenivasa, M.Y., Diwakar, B.T., Raj, A.P.C., Dass, R.S., Naidu, A., Janardhana, G.R., 2013. Toxigenic Fusarium species and Fumonisin B1 and B2 associated with freshly harvested sorghum and maize grains produced in Karnataka, India. Annals: Food Sci. Technol. 14, 100–107.

Stoloff, L., 1977. Aflatoxins – an overview. In: Rodncks, J.V., Hesseltine, C.W., Mehlman, M. A. (Eds.), Mycotoxins in Human and Animal Health. Pathotox, Park Forest South.

Streit, E., Schatzmayr, G., Tassis, P., Tzika, E., Marin, D., Taranu, I., et al., 2012. Current situation of mycotoxin contamination and co-occurrence in animal feed-focus on Europe. Toxins 4, 788–809.

Sundaramoorthy, S., Raguchander, T., Ragupathi, N., Samiyappan, R., 2012. Combinatorial effect of endophytic and plant growth promoting rhizobacteria against wilt disease of *Capsicum annum* L. caused by *Fusarium solani*. Biol. Control. 60, 59–67.

Tan, D., 2006. OTA Pollution Level Survey of Grain Food of 6 Provinces in China and Dietary Exposure Assessment Survey in Our Country (Thesis), Shijiazhuang, Hebei Medical University.

Toth, B., Baranyi, N., Berki, A., Torok, O., Kotai, E., Mesterhazy, A., et al., 2012. Occurrence of *Aspergillus flavus* on cereals in Hungary. Rev. Agric. Rural. Dev. 1, 446–451.

Udomkun, P., Wiredu, A.N., Nagle, M., Bandyopadhyay, R., Müller, J., Vanlauwe, B., 2017. Mycotoxins in Sub-Saharan Africa: present situation, socio-economic impact, awareness, and outlook. Food Control. 72, 110–122. Available from: https://doi.org/10.1016/j.foodcont.2016.07.039.

Venter, A.C., 2014. Glycerol Compositions and Solutions. US Patent Application Publication. No. 20140106008, A1.

Weidenbörner, M., 2001. Foods and fumonisins. Eur. Food Res. Technol. 212, 262–273.

Wichmann, G., Herbarth, O., Lehmann, I., 2002. The mycotoxins citrinin, gliotoxin, and patulin affect interferon-gamma rather than interleukin-4 production in human blood cells. Env. Toxicol. 17 (3), 211–218.

Williams, J.H., Phillips, T.D., Jolly, P.E., Stiles, J.K., Jolly, C.M., Aggarwal, D., 2004. Human aflatoxicosis in developing countries, a review of toxicology, exposure, potential health consequences, and interventions. Am. J. Clin. Nutr. 80, 1106–1122.

Wu, F., Groopman, J.D., Pestka, J.J., 2014. Public health impacts of food borne mycotoxins. Annu. Rev. Food Sci. Technol. 5 (1). Available from: https://doi.org/10.1146/annurev-food-030713-092431.

Luo, Y., Liu, X., Li, J., 2018. Updating techniques on controlling mycotoxins - a review. Food Control. Available from: https://doi.org/10.1016/j.foodcont.2018.01.016.

Yang, G., Bai, Y., Wu, X., Sun, X., Sun, M., Liu, X., et al., 2018. Patulin induced ROS-dependent autophagic cell death in human hepatoma G2 cells. Chem.-Biol. Interact. 288, 24–31.

Yang, J., Li, J., Jiang, Y., Duan, X., Qu, H., Yang, B., et al., 2014. Natural occurrence, analysis, and prevention of mycotoxins in fruits and their processed products. Crit. Rev. Food Sci. Nutr. 54, 64–83.

Yassin, M.A., El-Samawaty, A., Moslem, M.A., El-Naggar, M.A., 2013. Mycobiota of almond seeds and the toxigenicity of some involved genera. Life Sci. J. 10.

Zhang, J.B., Wang, J.H., Gong, A.D., Chen, F.F., Song, B., Li, X., et al., 2013. Natural occurrence of fusarium head blight, mycotoxins and mycotoxin-producing isolates of *Fusarium* in commercial fields of wheat in Hubei. Plant. Pathol. 62, 92–102.

Chapter 15

Bioprospection of endophytic fungi associated with ethnoveterinary plants for novel metabolites

Archana Nath[1] and S.R. Joshi[2,*]
[1]*Agricultural and Processed Food Products Export Development Authority, Housefed Complex, Guwahati, Assam, India,* [2]*Microbiology Laboratory, Department of Biotechnology and Bioinformatics, North-Eastern Hill University, Shillong, India*
**Corresponding author.* Email: srjoshi2006@yahoo.co.in

Introduction

India has one of the sophisticated medical cultures with a tradition of over 5000 years. Livestock keeping in ancient India was an inviolable tradition more than an economic enterprise. Atharvaveda, Mahabharata and many other old scriptures, and edicts provides ample evidence about ailments of animals, herbal medicines, and cure of diseases (Somvanshi, 2006). Ancient Indian literature in the form of the holy Vedas, Puranas, Brahmanas, epics, etc. are flooded with information on animal care, health management, and disease cure.

India is an agriculture-based country, whose majority of the population mostly depends on the agriculture and animal husbandry for their livelihood. Agriculture is the main source of Indian economy which also includes animal husbandry and their maintenance for obtaining good quality animal products. Since livestock can yield products throughout the year, they are the main source of income generation for farmers than that of agriculture which is mostly seasonal (Sansoucy, 1997). Thus, the health and maintenance of livestock and other animals are must important for the farmers. Due to the financial limitations, sometimes they are not able to afford modern medicines and facilities and thus, prefer traditional ethnoveterinary practices for the treatments of animals (Shrivastava et al., 2017).

Fungi Bio-Prospects in Sustainable Agriculture, Environment and Nano-Technology.
DOI: https://doi.org/10.1016/B978-0-12-821394-0.00015-9

Ethno-veterinary practice in modern India

The codified knowledge of ethnoveterinary practices still exists in the form of text manuscripts. The local folk of India are still practicing the ethnoveterinary medicines. The traditional knowledges are passed from one generation to another verbally and the folk ethnoveterinary practices mostly remain undocumented (Santhanakrishnan et al., 2008). There is a rich existence of these practices in the villages of India which becomes an integral part of a family which aids to their social, religious and economic values. There are local traditional healers who are experienced and knowledgeable in animal healthcare practices (Phondani et al., 2010). They use the locally available medicinal plants for the treatment of animals. Since the knowledge was passed verbally to the descendants and were not documented, many information gets extinct due to sudden demise of a generation.

Since 1980s, researchers got interested in documenting and validating ethnoveterinary practices (Eshetu et al., 2015). Several researches, conferences and workshops were held in this field which helped in the protection of traditional knowledge. Thus, the traditional knowledge on animal healthcare will now be available for all the practitioners, healers and researchers for several purposes in a documented form.

Ethno veterinary medicine

Ethnoveterinary medicine (EVM) is the holistic, interdisciplinary study of local knowledge and its associated skills, practices, beliefs, practitioners, and social structures pertaining to the healthcare of the animals (McCorkle, 1995). Ethnoveterinary practices are cost effective and dynamic (Warren, 1991). EVM is applicable for the treatment of almost all types of ailments and also for the maintenance of the reproductive health of the animals. The ethnoveterinary practices is most popular among the farmers and the traditional healers because its easily accessible, easy to prepare and administer at a cost-effective price.

Various types of EVM is practiced in different part of the country, some of them are highly effective against certain ailments and diseases, but few of them are based on superstitions and mythological myths which has hardly any basis to be considered as effective treatment (Sri Balaji and Chakravarthi, 2010).

Herbal remedies used for hundreds of years by traditional healers can be put to commercial use, but researchers are demanding that traditional knowledge should be validated, to verify the safety and efficacy of the treatments.

Threats to ethnoveterinary knowledge

Modernization in societies lifestyle and with the rapid technological progress, youngsters are becoming hesitant and refuses to learn the tradition knowledge from their parents and great grandparents (Martin, 1995; Longuefosse and Nossin, 1996). Since many of the information are not documented and passed verbally from one generation to another has greatly endangered its survival and sustainability into the future (Wanzala et al., 2005). Youngsters look tradition practices by a different perspective considering it as an evil, satanic, sinful and therefore ungodly. It is further anticipated that the rate of this knowledge's decay and disappearance is alarming worldwide and, if precautions are not taken, it may continue becoming worse with time.

Benefits of medicinal plants

The original source for both ancient and modern medicine is mostly the medicinal plant (Karimi and Jaafar, 2011). Traditional medicines are popular among the people because it is easily affordable, more closely corresponds to the patient's ideology, it eliminates the general concern of people about the side effects of chemical (synthetic) medicines and allows greater public access to health information (Wachtel-Galor and Benzie, 2011). A general perception has been made by the people about the herbal plants, that it is natural and safe and is not toxic. Recently in the health care sector, traditional medicine holds an important position as it is a flourishing global commercial enterprise though people have physical or financial access to allopathic medicine (Engebretson, 2002; Evans et al., 2007). The popularity of herbal medicines is growing faster among the people than any other alternative therapy (Eisenberg et al., 1998).

Medicinal plants are practically a mixture of over dozens or even hundreds of chemicals that act synergistically. Medicinal plants are considered as a rich resource of bioactive compounds which can be used in drug development either pharmacopoeial, non-pharmacopoeial or synthetic drugs. The bioactive compounds produced by plants are produced as secondary metabolites. The traditional plants play a significant role in the pharmacological research and drug development as not only the active compounds produced by them but also the plant itself as a starting material is used for the synthesis of drugs (Li and Vederas, 2009). About 20 decades ago, morphine, a pharmacologically active compound was derived from opium extracted from seeds pods of the *Papaver somniferum*. Discovery of Morphine followed by many other compounds like salicylic acid, a precursor of aspirin, derived from *Salix* sp.; reserpine, an antipsychotic and antihypertensive drug from *Rauwolfia* sp.; and antimalarials such as quinine from *Cinchona* bark etc. then contributed greatly to the commercial drug preparations (Wachtel-Galor and Benzie, 2011).

Endophytes

Endophytes are microorganisms of bacterial or fungal origin which resides in the healthy plant tissue intercellularly/intracellularly without causing any disease in the host plant (Wilson, 1995). Endophytes can colonize in all plant and are ubiquitous and their presence is evidenced in almost all the plants examined till date. They show three different types of association with the plants, commensal, parasitic or mutualistic. Host plants gets considerable benefits by the endophytes as they support the growth of the host plants and protects them from biotic and abiotic stresses. Endophytic fungi inhabiting the host plants has the potential to produce many bioactive metabolites with diverse structural features. These bioactive metabolites are known to possess pharmaceutical and ecological significance. According to chemical structures and biosynthetic pathways, secondary metabolites are mainly divided into four groups, including polyketide, non-ribosomal peptides, alkaloids, and terpenes (Liu and Liu, 2018). The characteristics of microbial adaptations in special and sequestered environments and their huge diversity makes endophytes an exciting field of study in the field of drug discovery. Drug resistance is becoming the world wide threat which is leading to the discovery of new drugs from the natural resources among which endophytes plays a significant role in drug discovery (Padhi et al., 2013).

Interaction of endophytes with medicinal plants

Genetic background of the concerned plants, ecological habitats where the plants live, and soil nutrients are some of the important factors which effects the quality and quantity of crude drugs originated from medicinal plants (Dai et al., 2003; Sherameti et al., 2005). In the past few years, endophytic fungi or endophytes are gaining interest as they are considered to played a very important role in affecting the quality and quantity of the crude drugs through a particular fungus-host interaction. There is an need to understand particular relationships between endophytic fungi and medicinal plants for promoting crude drug production (Faeth and Fagan, 2002). Although endophytic fungi play an important role in the plant micro-ecosystems which is important for the growth and development of the host plants but still very limited knowledge is available about the exact relationships between endophytic fungi and their host plants. Further detailed study on such relationships will lead to the production of better drugs by manipulating the growth conditions of medicinal plants (Jia et al., 2016). The diversity of the endophytic fungi in the medicinal host plants are mostly dependent on the genetic background, nutrient level, and ecological habitats of the host plants which in turn are beneficial to the host plant in terms of induced growth of host medicinal plants, increased resistance to biotic or abiotic stress (Rodriguez et al., 2009) and accumulation of secondary metabolites, including bioactive compounds (Firáková et al., 2007), some of which can be used by beneficial drugs. Thus, it is considered that the mutual interrelation between endophytic fungi and their host plants can influence the production of certain bioactive metabolites and a detailed study on such relationship may lead to some effective drug production. Species composition of the host plants are mostly influenced by the age of the host plant and tissue they reside (Rodrigues, 1994; Sieber, 2007) and also on their ability to utilize different substrates (Carroll and Petrini, 1983). Endophytic microbes from medicinal plants are good source of functional metabolites (Weber et al., 2004; Tejesvi et al., 2007; Huang et al., 2008). Endophyte infection was found to alter pattern of gene expression in the host plant (Bailey et al., 2006). Interaction between endophyte and plant is mainly controlled by the genes of both organism and host plant modulated by the environment. These findings have important practical implications for obtaining and producing drugs with improved quality and higher quantity.

Derivation of natural products from host plant

Several studies have reported that the medicinal plants producing pharmaceutically important secondary metabolites usually harbor endophytes which are capable of production of host-independent metabolites. Since the first report of endophyte *Taxomyces andreanae* producing same bioactive secondary metabolite taxol (paclitaxel) as its host *Taxus brevifolia* (Stierle et al., 1993) several other studies also revealed that plant derived secondary metabolites are also produced by endophytes (Zhao et al., 2011).

Different types of endophytes

Bacterial Endophytes: Bacteria that inhabits the plants tissues intercellularly/intracellulary are known as bacterial endophytes. For example: *Azorhizobium caulinodans* in rice (Engelhard et al., 2000), *Enterobacter sakazakii* in Soyabean (Kuklinsky-Sobral et al., 2004), *Pseudomonas fluorescens* and *Pseudomonas putida* in carrot (Surette et al., 2003), *Bacillus* spp. in citrus plants (Araújo et al., 2001, 2002) and *Streptomyces* in Wheat (Coombs and Franco, 2003).

Fungal Endophytes: Fungal microorganisms that inhabits the plants tissues intercellularly/intracellularly are known as fungal endophytes. For example, *Pestalotiopsis versicolor* and *Pestalotiopsis neglecta* are the fungal endophytes live inside *Taxus cuspidate* (Kumaran et al., 2009). Endophytic fungi have been classified into two broad groups based on their phylogeny and life history traits. These include the clavicipitaceous, which infect some grasses confined to cool regions and the non-clavicipitaceous endophytes, which are from asymptomatic tissues of non-vascular plants, ferns and allies, conifers and angiosperms and are limited to the Ascomycota or Basidiomycota group (Jalgaonwala et al., 2011; Bhardwaj and Agrawal, 2014).

Actinomycetes Endophytes: Actinomycetes that inhabits the plants tissues intercellularly/intracellularly are known as Actinomycete endophytes.

Traditionally, these actinomycetes were considered transitional forms between the fungi and bacteria (Barka et al., 2016). Endophytic actinomycetes are known to produce various chemical entities with unique structures of considerable medicinal importance (Gayathri and Muralikrishnan, 2013; Singh and Dubey, 2015).

Mycoplasma endophytes: Mycoplasma species are also reported as plant endophytes. Endophytic mycoplasma species are mostly found in a symbiotic relationship with some red algae, such as *Bryopsis pennata*, *B. hypnoides* and also in Arcobacter (Hollants et al., 2011). However, there importance to mankind is still is not confirmed.

Applications of endophytes

More than 70% of the antimicrobial and 60% of the anticancer drugs in medicinal uses are natural products or their derivatives (McAlpine et al., 2005). Microorganisms are inexpensive and easier form of natural products having various applications in drug discovery, agriculture and industries (Keller et al., 2005; Strobel, 2006; Porras-Alfaro and Bayman, 2011). Since last four decades, 40% of new chemical entities (NCEs) have been reported to be product of microorganisms (Khosla, 1997; Clardy and Walsh, 2004; Sieber and Marahiel, 2005). Bioactive natural compounds produced by endophytes have shown promising potentiality towards safety and human health concerns, although there is still a significant demand of drug industry for synthetic products due to economic and time-consuming reasons (Strobel et al., 2004).

Use of endophytes in agriculture

Endophytes play significant role in agriculture by offering beneficial support to the plants by plant growth promotion. Endophytes are involved in the synthesis of siderophores, in the production of plant hormones, in nitrogen fixation, solubilization of immobilized phosphorus, in nutrient cycling, production of volatile organic compounds, pathogenic resistance and stress tolerance (Firdous et al., 2019). The various mechanisms behind plant growth promotions are listed below.

Phosphate solubilization

Soil contains abundant amount of Phosphorus but it remains unavailable to the plant as it is present in insoluble form (Miller et al., 2010). Endophytes releases organic acids into the soil which convert insoluble phosphates into orthophosphates which is soluble and are easily available for plant uptake (Ramachandran et al., 2007).

Siderophore production

Siderophores are iron chelating agents produced by endophytes which can provide iron to the plants and make iron deficient to pathogens. There are two types of siderophores produced by endophytes – Hydroxamate-type and catecholate-type (Sharma and Johri, 2003).

Phytohormone production

Indole acetic acid, gibberellic acid (Khan et al., 2014), ethylene (Kang et al., 2012) and auxins (Dutta et al., 2014) are some of the important phytohormones produced by the endophytes which helps in the growth promotion of non-leguminous plants. For the proper growth of plants, all these phytohormones should be produced by plants and endophytes in optimal concentrations.

Nitrogen fixation

Plants cannot utilize atmospheric nitrogen as 80% of it is in dinitrogen form. Some of the endophytes can fix this nitrogen with the help of nitrogenase enzyme to reduce it to ammonium and nitrate which can easily be taken up by plants for their growth (Steenhoudt and Vanderleyden, 2000).

Bioremediation

Plants produces some toxic pollutants or metabolites which are non-degradable. Some engineered endophytes are now being used for degrading toxic pollutants, reduce heavy metal stress to plants, remove greenhouse gases from air and control pest growth on plants and thus helps in bioremediation process (Yadav and Yadav, 2017).

Phytoremediation

Heavy metal resistant endophytes enhances the plant growth by reducing metal phytotoxicity, affecting metal translocation and accumulation in plants by producing various enzymes which can degrade organic contaminants and reduce both the phytotoxicity and evapotranspiration of volatile contaminants (Li et al., 2012). Phytoremediation is useful for various pollutants like xenobiotics, petroleum, BTEX (benzene, toluene, ethyl benzene, xylene) and even heavy metals (Khan and Doty, 2011).

Biocontrol

Microorganisms are the important cause of plant diseases which affect agricultural practices and result in significant crop losses. Fungal pathogens are the major cause of plant diseases and infect most plants (De Silva et al., 2019). Agrochemicals play a significant role in plant disease management to ensure a sustainable and productive agricultural system. But, the extensive use of this chemicals will result in various biotic and abiotic stresses pertaining to the rise of pesticide resistant phytopathogens and decreased soil fertility. A sustainable agriculture is achieved through reduction or elimination of fertilizers and agrochemicals, resulting in minimal impact to the environment (Tewari et al., 2006). Recently, the use of antagonistic endophytes as biocontrol agents is drawing special attention as an attractive option for management of some plant diseases, resulting in minimal impact to the environment.

Benefits of endophytes in biopharmaceuticals

Researchers are gaining interest on endophytes producing novel structures for exploitation in medicine (Wang et al., 2000; Ezra et al., 2004; Gunatilaka, 2006). Many antibiotics, anti-cancer agents, biological control agents and other bioactive molecules having different functional roles are known to be produced by endophytes (Joseph and Priya, 2011). But still further studies are needed to explore endophytic bioactive compounds for therapeutic properties. The role of endophytes in biopharmaceuticals are described below.

Antimicrobial effect

Antimicrobial compounds of different structural classes like alkaloids, peptides, steroids, terpenoids, phenols, quinines and flavonoids produced by endophytes have been studied so far (Yu et al., 2010). Drug resistance is becoming an emerging public health concern worldwide and is creating havoc in plants, human and animals. Discovery of novel antimicrobial metabolites from endophytes can be a possible alternative to overcome against the resistant human and plant pathogens. Several other bioactive compounds, like sordaricin with antifungal activity (Pongcharoen et al., 2008), griseofulvin, a spirobenzofuran antifungal antibiotic agent, mellisol and 1,8- dihydroxynaphthol 1-O-aglucopyranoside with activity against herpes simplex virus type 1 (Pittayakhajonwut et al., 2005), Chaetomugilin A and D with antifungal activities (Qin et al., 2009) and Hypericin, a naphthodianthrone derivative and Emodin with antimicrobial activity (Joseph and Priya, 2011) etc. has been reported earlier which were produced by endophytes.

Anticancer agents

After the first report of Paclitaxel production by an endophytic fungus, there has been a significant increase in the anticancer compounds isolated from endophytes. Some of the important anticancer bioactive compounds produced by endophytes are-The taxol derived from an endophytic fungus, *Pestalotiopsis pauciseta* was isolated from a medicinal plant *Tabebuia pentaphylla* (Vennila et al., 2010), Beauvercin, isolated from *Fusarium oxysporum* (Wang et al., 2011), 5-methyl-8-(3-methylbut-2-enyl) furanocoumarin isolated from the mangrove endophytic fungus, *Penicillium* sp., three tremulane sesquiterpenes ceriponols F, G and K were isolated from *Ceriporia lacerate*, a fungal endophyte residing in the host plant *Huperzia serrata* (Ying et al., 2013), Two diterpenes, sphaeropsidins A and D, were isolated from an endophytic fungal strain, *Smardaea* sp., residing in *Ceratodon purpureus* (Wang et al., 2011), Alterporriol K and L, produced by endophytic fungus *Alternaria* sp. which was isolated from the *Aegiceras corniculatum* (Huang et al., 2011), Penicisteroids A and B, two new polyoxygenated steroids, obtained from *Penicillium chrysogenum*, an endophytic fungus isolated from Laurencia sp. (Gao et al., 2011). Furthermore, studies are being carried out by several researchers to discover more effective anticancer and antitumor drugs.

Antioxidant agents

Besides possessing a variety of potential, these endophytic fungi have become an alternative source for antioxidant moieties. Although synthetic antioxidants like Butylated hydroxyanisole (BHA), Butylated hydroxytoluene (BHT) and propyl gallate are being regularly used in the food and cosmetic industry but it has different side-effects like carcinogenicity, the concern of public safety and economic values. Therefore, compounds isolated from natural sources are gaining interest and endophytic fungi are one of most important.

Ethnoveterinary plants used in different parts of India and the fungal endophytes associated with them

A lot of work has been done to evaluate and document this hidden knowledge in different parts of India and abroad, but still the information is rare and unacquainted to others. Although a lot of work had been continuously taken place in this field from the different states of India but the present chapter emphasizes on the ethnoveterinary plants from the few states of Northern, Eastern, Western, Southern and Central part of the India depending on the importance of ethnoveterinary practices in those states. Also, the Chapter comprises the details of endophytic fungi which has been isolated from this plants along with their bioactivity.

Ethnoveterinary plants used in North India

This section comprises of different ethnoveterinary plant of Northern states of India which includes, Haryana, Uttarakhand, Jharkhand, Uttar Pradesh and Punjab and their medicinal uses in livestock sector. The fungal endophytes isolated from this ethnoveterinary plants along with their bioactivity are also shown in Table 15.1.

TABLE 15.1 Endophytic fungi and their bioactivity associated with the ethnoveterinary plants of Northern India.

Sl. no	Ethnoveterinary plant	Medicinal uses	State	Endophytes isolated with their bioactivity
1.	*Aanthus illicifolious*	Applied on the fracture area	Uttarakhand	Nodulisporium sp. and Xylaria sp. – Antibacterial activity (Chi et al., 2019)
2.	*Acacia nilotica* L.	Diarrhea, to kill the stomach worms and facilitating smooth delivery in cattle	Haryana	1. *Aspergillus awamori* – antidiabetic (Singh and Kaur, 2016) 2. *Aspergillus flavus* – antifungal activity (Tribhuvan and Tawre, 2019) 3. *Eupenicillium* sp. – Antimicrobial activity (Meenabiga and Rajagopal, 2016)
3.	*Acacia planiforns*	To treat flatulence and diarrhea	Haryana	No reports available
4.	*Aegle marmelos* L.	Dysentery and diarrhea	Haryana	1. *Aspergillus flavus* – antimicrobial activity, antioxidant (Patil et al., 2015) 2. *Bartaliniarobillardoides tassi* – anticancer agent (Gangadevi and Muthumary, 2008)
5.	*Allium sativum* L.	To treat cold and fever in cattle	Haryana	*Trichoderma brevicompactum* – Antifungal (Shentu et al., 2014)
6.	*Andrographis paniculata*	To treat worm infestation	Jharkhand	*Aspergillus sclerotiorum* – Antibacterial and Guignardia *sp.*- Antioxidant (Shang, 2018)
7.	*Argemone mexicana* L.	Constipation in camels, removal of retained placenta in cows and buffaloes	Haryana	No detailed study available
8.	*Barberis aritata*	To cure eye ailments and poisonous grasses effect	Uttarakhand	*Alternaria macrospora* – Antibacterial activity; *Alternaria alternata* and *Aspergillus flavus* – Antioxidant activity (Sharma et al., 2018)

9.	*Bauhina variegata* L.	To treat eye ailments	Uttar Pradesh	No detailed study available
10.	*Butea monosperma* L.	Deworming	Uttar Pradesh	*Fusarium* sp., *Colletotrichum* sp. *and Sclerotium sp.* – Antifungal and *Aspergillus fumigatus* – antibacterial activity (Shantaraja et al., 2014)
11.	*Bryophyllum daigremontianum*	Urinary problem in cattle	Haryana	No detailed study available
12.	*Calligonum comosum* L.	To treat stomach ache in cows and buffaloes	Haryana	No reports available
13.	*Carissa carandus* L.	Treat maggot wounds	Uttar Pradesh	*Nigrospora guilinensis* – Antifungal (Supaphon and Preedanon, 2019)
14.	*Cereus pterogonus* L.	Muscular pain	Jharkhand	No reports available
15.	*Citrullus colocynthis*	Tympany	Punjab	No detailed study available
16.	*Croton oblongifolia* Roxb.	Health supplement	Jharkhand	No reports available
17.	*Cyperus rotundus*	Remove pain and restlessness	Jharkhand	*Rhizoctonia solani* – Antibacterial activity (Ratnaveera et al., 2015)
18.	*Dalbergia sissoo* R.	To cure blisters and leg sore	Uttar Pradesh	No reports available.
19.	*Elettaria cardamomum* L.	To treat cold in horse	Haryana	No reports available.
20.	*Eruca vesicaria* L.	To treat mastitis in buffaloes and cows	Haryana	No reports available.

(*Continued*)

TABLE 15.1 (Continued)

Sl. no	Ethnoveterinary plant	Medicinal uses	State	Endophytes isolated with their bioactivity
21.	*Euphorbia fusiformis*	For poor lactation	Jharkhand	No reports available
22.	*Fumaria indica*	Constipation	Punjab	No reports available
23.	*Gymnema sylvestre* R.	To cure cataract	Uttar Pradesh	*Cladosporium* sp., *Mycelia sterilia* and *Acremonium* sp. – Antibacterial activity (Darbha and Tikole, 2013)
24.	*Helianthus annuus* L.	To pregnant cattle for smooth delivery	Haryana	*Nigrospora sphaerica* – antifungal activity, antioxidant activity, and cytotoxicity (Supaphon et al., 2019)
25.	*Helicteres isora*	Cure stomach ache and body ache	Jharkhand	No detailed study done on bioactivity.
26.	*Hyoscyamus niger* L.	Remove maggots from the wound	Uttar Pradesh	No reports available
27.	*Perilla frutescens*	Cure of diarrhea	Uttarakhand	No detailed study done on bioactivity.
28.	*Plumbago zeylanica* L.	To improve appetite	Haryana	No detailed study done on bioactivity.
29.	*Premna serratifolia* L	Applied on the wound to kill the germs	Haryana	*Hypocrea virens*-antimicrobial (Sahu et al., 2016)
30.	*Ricinus communis* L.	General gastric problems (to clean the stomach)	Haryana	*Phoma hedericola* (Sandhu et al., 2016)
31.	*Salmelia malabarica*	For easy expulsion of placenta	Uttarakhand	No reports available

32.	*Salvadora persica* L.	Increase the milk production	Haryana	1. *Alternaria* sp. – Antibacterial (Elgoraban et al., 2019) 2. *Penicillium* sp. – Antibacterial and antifungal (Korejo et al., 2014)
33.	*Sterculia urens* Roxb.	Stomach ache	Jharkhand	No reports available
34.	*Terminalia chebula* Retz.	For better digestion and stomach ache	Haryana	1. *Penicillium thiomii* – Anti colon cancer activity (Shoeb et al., 2013) 2. *Alternaria alternata* – Anticandidal activity (Phaopongthai et al., 2013)
35.	*Urtica dioica*	To enhance milk production, to treat urinary disorders	Uttarakhand	No detailed study done
36.	*Verginia ciliate*	To treat urinary disorders	Uttarakhand	No reports available
37.	*Withania somnifera* L.	To treat cold and cough	Haryana	*Taleromyces pinophilus*- Production of Withanolides, used to treat cardiovascular, Alzheimer's disease, etc. (Sathiyabama and Parthasarathy, 2019)
38.	*Ziziphus nummularia*	To remove intestinal worms, to treat cold and cough and for treating diarrhea	Haryana	No detailed study on bioactivity of endophytes
39.	*Ziziphus jujuba*	Prolapse and external wound	Punjab	*Alternaria* sp. was isolated but no reports on bioactivity (Orfali et al., 2017)

Ethnoveterinary plants used in Western India

This section comprises of different ethnoveterinary plant of Western states of India which includes, Rajasthan, Gujarat, Maharashtra and their medicinal uses in livestock sector. The fungal endophytes isolated from this ethnoveterinary plants along with their bioactivity are also shown in Table 15.2.

Ethnoveterinary plants used in Southern India

This section comprises of different ethnoveterinary plant of Southern states of India which includes, Andhra Pradesh, Tamil Nadu, Kerala, Karnataka and their medicinal uses in livestock sector. The fungal endophytes isolated from this ethnoveterinary plants along with their bioactivity are also shown in Table 15.3.

Ethnoveterinary plants used in Eastern India

This section comprises of different ethnoveterinary plant of Eastern states of India which includes, North eastern states, West Bengal, Orissa and their medicinal uses in livestock sector. The fungal endophytes isolated from this ethnoveterinary plants along with their bioactivity are also shown in Table 15.4.

Ethnoveterinary plants used in Central India

This section comprises of different ethnoveterinary plant of Central India which includes, Madhya Pradesh, Chhattisgarh and their medicinal uses in livestock sector. The fungal endophytes isolated from this ethnoveterinary plants along with their bioactivity are also shown in Table 15.5.

Conclusion

Our Chapter demonstrate the importance of traditional knowledge on medicinal plants and their uses in the treatment of livestock among the people of various states of India. Preservation of fungal endophytes from the ethnoveterinary plants will be more beneficial as the plants may get extinct due to the rapid deforestation, so the fungal endophytes may be more convenient source of drugs as they can be preserved for longer duration of time. As per Tables 15.1–15.5, it can be observed that huge number of ethnoveterinary plants are being used by the people of different regions of India, but only few studies have been conducted on the bioactive potential of fungal isolates derived from these plants. Thus, an urgent need is there to explore the ethnoveterinary plants from the different region of India for the isolation of more potential fungal endophytes in the field of biopharmaceuticals.

TABLE 15.2 **Endophytic fungi and their bioactivity associated with the ethnoveterinary plants of Western India.**

Sl. no.	Ethnoveterinary plant	Medicinal uses	State	Endophytes isolated
1.	*Abrus precatorius* L.	Constipation	Rajasthan	No detailed study on bioactivity
2.	*Agave americana* L.	Broken horns	Rajasthan	No reports available
3.	*Aloe vera* L.	Burn	Rajasthan	*Gliocladium roseum* and *Nigrospora* sp. – Antibacterial activity (Sahani et al., 2017)
4.	*Balanites aegytiaca*	Trypanosomiasis	Maharashtra	No reports available
5.	*Blumia wightiana*	Used to kill lice stuck on animals	Gujarat	No reports available
6.	*Bombax ceiba* L.	Bone fracture	Rajasthan	No reports available
7.	*Butea monosperma*	Fracture, Anthrax	Maharashtra	No detailed study on Bioactivity
8.	*Capparis sepiaria*	Topical healing of wounds	Gujarat	No detailed study on Bioactivity
9.	*Cassia fistula* L.	Constipation	Rajasthan	*Thelioviopsis periconia*, *Papulospora* sp. and Mycelia sterilia – Antimicrobial activity (Tapwal et al., 2015)
10.	*Celosia argentea* L.	Food poisoning	Rajasthan	No reports available
11.	*Corallocarpus epigaeus*	Tonsillitis	Rajasthan	No reports available
12.	*Curculigo orchioides* Gaertn.	Impaction	Rajasthan	No detailed study on Bioactivity
13.	*Daemia extensa*	Ulcers and warts	Gujarat	No reports available
14.	*Dalbergia sisso* Roxb.	Diarrhea	Rajasthan	No detailed study on Bioactivity

(*Continued*)

TABLE 15.2 (Continued)

Sl. no.	Ethnoveterinary plant	Medicinal uses	State	Endophytes isolated
15.	*Gardenia gummifera*	For traumatic wounds and ulcers	Gujarat	No detailed study on Bioactivity
16.	*Grewia damine* Gaertn.	Bone fracture	Rajasthan	No reports available
17.	*Hiptage madablota*	Maggots	Gujarat	No reports available
18.	*Hydrophylax mari-time*	Used as disinfectant in ulcers, wounds and also cures itching	Gujarat	No reports available
19.	*Indigofera cordifolia*	Used for topical application in wounds and ulcers	Gujarat	No reports available
20.	*Ipomea hederacea* J	Maggoted wound	Maharashtra	No reports available
21.	*Melia azedarach* L.	Removal of intestinal parasites	Rajasthan	No detailed study on Bioactivity
22.	*Momordica dioica* R.	Black quarter	Maharashtra	No detailed study on Bioactivity
23.	*Odina wodier*	Used as disinfectant for washing wounds and ulcers	Gujarat	No reports available
24.	*Plumbago zeylanica* L.	Flatulence	Rajasthan	No detailed study on Bioactivity
25.	*Pongamia pinnata*	Kills insects on wound	Gujarat	No detailed study on Bioactivity
26.	*Premna herbacea*	Used in non healing wound and ulcers	Gujarat	No reports available
27.	*Saccharum officinarum* L.	Infertility and wounds	Rajasthan	Endophytes showed antibacterial activity but the identification is not

(Continued)

TABLE 15.2 (Continued)

Sl. no.	Ethnoveterinary plant	Medicinal uses	State	Endophytes isolated
				done (Ogu and Ehiobu, 2016)
28.	*Sapindus trifolius*	To kill lice stuck on bodies of animals	Gujarat	No reports available
29.	*Securinego virosa*	Fracture, Dysentery	Maharashtra	No reports available
30.	*Soymida febrifuga* R.	Diarrhea	Rajasthan	No reports available
31.	*Trianthema portulacastrum* L.	Impaction and Foot and Mouth Disease	Rajasthan	No reports available
32.	*Truimfetta rhomboida*	For non healing traumatic wound and ulcers	Gujarat	No reports available
33.	*Vahlia voscosa*	Removes its maggots and heals the wound	Gujarat	No reports available
34.	*Ventilago denticulate* G	Enteritis	Maharashtra	No reports available
35.	*Vernornia anthelmin-tica*	Anthelmintic in horses	Gujarat	No reports available
36.	*Ziziphus nummularia*	Retention of placenta	Rajasthan	*Aspergillus calidoustus* -Lipase production (Fareed et al., 2017)

TABLE 15.3 Endophytic fungi and their bioactivity associated with the ethnoveterinary plants of Southern India.

Sl. no.	Ethnoveterinary plant	Medicinal uses	State	Endophytes isolated
1.	*Abrus precatorius*	Dysentry	Tamil Nadu	No detailed study on bioactivity
2.	*Acorus calamus* L.		Tamil Nadu	*Penicillium citrinum* – Plant growth

(*Continued*)

TABLE 15.3 (Continued)

Sl. no.	Ethnoveterinary plant	Medicinal uses	State	Endophytes isolated
		To remove internal parasites		promoting (Mani and Audipudi, 2016)
3.	*Alternanthera sessilis*	Mastitis	Tamil Nadu	*Fusarium oxysporum* – antimicrobial activity (Srinivas et al., 2015)
4.	*Strobilanthes foliosus* (wight.) Anders	Increase cattle lactation	Tamil Nadu	No reports available
5.	*Andrographis paniculata*	Foot and Mouth Diseases	Andhra Pradesh	*Aspergillus tubingensi* – Antioxidant and Hepatoprotective activity (Puri et al., 2018)
6.	*Artemisia absinthum*	Kill worms	Kerala	No reports available
7.	*Balanites aegyptiaca* L.	Eye infection	Andhra Pradesh	No reports available
8.	*Boerhavia diffusa* L.	Blood dysentery and dropsy	Karnataka	No detailed study on bioactivity
9.	*Ceiba pentandra* (L.)	Trypanosomiasis	Andhra Pradesh	No detailed study on bioactivity
10.	*Cinchona officinalis*	Dysentry	Kerala	No reports available
11.	*Coccinia grandis*	Mastitis	Tamil Nadu	*Nigrospora oryzae* – Antifungal activity (Thanabalasingam et al., 2015)
12.	*Cryptolepis dubia*	Snakebite in cattle	Karnataka	No reports available
13.	*Cyperus malaccensis*	increase milk secretion and cures scorpion bites	Kerala	No detailed study on bioactivity
14.	*Dendrocalamus strictus*	Frequent coughing	Karnataka	No reports available
15.	*Dodonaea viscose* L.	Bone fracture	Tamil Nadu	No reports available

(*Continued*)

TABLE 15.3 (Continued)

Sl. no.	Ethnoveterinary plant	Medicinal uses	State	Endophytes isolated
16.	*Elephantopus scaber* Linn.	Maggot wounds, Diarrhea	Andhra Pradesh	No reports available
17.	*Euphorbia rothianaspren*	Insect repellent	Tamil Nadu	No reports available
18.	*Gloriosa superba*	Treatment of skin diseases	Kerala	*Aspergillus* sp. – Antimicrobial and cytotoxic activity (Budhiraja et al., 2015)
19.	*Justicia adhatoda*	To cure dysentry	Karnataka	*Alternaria alternata*- Antimicrobial activity (Barman et al., 2016)
20.	*Kalanchoe pinnata*	Insect bite	Andhra Pradesh	No detailed study on bioactivity
21.	*Lobelia leschenaultiana*	Maggoted wound	Tamil Nadu	No reports available
22.	*Mallotus philippensis*	Intestinal worms	Karnataka	*Pestalotiopsis* sp. – Antifungal; *Pestalotiopsis* sp., and *Phomopsis* sp. – antimicrobial (Gangwar et al., 2015)
23.	*Mussaenda frondosa L.*	Treatment of poisonous bites in cattle	Karnataka	No reports available
24.	*Oxalis tuberosa*	To cure worms in wound and also bleeding from wound	Kerala	No reports available
25.	*Parmelia* sp.	Blood clotting	Tamil Nadu	No reports available
26.	*Pavonia zeylanica*	Constipation	Tamil Nadu	No reports available
27.	*Pergularia daemia*	To treat intestinal worms and Ranikhet diseases	Tamil Nadu	No reports available

(*Continued*)

TABLE 15.3 (Continued)

Sl. no.	Ethnoveterinary plant	Medicinal uses	State	Endophytes isolated
28.	*Sida rhombifolia* L.	Throat swelling	Tamil Nadu	No reports available
29.	*Solanum sisymbrifolium* L.	Deworming & diarrhea	Tamil Nadu	No reports available
30.	*Solanum torvum* L.	Used to reduce digestive problems	Kerala	No detailed study on bioactivity
31.	*Streblus asper* L.	To treat mouth disease	Karnataka	No reports available
32.	*Syzygium cumini* L.	Diarrhea	Karnataka	Endophytes showed antimicrobial activity but the endophytes were not identified. (Jain and Sharma, 2015)
33.	*Tephrosia purpurea*	Anorexia	Tamil Nadu	*Penicillium griseofulvum*- Antifungal activity (Luo et al., 2015)
34.	*Trianthema portulacastrum* L.	Eye problem and sore	Andhra Pradesh	No reports available
35.	*Tylophora indica*	Dysentry	Karnataka	*Dothideomycete* sp. and *Thielavia subthermophila* showed high activity against (Kumar et al., 2011)
36.	*Vitex negundo* L.	Used as an antibacterial and snake bite	Tamil Nadu	*Phomopsis* sp.- Antimicrobial activity (Desale and Bodhankar, 2013)
37.	*Xanthium indicum*	Maggots	Andhra Pradesh	No reports available
38.	*Ziziphus mauritiana*	To heal scorpion bite	Kerala	*Fusarium* sp.- Biocontrol (Sheeba et al., 2019)

TABLE 15.4 Endophytic fungi and their bioactivity associated with the ethnoveterinary plants of Eastern part of India.

Sl. no.	Ethnoveterinary plant	Medicinal uses	State	Endophytes isolated
1.	*Acacia nilotica* L.	Colic pain	Orissa	No reports available
2.	*Ampelocissus tomentosa* R	Pain	West Bengal	No reports available
3.	*Arundinaria maling*	To treat diarrhea and dysentery	Sikkim	No reports available
4.	*Bergenia ciliata* Stern.	To treat diarrhea and dysentery	Sikkim	No detailed study on bioactivity.
5.	*Calotropis gigantea* L.	Body swelling of cattle	West Bengal	*Phomopsis asparagi* – Anticandidal activity (Nath and Joshi, 2017)
6.	*Cassia fistula* L.	Throat swelling	Orissa	*Thelioviopsis, Periconia, Papulospora* sp. – Antimicrobial (Tapwal et al., 2015)
7.	*Cissampelos pareira* L.	Antidote on insect bite and scorpion sting	Sikkim	No reports available
8.	*Clerodendrum viscosum* V.	Constipation and Digestion promoter	Assam	No detailed study on bioactivity
9.	*Corchorus capsularis* L.	Dysentry	Assam	*Aspergillus flavus-* –Antimicrobial activity (Chowdhury et al., 2018)
10.	*Eclipta alba* L.	Cure swelling in the ear of cattle	West Bengal	No detailed study on bioactivity.
11.	*Elatostema ficoides*	Fever	Sikkim	No reports available
12.	*Euphorbia antiquorum*	Heal yoke wounds	Orissa	No reports available
13.	*Grewia* multiflora J.	Bone fracture	Assam	No reports available
14.	*Holarrhena antidysenterica* L.	Dysentery	Sikkim	No detailed study on bioactivity

(Continued)

TABLE 15.4 (Continued)

Sl. no.	Ethnoveterinary plant	Medicinal uses	State	Endophytes isolated
15.	*Impatiens balsamina* L.	Stomach ache	Assam	*Gibberella moniliformis* was isolated but no report on their Bioactivity (Sarang et al., 2017)
16.	*Justicia gendarussa*	Dysentry	Assam	*Colletotrichum gloeosporioides* – Antimicrobial (Gangadevi and Johnpaul, 2008)
17.	*Lawsonia inermis*	Foot and Mouth Disease		*Colletotrichum gloeosporioides* – Antimicrobial (Lakshmi and Selvi, 2013)
18.	*Litsea monopetala* R.	Bone fracture	West Bengal	No reports available
19.	*Lycopodium clavatum* L.	For treatment of muscle contraction in cattle.	Sikkim	No detailed study on bioactivity
20.	*Manihot esculenta*	Stomach ache	Assam	No detailed study on bioactivity.
21.	*Plumbago zeylanica* L.	Diarrhea	Orissa	Reports of isolation of *Aspergillus, Curvularia, Fusarium, Penicillium, Pythium, Rhizopus, Sporotrichum* and *Trichoderma* are there but no reports on bioactivity is available (Sagar et al., 2017).
22.	*Rauwolfia serpentina L.*	Snake bite	Sikkim	*Penicillium* sp. – hypocholesterolemic activity and *Aspergillus awamori*- antioxidant (Nath et al., 2015).
23.	*Rhododendron campanulatum* D	Chronic fever	Sikkim	*Penicillium* sp. and *Trichoderma* sp. – Antibacterial and antifungal (Baral et al., 2011)
24.	*Schrebera swientenioides R.*	To kill wound worms	West Bengal	No reports available

(*Continued*)

TABLE 15.4 (Continued)

Sl. no.	Ethnoveterinary plant	Medicinal uses	State	Endophytes isolated
25.	*Streblus asper* L.	Stomach ache	Assam	No reports available
26.	*Swertia chirata*	Fever	Sikkim	No reports available
27.	*Typhonium trilobatum* Schott.	Maggots on wounds	Assam	No reports available
28.	*Xanthium indicum*	Chronic wounds	Orissa	No reports available
29.	*Zanthoxyllum alatum* R.	Appetizer	Assam	No reports available
30.	*Zingiber officinale* R.	To treat cough	Sikkim	*Paraconiothyrium* sp.- Antimicrobial (Anisha et al., 2018)
31.	*Zehneria umbellata* T.	To cure cuts and bruises of cattle	West Bengal	No reports available

TABLE 15.5 Endophytic fungi and their bioactivity associated with the ethnoveterinary plants of Western India.

Sl. no.	Ethnoveterinary plant	Medicinal uses	State	Endophytes isolated
1.	*Abutilon indicum* L.	Dysentry	Chattisgarh	Endophytes have been isolated but no reports on bioactivity
2.	*Achyranthes aspera*	To get relief from watering in eyes	Chattisgarh	*Aspergillus terreus* – Antibacterial, Antifungal and anti-oxidant activity (Goutam et al., 2016)
3.	*Adhatoda vasica*	Cough	Madhya Pradesh	No detailed study on bioactivity
4.	*Andrographis paniculata*	Fever	Madhya Pradesh	*Aspergillus tubingensi* – Antioxidant and Hepatoprotective activity (Puri et al., 2018)
5.	*Aristolochia indica*	To cure insect bite	Chattisgarh	No detailed study on bioactivity

(*Continued*)

TABLE 15.5 (Continued)

Sl. no.	Ethnoveterinary plant	Medicinal uses	State	Endophytes isolated
6.	*Bombax ceiba*	Bone fracture	Madhya Pradesh	No detailed study on bioactivity
7.	*Cissus quardangularis*	Bone fracture	Madhya Pradesh	*Asclepias curassavica* and *Alternaria alternata*-cytotoxic activity (Tenguria and Firodiya, 2016)
8.	*Elephantopus scaber* L.	Dysentery	Chattisgarh	No reports available
9.	*Helicteres isora* L	Diarrhea	Madhya Pradesh	No detailed study on bioactivity
10.	*Holarrhena antidysenterica*	Dysentry	Chattisgarh	Reports on isolation of *Penicillium citrinum*, *Aspergillus niger*, *Curvularia vermiformis*, *Rhizopus* sp., *Cladosporium* sp., *Alternaria alternata* and *Fusarium* sp. is there but no reports on their bioactivity.
11.	*Nyctanthes arbor-tristis*	Fever	Madhya Pradesh	*Aspergillus nidulans* – Antiproliferative activity (Sana et al., 2019)
12.	*Porana paniculata*	Bone fracture	Chattisgarh	No reports available
13.	*Scindapsus officinalis*	Bone fracture	Chattisgarh	No reports available
14.	*Semecarpus anacardium*	Infertility	Madhya Pradesh	*Chaetomium arcuatum*-acetylcholinesterase inhibitory and anti-cancer activities (Bhukya and Tangutur, 2016)
15.	*Sida acuta*	Wounds	Madhya Pradesh	*Aspergillus sulphureus* – Antibacterial and antioxidant (Murali et al., 2017)
16.	*Solanum nigrum* L	Fever	Madhya Pradesh	Endophyte was isolated which produced a cytotoxic compound, Solamargine (El-Hawary et al., 2016).

(*Continued*)

TABLE 15.5 (Continued)

Sl. no.	Ethnoveterinary plant	Medicinal uses	State	Endophytes isolated
17.	*Syzygium cumini*	Dysentry	Chattisgarh	Endophytes have been isolated but no reports on bioactivity
18.	*Terminalia bellirica*	Wounds	Madhya Pradesh	No reports available
19.	*Tridax procumbens* L.	Cut wounds	Chattisgarh	*Aspergillus flavus* – antimicrobial (Deepake et al., 2012)
20.	*Vitex. negundo* L.	Dysentry	Chattisgarh	*Phomopsis* sp. – Antimicrobial (Desale and Bodhankar, 2013)
21.	*Woodfordia floribunda*	Bone fracture	Chattisgarh	No reports available
22.	*Ziziphus jujube*	Cough, fever	Madhya Pradesh	No detailed study on bioactivity

References

Anisha, C., Sachidanandan, P., Radhakrishnan, E.K., 2018. Endophytic Paraconiothyrium *sp.* from *Zingiber officinale* Rosc. displays broad-spectrum antimicrobial activity by production of danthron. Curr. Microbiol. 75 (3), 343–352.

Araújo, W.L., Maccheroni Jr., W., Aguilar-Vildoso, C.I., Barroso, P.A.V., Saridakis, H.O., Azevedo, J.L., 2001. Variability and interactions between endophytic bacteria and fungi isolated from leaf tissues of citrus rootstocks. Can. J. Microbiol. 47, 229–236.

Araújo, W.L., Marcon, J., Maccheroni Jr., W., Van Elsas, J.D., Azevedo, J.L., 2002. Diversity of endophytic bacterial populations and their interaction with *Xylella fastidiosa* in citrus plants. Appl. Environ. Microbiol. 68, 4906–4914.

Bailey, B.A., Bae, H., Strem, M.D., Roberts, D.P., Thomas, S.E., Crozier, J., et al., 2006. Fungal and plant gene expression during the colonization of cacao seedlings by endophytic isolates of four *Trichoderma species*. Planta 224, 1449–1464.

Baral, B., Rana, P., Maharjan, B.J., 2011. Antimicrobial potentials of endophytic fungi inhabiting *Rhododendron anthopogon* D. Ecoprint 18, 39–44.

Barka, E.A., Vatsa, P., Sanchez, L., Gaveau-Vaillant, N., Jacquard, C., Meier-Kolthoff, J.P., et al., 2016. Taxonomy, physiology, and natural products of actinobacteria. Microbiol. Mol. Biol. Rev. 80, 1–43.

Barman, S., Ghosh, R., Mandal, N., Gond, S.K., Mandal, N.C., 2016. Assessment of Antimicrobial Activity of Endophytic Fungi Isolated from Justicia adhatoda L. UGB J. Plant. Biol. Biotech 1 (1), 19–23.

Bhardwaj, A., Agrawal, P., 2014. A review fungal endophytes: as a store house of bioactive compound. World J. Pharm. Pharm. Sci. 3, 228–237.

Bhukya, B., Tangutur, A.D. (Eds.), 2016. Microbial Biotechnology: Technological Challenges and Developmental Trends. Apple Academy Press, Waretown.

Budhiraja, A., Nepali, K., Kaul, S., Dhar, K.L., 2012. Antimicrobial and cytotoxic activities of fungal isolates of medicinal plant Gloriosa superba. Int. J. Recent Adv. Pharm. Res. 2 (1), 37–45.

Dutta, D., Puzari, K.C., Gogoi, R., Dutta, P., 2014. Endophytes: exploitation as a tool in plant protection. Braz. Arch. Biol. Technol. 57 (5), 621–629.

Carroll, G.C., Petrini, O., 1983. Patterns of substrate utilization by some fungal endophytes from coniferous foliaee. Mycologia 75, 53–63.

Chi, W.C., Pang, K.L., Chen, W.L., Wang, G.J., Lee, T.H., 2019. Antimicrobial and iNOS inhibitory activities of the endophytic fungi isolated from the mangrove plant *Acanthus ilicifolius* var. xiamenensis. Bot. Stud. 60 (1), 4.

Chowdhury, F.T., Sarker, M., Islam, M.S., Nur, H.P., Islam, M.R., Khan, H., 2018. Investigation of antimicrobial activity and identification of bioactive volatile metabolites of jute endophytic fungus Aspergillus flavus, Biores. Commun 4 (1), 476–482.

Clardy, J., Walsh, C., 2004. Lessons from natural molecules. Nature 432, 829–837.

Coombs, J.T., Franco, C.M.M., 2003. Isolation and identification of actinobacteria from surface sterilized wheat roots. Appl. Environ. Microbiol. 69, 5603–5608.

Dai, C.C., Yu, B.Y., Xu, Z.L., Yuan, S., 2003. Effect of environmental factors on growth and fatty acid composition of five endophytic fungi from *Sapium sebiferum*. J. Appl. Ecol. 14, 1525–1528.

Darbha, N.S., Tikole, S., 2013. Studies on endophytic fungi of ayurvedic medicinal plant Gymnema sylvestre. Int. J. Curr. Sci 7, 118–127.

Deepake, U., Komple, T., Garimella, D., Garimella, G., 2012. Preliminary screening of endophytic fungi from *Tridax procumbens* Linn. and *Argemone mexicana* Linn. for their antimicrobial activity. Int. J. Curr. Res. 4 (2), 93–96.

Desale, S., Bodhankar, M.G., 2013. Antimicrobial activity of endophytic fungi isolated from *Vitex negundo* Linn. Int. J. Curr. Microbiol. Appl. Sci. 2 (12), 389–395.

De Silva, D.D., Groenewald, J.Z., Crous, P.W., Ades, P.K., Nasruddin, A., Mongkolporn, O., et al., 2019. Identification, prevalence and pathogenicity of Colletotrichum species causing anthracnose of *Capsicum annuum* in Asia. IMA Fungus 10, 1.

Eisenberg, D.M., Davis, R.B., Ettner, S.L., Appel, S., Wilkey, S., Rompay, M., et al., 1998. Trends in alternative medicine use in the United States, 1990-1997: results of a followup national survey. J. Am. Med. Assoc. 280, 1569–1575.

Elgorban, A.M., Bahkali, A.H., Al Farraj, D.A., Abdel-Wahab, M.A., 2019. Natural products of Alternaria sp., an endophytic fungus isolated from Salvadora persica from Saudi Arabia. Saudi J. Biol. Sci. 26 (5), 1068–1077.

El-Hawary, S.S., Mohammed, R., Abouzid, S.F., Bakeer, W., Ebel, R., Sayed, A.M., et al., 2016. Solamargine production by a fungal endophyte of *Solanum nigrum*. J. Appl. Microbiol. 120, 900–911.

Engebretson, J., 2002. Culture and complementary therapies, Complement. Ther. Nurs. Midwif 8 (4), 177–184.

Engelhard, M., Hurek, T., Reinhold, B., 2000. Preferential occurrence of diazotrophic endophytes, Azoarcus spp., in wild rice species and land races of Oryza sativa in comparison with modern races. Environ. Microbiol. 2 (2), 131–141.

Eshetu, G.R., Dejene, T.A., Telila, L.B., Bekele, D.F., 2015. Ethnoveterinary medicinal plants: preparation and application methods by traditional healers in selected districts of southern Ethiopia. Vet. World 8 (5), 674–684.

Evans, M., Shaw, A., Thompson, E., Falk, S., et al., 2007. Decisions to use complementary and alternative medicine (CAM) by male cancer patients: information-seeking roles and types of evidence used. BMC Complement. Altern Med. 7 (25), 1–13.

Ezra, D., Castillo, U.F., Strobel, G.A., Hess, W.M., Porter, H., et al., 2004. Coronamycins, peptide antibiotics produced by a verticillate Streptomyces *sp*. (MSU-2110) endophytic on *Monstera* sp. Microbiology 150, 785–793.

Faeth, S.H., Fagan, W.F., 2002. Fungal endophytes: common host plant symbionts but uncommon mutualists. Integr. Comp. Biol. 42, 360–368.

Fareed, S., Jadoon, U.N., Ullah, I., Jadoon, M.A., Rehman, M.U., Bibi, Z., Waqas, M., Nisa, S., 2017. Isolation and biological evaluation of endophytic fungus from Ziziphus Nummularia. J. Entomol. Zool. Stud. 5 (3), 32–38.

Firáková, S., Šturdíková, M., Mŭcková, M., 2007. Bioactive secondary metabolites produced by microorganisms associated with plants. Biologia 62, 251–257.

Firdous, J., Lathif, N.Ab., Resni, M., Muhamad, N., 2019. Endophytic bacteria and their potential application in agriculture: a review. Agric. Res. 53 (1), 1–7.

Ganga Mani, P., Audipudi, A.V., 2016. Penicillium citrinum AVGE1 an Endophyte of Acorus calamus its Role in bio Control and PGP in Chilli seedlings. Int. J. Curr. Microbiol. App. Sci 5 (5), 657–667.

Gangadevi, V., Muthumary, J., 2008. Isolation of *Colletotrichum gloeosporioides*, a novel endophytic taxol-producing fungus from the leaves of a medicinal plant, *Justicia gendarussa*. Mycol. Balcan. 5, 1–4.

Gangadevi, V., Muthumary, J., 2018. Isolation of Colletotrichum gloeosporioides, a novel endophytic taxol-producing fungus from the leaves of a medicinal plant, Justicia gendarussa. Mycologia Balc 5, 1–4.

Gangwar, M., Verma, V.C., Gautam, M.K., Nath, G., 2015. Isolation and evaluation of antimicrobial activities of endophytic fungal extract from *Mallotus philippinensis* Muell. Appl. Microbiol. 1, 103.

Gao, S.S., Li, X.M., Du, F.Y., Li, C.S., Proksch, P., Wang, B.G., 2011. Secondary metabolites from a marine-derived endophytic fungus Penicillium chrysogenum QEN-24S. Mar. Drugs. 9, 59–70.

Gayathri, P., Muralikrishnan, V., 2013. Isolation and characterization of endophytic actinomycetes from mangrove plants for antimicrobial activity. Int. J. Curr. Microbiol. Appl. Sci. 2, 78–79.

Goutam, J., Singh, S., Kharwar, R.N., Ramraj, V., 2016. *In-vitro* potential of endophytic fungus *Aspergillus terreus* (JAS-2) associated with *Achyranthus aspera* and study on its culture conditions. Biol. Med. 8, 349.

Gunatilaka, A.A.L., 2006. Natural products from plant-associated microorganisms: distribution, structural diversity, bioactivity and implications of their occurrence. J. Nat. Prod. 69, 509–526.

Hollants, J., Leroux, O., Leliaert, F., Decleyre, H., De Clerck, O., Willems, A., 2011. Who is in there? Exploration of endophytic bacteria within the siphonous green seaweed *Bryopsis* (Bryopsidales, Chlorophyta). PLoS One 6, e26458. Available from: https://doi.org/10.1371/journal.pone.0026458.

Huang, C.H., Pan, J.H., Chen, B., Yu, M., Huang, H.B., Zhu, X., et al., 2011. Three Bianthraquinone derivatives from the Mangrove endophytic fungus Alternaria sp. ZJ9-6B from the South China, Sea. Mar. Drugs. 9, 832–843.

Huang, W.Y., Cai, Y.Z., Hyde, K.D., Corke, H., Sun, M., 2008. Biodiversity of endophytic fungi associated with 29 traditional Chinese medicinal plants. Fungal Divers. 33, 61–75.

Jain, P., Sharma, P., 2015. Antagonistic activity of endophytic fungi isolated from *Syzygium cumini* (L.) Skeels. Int. J. Life Sci. Res. 3 (2), 59–63.

Jalgaonwala, R.E., Mohite, B.V., Mahajan, R.T., 2011. Natural products from plant associated endophytic fungi. J. Microbiol. Biotechnol. Res. 1, 21–32.

Jia, M., Chen, L., Xin, H.L., Zheng, C.J., Rahman, K., Han, T., et al., 2016. A friendly relationship between endophytic fungi and medicinal plants: a systematic review. Front. Microbiol. 7, 906.

Joseph, B., Priya, R.M., 2011. Bioactive compounds from endophytes and their potential in pharmaceutical effect: a review. Am. J. Biochem. Mol. Biol. 1, 291–309.

Kang, N.Y., Cho, C., Kim, N.Y., Kim, J., 2012. Cytokinin receptor-dependent and receptor-independent pathways in the dehydration response of Arabidopsis thaliana. J. Plant Physiol. 169, 1382–1391.

Karimi, E., Jaafar, H.Z., 2011. HPLC and GC–MS determination of bioactive compounds in microwave obtained extracts of three varieties of *Labisia pumila* Benth. Molecules 16 (8), 6791–6805.

Keller, N.P., Turner, G., Bennett, J.W., 2005. Fungal secondary metabolism from biochemistry to genomics. Nat. Rev. Microbiol. 3, 937–947.

Khan, M.I.R., Asgher, M., Khan, N.A., 2014. Alleviation of salt-induced photosynthesis and growth inhibition by salicylic acid involves glycine betaine and ethylene in mungbean (Vigna radiata L.). Plant Physiol. Biochem. 80, 67–74.

Khan, Z., Doty, S.L., 2011. Endophyte-assisted phytoremediation. Curr. Opin. Plant. Biol. 12, 97–105.

Khosla, C., 1997. Harnessing the biosynthetic potential of modular polyketide synthases. Chem. Rev. 97, 2577–2590.

Korejo, F., Ali, A.A., Ali, S.A., Shafique, H.A., Viqar, S., Ara, J., et al., 2014. Antifungal and antibacterial activity of endophytic Penicillium species isolated from Salvadora species. Pak. J. Bot. 46 (6), 2313–2318.

Kuklinsky-Sobral, J., Araújo, W.L., Mendes, R., Gerardi, I.O., Pizzirani-Kleiner, A.P., Azevedo, J.L., 2004. Isolation and characterization of soybean-associated bacteria and their potential for plant growth promotion. Environ. Microbiol. 6, 1244–1251.

Kumar, S., Kaushik, N., Edrada-Ebel, R., Ebel, R., Proksch, P., 2011. Isolation, characterization and bioactivity of endophytic fungi of Tylophora indica. World J. Microbiol. Biotechnol. 27 (3), 571–577.

Kumaran, R.S., Muthumary, J., Hur, B.K., 2009. Isolation and identification of an anticancer drug, taxol from *Phyllosticta tabernaemontanae*, a leaf spot fungus of an angiosperm, *Wrightia tinctoria*. J. Microbiol. 1, 40–49.

Lakshmi, P.J., Selvi, K.V., 2013. Anticancer potentials of secondary metabolites from endophytes of *Barringtonia acutangula* and its molecular characterization. Int. J. Curr. Microbiol. Appl. Sci. 2 (2), 44–45.

Li, J.W.H., Vederas, J.C., 2009. Drug discovery and natural products: End of an era or an endless frontier? Science 325, 161–165.

Li, H.Y., Wei, D.Q., Shen, M., Zhou, Z.P., 2012. Endophytes and their role in phytoremediation. Fungal Divers. 54, 11–18.

Liu, J., Liu, G., 2018. Analysis of secondary metabolites from plant endophytic fungi. Methods Mol. Biol. 1848, 25–38.

Longuefosse, J.L., Nossin, E., 1996. Medical ethno botany survey in Martinique. J. Ethnopharmacol. 53, 117–142.

Luo, Z.P., Lin, H.Y., Ding, W.B., He, H.L., You-Zh, L., 2015. Phylogenetic diversity and antifungal activity of endophytic fungi associated with *Tephrosia purpurea*. Mycobiology 43 (4), 435–443.

Martin, G., 1995. Ethnobontay. A Methods Manual. Chapman & Hall.

McAlpine, J.B., Bachmann, B.O., Piraee, M., Tremblay, S., Alarco, A.M., Zazopoulos, E., et al., 2005. Microbial genomics as a guide to drug discovery and structural elucidation: ECO-02301, a novel antifungal agent, as an example. J. Nat. Prod. 68 (4), 493–496.

McCorkle, C.M., 1995. Back to the future: lessons from ethnoveterinary RD&E for studying and applying local knowledge. Agric. Hum. Values 12 (2), 52–80.

Meenambiga, S.S., Rajagopal, K., 2016. Medium Optimization of the endophytic fungus Eupenicillium Sp. isolated from Acacia nilotica L. for its antimicrobial activity against oral pathogens. Int. J. Pharm. Sci. Rev. Res 41 (2), 6–12.

Miller, S.,H., Browne, P., Prigent-Cambaret, C., Combes-Meynet, E., Morrissey, J.P., O'Gara, F., 2010. Biochemical and genomic comparison of inorganic phosphate solubilisation in *Pseudomonas* species. Environ. Microbiol. Rep. 2, 403–411.

Murali, M., Mahendra, C., Hema, P., Rajashekar, N., Nataraju, A., Sudarshana, M., et al., 2017. Molecular profiling and bioactive potential of an endophytic fungus *Aspergillus sulphureus* isolated from *Sida acuta*: a medicinal plant. Pharm. Biol. 55 (1), 1623–1630.

Nath, A., Chattopadhyay, A., Joshi, S.R., 2015. Biological activity of endophytic fungi of *Rauwolfia serpentina* Benth: an ethnomedicinal plant used in folk medicines in Northeast India. Proc. Natl Acad. Sci. India Sect. B. Biol. Sci. 85, 233–240.

Nath, A., Joshi, S.R., 2017. Anti-candidal effect of endophytic fungi isolated from *Calotropis gigantea*. Rev. Biol. Trop. 65 (4), 1437–1447.

Ogu, G.I., Ehiobu, J.M., Ukpai, E.G., 2016. Efficacy of metabolites from Saccharum officinarum Linn. endophytic fungi against some uropathogenic bacteria. Int. J. Res. Med. Sci. 2 (1), 1–9.

Orfali, R., Ebrahim, W., El-Shafae, A.M., 2017. Secondary Metabolites from Alternaria sp., a Fungal Endophyte Isolated from the Seeds of Ziziphus jujube. Chem. Nat. Comp 53 (6).

Patil, M., Patil, R., Maheswari, V., 2015. Biological activities and identification of bioactive metabolite from endophytic *Aspergillus flavus* L7 Isolated from *Aegle marmelos*. Curr. Microbiol. 71 (1), 39–48.

Padhi, L., Mohanta, Y.K., Panda, S.K., 2013. Endophytic fungi with great promises: a review. J. Adv. Pharm. Educ. Res. 3 (3), 152–170.

Phaopongthai, J., Wiyakrutta, S., Meksuriyen, D., et al., 2013. Azole-synergistic anti-candidal activity of altenusin, a biphenyl metabolite of the endophytic fungus Alternaria alternata isolated from Terminalia chebula Retz. J. Microbiol. 51, 821–828.

Phondani, P.C., Maikhuri, K., Kala, P., 2010. Ethno veterinary uses of medicinal plants among traditional herbal healers in Alaknanda catchment of Uttarakhand, India. Afr. J. Trad. Complement. Med. 7, 195–206.

Pittayakhajonwut, P., Suvannakad, R., Thienhirun, S., Prabpai, S., Kongsaeree, P., Tanticharoen, M., 2005. An antiherpesimplex virus-type 1 agent from *Xylaria mellisii* BCC 1005. Tetrahedron Lett. 46, 1341–1344.

Pongcharoen, W., Rukachaisirikul, V., Phongpaichit, S., Kühn, T., Pelzing, M., Sakayaroj, J., et al., 2008. Metabolites from the endophytic fungus Xylaria *sp.* PSU-D14. Phytochemistry 69, 1900–1902.

Porras-Alfaro, A., Bayman, P., 2011. Hidden fungi, emergent properties: endophytes and microbiomes. Annu. Rev. Phytopathol. 49, 291–315.

Puri, S.K., Habbu, P.V., Kulkarni, P.V., Kulkarni, V.H., 2019. Characterization, in vitro antioxidant and hepatoprotective activity of fungal endophytic extracts of Andrographis Paniculata leaves in ccl4 induced hepatotoxicity. Int. J. Pharm. Pharm. Sci. 11 (1), 44.

Qin, J.C., Zhang, Y.M., Gao, J.M., Bai, M.S., Yang, S.X., Laatschb, H., et al., 2009. Bioactive metabolites produced by *Chaetomium globosum*, an endophytic fungus isolated from *Ginkgo biloba*. Bioorgan. Med. Chem. Lett. 19 (6), 1572–1574.

Ramachandran, S., Singh, S.K., Larroche, C., Soccol, C.R., Pandey, A., 2007. Oil cakes and their biotechnological applications – a review. Bioresour. Technol. 98, 2000–2009.

Ratnaweera, P.B., Williams, D.E., Patrick, B.O., de Silva, E.D., Andersen, R.J., 2015. Solanioic acid, an antibacterial degraded steroid produced in culture by the fungus rhizoctonia solani isolated from tubers of the medicinal plant cyperus rotundus. Org Lett. 17 (9), 2074–2077.

Rodrigues, K.F., 1994. The foliar fungal endophytes of the Amazonian palm *Euterpe oleracea*. Mycologia 86, 376–385.

Rodriguez, R.J., White, J.F.J., Arnold, A.E., Redman, R.S., 2009. Fungal endophytes: diversity and functional roles. N. Phytol.

Sagar, A., Bala, M., Prakash, V., 2017. Studies on endophytes and antibacterial activity of *Plumbago zeylanica* L. Int. J. Multidiscip. Res. Dev. 4 (1), 01–04.

Sahani, K., Thakur, D., Hemalatha, K.P.J., Ganguly, A., 2017. Antiglycemic activity of endophytic fungi from selected medicinal plants by alpha-amylase inhibition method. Int. J. Sci. Res. 6 (3), 6–391.

Sahu, R., Kumar, S., Aharwal, R.P., Sandhu, S.S., 2016. Antibacterial activity of isolated endophytic fungi from Rauvolfia serpentina (l.) benth. ex Kurz. Int. J. Pharm. Pharm. Sci. 8 (11), 38–42.

Sana, T., Siddiqui, B.S., Shahzad, S., Farooq, A.D., Siddiqui, F., Sattar, S., et al., 2019. Antiproliferative activity and characterization of metabolites of *Aspergillus nidulans*: An endophytic fungus from *Nyctanthes arbor-tristis* Linn. against three human cancer cell lines. Med. Chem. 15 (4), 352–359.

Sandhu, S.S., Kumar, S., Aharwal, R.P., Shukla, H., Rajak, R.C., 2014. Endophytic Fungi: As a Source of Antimicrobials Bioactive Compounds. World J. Pharm. Pharma. Sci. 3 (2), 1179–1197.

Santhanakrishnan, R., Hafeel, A., Hariramamurthi, B.A., Unnikrishnan, P.M., 2008. Documentation and participatory rapid assessment of ethnoveterinary practices. Indian J. Trad. Knowl. 7 (2), 360–364.

Sansoucy, R., 1997. Livestock – a driving force for food security and sustainable development. World Anim. Rev. 1995, 5–17.

Sarang, H., Rajani, P., Vasanthakumari, M.M., Kumara, P.M., Siva, R., Ravikanth, G., et al., 2017. An endophytic fungus, Gibberella moniliformis from *Lawsonia inermis* L. produces lawsone, an orange-red pigment. Antonie van Leeuwenhoek . Available from: https://doi.org/10.1007/s10482-017-0858-y.

Sathiyabama, M., Parthasarathy, R., 2018. Withanolide production by fungal endophyte isolated from Withania somnifera. Nat. Prod. Res 32 (13), 1573–1577.

Shang, T.W., 2017. Diversity and Bioactivities of Endophytic Fungi from Medicinal Plant Andrographis paniculata (Hempedu Bumi). Thesis, Monash University, Malaysia.

Shantharaja, S., 2014. Evaluation of endophytic fungi from Butea monosperma for antimicrobial and enzyme activity. J. Med. Plants. Stud. 2 (4), 38–45.

Sharma, A., Johri, B.N., 2003. Combat of iron-deprivation through a plant growth promoting fluorescent Pseudomonas strain GRP3A in mung bean. Microbiol. Res. 158, 77–81.

Sharma, R.K., Jindal, M., Kunwar, R.S., Lata, S., 2018. Blood sugar lowering effect of stem part of berberis aristata as add on therapy in type-2 diabetes mellitus patients. J. Appl. Med. Sci. Available from: https://doi.org/10.21276/sjams.2018.6.5.25.

Sheeba, H., Syed Ali, M., Anuradha, V., 2019. Bioactive compounds and antimicrobial activity of fungal crude extract from medicinal plants. J. Pharm. Sci. Res. 11 (5), 1826–1833.

Shentu, X., Zhan, X., Ma, Z., Yu, X., Zhang, C., 2014. Antifungal activity of metabolites of the endophytic fungus Trichoderma brevicompactum from garlic. Braz. J. Microbiol 45, 248–254.

Sherameti, I., Shahollari, B., Venus, Y., Altschmied, L., Varma, A., Oelmüller, R., 2005. The endophytic fungus *Piriformospora indica* stimulates the expression of nitrate reductase and the starch-degrading enzyme glucan-water dikinase in tobacco and Arabidopsis roots through a homeodomain transcription factor that binds to a conserved motif in their promoters. J. Biol. Chem. 280, 26241–26247.

Shoeb, M., Lin, P.K.T., Nahar, N., 2012. Anti-colon cancer activity of endophytic fungal strains from Terminalia chebula Rezt. Bangladesh. J. Pharmacol. 7 (1), 47–49.

Shrivastava, S., Jain, A.K., Tomar, R.S., 2017. Ethnoveterinary practices – a review on phytotherapeutical approaches in treatment of animals. World J. Pharm. Med. Res. 3, 96–100.

Sieber, S.A., Marahiel, M.A., 2005. Molecular mechanisms underlying nonribosomal peptide synthesis: approaches to new antibiotics. Chem. Rev. 105, 715–738.

Sieber, T.N., 2007. Endophytic fungi in forest trees: are they mutualists? Fungal Biol. Rev. 21, 75–89.

Singh, B., Kaur, A., 2016. Antidiabetic potential of a peptide isolated from an endophytic Aspergillus awamori. J. Appl. Microbiol. 120 (2), 301–311.

Singh, R., Dubey, A.K., 2015. Endophytic actinomycetes as emerging source for therapeutic compounds. Indo Glob. J. Pharm. Sci. 5, 106–116.

Somvanshi, R., 2006. Veterinary medicine and animal keeping in ancient India. Asian Agri-History 10 (2), 133–146.

Sri Balaji, N., Chakravarthi, R., 2010. Ethnoveterinary practices in India: a review. Vet. World 3 (12), 549–551.

Srinivas R.P., Nigam A., Aruna J., Alam M.A., Ishara L., Chamith Y.H. and Chikkaswamy B. K., Antimicrobial activity in cultures of endophytic fungi isolated in some medicinal plant species, Int. J. Adv. Res. 4(2), 1-24.

Steenhoudt, O., Vanderleyden, J., 2000. Azospirillum, a free living nitrogen-fixing bacterium closely associated with grasses: genetic, biochemical and ecological aspects. Fems Microbiol. Rev. 24, 487–506.

Stierle, A., Strobel, G., Stierle, D., 1993. Taxol and taxane production by *Taxomyces andreanae*, an endophytic fungus of Pacific yew. Science 260, 214–216.

Strobel, G., Daisy, B., Castillo, U., Harper, J., 2004. Natural products from endophytic microorganisms. J. Nat. Prod. 67 (2), 257–268.

Strobel, G., 2006. Harnessing endophytes for industrial microbiology. Curr. Opin. Microbiol. 9, 240–244.

Supaphon, P., Preedanon, S., 2019. Antimicrobial and antioxidant activities of endophytic fungi extracts isolated from Carissa carandas. Afr. J. Microbiol. Res. 13 (27), 464–473.

Supaphon, P., Preedanon, S., 2019. Evaluation of in vitro alpha-glucosidase inhibitory, antimicrobial, and cytotoxic activities of secondary metabolites from the endophytic fungus, Nigrospora sphaerica, isolated from Helianthus annuus. Ann. Microbiol. 69, 1397–1406.

Surette, M.A., Sturz, A.V., Lada, R.R., Nowak, J., 2003. Bacterial endophytes in processing carrots (Daucus carota L. var. sativus): their localization, population density, biodiversity and their effects on plant growth. Plant Soil 253, 381–390.

Tapwal, A., Pandey, P., Chandra, S., Rashmi, 2015. Antimicrobial activity and phytochemical screening of endophytic fungi associated with Cassia fistula. Int. j. biol. chem. sci. 2 (7), 15–21.

Tejesvi, M.V., Kini, K.R., Prakash, H.S., Subbiah, V., Shetty, H.S., 2007. Genetic diversity and antifungal activity of species of Pestalotiops is isolated as endophytes from medicinal plants. Fungal Divers. 24, 37–54.

Tenguria, R.K., Firodiya, A., 2016. Cytotoxic activity of endophytic fungi isolated from central region of Madhya Pradesh. J. Innov. Pharm. Biol. Sci. 3, 29–38.

Tewari, K., Onda, M., Ito, S., et al., 2006. Effect of deep placement of slow-release fertilizer (lime nitrogen) applied at different rates on growth, N2 fixation and yield of soya bean (*Glycine max* [L.] Merr.). J. Agron. Crop Sci. 192 (6), 417–426.

Thanabalasingam, D., Kumar, N.S., Jayasingh, L., Fujimoto, Y., 2015. Endophytic fungus Nigrospora oryzae from a medicinal plant Coccinia grandis, a high yielding new source of Phenazine-1-carboxamide. Nat. Prod. Commun. 10 (10), 1659–1660.

Tribhuvan, D.B., Tawre, A.S., 2019. Antifungal activity of endophyte Aspergillus flavus isolated from Acacia nilotica. Int. J. Curr. Res. 11 (7), 5139–5140.

Vennila, S., Deshmukh, A.J., Pinjarkar, D., Agarwal, M., Ramamurthy, V.V., Joshi, S., et al., 2010. Biology of the mealybug, *Phenacoccus solenopsis* on cotton in the laboratory. J. Insect Sci. 10, 115.

Wachtel-Galor, S., Benzie, I., 2011. Herbal medicine: an introduction to its history, usage, regulation, current trends, and research needs. Herb. Med.: Biomol. Clin. Asp. 2, 234–266.

Wang, J., Li, G., Lu, H., Zheng, Z., Huang, Y., Su, W., 2000. Taxol from Tubercularia *sp.* strain TF5, an endophytic fungus of *Taxus mairei*. FEMS Microbiol. Lett. 193, 249–253.

Wang, X.P., Fang, Y.L., Zhang, Z.N., 2011. Effects of delayed mating on the fecundity, fertility and longevity of females of diamondback moth, *Plutella xylostella*. Insect Sci. 18, 305–310.

Wanzala, W., Zessin, K.H., Kyule, N.M., Boumann, M.P.O., Mathias, E., Hassanali, A., 2005. Ethnoveterinary medicine: a critical review of its evolution, perception, understanding and the way forward. Livest. Res. Rural Dev. 17 (11), 1–134.

Warren, D.M., 1991. Using indigenous knowledge in agricultural development. World Bank Discussion Paper No.127.

Weber, A.P.M., Schneidereit, J., Voll, L.M., 2004. Using mutants to probe the in vivo function of plastid envelope membrane metabolite transporters. J. Exp. Bot. 55, 1231–1244.

Wilson, D., 1995. Endophyte: the evolution of a term, and clarification of its use and definition. Oikos 1995 (73), 274–276.

Yadav, A., Yadav, K., 2017. Exploring the potential of endophytes in agriculture: a minireview. Adv. Plants Agric. Res. 6, 00221.

Ying, Y.M., Shan, W.G., Zhang, L.W., Zhan, Z.J., 2013. Ceriponols A-K, tremulane sesquitepenes from Ceriporia lacerate HS-ZJUT-C13A, a fungal endophyte of Huperzia serrate. Phytochem. 95, 360–367.

Yu, H.S., et al., 2010. Recent development and future prospects of antimicrobial metabolites produced by endophytes. Microbiol. Res. 165, 437–449.

Zhao J., Zhou L., Wang J., Shan T., Zhong L., Liu X., et al., 2011. Endophytic fungi for producing bioactive compounds originally from their host plants. In: Mendez-Vilas A. (ed.), Current Research, Technology and Education Topics in Applied Microbiology and Microbial Biotechnology.

Chapter 16

Sustainable crop production and improvement through bio-prospecting of fungi

Mohammad Haris[1], Adnan Shakeel[1], Moh. Sajid Ansari[1], Touseef Hussain[1,*], Abrar Ahmad Khan[1] and Raunak Dhankar[2]
[1]*Department of Botany, Aligarh Muslim University, Aligarh, India,* [2]*Department of Biological Sciences, School of Basic and Applied Sciences, GD Goenka University, Gurugram, India*
**Corresponding author.* Email: hussaintouseef@yahoo.co.in

Introduction

Microorganisms present in the soil have a close symbiotic relationship with plants. They are most abundant in soil biota and are responsible for promoting nutrition and the organic cycle, fertilizing and restoring soil, plant growth as well as the primary production of the ecosystem. Two examples comprise mycorrhizae and rhizobia. Mycorrhizal fungi are very host-specific that create a symbiotic association with the roots of the host plant. Some of the beneficial fungi play a vital role in the improvement of growth and yield of the plant, involved in tolerance to biotic and abiotic stress, in the remediation of hazardous materials, in sustainable crop production and in food security (Borde et al., 2009). Some fungi do not permit the harmful ones to colonize the surface of root but can be used as biological control agents (Ha, 2010). In some studies, fungal inoculum applied as biofertilizers in the field trial and also in the greenhouse (Grigera et al., 2007; Rahi et al., 2009; Goetten et al., 2016; Zhang et al., 2016; Wang et al., 2018a).

In agriculture, mycorrhizal fungi are extensively used, because they create a symbiotic association in root and supply many beneficial substances to plant due to which plant show better plant growth and development, improve nutrient absorption and greater tolerance of plants from diseases (Whipps, 2004; Liu and Chen, 2007; El-Shaikh and Mohammed, 2009; Smith et al., 2010; Hernandez-Montiel et al., 2013; Janouskova et al., 2017). Biofertilizers improve the absorption of nutrients from the soil or atmosphere which increases the production of enzymes, hormones and bioactive

Fungi Bio-Prospects in Sustainable Agriculture, Environment and Nano-Technology.
DOI: https://doi.org/10.1016/B978-0-12-821394-0.00016-0

compounds that are able to improve plant growth (Chi et al., 2010; Abdel-Fattah et al., 2013; Pal et al., 2015). Symbiotic relationships of fungal endophytes show a range with their hosts and different lifestyles; For example, some of these kinds of interactions may be mutualistic in which the relationship of long-term is advantageous for both partners (Card et al., 2016). They are essential for the nutrient cycle and responsible for the degradation of cellulose and lignin (Pointing et al., 2001). It is an effective approach to improve the tolerance of plants to salt stress and this approach can play a role in the development of sustainable agricultural systems. Rhizospheric microorganisms are one of the most important groups that include *Trichoderma* spp which can cause some beneficial effects on plant growth and development (Harman et al., 2004; Qi and Zhao, 2013). This chapter will present an overview of the beneficial fungi and their role in sustainable agriculture production in various aspects.

Role of bioactive compounds in agriculture crop production

Fungi produce a variety of bioactive compounds that plays a major contribution to the physiological activities of the host plant, affecting host growth. This may lead to greater stress tolerance from biotic and abiotic plants (Pineda et al., 2010). Bioactive metabolites, isolated from medicinally used mushrooms and their mycelium, function as biological anti-hyperglycemic agents in the diabetes treatment (Table 16.1) (De Silva et al., 2012). *Ganoderma lingzhi* (Lingzhi) contains complex natural bioactive compounds, of low and high molecular weight that are mainly triterpenoids (Sandargo et al., 2019).

Many fungi produced 1-Octen-3-ol which contributes to improving the resistance of the plant from the necrotrophic fungal pathogen Botrytis cinerea through the inducing defense signaling cascades (Kishimoto et al., 2007; Contreras-Cornejo et al., 2014). The volatile profiles of *Alternaria alternata*, *Penicillium charlesii*, and *Penicillium aurantiogriseum* promote growth and starch accumulation in multiple plant species (Ezquer et al., 2010).

Endophytic fungi for improving plant performance and producing bioactive compounds

Endophytic plant fungi, a new and plentiful resource of microbes, which possess the special ability to release the same compounds, or similar compounds that originate from their host plants, as well as other bioactive compounds, have increased much interesting research both in basic research and applied fields. Endophytic plant fungi are a good source of naturally produce bioactive compounds regarding their potential approach in medicine, agriculture and the food industries. Two decades ago, endophytic fungi have successfully discovered many precious bioactive compounds including insecticidal, cytotoxic, antimicrobial and antitumor activities.

TABLE 16.1 Bioactive volatile compounds from different fungal species.

Fungal species	Volatile compound	References
Fusarium oxysporum	β-Caryophyllene	Minerdi et al. (2011)
Mold fungi	1-Octen-3-ol	Kishimoto et al. (2007)
Muscodor albus	Isoamyl acetate	Strobel et al. (2001)
Muscodor albus	2-Methyl butanol Isobutyric acid	Mercier and Jimenez (2004)
Muscodor crispans	Mixture of volatile compounds	Mitchell et al. (2010)
Muscodor yucatanensis	Mixture of VOCs	Macías-Rubalcava et al. (2010)
Phoma sp.	Unique mixture of VOCs, including a series of sesquiterpenoids, few alcohols and several reduced the naphthalene derivatives	Strobel et al. (2011)
Phomopsis sp.	Sabinene, isoamyl alcohol, 2-propanone, 2-methyl propanol	Singh et al. (2011)
Trametes gibbosa	1-Octen-3-ol	Thakeow et al. (2008)
Trametes versicolor	γ-Patchoulene, δ -cadinene, isoledene, β-guaiene	Drilling and Dettner (2009)
Trichoderma virens	β-Caryophyllene, β-elemene, germacrene D, δ-cadinene	Zhang et al. (2007)
Many species of bacteria, fungi, and plants	2-Butanone	Song and Ryu (2013), Farag et al. (2006), Wheatley (2002), Schöller et al. (2002)
Many species of bacteria, fungi, and plants	3-Pentanol	Song and Ryu (2013), Funes et al. (2009), Zerba and Gonzalez Audino (2009), Bukovinszky et al. (2005)

These fungi have a different kind of secondary metabolites (Tan and Zou, 2001; Gunatilaka, 2006), in which some are bioactive compounds that expressed as safeguard weapons to protect the host plant from diseases and pest but also act as metabolites which are responsible for the specific interactions with the host plant.

Other pronounced bioactive compounds that originate from host plants could also biosynthesize from their endophytic fungi, primarily huperzine A, α-irone, β-iron, diosgenin, hypericin and toosendanin (shown in Table 16.2). Wang et al. (2007) discovered three types of endophytic fungi which isolate from *Melia azedarach* that had the potential to produce toosendanin.

Fungi as an indicator of soil health

Soil is a non-renewable resource, a vital part of the environment and agriculture. Soil health term is widely used indicating to sustainable agriculture for discussion of widespread state or quality of soil resource. Concept of soil health includes soil biota, biotic interactions and also soil as a habitat (Young and Ritz, 2005). An combined approach to soil health presupposes that the soil is a living system and its health is determined by the synergy between different processes and features with a strong effect on the activity of soil microbiota (Frąc et al., 2018). By making primary participation in the process of decomposition, the soil fauna and microbial community are considered critical in any ecosystem (Cardoso et al., 2013). Management of soil is fundamental to agricultural systems as there is an extensive depletion of

TABLE 16.2 Endophytic fungi producing bioactive compounds from their host plants.

Endophytic fungus	Host plant	Bioactive compounds	Reference
Acremonium sp.	*Huperzia serrata*	Huperzine A	Li et al. (2007)
Blastomyces sp.	*Phlegmariurus cryptomerianus*	Huperzine A	Ju et al. (2009)
Botrytis sp.	*Phlegmariurus cryptomerianus*	Huperzine A	Ju et al. (2009)
Cephalosporium sp.	*Paris polyphylla* var. *yunnanensis*	Diosgenin	Zhou et al. (2004), Cao et al. (2007)
Chaetomium globosum	*Hypericum perforatum*	Hypericin, Emodin	Kusari et al. (2008)
Paecilomyces sp.	*Paris polyphylla* var. *yunnanensis*	Diosgenin	Zhou et al. (2004), Cao et al. (2007)
Penicillium chrysogenum	*Lycopodium serratum*	Huperzine A	Zhou et al. (2009)
Rhizopus oryzae	*Iris germanica*	α-Irone, β-Irone	Zhang et al. (1999)

agricultural soil by increased salinity, contamination, soil erosion, loss of organic matter and compaction (Council, the European Parliament, 2002). Soil-borne microbes are largely concerned in the release of soil mineral nutrients, plant disease resistance and plant disease (Zhang et al., 2013). Their diversity and activity play an important role in sustainability by maintaining essential functions in soil health and managing the carbon and nutrient cycling and other essential processes (Jeffries et al., 2003; Izquierdo et al., 2005). To various changes occurring in soil, the use of microbial indicators is more susceptible than chemical and physical, therefore an early and easily predict any interruption in the sustainability of an environment (MarchioriJúnior and Melo, 1999; Masto et al., 2009). Main biological attributes used in soil health studies are soil microorganisms that are living component of soil organic matter which include fungi, bacteria, algae and protozoa these are an important source of nutrients to fulfill plant demands (Sicardi et al., 2004)

Fungi are a distinct group of microorganisms and are successful inhabitants of different soil habitats (Frąc et al., 2017). The reason for their success to thrive better in adverse and unfavorable conditions is their plasticity and adaptability (Sun et al., 2005). They performs an important role in nutrient cycling as decomposers, managing carbon and nutrient balance by producing extracellular enzymes, that are able to breakdown organic matter (Sun et al., 2005; Zifcakova et al., 2016), their activity and diversity are regulated by various abiotic and biotic factors (López-Bucio et al., 2015; Rouphael et al., 2015). Fungal biodiversity determines plant diversity, productivity and ecosystem variability; they can also hinder or reduce the injury of root diseases and play a crucial role in the care of soil health and quality (Wagg et al., 2014).

Fungi make a considerable contribution to the list of bioindicators. Any living entity that displays information based on the environment is known as bioindicator (Wiłkomirski, 2013). The fungal community can be used as a bioindicator to reflect anthropogenic action in soil ecosystems as they interact with the ambient environment including soil. Due to these close interactions between fungal community disparity and anthropogenic activities, using the fungal community as a bioindicator is potentially feasible. Many studies have reported the impact of anthropogenic activities on the fungal community and found that community diversity and composition are influenced by such action (Artigas et al., 2008; Baudoin et al., 2008; Medeiros et al., 2009; Duarte et al., 2013). The lichens which also include fungi are considered as the most reliable biomonitors according to their specific physiological, morphological, and anatomical characteristics (Aslan et al., 2013). Keystone species are excellent indicators of soil health as they control the interactions of other organisms in an ecosystem (Bruggen and Semenov, 2000). *Rhizobium* species being keystone and highly sensitive to agrochemicals have been proposed as indicators of soil health (Visser and Parkinson, 1992; Anderson, 2003). Bergfur and Sundberg (2014) discover that agricultural land use can have symbolic effects on fungal communities. Pietryczuk et al. (2018) noticed that anthropogenically polluted rivers exhibit higher taxonomic diversity of fungi. Overall, most studies on river ecosystems have illustrated that anthropogenic activities (e.g., nutrient, micropollutant, and microplastic discharge) can shape fungal community arrangement (Jabiol et al., 2018). Fungi have been also reported as bioindicators of radioactive substances contaminated to the soil. They accumulate radionuclides on their fruiting bodies (Dighton et al., 2008) Hasel Examples include *Cortinarius armillatus* and *Rozites caperatus* (Haselwandter and Berreck, 1994).

The property of soil to suppress pathogenic microorganisms is also an attribute of healthy soil. Baker and Cook reported that suppressive soil has been defined as the soil in which the disease severity or incidence remains low in spite of the presence of pathogens, susceptible host plants, and climatic conditions favorable for disease development. Soil suppressiveness can be enhanced by the addition of biopolymers like chitin and its derivative chitosan which lead to changing the activity and structure of soil microorganisms (Kielak et al., 2013). *Trichoderma, Fusarium, Penicillium* belongs to non-pathogenic fungal groups which are antagonists of soil-borne plant pathogens. Arbuscular mycorrhizal fungi are found to reduce plant root diseases but development of disease may occur under unfavorable conditions and other conditions may cause soil suppressiveness (Garbeva et al., 2004; Gianinazzi et al., 2010).

Arbuscular mycorrhizal fungi (AMF) belongs to a class of beneficial microorganisms and in various experiments, it was found that AMF inoculated soil lead to the remarkable increase in the turnout of crop plants (Smith and Read, 2008; Thilagar and Bagyaraj, 2015; Bagyaraj and Ashwin, 2017). AMF symbiosis has a great impact on plant and soil health like, up-gradation of soil structure, stimulation of nutrient cycling, and development of rooting and plant establishment, enrichment of plant tolerance to stresses, enhancement of plant community diversity and expansion of low mobility ions (Azcon-Aguilar and Barea, 1997) Arbuscular mycorrhizal fungi also found to protect biological and physicochemical soil qualities, improved revegetation process for desertification ecosystem (Bagyaraj and Ashwin, 2017). Other microorganisms, like nitrogen-fixing and phosphorus solubilizing and inoculation of AMF in combining with plant growth-promoting rhizobacteria (PGPR) has great consequences on plant health and soil quality, the favorable impact of AMF together with PGPR on soil microbial diversity, activity and horticultural plant growth has also been reported (Azcon-Aguilar and Barea, 1997; De Coninck et al., 2015; Fernandez et al., 1997).

Biofertilizers as a key player in crop production

Biofertilizers are manufacture taken away organic matter or agricultural industrial wastes, which act as a substance for procreation of inoculum of chosen microorganisms (Kaewchai et al., 2009). Mycorrhizae make mutualistic symbiotic association with more than 80% roots of land plants including tree species of forest and many important crops (Gentili and Jumpponen, 2006; Rinaldi et al., 2018). There are seven categories of mycorrhiza: ectomycorrhiza (ECM), endomycorrhiza or arbuscular mycorrhiza (AM), arbutoid mycorrhizae, ect-endomycorrhiza, ericoid mycorrhiza, orchidoid mycorrhiza and monotropoid mycorrhizae (Tao et al., 2008; Zhu et al., 2008). Among them two dominant mycorrhizae are arbuscular mycorrhizae (AM) and ectomycorrhizae (ECM) that can increase nutrients, water uptake and provide safety against pathogen but some families of plants have potential to form functional relationship with both fungi (Siddiqui and Pichtel, 2008). However, most commonly, AM fungi found in the wide range of herbaceous and woody plants rhizospheric roots (Rinaldi et al., 2018). In this chapter, we focus on arbuscular mycorrhizal, Tricoderma spp. and some other fungi that have most widespread and economically important value.

Use of arbuscular mycorrhizal fungi (AMF) as a biofertilizers

Biofertilizers naturally occurring substances used to improve soil fertility. These fertilizers are increase to soil health as well as development of plant growth (Sadhana, 2014). Soil borne pathogens AMF significantly enhance absorption of plant nutrients and resistance from many abiotic stress factors (Sun et al., 2018). Majority of the AMF species belong to the phylum Mucoromycota and the sub-phylum Glomeromycotina (Spatafora et al., 2016). According to this AMF are important endosymbionts playing a useful role in the function of ecosystem and plant productivity? They perform key important role for sustainable crop enhancement (Gianinazzi et al., 2010). Water dynamics and improve aeration; permit the use of insoluble P sources (Smith and Read, 1997) and most higher tolerance capacity from heavy metal contamination or drought and a low susceptible to root pathogens or herbivores (Gentili and Jumpponen, 2006). Many research studies investigate that AMF has uncountable benefits for the improvement of soil health and productivity of crop from last two decades. Therefore, it is believed that AMF can be used as in place of inorganic fertilizers in future because application of mycorrhizal can diminish the high use of chemically manufactured fertilizer input mainly phosphorus (Ortas, 2012). Because inorganic fertilizers such as fungicide and herbicides have created different problems to human health, plants and soil by their harmful effect on the soil health, air quality, water system and many food products (Yang et al., 2004). AM used as an important phosphate mobilizing biofertilizers (Zhang et al., 2018).

Use of *Trichoderma* spp. as a biofertilizers

Trichoderma spp. It is a beneficial fungus that essentially lives in the soil or in plant tissues. They are found in many root ecosystems and it has been perceived that sucrose of plant origin plays an important role in facilitating root colonization by *Trichoderma* cells and in coordinating defense mechanisms (Vargas et al., 2008). In addition, some *Trichoderma* species also use other beneficial effects for plants, such as promoting the growth and absorption of nutrients, the efficiency of fertilizer use and tolerance to abiotic stress (Lorito et al., 2010). The use of natural sources is recommended to develop antagonistic action against pathogens and the development of harmful plants in viable agriculture to reduce the use of harmful chemical pesticides to the environment (Sadoma et al., 2011). The ability of *Trichoderma* to secrete antifungal metabolites in which caused the physical abnormality of *G. boninense* during the interaction has been reported (Siddiquee et al., 2009; Angel et al., 2018). In addition, the evaluation of the nursery and the field carried out by (Sundram et al., 2016) showed the potential of endophytic strains isolated from *Trichoderma* as suitable biofertilizers for disease management in crops.

Use of other fungi as a biofertilizers

Another fungal biofertilizer like *Penicillium* species that have been used to enhance plant growth and soil fertility. They are phosphate solubilizing microorganisms which increase absorption of phosphorus in plants that regulate plant growth (Pradhan and Sukla, 2005). Some of the commercial products like JumpStart® has been formulated by *Penicillium bilaiae* and comes in the market as a wettable powder in 1999 (Burton and Knight, 2005). In canola (*Brassica napus*) enhance seed yield, dry matter and phosphorus (P) absorbance by the use of *Penicillium bilaiae*. Burton and Knight (2005). *Penicillium italicum* and *P. radicum* are also phosphate-solubilizing taxa (El-Azouni, 2008). Isolation of *Penicillium radicum* from the rhizosphere of wheat roots has exhibited a benefit in the promotion of plant growth (Whitelaw et al., 1999) while *P. italicum* isolated from rhizospheric soil have ability to solubilize tri-calcium phosphate (TCP) and responsible for the increase growth of soybean crop (Akintokun et al., 2007).

Several *Aspergillus* species have been discovered which involved in the inorganic phosphate solubilization such as *A. niger, A. terreus* and *A. flavus* (Akintokun et al., 2007). These fungi are capable of solubilization of inorganic phosphate through the formation of acids, for example, succinic acid, citric acid, succinic acid oxalic acid and glycolic acid (Barroso et al., 2006). *Aspergillus fumigates* has been reported as a potassium releasing fungus that isolated from the compost (Lian et al., 2008). Treatments of fungal inocula used for plants which are shown in Table 16.3. With the help of fungal biofertilizers, inorganic and organic forms of phosphorus mobilize and solubilize into soluble forms, which are available to plants. For example, Mixture of *Aspergillus niger* and *Bacillus megaterium* to form such microorganisms that solubilize the phosphate. In India, these microorganisms have been used

TABLE 16.3 Use of different fungal inoculum as bio fertilizers on the performance of different plants.

Fungal inoculum	Treated plant	Result	References
Arbuscular mycorrhiza	*Zea mays*	Beneficial for increase productivity and reproductive growth of maize crops through enhancing phosphorus absorption	Grigera et al. (2007)
Arbuscular mycorrhiza	*Citrulls lanatus*	Low transplanting problems, change the structure of soil microbiota and increase the activity of soil enzymes	Zhao et al. (2010)
Arbuscular mycorrhiza (*Rhizophagus clarus* and *Claroideoglomus etunicatum*)	Seedling of various woody plants	1. Improved root colonization and seedlings of woody plants 2. Improved stem diameter and plant length of *C. robustum, C. fissilis* and *L. divaricata,* 3. Improved biomass of shoot and growth of *G. gardneriana, L. divaricata, C. fissilis* and *C. robustum* 4. Improved phosphorus availability in shoot of *G. gardneriana, S. terebinthifolius* and *C. robustum*	Goetten et al. (2016)
Aspergillus niger (CS-1)	*Triticum aestivum*	Regulate plant growth through improving the fresh and dry mass of wheat per plant	Wang et al. (2018b)
Discosia sp. (HKUCC 6626)	*Pisum sativum, Zea mays, Cicer arietinum*	Improved dry mass, shoot and root length in the test plants over the uninoculated control under the controlled environment	Rahi et al. (2009)
Endophytic strains of *Fusarium tricinctum* (RSF-4L) and *Alternaria alternata* (RSF-6L)	*Oryza sativa*	Improved growth attributes as well as enhanced the length of root and shoot, plant biomass and chlorophyll contents	Khan et al. (2015)
Penicillium sp., *Trichoderma* sp., (*Pythium* sp.)	*Pennisetum glaucum* seeds	1. *Penicillium* sp. at 5% (w/w) concentration observed maximum germination of seed 92% and seedling vigor 1701.9	Murali et al. (2012)

(*Continued*)

TABLE 16.3 (Continued)

Fungal inoculum	Treated plant	Result	References
		2. *Penicillium* sp. at 5% (w/w) and *Pythium* sp. at 10% (w/w) responsible for highest disease control of 67% and 61% respectively facing downy mildew disease of pearl millet 3. *Penicillium* sp. and *Trichoderma* sp. observed huge disease controlled at 71% and 66% respectively, under greenhouse experiments	
Purpureocillium lilacinum	*Solanum lycopersicum* seeds	Enhanced % seed germination of *Solanum lycopersicum* from 71% to 85% after 48 h	Cavello et al. (2015)
Trichoderma longibrachiatum	*Triticum aestivum* seeds	Enhanced root length and seedling height of wheat compared to the stress treatment of NaCl	Zhang et al. (2016)

as a biofertilizer (Pal et al., 2015). Biofertilizers as a key player in the salvage of plant nutrients and enhance the rate of fertilizer degradation (Pal et al., 2015).

Role of fungi in the fight with climate change

Recently, the world is facing a serious problem of global warming. Climate change and soil contamination due to human activating greatly contributed to increase the CO_2 concentration globally. CO_2 causes greenhouse gas effect. It is quite evident that global CO_2 concentration will reach to two-folds higher in the coming 100 years which will play a significant role in changing the biogeochemical reactions of the atmosphere. However, the heterogeneous influence of climate change is observed in a different part of the globe. Equatorial and temperate regions are less prone to change in atmospheric temperature and precipitation comparison to high latitude areas (Flombaum et al., 2016). The boreal peatland of these areas are dominated by peat moss which play great role in maintaining the balance of abiotic factors such as temperature, moisture, air and nutrients around plant roots (Turetsky et al., 2012; Quinton et al., 2007) by negatively influencing the microbial activities by keeping low pH keeping soil temperature low and by secreting antimicrobial compounds (Hájek et al., 2011). The increase in atmospheric temperature is positively correlated to the increase in concentration of CO_2 which indirectly increase decomposition rate in ecosystem. The high atmospheric CO_2 concentration help to increase plant growth, however it completely depends on the plant species and water availability to the plant (Gorham, 1991; Blankinship et al., 2011) and symbiotic relationship with fungi.

The predicted global temperature rise is average of 1.5°C (Allen et al., 2018) however it is expected that it will be two to four times greater than the predicted global average, which could potentially alter many ecosystem processes and influence the pattern of carbon cycling in these crucial carbon sequestering ecosystems (McGuire et al., 2009).

A huge variation in biodiversity is noted with changing climatic conditions. In Climatic change scenario, it is expected that as the temperature change a sharp shift in a plant community is observed which results in competitive outcomes within plant communities (Chen et al., 2011). The higher and dominated species of vascular plants are observed in boreal peatlands which is due to the effect of *Sphagnum* mosses and many fungal species has been seen in these areas (Chen et al., 2011). These scenarios may cause loss of few diversity and emergence of few new species of the fungi kingdom.

Out of these diverse, fungal is more diverse and perform a range of function which may change depending on the community structure. It's a great need to understand the exact effect of microbes which may be a driving factor in making the change in the reorganization of plant communities and overall ecosystem process. These changes can provide a great shift in few species of fungi According to climatic change agency IPCC more comprehensive, realistic study are needed for the relationships between the variations in the composition of fungal communities and reorganization of plant communities and ecosystem-level processes (Bardgett and van der Putten, 2014). These kinds of evidence will help in the prediction of climate change in the future.

A rapid change in the communities of fungi has been observed nowadays due to change in many abiotic factors of environment such as high a significant elevation in temperature, increased level of CO_2 and decreasing water table surroundings as well as varying amount of peat matrix (Andersen et al., 2012; Dieleman et al., 2015). Fungal communities of northern peatlands are likely currently undergoing a rapid change in their environment through the multiple drivers of climate change (i.e. increased temperature, elevated atmospheric CO_2 and lowered water table conditions) as well as through variations of plant inputs into peat matrix caused by shifts in aboveground communities under climate change scenarios (Andersen et al., 2012; Dieleman et al., 2015; Allison and Treseder, 2011).

Due to change in above-ground communities, the decomposition activities of different microbes change significantly in different layers of soil due to change in soil temperature and moisture. These changes significantly influence the metabolic rate of soil microbiota. The amount of CO_2 is higher in the soil in the atmosphere (Schwartz and Bazzaz, 1973), which can't exert any influence on soil fungal communities. Many studies have been shown a considerable influence on soil microbial respiration and decomposition of unmanageable material which has been very well represented by different carbon models (Todd-Brown et al., 2014). These kinds of studies have been shown a great association between carbon quality and temperature sensitivity in the process of decomposition (Tang and Riley, 2014).

Few studies have suggested the increased rate of decomposition by influencing the water table position and giving an appropriate habitat for aerobic decomposers (Allison and Treseder, 2011). Even in a few cases, elevation in temperature can accelerate anaerobic microbial decomposition, which releases methane gas. It is very well known that methane is twenty times more potent than CO_2 global warming gas (Godin et al., 2012). Many studies have been revealed that increased rate of CO_2 flux to the atmosphere occurs incidentally with lowering the water table in peatlands which ultimately happens due to the increased rate of evapotranspiration by plants (as would occur with increased air temperature) which further can reduce soil moisture.

Recently it is found that an increased rate of CO_2 flux to the atmosphere happens coincidently with lowering the water table in peatlands, and an increased rate of evapotranspiration by plants (as would occur with increased air temperatures) can further reduce soil moisture, potentially affecting the rate carbon turn over in lower peat horizons (Ostle et al., 2009). The changing climatic conditions can bring a sustainable change in microbial communities especially can influence fungal growth and activities, which significantly impact the plant productivity and total ecosystem processes (Andersen et al., 2012).

A cause and effect relationship was observed between climate change and variations in the composition and function of the fungal group on the basis of the projections made by IPCC (Laiho, 2006; Morgado et al., 2015). However, more comprehensive studies including the variations in the overall structure of microbial communities and their re-organization of plant communities and ecosystem-scale processes (Bardgett and van der Putten, 2014; Van der Heijden et al., 2008) will help to predict future climate change more accurately.

Fungi are a unique and fascinating group of living beings. They are distributed abundantly over the earth and due to many distinguishable properties make their distinguishable identity. There are approximately 5 million species of fungi are found over the earth. The largest single organism on the planet belongs to the fungus category i.e. *Armillaria ostoya* (the four sq. mi. (10 sq. km)) which is more than 8000 years old and found in the soil of Oregon's Blue Mountains. Fungi are the potential source of blue cheeses, beer and antibiotics and beyond. Fungi play a substantial role in the carbon cycle during the biogeochemical process it helps in the movement of carbon on the earth between air, soils and water (Bardgett and van der Putten, 2014). Fungi sequester a major portion of atmospheric carbon dioxide which ultimately enters as carbon in the soil. As the dead matter of plant material is broken down by fungi in the oil after that carbon is released back into the air. After releasing this air in the atmosphere a significant change has been made in the climate change (Lange, 2014).

The growth of these microbes is depending on the presence of the amount of nitrogen in the soil. Both plants, as well as fungi, compete for nitrogen. Lower the amount of the nitrogen available for decomposers slower will be the decomposition rate, lessor amount of carbon will be released in the atmosphere because a major portion of carbon will remain in the soil. Now the actual role of the fungus come in the frame (Bardgett and van der Putten, 2014). As we all know that fungi found with most of the plants in a symbiotic relationship. The fungi are known for the extraction of nitrogen from the soil which is available to the plants through their roots.

Researchers from the University of Texas, Boston University and the Smithsonian Tropical Research Institute ran computer models on data from more than 200 soil profiles from around the world. They compared the carbon capacity of ecto and ericoid mycorrhizal (EEM) fungi and arbuscular mycorrhizal (AM) fungi. They found that EEM fungi contain 70% more carbon than soils dominated by AM fungi. EEM is able to do this due to the presence of more nitrogen degrading enzymes. They further concluded that their results clearly indicated the dominant role of symbiotic fungi in the extraction of carbon globally. However, their role not fully recognized until now. According to researchers, the role of fungi cannot be forgotten in the carbon cycle because forest functionality depends to some extent on the mycorrhizal relationship between trees and decomposers. So its high time to think about this system holistically in a more scientific way because indirectly this group is 'running the world'.

Conclusion and Future prospects

In the global market, agriculture is increasingly integrated into the food chain, although environmental regulations, food safety and animal quality and welfare also increasingly affect the sector. It faces new challenges to meet the growing demand for food, is internationally competitive and produces high quality agricultural products. In addition, it should meet sustainability objectives in the context of ongoing agricultural policy reform. In order to better understand the effects of globalization on the entire food system and its outer regions, it is urgently necessary. World trade is necessary to balance supply and demand in all regions as food production and population growth are increasing at different speeds in different geographical regions. Fungi can be investigated by some of the strategies mentioned above and many others through the means of biotechnology to improve productivity and agricultural diversity. In many developed countries, some of these strategies have been in use for years. The problems of food insecurity due to climate change along with population explosion, industrialization (contamination of soil, water, air) and its related problems (invasive and pathogenic diseases, drought, etc.) are causing more and more stress are critical in agricultural production and productivity. Therefore, agricultural sustainability requires strategic implementation of scientific findings worldwide.

Conflict of interest

The authors declare no conflict of interest.

References

Abdel-Fattah, G.M., Ibrahim, A.H., Al-Amri, S.M., Shoker, A.E., 2013. Synergistic effect of arbuscular mycorrhizal fungi and spermine on amelioration of salinity stress of wheat ("*Triticum aestivum* L." cv. 9). Aust. J. Crop. Sci. 7, 1525.

Akintokun, A.K., Akande, G.A., Akintokun, P.O., Popoola, T.O.S., Babalola, A.O., 2007. Solubilization on insoluble phosphate by organic acid-producing fungi isolated from Nigerian soil. Int. J. Soil Sci. 2, 301–307.

Allen, M.R., O.P. Dube, W. Solecki, F. Aragón-Durand, W. Cramer, S. Humphreys, M. Kainuma, J. Kala, N. Mahowald, Y. Mulugetta, R. Perez, M. Wairiu, and K. Zickfeld, 2018: Framing and Context. In: *Global Warming of 1.5°C. An IPCC Special Report on the impacts of global warming of 1.5°C above pre-]industrial levels and related global greenhouse gas emission pathways, in the context of strengthening the global response to the threat of climate change, sustainable development, and efforts to eradicate poverty* [Masson-]Delmotte, V., P. Zhai, H.-]O. Pörtner, D. Roberts, J. Skea, P.R. Shukla, A. Pirani, W. Moufouma-] Okia, C. Péan, R. Pidcock, S. Connors, J.B.R. Matthews, Y. Chen, X. Zhou, M.I. Gomis, E. Lonnoy, T. Maycock, M. Tignor, and T. Waterfield (eds.)]. In Press.

Allison, S.D., Treseder, K.K., 2011. Climate change feedbacks to microbial decomposition in boreal soils. Sci. Direct 4, 362–374.

Andersen, R., Chapman, S.J., Artz, R.R.E., 2012. Microbial communities in natural and disturbed peatlands: a review. Soil. Biol. Biochem. 57, 979–994.

Anderson, T.H., 2003. Microbial eco-physiological indicators to asses soil quality. Agric. Ecosyst. Environ. 98 (1–3), 285–293.

Angel, L.P.L., Sundram, S., Tay, B.Y.P., Yusof, M.T., Ismail, I.S., 2018. Profiling of anti-fungal activity of *Trichoderma virens* 159c involved in biocontrol assay of *Ganoderma boninense*. J. Oil Palm. Res. 30 (1), 83–93.

Artigas, J., Romaní, A.M., Sabater, S., 2008. Relating nutrient molar ratios of microbial attached communities to organic matter utilization in a forested stream. Fundam. Appl. Limnol./Arch. Hydrobiol. 173 (3), 255–264.

Aslan, A., Gurbuz, H., Yazici, K., Cicek, A., Turan, M., Ercisli, S., 2013. Evaluation of lichens as bio-indicators of metal pollution. J. Elementol. 18 (3).

Azcon-Aguilar, C., Barea, J.M., 1997. Applying mycorrhiza biotechnology to horticulture: significance and potentials. Sci. Hortic. 68, 1–24. Available from: https://doi.org/10.1016/S0304-4238(96)00954-5.

Bagyaraj, D.J., Ashwin, R., 2017. Soil biodiversity: role in sustainable horticulture. Biodivers. Hortic. Crop. 5, 1–18. Available from: https://doi.org/10.1016/j.jenvman.2017.08.001.

Baker K.F., Cook R.J. Biological Control of plant Pathogens. WH Freeman and Company.

Bardgett, R.D., van der Putten, W.H., 2014. Belowground biodiversity and ecosystem functioning. Nature 515, 505–511.

Barroso, C.B., Pereira, G.T., Nahas, E., 2006. Solubilization of CAHPO4 and ALPO4 by *Aspergillus niger* in culture media with different carbon and nitrogen sources. Brazil. J. Microbiol. 37, 434–438.

Baudoin, J.M., Guérold, F., Felten, V., Chauvet, E., Wagner, P., Rousselle, P., 2008. Elevated aluminium concentration in acidified headwater streams lowers aquatic hyphomycete diversity and impairs leaf-litter breakdown. Microb. Ecol. 56 (2), 260–269.

Bergfur, J., Sundberg, C., 2014. Leaf-litter-associated fungi and bacteria along temporal and environmental gradients in boreal streams. Aquat. Microb. Ecol. 73 (3), 225–234.

Blankinship, J.C., Niklaus, P.A., Hungate, B.A., 2011. A meta-analysis of responses of soil biota to global change. Oecologia 165, 553–565.

Borde, M., Dudhane, M., Jite, P.K., 2009. Role of bioinoculant (AM fungi) increasing in growth, flavor content and yield in *Allium sativum* L. under field condition. Not. Bot. Horti Agrobot. Cluj. 37, 124–128.

Bruggen, A.H., Semenov, A.M., 2000. In search of biological indicators for soil health and disease suppression. Appl. Soil. Ecol. 15 (1), 13–24.

Bukovinszky, T., Gols, R., Posthumus, M.A., Vet, L.E.M., Van Leanteren, J.C., 2005. Variation in plant volatiles and attraction of the parasitoid Diadegma semiclausum (Hellen). J. Chem. Ecol. 31, 461–480.

Burton, E.M., Knight, S.D., 2005. Survival of *Penicillium bilaiae* inoculated on canola seed treated with Vitavax RS and Extender. Bio Fertil. Soils 42, 54–59.

Cao, X., Li, J., Zhou, L., Xu, L., Li, J., Zhao, J., 2007. Determination of diosgenin content of the endophytic fungi from *Paris polyphylla* var. *yunnanensis* by using an optimum ELISA. Nat. Prod. Res. Dev. 19, 1020–1023.

Card, S.D., Johnson, L.J., Teasdale, S., Caradus, J., 2016. Deciphering microbial behaviour – the link between endophyte biology and efficacious biological control agents. FEMS Microbiol. Ecol. 92.

Cardoso, E.J.B.N., Vasconcellos, R.L.F., Bini, D., Miyauchi, M.Y.H., dos Santos, C.A., Alves, P.R.L., 2013. Soil health: looking for suitable indicators. What should be considered to assess the effects of use and management on soil health? Sci. Agric. 70, 274–289. Available from: https://doi.org/10.1590/S0103-90162013000400009.

Cavello, I.A., Crespo, J.M., García, S.S., Zapiola, J.M., Luna, M.F., Cavalitto, S.F., 2015. Plant growth promotion activity of keratinolytic fungi growing on a recalcitrant waste known as "Hair Waste". Biotech. Res. Int. 1–10.

Chen, I.C., Hill, J.K., Ohlemüller, R., Roy, D.B., Thomas, C.D., 2011. Rapid range shifts of species associated with high levels of climate warming. Science 333, 1024–1026.

Chi, F., Yang, P., Han, F., Jing, Y., Shen, S., 2010. Proteomic analysis of rice seedlings infected by *Sinorhizobium meliloti* 1021. Proteomics 10, 1861–1874.

Contreras-Cornejo, H.A., Macías-Rodríguez, L., Herrera-Estrella, A., López-Bucio, J., 2014. The 4-phosphopantetheinyl transferase of *Trichoderma virens* plays a role in plant protection against *Botrytis cinerea* through volatile organic compound emission. Plant Soil 379, 261–274.

Council, The European Parliament. The Economic and Social Committee and the Committee of the Regions: towards a thematic strategy for soil protection. COM (2002) 179 final. Brussels, Belgium: European Commission.

De Coninck, B., Timmermans, P., Vos, C., Cammue, B.P.A., Kazan, K., 2015. What lies beneath: belowground defense strategies in plants. Trends Plant. Sci. 20, 91. Available from: https://doi.org/10.1016/j.tplants.2014.09.007.

De Silva, D.D., Rapior, S., Hyde, K.D., Bahkali, A.H., 2012. Medicinal mushrooms in prevention and control of diabetes mellitus, Fungal Divers., 56. pp. 1–29.

Dieleman, C.M., Branfireun, B.A., McLaughlin, J.W., Lindo, Z., 2015. Climate change drives a shift in peatland ecosystem plant community: implications for ecosystem function and stability. Glob. Change Biol. 21, 388–395.

Dighton, J., Tugay, T., Zhdanova, N., 2008. Fungi and ionizing radiation from radionuclides. FEMS Microbiol. Lett. 281 (2), 109–120.

Drilling, K., Dettner, K., 2009. Electrophysiological responses of four fungivorous Coleoptera to volatiles of Trametes versicolor: implications for host selection. Chemoecol 19, 109–115.

Duarte, S., Fernandes, I., Nogueira, M.J., Cássio, F., Pascoal, C., 2013. Temperature alters interspecific relationships among aquatic fungi. Fungal Ecol. 6 (3), 187–191.

El-Azouni, I.M., 2008. Effect of phosphate solubilizing fungi on growth and nutrient uptake of soybean (*Glycine Max* L.) plants. J. Appl. Sci. Res. 4, 592–598.

El-Shaikh, K.A.A., Mohammed, M.S., 2009. Enhancing fresh and seed yield of okra and reducing chemical phosphorus fertilizer via using VA-mycorrhizal inoculants. World J. Agric. Sci. 5 (S), 810–818.

Ezquer, I., Li, J., Ovecka, M., Baroja-Fernández, E., José Muñoz, F., Montero, M., et al., 2010. Microbial volatile emissions promote accumulation of exceptionally high levels of starch in leaves in mono- and dicotyledonous plants. Plant Cell Physiol. 51, 1674–1693.

Farag, M.A., Ryu, C.M., Sumner, L.W., Pare, P.W., 2006. GC–MS SPME profiling of rhizobacterial volatiles reveals prospective inducers of growth promotion and induced systemic resistance in plants. Phytochemistry 67, 2262–2268.

Fernandez, F., Ortiz, R., Martinez, M.A., CostalesA, Llonin, D., 1997. The effect of commercial arbuscular mycorrhizal fungi (AMF) inoculants on rice in different types of soil. Cultivo Stropicales 18 (1), 5–9.

Flombaum, P., Yahdjian, L., Sala, O.E., 2016. Global change drivers of ecosystem functioning modulated by natural variability and saturating responses. Glob. Change Biol. Available from: https://doi.org/10.1111/gcb.13441.

Frąc, M., Hannula, S.E., Bełka, M., Jędryczka, M., 2018. Fungal biodiversity and their role in soil health. Front. Microbiol. 9, 707.

Frąc, M., Weber, J., Gryta, A., Debicka, M., Kocowicz, A., Jamroz, E., 2017. Microbial functional diversity in podzolec to humus horizons affected by alkaline fly ash in the vicinity of electric power plant. J. Geomicrobiol. 34, 579–586.

Funes, H., et al., 2009. Comparison of three types of traps baited with sexual pheromones for ambrosia beetle Megaplatypus mutatus (Coleoptera: Platypodinae) in poplar plantations. J. Econ. Entomol. 102, 1546–1550.

Garbeva, P.V., Van Veen, J.A., Van Elsas, J.D., 2004. Microbial diversity in soil: selection of microbial populations by plant and soil type and implications for disease suppressiveness. Annu. Rev. Phytopathol. 42, 243–270.

Gentili, F., Jumpponen, A., 2006. Potential and possible uses of bacterial and fungal biofertilizers. In: Rai, M.K. (Ed.), Handbook of Microbial Biofertilizers. Food Products Press, pp. 1–28.

Gianinazzi, S., Golotte, A., Binet, M.N., Van Tuinen, D., Redecker, D., Wipf, D., 2010. Agroecology: the key role of arbuscular mycorrhizas in ecosystem services. Mycorrhiza 20 (8), 519–530.

Godin, A., McLaughlin, J.W., Webster, K.L., Packalen, M., Basiliko, N., 2012. Methane and methanogen community dynamics across a boreal peatland nutrient gradient. Soil. Biol. Biochem. 48, 96–105.

Goetten, L.C., Moretto, G., Sturmer, S.L., 2016. Influence of arbuscular mycorrhizal fungi inoculum produced on-farm and phosphorus on growth and nutrition of native woody plant species from Brazil. Acta Bot. Bras. 30, 9–16.

Gorham, E., 1991. Northern peatlands: role in the carbon cycle and probable responses to climatic warming. Ecol. Appl. 1, 182–195.

Grigera, M.S., Drijber, R.A., Wienhold, B.J., 2007. Increased abundance of arbuscular mycorrhizal fungi in soil coincides with the reproductive stages of maize. Soil. Biol. Biochem. 39, 1401–1409.

Gunatilaka, A.A.L., 2006. Natural products from plant-associated microorganisms: distribution, structural diversity, bioactivity, and implications of their occurrence. J. Nat. Prod. 69, 509–526.

Ha, T.N., 2010. Using *Trichoderma* species for biological control of plant pathogens in Vietnam. J. ISSAAS 16, 17–21.

Hájek, T., Ballance, S., Limpens, J., Zijlstra, M., Verhoeven, T.A., 2011. Cell-wall polysaccharides play an important role in decay resistance of Sphagnum and actively depressed decomposition *in vitro*. Biogeochemistry 103, 45–57.

Harman, G.E., Howell, C.R., Viterbo, A., Chet, I., Lorito, M., 2004. *Trichoderma* species-opportunistic, avirulent plant symbionts. Nat. Rev. Microbiol. 2, 43–56.

Haselwandter, K., Berreck, M., 1994. Accumulation of radionuclides in fungi. Met. Ions Fungi 259–277.

Hernandez-Montiel, L.G., Rueda-Puente, E.O., Cordoba-Matson, M.V., Holguın-Pena, J.R., Zulueta-Rodrıguez, R., 2013. Mutualistic interaction of rizobacteria with arbuscular mycorrhizal fungi and its antagonistic effect on *Fusarium oxysporum* in *Carica* papaya seedlings. Crop. Prot. 47, 61–66.

Izquierdo, I., Caravaca, I., Alguacil, M.M., Hernandez, G., Roldan, A., 2005. Use of microbiological indicators for evaluating success in soil restoration after revegetation of a mining area under subtropical conditions. Appl. Soil. Ecol. 30, 3–10.

Jabiol, J., Cornut, J., Tlili, A., Gessner, M.O., 2018. Interactive effects of dissolved nitrogen, phosphorus and litter chemistry on stream fungal decomposers. FEMS Microbiol. Ecol. 94 (10), fiy151.

Janouskova, M., Krak, K., Vosatka, M., Puschel, D., Storchova, H., 2017. Inoculation effects on root-colonizing arbuscular mycorrhizal fungal communities spread beyond directly inoculated plants. PLoS One 12, e0181525.

Jeffries, P., Gianinazzi, S., Perotto, S., Turnau, K., Barea, J.M., 2003. The contribution of arbuscularmycorrhizal fungi in sustainable maintenance of plant health and soil fertility. Biol. Fertil. Soils 37 (1), 1–16.

Ju, Z., Wang, J., Pan, S., 2009. Isolation and preliminary identification of the endophytic fungi which produce hupzine A from four species in Hupziaceae and determination of huperzine A by HPLC. Fudan Univ. J. (Med. Sci. Ed.) 36, 445–449.

Kaewchai, S., Soytong, K., Hyde, K.D., 2009. Mycofungicides and fungal biofertilizers. Fungal Divers. 38, 25–50.

Khan, A.L., Hussain, J., Al-Harrasi, A., Al-Rawahi, A., Lee, I.J., 2015. Endophytic fungi: resource for gibberellins and crop abiotic stress resistance. Crit. Rev. Biotechnol. 35, 62–74.

Kielak, A.M., Cretoiu, M.S., Semenov, A.V., Sørensen, S.J., van Elsas, J.D., 2013. Bacterial chitinolytic communities respond to chitin and pH alteration in soil. Appl. Environ. Microbiol. 79 (1), 263–272.

Kishimoto, K., Matsui, K., Ozawa, R., Takabayashi, J., 2007. Volatile 1-octen-3-ol induces a defensive response in Arabidopsis thaliana. J. Gen. Plant. Pathol. 73, 35–37.

Kusari, S., Lamshoft, M., Zuhlke, S., Spiteller, M., 2008. An endophytic fungus from *Hypericum perforatum* that produces hypericin. J. Nat. Prod. 71, 159–162.

Laiho, R., 2006. Decomposition in peatlands: reconciling seemingly contrasting results on the impacts of lowered water levels. Soil Biol. Biochem. 38, 2011–2024.

Lange, L., 2014. The importance of fungi and mycology for addressing major global challenges. IMA Fungus 5 (2), 463–471.

Li, W., Zhou, J., Lin, Z., Hu, Z., 2007. Study on fermentation condition for production of huperzine A from endophytic fungus 2F09P03B of *Huperzia serrata*. Chin. Med. Biotechnol. 2, 254–259.

Lian, B., Wang, B., Pan, M., Liu, C., Teng, H.H., 2008. Microbial release of potassium from Kbearing minerals by thermophlic fungus *Aspergillus fumigatus*. Geochim. Cosmochim. Acta 72, 87–98.

Liu, R.J., Chen, Y.L., 2007. Mycorrhizology. Science Press, Beijing.

López-Bucio, J., Pelagio-Flores, R., Herrera-Estrella, A., 2015. *Trichoderma* as biostimulant: exploiting the multilevel properties of a plant beneficial fungus. Sci. Hortic. 196, 109–123.

Lorito, M., Woo, S.L., Harman, G.E., Monte, E., 2010. Translational research on *Trichoderma*: From omics to the field. Ann. Rev. Phytopath. 48, 395–417.

Macías-Rubalcava, M.L., Hernández-Bautista, B.E., Oropeza, F., Duarte, G., González, M.C., Glenn, A.E., et al., 2010. Allelochemical effects of volatile compounds and organic extracts from *Muscodor yucatanensis*, a tropical endophytic fungus from *Bursera simaruba*. J. Chem. Ecol. 36, 1122–1131.

MarchioriJúnior, M., Melo, W.J., 1999. Carbon, microbial biomass carbon and enzyme activity of a soil under natural forest, grassland and cotton culture. Rev. Brasil. Cienc. Solo 23 (2), 257–263.

Masto, R.E., Chhonkar, P.K., Singh, D., Patra, A.K., 2009. Changes in soil quality indicators under long-term sewage irrigation in a sub-tropical environment. Environ. Geol. 56 (6), 1237–1243.

McGuire, A.D., Anderson, L.G., Christensen, T.R., et al., 2009. Sensitivity of the carbon cycle in the Arctic to climate change. Ecol. Monogr. 79, 523–555.

Medeiros, A.O., Pascoal, C., Graça, M.A.S., 2009. Diversity and activity of aquatic fungi under low oxygen conditions. Freshw. Biol. 54 (1), 142–149.

Mercier, J., Jimenez, J.I., 2004. Control of fungal decay of apples and peaches by the biofumigant fungus *Muscodor albus*. Postharvest Biol. Technol. 31, 1–8.

Minerdi, D., Bossi, S., Maffei, M.E., Gullino, M.L., Garibaldi, A., 2011. *Fusarium oxysporum* and its bacterial consortium promote lettuce growth and expansion A5 gene expression through microbial volatile organic compound (MVOC) emission. FEMS Microbiol. Ecol. 76, 342–351.

Mitchell, A.M., Strobel, G.A., Moore, E., Robison, R., Sears, J., 2010. Volatile antimicrobials from Muscodor crispans, a novel endophytic fungus. Microbiology 156, 270–277.

Morgado, L., Semenova, T.A., Welker, J.M., Walker, M.D., Smets, E., Geml, J.O., 2015. Summer temperature increase has distinct effects on the ectomycorrhizal fungal communities of moist tussock and dry tundra in Arctic Alaska. Glob. Change Biol. 21, 959–972.

Murali, M., Amruthesh, K.N., Sudisha, J., Niranjana, S.R., Shetty, H.S., 2012. Screening for plant growth promoting fungi and their ability for growth promotion and induction of resistance in pearl millet against downy mildew disease. J. Phytol. 4, 30–36.

Ortas, I., 2012. The effect of mycorrhizal fungal inoculation on plant yield, nutrient uptake and inoculation effectiveness under long-term field conditions. Field Crop. Res. 125, 35–48.

Ostle, N.J., Smith, P., Fisher, R., et al., 2009. Integrating plant-soil interactions into global carbon cycle models. J. Ecol. 97, 851–863.

Pal, S., Singh, H.B., Farooqui, A., Rakshit, A., 2015. Fungal biofertilizers in Indian agriculture: perception, demand and promotion. J. Ecofriendly Agric. 10, 101–113.

Pietryczuk, A., Cudowski, A., Hauschild, T., Świsłocka, M., Więcko, A., Karpowicz, M., 2018. Abundance and species diversity of fungi in rivers with various contaminations. Curr. Microbiol. 75 (5), 630–638.

Pineda, A., Zheng, S.J., van Loon, J.J.A., Pieterse, C.M.J., Dicke, M., 2010. Helping plants to deal with insects: the role of beneficial soilborne microbes. Trends Plant Sci. 15, 507–514.

Pointing, S.B., Pelling, A.L., Smith, G.J.D., Hyde, K.D., Reddy, C.A., 2001. Screening of basidiomycetes and xylariaceous fungi for lignin peroxidase and laccase gene-specific sequences. Mycol. Res. 109, 115–124.

Pradhan, N., Sukla, L.B., 2005. Solubilization of inorganic phosphates by fungi isolated from agriculture soil. Afr. J. Biotech. 5, 850–854.

Qi, W.Z., Zhao, L., 2013. Study of the siderophore-producing *Trichoderma asperellum* Q1 on cucumber growth promotion under salt stress. J. Basic Microbiol. 53, 355–364.

Quinton, W.L., Hayashi, M., Carey, S.K., Myers, T., 2007. Peat hydraulic conductivity in cold regions and its relation to pore size and geometry. Hydrol. Process. 22, 2829–2837.

Rahi, P., Vyas, P., Sharma, S., Gulati, A., Gulati, A., 2009. Plant growth promoting potential of the fungus Discosia sp. FIHB 571 from tea rhizosphere tested on chickpea, maize and pea. Indian J. Microbiol. 49, 128–133.

Rinaldi, A.C., Comandini, O., Kuyper, T.W., 2018. Ectomycorrhizal fungal diversity: separating the wheat from the chaff. Fungal Divers. 33, 1–45.

Rouphael, Y., Franken, P., Schneider, C., Schwarz, D., Giovannetti, M., Agnolucci, M., et al., 2015. Arbuscular mycorrhizal fungi act as biostimulants in horticultural crops. Sci. Hortic. 196, 91–108.

Sadhana, B., 2014. Arbuscular mycorrhizal fungi (AMF) as a biofertilizers—a review. Int. J. Curr. Microbiol. App. Sci. 3 (4), 384–400.

Sadoma, M.T., El-Sayed, A.B.B., El-Moghazy, S.M., 2011. Biological control of downy mildew disease of maize caused by *Peronosclerospora sorghi* using certain biocontrol agents alone or in combination. J. Agric. Res. 37, 1–11.

Sandargo, B., Chepkirui, C., Cheng, T., Chaverra-Munoz, L., Thongbai, B., Stadler, M., et al., 2019. Biological and chemical diversity go hand in hand: Basidomycota as source of new pharmaceuticals and agrochemicals. Biotechnol. Adv. 37 (6), 107344.

Schöller, C.E., Gürtler, H., Pedersen, R., Molin, S., Wilkins, K., 2002. Volatile metabolites from actinomycetes. J. Agric. Food Chem. 50, 2615–2621.

Schwartz, D.M., Bazzaz, F.A., 1973. *In situ* measurements of carbon dioxide gradients in a soil-plant-atmosphere system. Oecologia 12, 161–167.

Sicardi, M., García-Préchac, F., Frioni, L., 2004. Soil microbial indicators sensitive to land use conversion from pastures to commercial *Eucalyptus grandis* (Hill ex Maiden) plantations in Uruguay. Appl. Soil Ecol. 27 (2), 125–133.

Siddiquee, S., Yusuf, U.K., Hossain, K., Jahan, S., 2009. *In vitro* studies on the potential *Trichoderma harzianum* for antagonistic properties against *Ganoderma boninense*. J. Food Agric. Environ. 7, 970–976.

Siddiqui, Z.A., Pichtel, J., 2008. Mycorrhixae: an overview. In: Siddiqui, Z.A., Akhtar, M.S., Futai, K. (Eds.), Mycorrhizae: Sustainable Agriculture and Forestry. Springer, pp. 1–36.

Singh, S.K., Strobel, G.A., Knighton, B., Geary, B., Sears, J., Ezra, D., 2011. An endophytic Phomopsis sp. possessing bioactivity and fuel potential with its volatile organic compounds, Microb. Ecol., 61. pp. 729–739.

Smith, S.E., Facelli, E., Pope, S., Smith, F.A., 2010. Plant performance in stressful environments: interpreting new and established knowledge of the roles of arbuscular mycorrhizas, Plant Soil, 326. pp. 3–20.

Smith, S.E., Read, D.J., 2008. Mycorrhizal Symbiosis, third ed. Academic Press, London.

Smith, S.E., Read, D.J., 1997. Mycorrhizal Symbiosis. Academic Press, San Diego, p. 607.

Song, G.C., Ryu, C.M., 2013. Two volatile organic compounds trigger plant self-defense against a bacterial pathogen and a sucking insect in cucumber under open field conditions. Int. J. Mol. Sci. 14, 9803–9819.

Spatafora, J.W., Chang, Y., Benny, G.L., Lazarus, K., Smith, M.E., Berbee, M.L., et al., 2016. A phylum-level phylogenetic classification of zygomycete fungi based on genome-scale data. Mycologia 108, 1028–1046.

Strobel, G., Singh, S.K., Riyaz-Ul-Hassan, S., Mitchell, A.M., Geary, B., Sears, J., 2011. An endophytic/pathogenic *Phoma* sp. from creosote bush producing biologically active volatile compounds having fuel potential. FEMS Microbiol. Lett. 320, 87–94.

Strobel, G.A., Dirkse, E., Sears, J., Markworth, C., 2001. Volatile antimicrobials from Muscodor albus, a novel endophytic fungus. Microbiology 147, 2943–2950.

Sun, J.M., Irzykowski, W., Jędryczka, M., Han, F.X., 2005. Analysis of the genetic structure of *Sclerotinia sclerotiorum* (Lib.) de Bary populations from different regions and host plants by Random Amplified Polymorphic DNA markers. J. Integr. Plant. Biol. 47, 385–395. Available from: https://doi.org/10.1111/j.1744-7909.2005.00077.x.

Sun, Z., Song, J., Xin, X., Xie, X., Zhao, B., 2018. Arbuscular mycorrhizal fungal proteins 14-3-3- are involved in arbuscule formation and responses to abiotic stresses during AM symbiosis. Front. Microbiol. 5, 9–19.

Sundram, S., Angel, L.P.L., Tay, B.Y.P., Roslan, N.D., Mohamed Azni, I.N.A., Idris, A.S., 2016. *Trichoderma virens*, an effective biocontrol agent against *Ganoderma boninense*. MPOB Inf. Ser. 587.

Tan, R.X., Zou, W.X., 2001. Endophytes: a rich source of functional metabolites. Nat. Prod. Rep. 18, 448–459.

Tang, J., Riley, W.J., 2014. Weaker soil carbon–climate feedbacks resulting from microbial and abiotic interactions. Nat. Clim. Change 5, 56–60.

Tao, G., Liu, Z.Y., Hyde, K.D., Lui, X.Z., Yu, Z.N., 2008. Whole rDNA analysis reveals novel and endophytic fungi in *Bletilla ochracea (Orchidaceae)*. Fungal Divers. 33, 101–122.

Thakeow, P., Angeli, S., Weißbecker, Schutz, S., 2008. Antennal and behavioral responses of Cis boleti to fungal odor of *Trametes gibbosa*. Chem. Senses 33, 379–387.

Thilagar, G., Bagyaraj, D.J., 2015. Influence of different arbuscular mycorrhizal fungi on growth and yield of chilly. Proc. Natl. Acad. Sci. India Biol. Sci. 85, 71–75. Available from: https://doi.org/10.1007/s40011-013-0262-y.

Todd-Brown, K.E.O., Randerson, J.T., Hopkins, F., et al., 2014. Changes in soil organic carbon storage predicted by Earth system models during the 21st century. Biogeosciences 11, 2341–2356.

Turetsky, M.R., Bond-Lamberty, B., Euskirchen, E., Talbot, J., Frolking, S., McGuire, A.D., et al., 2012. The resilience and functional role of moss in boreal and arctic ecosystems. N. Phytol. 196, 49–67.

Van der Heijden, M.G.A., Bardgett, R.D., van Straalen, N.M., 2008. The unseen majority: soil microbes as drivers of plant diversity and productivity in terrestrial ecosystems. Ecol. Lett. 11, 296–310.

Vargas, W.A., Djonovic, S., Sukno, S.A., Kenerley, C.M., 2008. Dimerization controls the activity of fungal elicitors that trigger systemic resistance in plants. J. Biol. Chem. 283, 19804–19815.

Visser, S., Parkinson, D., 1992. Soil biological criteria as indicators of soil quality: soil microorganisms. Am. J. Altern. Agric. 7 (1–2), 33–37.

Wagg, C., Bender, S.F., Widmer, F., Van der Heijden, M.G.A., 2014. Soil biodiversity and soil community composition determine ecosystem multifunctionality. Proc. Natl. Acad. Sci. USA 111, 5266–5270. Available from: https://doi.org/10.1073/pnas.1320054111.

Wang, Q., Fu, Y., Gao, J., Wang, Y., Li, X., Zhang, A., 2007. Preliminary isolation and screening of the endophytic fungi from *Melia azedarach* L. Acta C. Boreali-Occidentalis Sin. 16, 224–227.

Wang, X., Wang, C., Sui, J., Liu, Z., Li, Q., Ji, C., et al., 2018a. Isolation and characterization of phosphofungi, and screening of their plant growth-promoting activities. AMB Expr. 8, 1–12.

Wang, X., Wang, C., Sui, J., Liu, Z., Li, Q., Ji, C., et al., 2018b. Isolation and characterization of phosphofungi, and screening of their plant growth promoting activities. AMB. Express 8 (1), 63.

Wheatley, R.E., 2002. The consequences of volatile organic compound mediated bacterial and fungal interactions. Anto van. Leeuwen 81, 357–364.

Whipps, J.M., 2004. Prospects and limitations for mycorrhizas in biocontrol of root pathogens. Can. J. Bot. 82, 1198–1227.

Whitelaw, M.A., Hardena, T.J., Helyar, K.R., 1999. Phosphate solubilisation in solution culture by the soil fungus *Penicillium radicum*. Soil Biol. Biochem. 3, 655–665.

Wiłkomirski, B., 2013. History of bioindication. Monit. SrodPrzyr 14, 137–142.

Yang, S., Li, F., Malhi, S.S., Wang, P., Dongrang, S., Wang, J., 2004. Long term fertilization effects on crop yield and nitrate nitrogen accumulation in soil in Northwestern China. Agron. J. 96, 1039–1049.

Young, I.M., Ritz, K., 2005. The habitat of soil microbes. In: Bardgett, R.D., Usher, M.B., Hopkins, D.W. (Eds.), Biological Diversity and Function in Soils. Cambridge University Press, Cambridge, UK, pp. 31–43.

Zerba, E., Gonzalez Audino, P., 2009. Comparison of three types of traps baited with sexual pheromones for ambrosia beetle *Megaplatypus mutatus* (Coleoptera: Platypodinae) in poplar plantations. J. Econ. Entomol. 102, 1546–1550.

Zhang, H., Kim, M.S., Krishnamachari, V., Payton, P., Sun, Y., Grimson, M., et al., 2007. Rhizobacterial volatile emissions regulate auxin homeostasis and cell expansion in Arabidopsis. Planta 226, 839–851.

Zhang, L., Gu, S., Shao, H., Wei, R., 1999. Isolation, determination and aroma product characterization of fungus producing irone. Mycosystema 18, 49–54.

Zhang, L., Shi, N., Fan, J., Wang, F., 2018. Arbuscular mycorrhizal fungi stimulate organic phosphate mobilization associated with changing bacterial community structure under field conditions. Environ. Microbiol. 20, 2639–2651.

Zhang, Q., Sun, J., Liu, S., Wei, Q., 2013. Manure refinement affectsapple rhizosphere bacterial community structure: a study in sandy soil. PLoS One 8, e76937.

Zhang, S., Gan, Y., Xu, B., 2016. Application of plant-growth promoting fungi *Trichoderma longibrachiatum* T6 enhances tolerance of wheat to salt stress through improvement of antioxidative defense system and gene expression. Front. Plant Sci. 7, 1–11.

Zhao, M., Li, M., Liu, R.J., 2010. Effects of arbuscular mycorrhizae on microbial population and enzyme activity in replant soil used for watermelon production. Int. J. Eng. Sci. Technol. 2, 17–22.

Zhou, L., Cao, X., Yang, C., Wu, X., Zhang, L., 2004. Endophytic fungi of *Paris polyphylla* var. *yunnanensis* and steroid analysis in the fungi. Nat. Prod. Res. Dev. 16, 198–200.

Zhou, S., Yang, F., Lan, S., Xu, N., Hong, Y., 2009. Huperzine A producing conditions from endophytic fungus in SHB *Huperzia serrata*. J. Microbiol. 29, 32–36.

Zhu, G.S., Yu, Z.N., Gui, Y., Liu, Z.Y., 2008. A novel technique for isolating orchid mycorrhizal fungi. Fungal Divers. 33, 123–137.

Zifcakova, L., Vetrovsky, T., Howe, A., Baldrian, P., 2016. Microbial activity in forest soil reflects the changes in ecosystem properties between summer and winter. Environ. Microbiol. 18, 288–301. Available from: https://doi.org/10.1111/1462-2920.13026.

Chapter 17

Yeast as plant growth promoter and biocontrol agent

Dhriti Kapoor[1] **and Arun Karnwal**[2,*]
[1]*Department of Botany, School of Bioengineering & Biosciences, Lovely Professional University, Phagwara, India,* [2]*Department of Microbiology, School of Bioengineering & Biosciences, Lovely Professional University, Phagwara, India*
**Corresponding author.* Email: *arunkarnwal@gmail.com*

Introduction

Yeast considered an illustrious eukaryotic living organism due to its incredible biodiversity and widespread industrial uses (Nandy and Srivastava, 2018). Among these organisms, *Saccharomyces cerevisiae* is applied majorly for industrial purposes, such as for the synthesis of metabolites, proteins, and other chemical products (Bennett and Whipps, 2008). Genetic manipulation of *S. cerevisiae* is comparatively a convenient process, and it is also a conventional technique for gathering genetic tools in comparison to several microorganisms. *S. cerevisiae* is a single-celled microorganism that is studied on a large scale and utilized for the generation of different protein expression and synthesis of certain biochemicals in the industries (Mattanovich et al., 2014). Apart from the production of various chemical products and protein expression, *S. cerevisiae* can be used in some traditional applications like the fermentation of alcohol, baking, or bio-ethanol processes (Nandy and Srivastava, 2018).

Certain microorganisms which contribute in growth promotion of the plants, also having some primary & secondary advantages such as nitrogen assimilation, degradation of toxic substances, production of antibiotics, vitamins and antifungal compounds, solubilization of minerals & nutrients like phosphorus along with certain plant growth regulators like gibberellic acid, auxins, and such microorganisms are termed as PGPM; plant growth-promoting microorganisms (Hu and Qi, 2013; Nutaratat et al., 2014). It is a significant fact that in natural conditions like in agricultural lands as well as in vitro conditions, the influence of PGPM can be conserved that effects of PGPM are preserved (Agamy et al., 2013; Ji et al., 2014). According to the

Fungi Bio-Prospects in Sustainable Agriculture, Environment and Nano-Technology.
DOI: https://doi.org/10.1016/B978-0-12-821394-0.00017-2

several reports, promotion in seed germination, root-shoot lengths of seedlings, their biomass, photosynthetic efficiency or overall enhancement in growth was observed when treated with the strains of yeast (Pérez-Montaño et al., 2014; Amprayn et al., 2012; Nassar et al., 2005). Besides, optimistic influences in context with the synthesis of bioactive compounds, enhancement in the product formation, and retardation in the activity of plant pathogens were observed with the yeast applications even in the early stages of the plant development (Agamy et al., 2013).

Plant growth promoting activities of microorganisms

Certain microorganisms like *Trichoderma longibrachiatum* T6 has the capacity to enhance the growth of agricultural plants like wheat, rice, tomato, and others has ability to suppress the effects of pathogen attack, i.e., they provide resistance against specific biotic and abiotic stresses such as nematodes, salinity stress, etc. and further leads to growth and yield promotion (Zhang et al., 2014). T6 and certain other microorganisms like yeast also having the potential to combat the deleterious effects produced by *Heterodera avenae* like pathogens simultaneously growth and yield can be regulated in the significantly important crops (Zhang et al., 2014). Reports suggested that these growth influencing features of endophytes can be understood by analyzing and comparing the activities of specific endophytes such as bacterial species and many yeast strains that can be helpful for the agricultural practices also (Nassar et al., 2005). The locale used by the microorganisms like yeast and bacteria and the phytosphere occupied by them gives the indications at the mark of their convergent evolution. We can get comprehensible ideas from the same lifestyles of these microorganisms in context with their physiology, their unicellular body by which they adjust in the aquatic habitats, their ecological positions, as well as the significance of biofilms, and their inclined growth rates in their lifecycle (Firrincieli et al., 2015).

Apart from this, certain reports indicated that yeasts and bacteria have the potential to form mixed biofilms (Bandara et al., 2006; Marcet-Houben and Gabaldón, 2010). Such concurrent behaviors, along with their physical or ecological vicinities, appear to specify the prospects of cross-kingdom gene transfer measures horizontally. Such activities were found to be implicated in the past; however, until now, these indications have proved very less convincing (Firrincieli et al., 2015). Along with various growth enhancing characteristics, endophytic yeasts may also possess capacities like providing protection from the attack of pathogens, encouraging adequate minerals accumulation, biosynthesis of plant growth regulators, and amelioration of both biotic as well as abiotic stresses. The different strains of yeast can possess these features, and few of them are considered as their peculiar features, which make them unique (Moller et al., 2016).

Applications of endophytes lead to the enhanced production of plant growth promoters, which acts as a primary method for enhancing plant growth, and these include the biosynthesis of indole-3-acetic acid (IAA), enhancement in the activity of ACC deaminase and siderophore synthesis (Doty, 2013). Various growth stimulation characteristics are possessed by the plant hormones like auxins and gibberellins, such as increased root and shoot growth along with the elongation of cells and its proliferation. Reports suggested that endophytic yeasts mainly lead to the synthesis of indole-3-acetic acid that has been extensively described by numerous groups and also has been reviewed widely (Moller et al., 2016; Nassar et al., 2005; Xin et al., 2009). Remarkably, this characteristic was also analyzed in the media in which tryptophan was supplied (Hardoim et al., 2008). Apart from yeast strains, this feature was also a characteristic of endophytic bacteria (Kandel et al., 2017). Such type of analysis gives a resilient foundation for the symbiotic relationship of plants and endophytes.

A multifarious amino acid, tryptophan, is endogenously present in the plants and their exudates, but it is expensive to synthesize it artificially for them (Kamilova et al., 2006). As tryptophan is a precursor of IAA, as it provided to the plants, they tend to synthesize more IAA, consequently stimulates the growth and yield of plants, though the exogenous application of this stimulator is too costly (Glick et al., 2007). Another role of this process is a diversion of nitrogen content in different parts of the plants, including roots, shoots, etc., when nitrogen content is in excess. Another stimulator known as 1-aminocyclopropane-1-carboxylate (ACC) is found in endophytes like yeasts or bacteria, which acts as a precursor for ethylene hormone synthesis (also acts as a stress ameliorator). ACC is transformed into a-ketobutyrate and ammonia by the help of enzyme ACC deaminases after cleavage (Glick et al., 2007; Nutaratat et al., 2014) and different strains of endophytic yeasts do this function, hence they act as a sink for ACC, consequently level of ethylene tend to decrease in the plants as ethylene is considered as a stress hormone and contributes in growth retardation in plants. Opposite to this, yeasts enhance the resistance of plants against stress conditions, hence play a role in preventing the reduction in growth. According to the previous studies, during this process, enzymes secreted by yeasts like ACC deaminases contribute to the release of ammonia, which can generate a mechanism of recycling nitrogen for the plants by the use of their symbiotic partners (Nutaratat et al., 2014).

The other fundamental characteristic of endophytic yeast is the synthesis of siderophore, which has been considerably studied in the case of arboreal endophytes and observed in the crops like rice and sugarcane in their leaf parts (Moller et al., 2016; Nutaratat et al., 2014). When plants face the deficiency of iron, then iron-chelating substances are synthesized by them, which are known as siderophores (Loaces et al., 2011). Such substances are released and again taken when they possess the chelated iron. Following this mechanism, plants get the chelated iron, which cannot be gathered from the soil and hence contributes to the stimulation of plant growth (Loaces et al., 2011). Moreover, for colonization of plants and excluding different microorganisms, such compounds may help the endophytic yeasts colonize the plant, which justifies the antifungal properties of particular yeast. Apart from this, avoiding the spoilage of fruit is also one of the peculiar features of endophytic yeasts (Nutaratat et al., 2014).

Though various other strategies of growth stimulation in plants were observed, those have not been used generally most probably due to scarcity of possibly as a result of a lack of knowledge in the detailed description of such endophytic yeasts. For example, an antioxidative enzyme, namely catalase (CAT) activity, was found to enhance during stress conditions as this enzyme causes scavenging of reactive oxygen species generated due to stress (Khalifa et al., 2016). Recently, the activity of CAT was also observed in various strains of yeast endophytes in the leaves of some crops such as rice and sugar cane (Nutaratat et al., 2014). It has also been found out that the release of ammonium is independent of the activity of ACC deaminase and also from the enhanced solubilization of phosphate by different yeasts.

Besides, the synthesis of plant growth regulators like polyamines has also been analyzed in the case of yeast. Such compounds have a vital contribution in a plant for the stimulation of plant growth, and these were also found to be involved in different cellular processes, which includes the production of macromolecules, cell stress resistance along with growth and survival. Conversely, these growth regulators usually play the role of growth enhancement in yeasts (Takahashi and Kakehi, 2009).

Yeast as a biofertilizers (nutritional value enhancement)

Though in both vegetative and the reproductive phases, yeast suspensions play an essential role as natural biostimulants in plants (Ibraheim, 2014). A few detailed studies are there, where yeast had been applied as an elicitor. There is variation in the degree of elicitation amongst plant species; hence, research at the minimum species level is required in this area (Złotek and Świec, 2016). Many reports suggested that in crops, the nutritional value is increased by the application of yeast suspension. For elicitation, yeast suspension is suggested to be applied by Gawlik-Dziki et al. (2016), and hence it promotes the more advantageous nutritional level of the products in place of transgenic foods that are having less consumer recognition.

Apart from this, presence of certain macronutrients like NPK (nitrogen, phosphorus and potassium) along with magnesium and calcium in plants caused enhancement in growth in the plants like tomato, pepper, eggplant, lettuce, bean, etc. (El-Tohamy et al., 2015; Farrag et al., 2016; Fathy and Farid, 1996). The application of yeast suspension decreased the concentration of nitrates Zn, Fe, and Mn content in lettuce (Fawzy, 2010). Similar results were observed in caraway shoots (Ramadan and Ragab, 2015). An enhancement in protein, free amino acid, and carbohydrate content was observed in caraway, wheat, sugarbeet, and faba bean plants, treated with yeast extracts (Neseim et al., 2014; Abdallah et al., 2016; Mady, 2009; Agamy et al., 2013; Fathy and Farid, 1996; Abbas 2013; Hammad and Ali, 2014; Ramadan and Ragab, 2015) whereas opposite results were analyzed in milk thistle seedlings where reduction in amino acids was observed by Saad-Allah et al. (2017) under yeast suspension treatment.

In the leaves of wheat plants, activities of peroxidases and phenoloxidases were found to enhance when treated with yeast suspension (Hammad and Ali, 2014). On the other hand, a reduction in the activity of phenylalanine ammonia-lyase and enhancement in tyrosine aminotransferase was analyzed in red sage leaves (Yan et al., 2006). According to Fawzy (2010), in lettuce, a substantial rise in ascorbic acid content was observed. Enhancement in the level of auxins and gibberellic acid was analyzed in bean shoot after the application of yeast (El-Tohamy and El-Greadly, 2007). Similarly, indole acetic acid content was also enhanced in the quinoa leaf part (Abdallah et al., 2016).

On the other hand, a collective rise in abscisic acid and cytokinin was measured by Abbas (2013) and El-Shafey et al. (2016) in bean plants. Increased content of cytokinins was also analyzed by El-Tohamy and El-Greadly (2007) on bean and by El-Tohamy et al. (2015) on eggplant. Whereas in the leaves of faba bean, specific contradictory results were obtained, where retardation in the content of ABA and rise in auxins were analyzed (Mady 2009). Similarly, Abou El-Yazied and Mady (2011, 2012) analyzed the increase in auxin, gibberellins, and cytokinins levels and reduction in abscisic acid in the tomato and broad bean plants. In the leaves of lettuce, alleviated content of caffeic acid, ferulic acid, and chicoric acid was observed with yeast treatment by Złotek and Świeca (2016).

In the xerophytic plants, the concentration of specific ions like potassium, nitrogen, and phosphorus was found to increase in their seeds, when yeast suspension was applied (Mady, 2009; Mahmoud et al., 2016). On the other hand, findings of Agamy et al. (2013) suggested the enhanced level of sucrose in sugar beetroots with increasing the concentration of yeast extracts. When foliar spray of yeast suspension was given to the onion bulbs, total soluble solids content was found to be enhanced drastically (Abdel-Moneim et al., 2015). Similar results were obtained after the experiments performed by El-Tohamy et al. (2015) and Hernández et al. (2007) in sweet potato tubers and potato, respectively.

Application of yeast suspension on potato foliage leads to enhancement in the percentage of starch Apart from this, in onion plants, level of Cu, Mn and Fe and in sugar beetroots, protein content, phenols and free amino acids level was elevated after yeast treatment (Agamy et al., 2013; Ahmed et al., 2011). Whereas foliar spray of yeast extract on onion bulbs caused an increase in the level of macronutrients along with nitrate and nitrite content (Ahmed and Farm, 2015; Abdel-Moneim et al., 2015). Ahmed et al. (2011) also analyzed the same results on potato plants due to yeast application, where macronutrients, along with the Zn content, were raised (Ahmed et al., 2011). According to the reports of Ahmed et al. (2011), enhanced protein content was found in potato tubers after yeast treatment. Whereas, onion bulbs showed a rise in total carbohydrates and volatile oil content on the application of yeasts (Ahmed and Farm, 2015). A positive influence was analyzed in context with the ascorbic acid level of tomato plants and the fruits of sweet pepper after treatment of yeast suspension (Abou El-Yazied and Mady, 2011; Ghoname et al., 2010). Additionally, in the inflorescence of *Cynara scolynus*, alteration at a nutritional level was found after following yeast applications, hence drastically stimulated sodium, inulin, and total carbohydrate (Hafez, 2013).

Stress tolerance

In the natural environmental conditions, living organisms are exposed to usual fluctuations in conditions such as temperature, humidity, pH, mineral/nutrient availability, and heavy metals. These adverse modifications necessitate an adequate response. Suboptimal conditions can be avoided by the organisms which are multicellular and motile in nature as they can move from place to another. Whereas, unicellular organisms do not possess this ability, but they can carry different strategies that make them to identify such conditions immediately and to accord the reactions against them except for the extremely hazardous conditions due to overdoses of toxic substances. One of the strategies among them is gene rearrangement or stochastic switching, in which the organisms tend to make certain modifications in their cells so that they can become resistant towards the unfavorable conditions called stress. Boosting up specific defensive strategies inside the organism's body is another way to resist the environmental stresses; the mechanism of the organism's responses (Okada et al., 2014).

Yeasts can survive almost in every type of environment, and these are also considered as persuasive antagonists of a large number of plant pathogens. Unchallenging cultivation necessities and restricted biosafety alarms like such antagonistic features of yeasts, various strains of yeasts have been used for the biocontrol applications (Freimoser et al., 2019).

There are numerous reports, where the responses of organisms against stresses have been studied, and reports on *Saccharomyces cerevisiae* are among them, which is also called baker's yeast. These organisms are single-celled, eukaryotes and not capable of causing any disease and reside in water, soil, and different plant parts like flowers, leaves, and fruits. Due to specific anthropogenic reasons, these are also related to human life. These organisms have been utilized from ancient times for the production of bread and beer, and nowadays also been utilized in various industrial processes like fermentation which includes baking, distillation, beer & wine production. Further, another role of this organism includes the generation of biomass involving baker's yeast, fodder yeast, yeast extracts, or single-cell proteins (SCP) (Gasch et al., 2000).

In yeast, numerous strategies are functional against different types of stress. Among them, individual responses are initiated by different stress factors called 'Environmental Stress Response' (ESR), also termed as 'common environmental response' (CSR) and to fight against any specific type of stress, there are particular precise stress factor (Causton et al., 2001; Gasch et al., 2000). Different environmental factors have their distinctive response against stresses, as they possess essentials of specific as well as the environmental stress response. One specific process called 'systemic acquired resistance' is widespread, also known as 'adaptive responses.' Such responses are present in budding yeast namely *S. cerevisiae* along with various microorganisms as well as individual multicellular organisms, for example plants, mammals and flatworms including humans (Durrant and Dong, 2004; Kensler et al., 2007; Mitchell et al., 2009; Scholz et al., 2005). This strategy contributes significantly to the survival of different organisms in the changing environmental conditions. This mechanism includes the attainment of resistance against stress conditions in the former incubation under mild stress specifically of the same type or the different type, which is further called same and cross stress resistance, respectively. Yeast cells, when exposed to 0.7 M NaCl stress conditions for a short period, triggers the survival of more cells under a high concentration of NaCl, i.e., 1.4 M NaCl (Trollmo et al., 1988; Varela et al., 1992). Similar results were obtained when the cells were subjected to heat stress conditions such as 45 °C succeeding earlier slight advancement of the culture temperature from 28 to 38 °C (Davies et al., 1995).

Several facets must be compared while exploring the molecular aspect of stress resistance strategies. The basis of stress resistance depends upon the potential of an organism that how strong cellular defense and the repair mechanism it is having (Hillenmeyer et al., 2008). In the yeast S. cerevisiae, degradation of vacuoles, vesicular transport related to endosomes synthesis of RNA like functions of nucleus responsible for showing the susceptibility towards stress (Dudley et al., 2005; Parsons et al., 2006; Okada et al., 2014).

For protecting against stress in plants, *C. oleophila* was used as the first agent among yeast, and also, this research is considered an initial step to make it commercialized. There are different reports which also indicated the antifungal strategies of the yeast cell. Though yeasts play an antagonistic role against fungal species, which are pathogenic and these two have intense competition with each other for food and space. In some species like *C. oleophila* and other *Candida* species possess certain hydrolytic enzymes, including proteases, chitinases, and glucanases, along with some volatile compounds that possess antifungal activity (Huang et al., 2011; Segal et al., 2002). Besides, the yeast species such as *Candida* considered as biocontrol agents as they have potential to form biofilms, stimulate stress tolerance in different plant parts, contributes in osmoregulation and shows parasitism of hyphae (Droby et al., 2002; El Ghaouth et al., 2003; Wisniewski et al., 2007).

Various reports (Jijakli and Lepoivre, 1993; Tian et al., 2004; Zhang et al., 2007; Dudley et al., 2005; Parsons et al., 2006; Okada et al., 2014) suggested that *Cryptococcus albidus* can lead to high lipid yields for the production of biodiesel, Whereas some other yeast strains like *C. albidus*, *C. laurentii* and *C. flavus* provide protection to peach, cherry, citrus, strawberry, tomato and pome fruits against postharvest decay (Tian et al., 2004; Zhang et al., 2007). In one of the products, namely Yieldplus® (registered in 1997) and marketed in South Africa by Anchor Bio-Technologies, *C. albidus* was applied as a biocontrol agent. For a long period of 15 years, this product was sold in the market, but now it has been introverted (Mbili, 2012). This product was developed and used for protecting citrus fruits and pomegranates against the attack of *B. cinerea* and *P. expansum* and also been applied in strawberries against Botrytis at the time of post-harvest cold storage (Kowalska et al., 2012).

Metschnikowia genera are having majorly nectar yeasts and phyllosphere, which are ubiquitous in nature (Chappell and Fukami, 2018; Pozo et al., 2012; Vadkertiová et al., 2012). In this particular genera, *M. fructicola* and *M. pulcherrima* species are the most studied in context with biocontrol, which is having ability to reduce a range of postharvest and plant rot diseases, including the most persuasive antagonistic yeasts that have been recognized till date (Hilber-Bodmer et al., 2017; Parafati et al., 2015; Türkel et al., 2014). In these two genera, entire genomes are existing in various species, including *M. fructicola* and *M. pulcherrima* (Gore-Lloyd et al., 2019; Piombo et al., 2018). Additionally, transformation practices have also been developed, which can be expressed as a green fluorescent protein and accompaniment, a naturally existing mutant (Gore-Lloyd et al., 2019).

Different strains of *S. cerevisiae* like DISAABA1182, RC008, RC009, RC012, and RC016 caused the inhibition in the growth of plant pathogens like *A. parasiticus*, *A. carbonarius*, *A. ochraceus* or *Fusarium graminearum* and also leads to reduction in the synthesis of mycotoxins such as zearalenone, aflatoxin, deoxynivalenol, ochratoxin A (Armando et al., 2013; Cubaiu et al., 2012). Against the different stress responses, the removal of mycotoxins occur by the S. cerevisiae cell walls by adsorption process, by alterations in the composition of plasma membrane, followed by petulin revelation and transcriptional downregulation of polyketide formation (Armando et al., 2012; Cubaiu et al., 2012; Oporto et al., 2019). Emancipating killer activity and certain volatile compounds along with some hydrolytic enzymes can be described as another strategy as biocontrolling agents in yeast cells mainly against antifungal activities such as *C. acutatum* in citrus species (Lopes et al., 2015).

Comprehending the strategies of salt stress tolerance, eukaryotic microorganisms play a significant role as model organisms, which are halotolerant in nature. The generation of oxidative burst due to reactive oxygen species (ROS), can be ameliorated by black yeast *Hortaea werneckii*, which is highly saline resistant (Petrovič, 2006). These saline stress tolerant microorganisms are significantly important model organisms as they have evolutionary achieved strategies to cope with saline stress (Prista et al., 2002). As the problem of salinity stress is the primary agricultural issue, these days worldwide, so considering such strategies may contribute to ameliorating salinity stress in crops, higher eukaryotic species (Zhu, 2001; Wang et al., 2003). Therefore, these eukaryotic organisms which are having the potential of stress resistance have sought such great attention these days due to increasing salt stress conditions in the environment (Gunde-Cimerman et al., 2004). *Debaryomyces hansenii*, *Hortaea werneckii*, and *C. halophile* are few examples of extremely halotolerant yeasts; thus, these species are considered as model organisms to study the physiological and molecular aspects in eukaryotic organisms for salt stress resistance (Petrovič et al., 2002; Silva-Graça and Lucas, 2003).

After the heat stress conditions, plant parts can be recovered by soaking them in yeast suspension in NFP emulsion. After this treatment, certain features of the fruits like acidity, skin color, mass, and insistence are not adversely affected (Jijakli and Lepoivre, 1993). It has been reported that exposure to high-temperature conditions along with the application of yeast suspension can contribute to regulating the infections which are caused by *B. cinerea* and *Gloeosporium* rot and several apples rotted (in percentage) by *Penicillium* spp. has been observed to reduce from 18.2% to 3.8% without causing any adverse effect on any quality parameters. Hence, against the various fungal treatments, yeast suspensions have proved beneficial to overcome their adverse effects. However, thermotherapy and the biological control might be further used to analyze as an integrated approach for the assessment of the reasonable probability with a low exhaustive pattern of chemical treatment (Jijakli et al., 1999).

Mechanism of stress resistance

The organism which is used as a biocontrol agent that must be effective to prevent any disease caused by pathogens along with certain characters is also significant like conditions & requirement of products, biosafety and registration issues, preparation selections, and the prerequisite solicitation equipment, etc. Though the morphology of yeast, including its filamentous growth and absence of intrusive, make some hindrance but precisely its external appearance, only favors the manageable in the fermenters, beneficial invention features, and sufficient treatment selections. Besides, the unicellular structure of yeast helps in the creation of biofilm and also leads to adhesion and hence affects the environmental perseverance, affordability and thus upgraded its biocontrol activity (Fanning and Mitchell, 2012; Pandin et al., 2017; Rossouw et al., 2018; Verstrepen and Klis, 2006).

Although non-conventional yeast does not possess plasmid, they can maintain extra-chromosomal and foreign DNA with the help of plasmid vector, whereas *S. cerevisiae* possesses approximately 650–1000 sequences plasmid or can copy 2-micrometer plasmid (Peter et al., 2018; Cao et al., 2017). That is why yeast, along with certain bacteria, can contribute to growth promotion without passing any toxin genes or antibiotic resistance, which is plasmid-based, and hence, they also possess specific biocontrol activity. Additionally, horizontal gene transfer is not very common in yeast as in fungi, as they possess the complex genome structure (Fitzpatrick, 2012; Moriguchi et al., 2013; Richards et al., 2011).

In the yeast, there are different strategies which make them biocontrol agents such as they secrete certain enzymes, toxins, compete for space and nutrients, release certain volatile compounds, mycoparasitism and can stimulates resistance against different stress conditions in plants (Fig. 17.1) (Droby et al., 2009; Punja and Utkhede, 2003; Wisniewski and Droby, 2012).

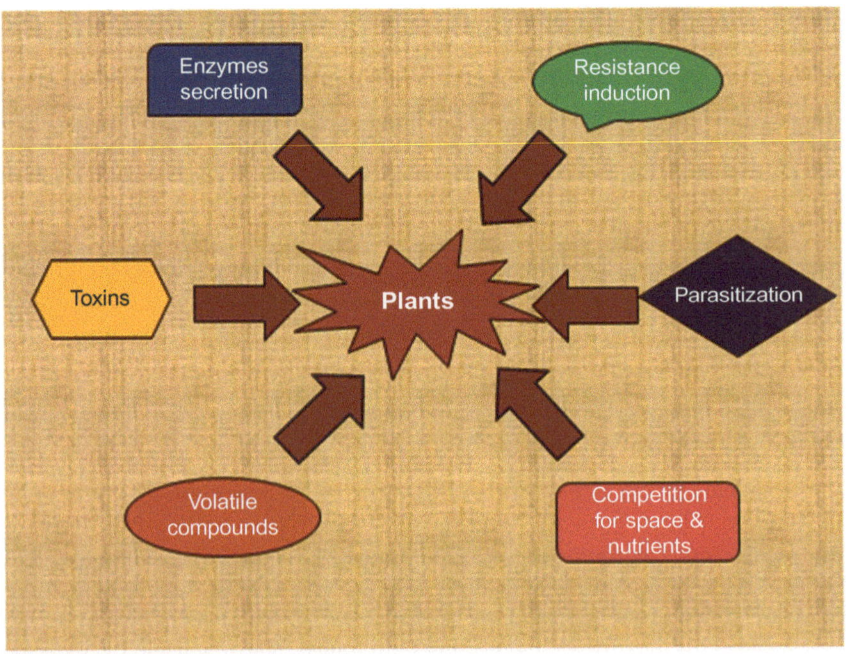

FIGURE 17.1 Mechanisms involved in biocontrol activity to yeasts.

Competition for nutrients and space

Different microorganisms have competition with each other for nutrition and food (Spadaro and Droby, 2016). Acting as a biocontrol agent, they have to do struggle for these primary requirements (Schaible and Kaufmann, 2005; Wisniewski et al., 2007). Mechanically, competition is challenging to understand, specifically in the natural conditions of the environment, where fewer resources are available, and a large number of organisms are there to compete. Competition for habitat and food in community ecology was understood penetratingly as elements of species diversity. *Metschnikowia reukaufii*, nectar yeast, plays an essential role as teamster of competitiveness for various species that are very similar to each other in context with biocontrol agent leads to some prior influences like effectual resource exhaustion as a result of repetition of nitrogen transporter and metabolism genes (Dhami et al., 2016).

To compete for the habitat, biofilm creation is also measured as an explicit and extremely efficacious mechanism. Biofilms are the population of microorganisms, made up of single or multiple numbers of species conglomerates, where these species grow on the surfaces (Costa-Orlandi et al., 2017). These biofilms immensely carry distinguishing features in comparison to free-floating cells and are also deliberated a virulence factor for pathogenic microbes (Costa-Orlandi et al., 2017; Davey and O'toole, 2000; Desai et al., 2014). In case of yeast strains, the formation of biofilms start with the attachment of a cell with the surface and which further leads to release of the extracellular matrix, alterations in the cell wall and creation of hyphal structure or pseudohyphae (Cavalheiro and Teixeira, 2018; Costa-Orlandi et al., 2017). Even in yeast strain, the development of biofilms has been understood at the molecular level (Cabral et al., 2014; Cavalheiro and Teixeira, 2018; d'Enfert and Janbon, 2016; Lohse et al., 2018; Reynolds and Fink, 2001). Conversely, the molecular reinforcements of the procedure and the configuration of various biofilms such as cell differentiation and multispecies biofilms have only been observed in detail for *Pichia fermentans* (Giobbe et al., 2007; Maserti et al., 2015). Formation of biofilm in case of apple wounds helps in providing protection against the postharvest diseases; thus this species has explicitly proven captivating regarding this, whereas *P. fermentans* shifts from the yeast, elongates as hyphal structure and rapid decay is caused in the absence of any pathogen in inoculated fruits (Fiori et al., 2012; Sanna et al., 2012, 2014).

Secreted enzymes

Reports suggested that in all types of host-pathogen interactions, certain enzymes are secreted by them, which causes degradation of all the cells and their components. These enzymes are secreted by prey cells that cause the killing of cells. Generally, these enzymes get upregulated when fewer nutrients are available and also act as a source of carbon and amino acids, which act as nutrients. Many enzymes are released, activated, and usually testified in yeasts such as proteases, chitinases, glucanases, etc. and can be utilized their biocontrol activity. One of the enzymes, namely chitinolytic, causes the degradation of cell walls and possesses the required features for acting as a biocontrol agent (Zajc et al., 2019). Chitinase enzyme activity was found in various genera of yeasts, which act as biocontrol agents like *Aureobasidium*, *Candida*, *Metschnikowia*, *Meyerozyma*, *Pichia*, *Saccharomyces*, *Tilletiopsis*, and in *Saccharomycopsis*, etc. in the existence of prey (Bar-Shimon et al., 2004; Junker et al., 2019; Lopes et al., 2015; Pretscher et al., 2018; Saravanakumar et al., 2009; Urquhart and Punja, 2002; Zajc et al., 2019; Zhang et al., 2011).

Major cell wall components of fungi are glucans, and for attachment to the cells, their alterations, and tolerance against toxins, exoglucanases are present (Tsai et al., 2011; Xu et al., 2013). The first gene, namely 1,3-β-glucanase (CoEXG1), was isolated from the yeast *Candida oleophila* that was used in the cloning purpose (Segal et al., 2002). Initially, spore germination of *Penicillium digitatum* was not severely influenced by the overexpression or removal of CoEXG1 (Yehuda et al., 2003), but later on, reduction in the activity of β-exoglucanase deletion mutant was analyzed in comparison to the wild type. Hence, the overexpression of yeast strain has proven the significant importance of glucanases as biocontrol activity (Bar-Shimon et al., 2004). In *Pichia anomala* (*Wickerhamomyces* sp.), removal of 2 exo-β-glucanases (PaEXG1 and PaEXG2) drastically retard biocontrol activity on different parts like fruits against *Botrytis cinerea* (Friel et al., 2007), however biocontrol activity was not found to decrease by the single omission of PaEXG2 (Grevesse et al., 2003). The activity of exoglucanase without representing an underlying engrossment, associated with the incompatible role, was identified in several biocontrol yeasts (Chan and Tian, 2005). Activities of β-1,3-glucanases were not linked to its particular retarding activity against *B. cinerea* in *Rhodotorula glutinis* and *Cryptococcus laurentii* (Castoria et al., 1997). In *S. cerevisiae*, six isolates showed antifungal activities against exoglucanases secreted by *Colletotrichum acutatum* as well as by the *Pichia guilliermondi* (Lopes et al., 2015; Zhang et al., 2011).

In different strains of yeasts, a lipolytic activity exists typically at the time of the screening for the extracellular activities of enzymes (Arroyo-López et al., 2008; Hernández et al., 2007). This feature of yeast is correlated with the ingestion of pre-accumulated lipids and also for providing resistance to chilling stress (Białkowska and Turkiewicz, 2014; Breuer and Harms, 2006; Papanikolaou and Aggelis, 2011; Szczęsna-Antczak et al., 2014). Apart from this, lipase activity was also identified to be convoluted in the pathogenicity of yeasts like *Candida*, *Cryptococcus*, or many other species (Mayer et al., 2013; Park et al., 2013; Sommer et al., 2016). Various reports (Keyhani, 2018; Manuel et al., 2012; de Carolina Sánchez-Pérez et al., 2014; Zha et al., 2014) suggested the functions of lipase enzymes to regulate the growth of fungi and viruses. They hence proved their role as a strong biocontrol potential against different types of the pathogen (Keyhani, 2018; Manuel et al., 2012; de Carolina Sánchez-Pérez et al., 2014; Zha et al., 2014), thus signify an encouraging mark for advanced learning on several control applications.

In some genera of yeasts, protease activity was also found, such as in *Wickerhamomyces*, *Metschnikowia*, *Pichia*, etc. (Pretscher et al., 2018). Lastly, Saccharomycopsis protease records were ominously enhanced at the time of predation, while it is never considered in context with their roles (Junker et al., 2019). From *A. pullulans*, Alp5 (alkaline serine protease) caused decrease in the germination of spores and pollen tube length in reduced spore germination and germ tube length in different fungal species such as *Penicillium* spp., *M. fructicola*, and *Alternaria* sp. in vitro in a dose-dependent manner (Banani et al., 2014; Zhang et al., 2012).

Induction of resistance

Plants possess a unique defense system by which they can identify the existence of pathogens and respond accordingly (Chisholm et al., 2006). The defensive system of the plants provides tolerance against various types of stresses, and biofertilizers also contribute in this context, where microorganisms are used (Gozzo and Faoro, 2013; Pieterse et al., 2014). Systemic resistance is provided to the plants against comprehensive array of pathogens by the application of various strains of yeast (Barda et al., 2015; Lee et al., 2017; Liu et al., 2017). Significant examples of such yeast strains are S. *cerevisiae*, *Rhodosporidium paludigenum*, *Candida saitoana*, *Cryptococcus* spp. and *Metschnikowia* and ultimately, tolerance against phyllosphere pathogens can also be induced (De Miccolis et al., 2019; Hadwiger et al., 2015; Sun et al., 2018).

Due to the attack of any type of stress in plants, reactive oxygen species are produced, and overexpression of yeast strain such as *C. oleophila* triggers the systemic resistance mechanism (Macarisin et al., 2010; De Miccolis et al., 2019). Apart from this, certain other yeast strains, including *C. laurentii*, *Cryptococcus flavescens*, and *R. glutinis*, also play their role in ameliorating the toxins from the plant cells produced by different types of pathogens; hence yeast cells act as a potential biocontrol agent by following several mechanisms.

Yeast as biocontrol agent

Increase incidences of plant diseases are of great concern for agriculture and mainly responsible for extensive crop damage with significant Socio-economic loss that was leading to more than 40 billion dollars loss worldwide, annually (Droby et al., 2003; Abadias et al., 2001; Prista et al., 2002). With time, scientists have made many discoveries and also do improves the pre-discovered technologies too (Silva-Graça and Lucas, 2003; Petrovič et al., 2002; Zhu, 2001; Cubaiu et al., 2012). In the field of agriculture, different researches (Parafati et al., 2015; Türkel et al., 2014; Kowalska et al., 2012) have been made to control plant diseases and also increase the shelf life of the plant products. In the conventional agricultural system, tons of chemicals were used every year in order to control plant diseases and maintaining good plant health (Fanning and Mitchell, 2012; Rossouw et al., 2018). These chemicals might give instant results, but long term usage of these cause environmental pollution, i.e., bioaccumulation, antibiotic resistance, and carcinogenicity.

Biological control of plant diseases came into light in the year 1890, triggered by the idea of using the natural microflora, which produces some biologically active compounds those act as a natural biocontrol agent for a specific group of pathogenic microbes (Bosch Van Den et al., 1982). These bio-active molecules worked as the backbone of biological control (Butt and Copping, 2000; Miles et al., 2012). Some of these controlling management can be naturally seen in wild yeast present on the surface of grapes (Pretscher et al., 2018). Though, there are other microbes available in natural agricultural system, apart from yeast which can show suitable biocontrol property, i.e., bacteria (Carvalho et al., 2010; Ongena and Jacques, 2008), fungi (Butt et al., 2001; Goettel et al., 2001), and actinomycetes (El-Tarabily and Sivasithamparam, 2006; Doumbou et al., 2001). However, the genetic stability of these microbes under environmental stress conditions is uncertain, whereas the yeast genome is far more stable than other microorganisms as biological agents (Devin et al., 1990). Yeast is not only controlled diseases during cultivation but also controls post-harvest diseases, especially in fruits, since they can tolerate higher sugar concentration than bacteria. (Ziedan and Farrag, 2011; Martini et al., 2016; Godoy and Ganga, 2017).

Numerous researches (Abadias et al., 2001; Droby et al., 2003; Godoy and Ganga, 2017) have been performed, and many more works are going on where the yeast is used as a biocontrol agent. The effectivity of yeast biocontrol mechanism is better than others because it has specific transcription factors such as heat shock proteins regulated by Msn2 and Msn4 genes (Nicholls et al., 2004); it controls the heat shock responses, which help it to survive in cold temperature. Apart from heat shock, the other stresses include, i.e., oxidative stress, which makes transcriptional changes into yeast cell by hindering the DNA, proteins, and lipids (Jamieson, 1998; Herrero et al., 2008). To fight against the oxidative stress, yeast made changes in the transcription level, which induced by few transcriptional factors, i.e., Yap1p, Skn7p, Snt2, Msn2p, and Msn4p (Qiu et al., 2019). These genes help in the production of peroxiredoxin Tsa1, which prevents the cell from forming protein aggregate under oxidative stress. This kind of approaches not only helps in good plant health but also helps in preventing environment pollution (Joubert and Doty, 2018)

Yeast against phytopathogens

Many workers (Parafati et al., 2015; Türkel et al., 2014; Kowalska et al., 2012; Joubert and Doty, 2018) claimed the potential use of yeast as a biocontrol agent which set new trends in the field of plant disease control. Yeast is the general resident flora of all types of plants and plant products. Availability of different yeast species on plant surface decreased host tissue colonization by plant pathogenic bacteria, and reduced pathogen survival and sporulation with varying degrees of efficiency (Jijakli and Lepoivre, 1993). This also includes the competition for nutrition between yeast and pathogen, which made it quite difficult for another organism to survive (Prista et al., 2002).

Various In-vitro and In-vivo researches has been conducted across the globe to overcome the adverse effect of fungal plant diseases, i.e., Botrytis rots (Jijakli and Lepoivre, 1993), Downy mildews (Hindumathy, 2012), *Fusarium* rots (Schisler et al., 1995), Powdery mildews (Ng et al., 1997), Rusts, *Rhizoctonia* rots, *Sclerotinia* rots, Gray mold etc. (Lee et al., 2012; Jedryczka et al., 2002) by using yeast as biocontrol agent. It has been found that yeast having multiple bio-controlling mechanisms or produce various natural antagonistic compounds, i.e., mycoparasitism and induction of resistance in plants, diffusible metabolites, secretion of hydrolytic enzymes, toxin production, releasing volatile compounds, those have antifungal activity for the molds (Jijakli and Lepoivre, 1993; Lee et al., 2012; Jedryczka et al., 2002). These metabolic compounds and hydrolytic enzymes, such as β-glucanases and chitinases, destroy the cell walls of phytopathogenic fungi and eventually leads to the death of the diseases causing fungal spp. Apart from the disease control, yeast also helps in the better plant growth and may work as plant growth stimulator (Jijakli et al., 1999; Ziedan and Farrag, 2011; Kheireddine et al., 2018; Martini et al., 2016; Köhl et al., 2019).

It was reported by Droby et al. (1989) that the presence of biocontrol yeast *Candida guilliermondii* (strain US 7) increased the nutrient competition with phytopathogenic *Penicillium digitatum*, a causative organism of green rot of citrus fruits, which caused a negative impact on survival of *Penicillium digitatum*. Gray *mold* (*Botrytis cinerea*) is one of the most common diseases of strawberries (*Fragaria* × *ananassa* Duchesne) worldwide. Continuous use of chemical fungicides increased the probability of resistance in *Botrytis cinerea* and made it challenging to control gray mold infection in strawberries. Chen et al. (2018) reported the application of yeast as an alternative approach for naturally controlling of this fungal pathogen. In their study, Chen et al. (2018) screened 24 yeast strains, with antagonistic potential against *Botrytis cinerea*, and analyzed the effect of volatile and diffusible metabolites for biocontrol potential against a fungal pathogen. They reported that out of 24 yeast strains, volatile organics compound synthesized by *Galactomyces candidum* JYC1146 shown significant higher antifungal potential against *Botrytis cinerea* and may be useful in the biological control of plant pathogens and may become an alternative of chemical fungicide against gray mold disease in plants.

Biological control mechanism

Competition

Yeast produces highly dense colonies as compared to filamentous fungi (Armando et al., 2013). This growth pattern generated a struggle for oxygen, available nutrition (such as carbohydrates, vitamins, minerals, and amino acids) and space for other available microflora. It assumed as the first approach of bio-controlling phytopathogens by yeasts (Cubaiu et al., 2012; Parafati et al., 2015). The massive number of yeast populations could reduce the availability of resources, i.e., amount of nutrients, space for fungal growth, and generate an adverse situation of growth for fungal pathogens. Iron is one of the essential nutritional requirements for all living organisms. Non-availability of iron may cause a reduction in growth, development, and virulence of the fungal pathogens. Yeasts secrete iron chelators (siderophores) that deficit the iron availability in the nearby areas. Saravanakumar et al. (2008) reported the effect of siderophores (pulcherrimin) secreted by *Metschnikowia pulcherrima* on the growth and development of *Botrytis cinerea, Alternaria alternata*, and *Penicillium expansum* on Apples. In their study, they observed that iron deficiency caused by the yeast strain *Metschnikowia pulcherrima* MACH1 could negatively affect the development of *Botrytis cinerea, Alternaria alternata*, and *Penicillium expansum* pathogens on Apples In-vitro and In-vivo.

Many workers reported that various pathogenic fungi could attack sugar beet plant grown under field conditions (*Cercospora* leaf spot: *Cercospora beticola*, Powdery mildew: *Erysiphe betae*, rust: *Uromyces betae*) and cause serious socio-economically concern for the nation and worldwide (Freimoser et al., 2019; Ferraz et al., 2016). Foliar diseases caused by *Cercospora beticola* and *Erysiphe betae* are the main socio-economic concern in many sugar beet growing regions of the world. Ziedan and Farrag (2011) reported the reduction in the growth of phytopathogens (*Cercospora beticola, Erysiphe betae*, and *Uromyces betae*) caused due to nutrition and space competition initiated by the presence of yeast spp (*Saccharomyces cerevisiae, Pichia albicans, Candida sake*). In their study, results of SEM revealed that high colonization of *S. cerevisiae* has occurred on the leaf surface of sugar beet and nearby area of *Erysiphe betae* mycelium, which inhibits the development of *Erysiphe betae* fungi.

Antagonism

The primary mode of action in antagonism is predation that can be seen when the yeast starts secreting enzymes during the nutrient deficient condition. Though they require some amount of nutrients to survive, in the scarcity periods they started secreting enzymes such as chitinases, glucanases, lipases, proteases, etc. these enzymes affect the "Prey cells", kills the cells and use the nutrients obtained from the prey cells for its survival (Freimoser et al., 2019)

Ferraz et al. (2016) reported the antagonistic killing behavior of metabolic products (hydrolytic enzyme) secreted by various yeast and primarily responsible for the bio-controlling of phytopathogens. They were isolated and screened three yeast spp. *Rhodotorula minuta*, *Candida azyma*, and *Aureobasidium pullulans* responsible for the bio-controlling of sour rot in the various plants.

The release of the chitin hydrolysis enzyme is another primary biocontrolling mechanism that allows degradation of the fungal cell wall and resembles yeast as beneficial biocontrol agents against phytopathogenic fungi. There are various yeast species, i.e., *Aureobasidium*, *Candida*, *Debaryomyces*, *Metschnikowia*, *Meyerozyma*, *Pichia*, *Saccharomyces*, *Tilletiopsis*, and *Wickerhamomyces*, those potentially secrete Chitinolytic enzymes and can be utilized as biocontrolling agent.

Another important class of enzymes that are secreted by yeast and may be utilized for biocontrolling of fungal pathogens is glucanases. Glucans (1,3-b- and 1,6-b-glucan) are an essential polysaccharide that is primarily available in the fungal cell wall. It was reported in earlier studies that *S. cerevisiae* can code for the production of a versatile range of endo- and Exo-1,3-b-glucanases. Nearly 15 genes control the production of endo- and Exo-1,3-b-glucanases in *S. cerevisiae* that encode polypeptides with glucanase or related enzymatic activities. These glucanases have various roles in cell separation cell wall modification, cell adhesion, and killer toxin resistance, whereas another type of glucanases have transglycosylase activity and primarily responsible for crosslinking and modification of 1,3-b-glucan chains with other cell wall components. Glucanases based biocontrol of filamentous fungal pathogens is studied as mycoparasitism type antagonistic behavior. Wisniewski et al. (1991) reported the use of *Pichia guilliermondii* as mycoparasitic yeast against *Botrytis cinerea* phytopathogen. They observed that available glucans in the fungal cell wall and glucanases (β-(1−3) glucanases) secreted by yeast produce a lectin-like linking between antagonist and *Botrytis cinerea* which leads to the lysis of fungal cells.

Parasitism

It is a unique relationship that exists between two organisms, in which one got benefited, and the other got harm. The parasitic relation can be even dangerous because the parasites feed upon host nutrients and eventually lead to host death. The parasitic effect does not only come from other microbes but also other parasitic plants too. The parasitic effect caused by specific sorts of microbes can be controlled by yeast to some extent. The main focus over here is the allowance of other microbes to colonize on the surface of the plants, and this interaction allows exploiting the plant and hence cause diseases. However, as previously discussed that yeast can produce specific sort of hydrolytic enzymes, volatile organic compounds (ex *A. pullulans*) that can control the growth of other microbes. Furthermore, this mechanism can be used for controlling parasitic effect of specific disease-causing microbes (Chen et al., 2018).

Conclusion

The necessity and demand of today's generation towards organic products make a tremendous movement to the new revolution in the field of agriculture. Today's trends follow the use of products free from chemicals. The usage of chemicals leaves a negative impact on not only our health but also the environment. This may include antibiotic resistance in microbes, carcinogenicity, soil fertility loss, etc. biocontrol sets new trends in the field of controlling diseases to plants. Numerous researches have been conducted since 1890 for the same.

Searching for a good biocontrol microbe was a tough job as many microbes (both prokaryotes and eukaryotes) can exhibit the biocontrol property. Yeast is selected as a right biocontrol agent because of its high stability, both climate basis and mode of nutrition basis. It has stable heat shock proteins, which makes it resistant to the cold environment, individual transcription factors that support this need. Due to its antagonistic property, it may be served as an excellent biocontrol agent. It produces exoenzymes such as protease, lipases, chitinases, volatile organic compounds, toxins, biofilm production, etc. that helps in killing the disease-causing microbes. To support the fact that different researches have been conducted in different plants around the globe, where the disease control property has been checked. Such as in controlling foliar disease in sugar beet plant, where along with chemical products, yeast is also used, which shows almost equal effect in disease control. Apart from that, it shows good results in controlling post-harvest damage in apples, lemons, controlling diseases in strawberry, etc. it not only controls diseases but also improves the quality of the product. It improves the root weight of sugar beet. It is an excellent competition organism with the antagonistic property.

References

Abadias, M., Benabarre, A., Teixido, N., Usall, J., Vinas, I., 2001. Effect of freeze drying and protectants on viability of the biocontrol yeast Candida sake. Int. J. Food Microbiol. 65 (3), 173–182.

Abbas, S.M., 2013. The influence of biostimulants on the growth and on the biochemical composition of Vicia faba CV. Giza 3 beans. Roman. Biotechnol. Lett. 18 (2), 8061–8068.

Abdallah, M.M.S., El Habbasha, S.F., El Sebai, T., 2016. Comparison of yeast extract and Nicotinaminde foliar applications effect on quinoa plants grown under sandy soil condition. Int. J. PharmTech Res. 9, 24–32.

Abdel-Moneim, M.M., El-Mazny, M.Y., Abdel-Mageed, Y.T., Moustafa, Y.M.M., Yamani, S.H. S., 2015, March. Effect of some natural antioxidants on the productivity and storage ability of Egyptian onion grown in sandy soil. In: Book of Proceedings 2nd International Conference for Agriculture and Irrigation in the Nile Basin Countries. pp. 23–25.

Abou El-Yazied, A., Mady, M.A., 2011. Effect of naphthalene acetic acid and yeast extract on growth and productivity of tomato (Lycopersicon esculentum Mill.) plants. Res. J. Agric. Biol. Sci. 7 (2), 271–281.

Abou El-Yazied, A., Mady, M.A., 2012. Effect of boron and yeast extract foliar application on growth, pod setting and both. J. Appl. Sci. Res. 8 (2), 1240–1251.

Agamy, R., Hashem, M., Alamri, S., 2013. Effect of soil amendment with yeasts as biofertilizers on the growth and productivity of sugar beet. Afr. J. Agric. Res. 8 (1), 46–56.

Ahmed, A.A., El-Baky, M.A., Zaki, M.F., El-Aal, F.S.A., 2011. Effect of foliar application of active yeast extract and zinc on growth, yield and quality of potato plant (Solanum tuberosum L.). J. Appl. Sci. Res. 7 (12), 2479–2488.

Ahmed, M.E.M., Farm, E., 2015. Response of garlic plants (Allium sativum L.) to foliar application of some bio-stimulants. Egypt. J. Hort. 42, 613–625.

Amprayn, K.O., Rose, M.T., Kecskés, M., Pereg, L., Nguyen, H.T., Kennedy, I.R., 2012. Plant growth promoting characteristics of soil yeast (Candida tropicalis HY) and its effectiveness for promoting rice growth. Appl. Soil. Ecol. 61, 295–299.

Armando, M.R., Dogi, C.A., Poloni, V., Rosa, C.A.R., Dalcero, A.M., Cavaglieri, L.R., 2013. In vitro study on the effect of Saccharomyces cerevisiae strains on growth and mycotoxin production by Aspergillus carbonarius and Fusarium graminearum. Int. J. Food Microbiol. 161 (3), 182–188.

Armando, M.R., Dogi, C.A., Rosa, C.A.R., Dalcero, A.M., Cavaglieri, L.R., 2012. Saccharomyces cerevisiae strains and the reduction of Aspergillus parasiticus growth and aflatoxin B1 production at different interacting environmental conditions, in vitro. Food Add. Contamin.: Part. A 29 (9), 1443–1449.

Arroyo-López, F.N., Querol, A., Bautista-Gallego, J., Garrido-Fernández, A., 2008. Role of yeasts in table olive production. Int. J. Food Microbiology 128 (2), 189–196.

Banani, H., Spadaro, D., Zhang, D., Matic, S., Garibaldi, A., Gullino, M.L., 2014. Biocontrol activity of an alkaline serine protease from Aureobasidium pullulans expressed in Pichia pastoris against four postharvest pathogens on apple. Int. J. Food Microbiol. 182, 1–8.

Bandara, W.M.M.S., Seneviratne, G., Kulasooriya, S.A., 2006. Interactions among endophytic bacteria and fungi: effects and potentials. J. Biosci. 31 (5), 645–650.

Barda, O., Shalev, O., Alster, S., Buxdorf, K., Gafni, A., Levy, M., 2015. Pseudozyma aphidis induces salicylic-acid-independent resistance to Clavibacter michiganensis in tomato plants. Plant Dis. 99 (5), 621–626.

Bar-Shimon, M., Yehuda, H., Cohen, L., Weiss, B., Kobeshnikov, A., Daus, A., et al., 2004. Characterization of extracellular lytic enzymes produced by the yeast biocontrol agent Candida oleophila. Curr. Genet. 45 (3), 140–148.

Bennett, A.J., Whipps, J.M., 2008. Beneficial microorganism survival on seed, roots and in rhizosphere soil following application to seed during drum priming. Biol. Control. 44 (3), 349–361.

Białkowska, A., Turkiewicz, M., 2014. Miscellaneous cold-active yeast enzymes of industrial importance. Cold-Adapted Yeasts. Springer, Berlin, Heidelberg, pp. 377–395.

Bosch Van Den, R., Messenger, P.S., Gutierrez, A.P., 1982. An Introduction to Biological Control. Plenum Press, New York (No. SB 975. V36 1982).

Breuer, U., Harms, H., 2006. Debaryomyces hansenii—an extremophilic yeast with biotechnological potential. Yeast 23 (6), 415–437.

Butt, T.M., Copping, L.G., 2000. Fungal biological control agents. Pesticide Outlook 11 (5), 186–191.

Butt, T.M., Jackson, C., Magan, N. (Eds.), 2001. Fungi as Biocontrol Agents: Progress Problems and Potential. CABI.

Cabral, V., Znaidi, S., Walker, L.A., Martin-Yken, H., Dague, E., Legrand, M., et al., 2014. Targeted changes of the cell wall proteome influence Candida albicans ability to form single-and multi-strain biofilms. PLoS Pathog. 10 (12), e1004542.

Cao, M., Gao, M., Lopez-Garcia, C.L., Wu, Y., Seetharam, A.S., Severin, A.J., et al., 2017. Centromeric DNA facilitates nonconventional yeast genetic engineering. ACS Synth. Biol. 6 (8), 1545–1553.

Carvalho, A.L.U.D., Oliveira, F.H.P.C.D., Mariano, R.D.L.R., Gouveia, E.R., Souto-Maior, A. M., 2010. Growth, sporulation and production of bioactive compounds by Bacillus subtilis R14. Braz. Arch. Biol. Technol. 53 (3), 643–652.

Castoria, R., De Curtis, F., Lima, G., De Cicco, V., 1997. β-1, 3-Glucanase activity of two saprophytic yeasts and possible mode of action as biocontrol agents against postharvest diseases. Postharvest Biol. Technol. 12 (3), 293–300.

Causton, H.C., Ren, B., Koh, S.S., Harbison, C.T., Kanin, E., Jennings, E.G., et al., 2001. Remodeling of yeast genome expression in response to environmental changes. Mol. Biol. Cell 12 (2), 323–337.

Cavalheiro, M., Teixeira, M.C., 2018. Candida biofilms: threats, challenges, and promising strategies. Front. Med. 5, 28.

Chan, Z., Tian, S., 2005. Interaction of antagonistic yeasts against postharvest pathogens of apple fruit and possible mode of action. Postharvest Biol. Technol. 36 (2), 215–223.

Chappell, C.R., Fukami, T., 2018. Nectar yeasts: a natural microcosm for ecology. Yeast 35 (6), 417–423.

Chen, P.H., Chen, R.Y., Chou, J.Y., 2018. Screening and evaluation of yeast antagonists for biological control of Botrytis cinerea on strawberry fruits. Mycobiology 46 (1), 33–46.

Chisholm, S.T., Coaker, G., Day, B., Staskawicz, B.J., 2006. Host-microbe interactions: shaping the evolution of the plant immune response. Cell 124 (4), 803–814.

Costa-Orlandi, C.B., Sardi, J.C., Pitangui, N.S., De Oliveira, H.C., Scorzoni, L., Galeane, M.C., et al., 2017. Fungal biofilms and polymicrobial diseases. J. Fungi 3 (2), 22.

Cubaiu, L., Abbas, H., Dobson, A., Budroni, M., Migheli, Q., 2012. A Saccharomyces cerevisiae wine strain inhibits growth and decreases ochratoxin A biosynthesis by Aspergillus carbonarius and Aspergillus ochraceus. Toxins 4 (12), 1468–1481.

Davey, M.E., O'toole, G.A., 2000. Microbial biofilms: from ecology to molecular genetics. Microbiol. Mol. Biol. Rev. 64 (4), 847–867.

Davies, J.M., Lowry, C.V., Davies, K.J., 1995. Transient adaptation to oxidative stress in yeast. Arch. Biochem. Biophys. 317 (1), 1–6.

de Carolina Sánchez-Pérez, L., Barranco-Florido, J.E., Rodríguez-Navarro, S., Cervantes-Mayagoitia, J.F., Ramos-López, M.Á., 2014. Enzymes of entomopathogenic fungi, advances and insights. Adv. Enzyme Res. 2 (02), 65.

De Miccolis, A.R.M., Rotolo, C., Gerin, D., Abate, D., Pollastro, S., Faretra, F., 2019. Global transcriptome analysis and differentially expressed genes in grapevine after application of the yeast-derived defense inducer cerevisane. Pest. Manag. Sci. 75 (7), 2020–2033.

d'Enfert, C., Janbon, G., 2016. Biofilm formation in Candida glabrata: What have we learnt from functional genomics approaches? FEMS Yeast Res. 16 (1).

Desai, J.V., Mitchell, A.P., Andes, D.R., 2014. Fungal biofilms, drug resistance, and recurrent infection. Cold Spring Harb. Perspect. Med. 4 (10), a019729.

Devin, A.B., Prosvirova, T.Y., Peshekhonov, V.T., Chepurnaya, O.V., Smirnova, M.E., Koltovaya, N.A., et al., 1990. The start gene CDC28 and the genetic stability of yeast. Yeast 6 (3), 231–243.

Dhami, M.K., Hartwig, T., Fukami, T., 2016. Genetic basis of priority effects: insights from nectar yeast. Proc. R. Soc. B: Biol. Sci. 283 (1840), 20161455.

Doty, S.L., 2013. Endophytic yeasts: biology and applications. Symbiotic Endophytes. Springer, Berlin, Heidelberg, pp. 335–343.

Doumbou, C.L., Hamby Salove, M.K., Crawford, D.L., Beaulieu, C., 2001. Actinomycetes, promising tools to control plant diseases and to promote plant growth. Phytoprotection 82 (3), 85–102.

Droby, S., Chalutz, E., Wilson, C.L., Wisniewski, M., 1989. Characterization of the biocontrol activity of Debaryomyces hansenii in the control of Penicillium digitatum on grapefruit. Can. J. Microbiol. 35 (8), 794–800.

Droby, S., Vinokur, V., Weiss, B., Cohen, L., Daus, A., Goldschmidt, E.E., et al., 2002. Induction of resistance to Penicillium digitatum in grapefruit by the yeast biocontrol agent Candida oleophila. Phytopathology 92 (4), 393–399.

Droby, S., Wisniewski, M., El Ghaouth, A., Wilson, C., 2003. Influence of food additives on the control of postharvest rots of apple and peach and efficacy of the yeast-based biocontrol product Aspire. Postharvest Biol. Technol. 27 (2), 127–135.

Droby, S., Wisniewski, M., Macarisin, D., Wilson, C., 2009. Twenty years of postharvest biocontrol research: is it time for a new paradigm? Postharvest Biol. Technol. 52 (2), 137–145.

Dudley, A.M., Janse, D.M., Tanay, A., Shamir, R., Church, G.M., 2005. A global view of pleiotropy and phenotypically derived gene function in yeast. Mol. Syst. Biol. 1 (1).

Durrant, W.E., Dong, X., 2004. Systemic acquired resistance. Annu. Rev. Phytopathol. 42, 185–209.

El Ghaouth, A., Wilson, C.L., Wisniewski, M., 2003. Control of postharvest decay of apple fruit with Candida saitoana and induction of defense responses. Phytopathology 93 (3), 344–348.

El-Shafey, A.I., El-Feky, S.S., Abo-Hamad, S.A., 2016. Effect of sowing time and foliar application of yeast extract on growth and productivity of different cultivars of faba bean (Vicia faba L). Egypt. J. Bot. 56 (1), 35–48.

El-Tarabily, K.A., Sivasithamparam, K., 2006. Non-streptomycete actinomycetes as biocontrol agents of soil-borne fungal plant pathogens and as plant growth promoters. Soil Biol. Biochem. 38 (7), 1505–1520.

El-Tohamy, W.A., El-Greadly, N.H.M., 2007. Physiological responses, growth, yield and quality of snap beans in response to foliar application of yeast, vitamin E and zinc under sandy soil conditions. Aust. J. Basic Appl. Sci. 1 (3), 294–299.

El-Tohamy, W.A., El-Abagy, H.M., Badr, M.A., Abou-Hussein, S.D., Helmy, Y.I., Shafeek, M. R., 2015. Effects of yeast extract and GA3 on water status, growth, productivity and quality of sweet potato grown in sandy soils. Int. J. Env. 4 (1), 256–261.

Fanning, S., Mitchell, A.P., 2012. Fungal biofilms. PLoS Pathog. 8 (4), e1002585.

Farrag, D.K., El-Dein, O.A., Khafagy, I.F., 2016. Impact of spraying with Saccharomyces cerevisiae and some commercial nutrients on lettuce plant (Lactuca sativa L.) productivity and prevention of some insect pests. Environment 2, 3.

Fathy, E.S., Farid, S., 1996. The possibility of using vitamin Bs and yeast to delay senescence and improve growth and yield of common beans (Phaseolus vulgaris L.). J. Agric. Sci. Mansoura Univ. 21 (4), 1415–1423.

Fawzy, Z.F., 2010. Increasing productivity of head lettuce by foliar spraying of some bio and organic compounds. Mesopotamia J. Agric. 38, 20–28 (عدد خاص بمؤتمر قسم البستنة).

Ferraz, L.P., da Cunha, T., da Silva, A.C., Kupper, K.C., 2016. Biocontrol ability and putative mode of action of yeasts against Geotrichum citri-aurantii in citrus fruit. Microbiol. Res. 188, 72–79.

Fiori, S., Scherm, B., Liu, J., Farrell, R., Mannazzu, I., Budroni, M., et al., 2012. Identification of differentially expressed genes associated with changes in the morphology of Pichia fermentans on apple and peach fruit. FEMS Yeast Res. 12 (7), 785–795.

Firrincieli, A., Otillar, R., Salamov, A., Schmutz, J., Khan, Z., Redman, R.S., et al., 2015. Genome sequence of the plant growth promoting endophytic yeast Rhodotorula graminis WP1. Front. Microbiol. 6, 978.

Fitzpatrick, D.A., 2012. Horizontal gene transfer in fungi. FEMS Microbiol. Lett. 329 (1), 1–8.

Freimoser, F.M., Rueda-Mejia, M.P., Tilocca, B., Migheli, Q., 2019. Biocontrol yeasts: mechanisms and applications. World J. Microbiol. Biotechnol. 35 (10), 154.

Friel, D., Pessoa, N.M.G., Vandenbol, M., Jijakli, M.H., 2007. Separate and combined disruptions of two exo-β-1, 3-glucanase genes decrease the efficiency of Pichia anomala (strain K) biocontrol against Botrytis cinerea on apple. Mol. Plant–Microbe Interact. 20 (4), 371–379.

Gasch, A.P., Spellman, P.T., Kao, C.M., Carmel-Harel, O., Eisen, M.B., Storz, G., et al., 2000. Genomic expression programs in the response of yeast cells to environmental changes. Mol. Biol. Cell 11 (12), 4241–4257.

Gawlik-Dziki, U., Dziki, D., Nowak, R., Świeca, M., Olech, M., Pietrzak, W., 2016. Influence of sprouting and elicitation on phenolic acids profile and antioxidant activity of wheat seedlings. J. Cereal Sci. 70, 221–228.

Ghoname, A.A., El-Nemr, M.A., Abdel-Mawgoud, A.M.R., El-Tohamy, W.A., 2010. Enhancement of sweet pepper crop growth and production by application of biological, organic and nutritional solutions. Res. J. Agric. Biol. Sci. 6 (3), 349–355.

Giobbe, S., Marceddu, S., Scherm, B., Zara, G., Mazzarello, V.L., Budroni, M., et al., 2007. The strange case of a biofilm-forming strain of Pichia fermentans, which controls Monilinia brown rot on apple but is pathogenic on peach fruit. FEMS Yeast Res. 7 (8), 1389–1398.

Glick, B.R., Cheng, Z., Czarny, J., Duan, J., 2007. Promotion of plant growth by ACC deaminase-producing soil bacteria. New Perspectives and Approaches in Plant Growth-Promoting Rhizobacteria Research. Springer, Dordrecht, pp. 329–339.

Godoy, L., Ganga, A., 2017. Use of yeast as a method of biological control. Appl. Food Sci. J. 1 (1).

Goettel, M.S., Hajek, A.E., Siegel, J.P., Evans, H.C., 2001. 13 Safety of fungal biocontrol agents. Fungi Biocontrol Agents 347.

Gore-Lloyd, D., Sumann, I., Brachmann, A.O., Schneeberger, K., Ortiz-Merino, R.A., Moreno-Beltrán, M., et al., 2019. Snf2 controls pulcherriminic acid biosynthesis and antifungal activity of the biocontrol yeast Metschnikowia pulcherrima. Mol. Microbiol.

Gozzo, F., Faoro, F., 2013. Systemic acquired resistance (50 years after discovery): moving from the lab to the field. J. Agric. Food Chem. 61 (51), 12473–12491.

Grevesse, C., Lepoivre, P., Jijakli, M.H., 2003. Characterization of the exoglucanase-encoding gene PaEXG2 and study of its role in the biocontrol activity of Pichia anomala strain K. Phytopathology 93 (9), 1145–1152.

Gunde-Cimerman, N., Zalar, P., Petrovic, U., Turk, M., Kogej, T., De Hoog, G.S., et al., 2004. Fungi in salterns. In: Ventosa, A. (Ed.), Halophilic Microorganisms.

Hadwiger, L.A., McDonel, H., Glawe, D., 2015. Wild yeast strains as prospective candidates to induce resistance against potato late blight (Phytophthora infestans). Am. J. Potato Res. 92 (3), 379–386.

Hafez, M.R., 2013. Effect of some biological components on Jerusalem Artichoke (Helianthus tuberosus L.) productivity under north sinai conditions. J. Appl. Sci. Res. 9 (1), 804–810.

Hammad, S.A., Ali, O.A., 2014. Physiological and biochemical studies on drought tolerance of wheat plants by application of amino acids and yeast extract. Ann. Agric. Sci. 59 (1), 133–145.

Hardoim, P.R., van Overbeek, L.S., van Elsas, J.D., 2008. Properties of bacterial endophytes and their proposed role in plant growth. Trends Microbiol. 16 (10), 463–471.

Hernández, A., Martín, A., Aranda, E., Pérez-Nevado, F., Córdoba, M.G., 2007. Identification and characterization of yeast isolated from the elaboration of seasoned green table olives. Food Microbiol. 24 (4), 346–351.

Herrero, E., Ros, J., Bellí, G., Cabiscol, E., 2008. Redox control and oxidative stress in yeast cells. Biochim. Biophys. Acta (BBA) – Gener. Subj. 1780 (11), 1217–1235.

Hilber-Bodmer, M., Schmid, M., Ahrens, C.H., Freimoser, F.M., 2017. Competition assays and physiological experiments of soil and phyllosphere yeasts identify Candida subhashii as a novel antagonist of filamentous fungi. BMC Microbiol. 17 (1), 4.

Hillenmeyer, M.E., Fung, E., Wildenhain, J., Pierce, S.E., Hoon, S., Lee, W., et al., 2008. The chemical genomic portrait of yeast: uncovering a phenotype for all genes. Science 320 (5874), 362–365.

Hindumathy, C.K., 2012. The defense activator from yeast for rapid induction of resistance in susceptible pearl millet hybrid against downy mildew disease. Int. J. Agric. Sci. 4 (2), 196.

Hu, C., Qi, Y., 2013. Long-term effective microorganisms application promote growth and increase yields and nutrition of wheat in China. Eur. J. Agron. 46, 63–67.

Huang, R., Li, G.Q., Zhang, J., Yang, L., Che, H.J., Jiang, D.H., et al., 2011. Control of postharvest Botrytis fruit rot of strawberry by volatile organic compounds of Candida intermedia. Phytopathology 101 (7), 859–869.

Ibraheim, S., 2014. Effect of foliar spray with some biostimulants on growth, yield and seeds quality of pea plants grown in sandy soil. J. Appl. Sci. Res. 10 (5), 400–407.

Jamieson, D.J., 1998. Oxidative stress responses of the yeast Saccharomyces cerevisiae. Yeast 14 (16), 1511–1527.

Jedryczka, M., Nikonorenkov, V.A., Levitin, M., Gasich, E., Lewartowska, E., Portenko, L., 2002. Spectrum and severity of fungal diseases on spring oilseed rape in rus sia. IOBC WPRS Bull. 25 (2), 13–20.

Ji, S.H., Gururani, M.A., Chun, S.C., 2014. Isolation and characterization of plant growth promoting endophytic diazotrophic bacteria from Korean rice cultivars. Microbiol. Res. 169 (1), 83–98.

Jijakli, H., Lepoivre, P., 1993. Biological control of postharvest Botrytis cinerea and Penicillium on apples. Bull. OILB/SROP (16), 106–110.

Jijakli, M.H., Lepoivre, P., Grevesse, C., 1999. Yeast species for biocontrol of apple postharvest diseases: an encouraging case of study for practical use. Biotechnological Approaches in Biocontrol of Plant Pathogens. Springer, Boston, MA, pp. 31–49.

Joubert, P.M., Doty, S.L., 2018. Endophytic yeasts: biology, ecology and applications. Endophytes of Forest Trees. Springer, Cham, pp. 3–14.

Junker, K., Chailyan, A., Hesselbart, A., Forster, J., Wendland, J., 2019. Multi-omics characterization of the necrotrophic mycoparasite Saccharomycopsis schoenii. PLoS Pathog. 15 (5), e1007692.

Kamilova, F., Kravchenko, L.V., Shaposhnikov, A.I., Azarova, T., Makarova, N., Lugtenberg, B., 2006. Organic acids, sugars, and L-tryptophane in exudates of vegetables growing on stonewool and their effects on activities of rhizosphere bacteria. Mol. Plant–Microbe Interact. 19 (3), 250–256.

Kandel, S.L., Firrincieli, A., Joubert, P.M., Okubara, P.A., Leston, N.D., McGeorge, K.M., et al., 2017. An in vitro study of bio-control and plant growth promotion potential of Salicaceae endophytes. Front. Microbiol. 8, 386.

Kensler, T.W., Wakabayashi, N., Biswal, S., 2007. Cell survival responses to environmental stresses via the Keap1-Nrf2-ARE pathway. Annu. Rev. Pharmacol. Toxicol. 47, 89–116.

Keyhani, N.O., 2018. Lipid biology in fungal stress and virulence: entomopathogenic fungi. Fungal Biol. 122 (6), 420–429.

Khalifa, A.Y., Alsyeeh, A.M., Almalki, M.A., Saleh, F.A., 2016. Characterization of the plant growth promoting bacterium, Enterobacter cloacae MSR1, isolated from roots of non-nodulating Medicago sativa. Saudi J. Biol. Sci. 23 (1), 79–86.

Kheireddine, A., Essghaier, B., Hedi, A., Dhieb, C., Zouaoui, N.S., 2018. New epiphytic yeasts able to reduce grey mold disease on apples. Plant Prot. Sci. 54 (4), 248–257.

Köhl, J., Kolnaar, R., Ravensberg, W.J., 2019. Mode of action of microbial biological control agents against plant diseases: relevance beyond efficacy. Front. Plant Sci. 10, 845.

Kowalska, J., Rożdżyński, D., Remlein-Starosta, D., Sas-Paszt, L., Malusá, E., 2012. Use of Cryptococcus albidus for controlling grey mould in the production and storage of organically grown strawberries. J. Plant Dis. Prot. 119 (5–6), 174–178.

Lee, G., Lee, S.H., Kim, K.M., Ryu, C.M., 2017. Foliar application of the leaf-colonizing yeast Pseudozyma churashimaensis elicits systemic defense of pepper against bacterial and viral pathogens. Sci. Rep. 7, 39432.

Lee, Y.H., Cho, Y.S., Lee, S.W., Hong, J.K., 2012. Chemical and biological controls of balloon flower stem rots caused by Rhizoctonia solani and Sclerotinia sclerotiorum. Plant Pathol. J. 28 (2), 156–163.

Liu, J., Li, G., Sui, Y., 2017. Optimization of culture medium enhances viable biomass production and biocontrol efficacy of the antagonistic yeast, Candida diversa. Front. Microbiol. 8, 2021.

Loaces, I., Ferrando, L., Scavino, A.F., 2011. Dynamics, diversity and function of endophytic siderophore-producing bacteria in rice. Microb. Ecol. 61 (3), 606–618.

Lohse, M.B., Gulati, M., Johnson, A.D., Nobile, C.J., 2018. Development and regulation of single-and multi-species Candida albicans biofilms. Nat. Rev. Microbiol. 16 (1), 19–31.

Lopes, M.R., Klein, M.N., Ferraz, L.P., da Silva, A.C., Kupper, K.C., 2015. Saccharomyces cerevisiae: a novel and efficient biological control agent for Colletotrichum acutatum during pre-harvest. Microbiological Res. 175, 93–99.

Macarisin, D., Droby, S., Bauchan, G., Wisniewski, M., 2010. Superoxide anion and hydrogen peroxide in the yeast antagonist–fruit interaction: a new role for reactive oxygen species in postharvest biocontrol? Postharvest Biol. Technol. 58 (3), 194–202.

Mady, M.A., 2009. Effect of foliar application with yeast extract and zinc on fruit setting and yield of faba bean (Vicia faba L.). J. Biol. Chem. Environ. Sci. 4 (2), 109–127.

Mahmoud, H.I., Azzaz, N.A., Khalifa, Y.A., Mahmoud, M.A., Fakhry, G., 2016. Effect of foliar application with active yeast extract and benzyladenine on some vegetative growth criteria and chemical composition of lupine (Lupinus termis L.) plants. Minia J. Agric. Res. Dev. 36, 193–214.

Manuel, J., Selin, C., Fernando, W.D., de Kievit, T., 2012. Stringent response mutants of Pseudomonas chlororaphis PA23 exhibit enhanced antifungal activity against Sclerotinia sclerotiorum in vitro. Microbiology 158 (1), 207–216.

Marcet-Houben, M., Gabaldón, T., 2010. Acquisition of prokaryotic genes by fungal genomes. Trends Genet. 26 (1), 5–8.

Martini, C., Tauk-Tornisielo, S.M., Codato, C.B., Bastos, R.G., Ceccato-Antonini, S.R., 2016. A strain of Meyerozyma guilliermondii isolated from sugarcane juice is able to grow and ferment pentoses in synthetic and bagasse hydrolysate media. World J. Microbiol. Biotechnol. 32 (5), 80.

Maserti, B., Podda, A., Giorgetti, L., Del Carratore, R., Chevret, D., Migheli, Q., 2015. Proteome changes during yeast-like and pseudohyphal growth in the biofilm-forming yeast Pichia. fermentans. Amino Acids 47 (6), 1091–1106.

Mattanovich, D., Sauer, M., Gasser, B., 2014. Yeast biotechnology: teaching the old dog new tricks. Microb. Cell Factories 13 (1), 34.

Mayer, F.L., Wilson, D., Hube, B., 2013. Candida albicans pathogenicity mechanisms. Virulence 4 (2), 119–128.

Mbili, N.C., 2012. Evaluation of integrated control of postharvest grey mould and blue mould of pome fruit using yeast, potassium silicate and hot water treatments (Doctoral dissertation).

Miles, L.A., Lopera, C.A., González, S., de García, M.C., Franco, A.E., Restrepo, S., 2012. Exploring the biocontrol potential of fungal endophytes from an Andean Colombian Paramo ecosystem. BioControl 57 (5), 697–710.

Mitchell, A., Romano, G.H., Groisman, B., Yona, A., Dekel, E., Kupiec, M., et al., 2009. Adaptive prediction of environmental changes by microorganisms. Nature 460 (7252), 220.

Moller, L., Lerm, B., Botha, A., 2016. Interactions of arboreal yeast endophytes: an unexplored discipline. Fungal Ecol. 22, 73–82.

Moriguchi, K., Yamamoto, S., Tanaka, K., Kurata, N., Suzuki, K., 2013. Trans-kingdom horizontal DNA transfer from bacteria to yeast is highly plastic due to natural polymorphisms in auxiliary nonessential recipient genes. PLoS One 8 (9), e74590.

Nandy, S.K., Srivastava, R.K., 2018. A review on sustainable yeast biotechnological processes and applications. Microbiol. Res. 207, 83–90.

Nassar, A.H., El-Tarabily, K.A., Sivasithamparam, K., 2005. Promotion of plant growth by an auxin-producing isolate of the yeast Williopsis saturnus endophytic in maize (Zea mays L.) roots. Biol. Fertil. Soils 42 (2), 97–108.

Neseim, M.R., Amin, A.Y., El-Mohammady, M.M.S., 2014. Effect of potassium applied with foliar spray of yeast on sugar beet growth and yield under drought stress. Glob. Adv. Res. J. Agric. Sci. 3 (8), 211–222.

Ng, K.K., MacDonald, L., Punja, Z.K., 1997. Biological control of rose powdery mildew with the antagonist yeast Tilletiopsis pallescens. HortScience 32 (2), 262–266.

Nicholls, S., Straffon, M., Enjalbert, B., Nantel, A., Macaskill, S., Whiteway, M., et al., 2004. Msn2-and Msn4-like transcription factors play no obvious roles in the stress responses of the fungal pathogen Candida albicans. Eukaryot. Cell 3 (5), 1111–1123.

Nutaratat, P., Srisuk, N., Arunrattiyakorn, P., Limtong, S., 2014. Plant growth-promoting traits of epiphytic and endophytic yeasts isolated from rice and sugar cane leaves in Thailand. Fungal Biol. 118 (8), 683–694.

Okada, N., Ogawa, J., Shima, J., 2014. Comprehensive analysis of genes involved in the oxidative stress tolerance using yeast heterozygous deletion collection. FEMS Yeast Res. 14 (3), 425–434.

Ongena, M., Jacques, P., 2008. Bacillus lipopeptides: versatile weapons for plant disease biocontrol. Trends Microbiol. 16 (3), 115–125.

Oporto, C.I., Villarroel, C.A., Tapia, S.M., García, V., Cubillos, F.A., 2019. Distinct transcriptional changes in response to patulin underlie toxin biosorption differences in Saccharomyces cerevisiae. Toxins 11 (7), 400.

Pandin, C., Le Coq, D., Canette, A., Aymerich, S., Briandet, R., 2017. Should the biofilm mode of life be taken into consideration for microbial biocontrol agents? Microb. Biotechnol. 10 (4), 719–734.

Papanikolaou, S., Aggelis, G., 2011. Lipids of oleaginous yeasts. Part I: Biochemistry of single cell oil production. Eur. J. Lipid Sci. Technol. 113 (8), 1031–1051.

Parafati, L., Vitale, A., Restuccia, C., Cirvilleri, G., 2015. Biocontrol ability and action mechanism of food-isolated yeast strains against Botrytis cinerea causing post-harvest bunch rot of table grape. Food Microbiol. 47, 85–92.

Park, M., Do, E., Jung, W.H., 2013. Lipolytic enzymes involved in the virulence of human pathogenic fungi. Mycobiology 41 (2), 67–72.

Parsons, A.B., Lopez, A., Givoni, I.E., Williams, D.E., Gray, C.A., Porter, J., et al., 2006. Exploring the mode-of-action of bioactive compounds by chemical-genetic profiling in yeast. Cell 126 (3), 611–625.

Pérez-Montaño, F., Alías-Villegas, C., Bellogín, R.A., Del Cerro, P., Espuny, M.R., Jiménez-Guerrero, I., et al., 2014. Plant growth promotion in cereal and leguminous agricultural important plants: from microorganism capacities to crop production. Microbiol. Res. 169 (5–6), 325–336.

Peter, J., De Chiara, M., Friedrich, A., Yue, J.X., Pflieger, D., Bergström, A., et al., 2018. Genome evolution across 1,011 Saccharomyces cerevisiae isolates. Nature 556 (7701), 339.

Petrovič, U., 2006. Role of oxidative stress in the extremely salt-tolerant yeast Hortaea werneckii. FEMS Yeast Res. 6 (5), 816–822.

Petrovič, U., Gunde-Cimerman, N., Plemenitaš, A., 2002. Cellular responses to environmental salinity in the halophilic black yeast Hortaea werneckii. Mol. Microbiol. 45 (3), 665–672.

Pieterse, C.M., Zamioudis, C., Berendsen, R.L., Weller, D.M., Van Wees, S.C., Bakker, P.A., 2014. Induced systemic resistance by beneficial microbes. Annu. Rev. Phytopathol. 52, 347–375.

Piombo, E., Sela, N., Wisniewski, M., Hoffmann, M., Gullino, M.L., Allard, M.W., et al., 2018. Genome sequence, assembly and characterization of two Metschnikowia fructicola strains used as biocontrol agents of postharvest diseases. Front. Microbiol. 9, 593.

Pozo, M.I., Lachance, M.A., Herrera, C.M., 2012. Nectar yeasts of two southern Spanish plants: the roles of immigration and physiological traits in community assembly. FEMS Microbiol. Ecol. 80 (2), 281–293.

Pretscher, J., Fischkal, T., Branscheidt, S., Jäger, L., Kahl, S., Schlander, M., et al., 2018. Yeasts from different habitats and their potential as biocontrol agents. Fermentation 4 (2), 31.

Prista, C., Soeiro, A., Vesely, P., Almagro, A., Ramos, J., Loureiro-Dias, M.C., 2002. Genes from Debaryomyces hansenii increase salt tolerance in Saccharomyces cerevisiae W303. FEMS Yeast Res. 2 (2), 151–157.

Punja, Z.K., Utkhede, R.S., 2003. Using fungi and yeasts to manage vegetable crop diseases. Trends Biotechnol. 21 (9), 400–407.

Qiu, X., Zhang, J., Zhou, J., Fang, Z., Zhu, Z., Li, J., et al., 2019. Stress tolerance phenotype of industrial yeast: industrial cases, cellular changes, and improvement strategies. Appl. Microbiol. Biotechnol. 103 (16), 6449–6462.

Ramadan, A.M., Ragab, S.T., 2015. Improving growth and yield of caraway (Carum carvi L.,) plants by decapitation and/or active dry yeast application. Int. J. Curr. Microbiol. Appl. Sci. 4 (9), 47–60.

Reynolds, T.B., Fink, G.R., 2001. Bakers' yeast, a model for fungal biofilm formation. Science 291 (5505), 878–881.

Richards, T.A., Leonard, G., Soanes, D.M., Talbot, N.J., 2011. Gene transfer into the fungi. Fungal Biol. Rev. 25 (2), 98–110.

Rossouw, D., Meiring, S.P., Bauer, F.F., 2018. Modifying Saccharomyces. cerevisiae adhesion properties regulates yeast ecosystem dynamics. mSphere 3 (5), e00383–18.

Saad-Allah, K.M., Fetouh, M.I., Elhaak, M.A., 2017. Induction of milk thistle (Silybum marianum L. Gaertn) growth and phytochemicals production by natural stimulants. J. Appl. Res. Med. Aromatic Plants 6, 101–110.

Sanna, M.L., Zara, G., Zara, S., Migheli, Q., Budroni, M., Mannazzu, I., 2014. A putative phospholipase C is involved in Pichia fermentans dimorphic transition. Biochim. Biophys. Acta (BBA) – Gener. Subj. 1840 (1), 344–349.

Sanna, M.L., Zara, S., Zara, G., Migheli, Q., Budroni, M., Mannazzu, I., 2012. Pichia fermentans dimorphic changes depend on the nitrogen source. Fungal Biol. 116 (7), 769–777.

Saravanakumar, D., Ciavorella, A., Spadaro, D., Garibaldi, A., Gullino, M.L., 2008. Metschnikowia pulcherrima strain MACH1 outcompetes Botrytis cinerea, Alternaria alternata and Penicillium expansum in apples through iron depletion. Postharvest Biol. Technol. 49 (1), 121–128.

Saravanakumar, D., Spadaro, D., Garibaldi, A., Gullino, M.L., 2009. Detection of enzymatic activity and partial sequence of a chitinase gene in Metschnikowia pulcherrima strain MACH1 used as post-harvest biocontrol agent. Eur. J. Plant Pathol. 123 (2), 183–193.

Schaible, U.E., Kaufmann, S.H., 2005. A nutritive view on the host–pathogen interplay. Trends Microbiol. 13 (8), 373–380.

Schisler, D.A., Kurtzman, C.P., Bothast, R.J., Slininger, P.J., 1995. Evaluation of yeasts for biological control of Fusarium dry rot of potatoes. Am. Potato J. 72 (6), 339–353.

Scholz, H., Franz, M., Heberlein, U., 2005. The hangover gene defines a stress pathway required for ethanol tolerance development. Nature 436 (7052), 845.

Segal, E., Yehuda, H., Droby, S., Wisniewski, M., Goldway, M., 2002. Cloning and analysis of CoEXG1, a secreted 1, 3-β-glucanase of the yeast biocontrol agent Candida oleophila. Yeast 19 (13), 1171–1182.

Silva-Graça, M., Lucas, C., 2003. Physiological studies on long-term adaptation to salt stress in the extremely halotolerant yeast Candida versatilis CBS 4019 (syn. C. halophila). FEMS Yeast Res. 3 (3), 247–260.

Sommer, B., Overy, D.P., Haltli, B., Kerr, R.G., 2016. Secreted lipases from Malassezia globosa: recombinant expression and determination of their substrate specificities. Microbiology 162 (7), 1069–1079.

Spadaro, D., Droby, S., 2016. Development of biocontrol products for postharvest diseases of fruit: the importance of elucidating the mechanisms of action of yeast antagonists. Trends Food Sci. Technol. 47, 39–49.

Sun, C., Fu, D., Lu, H., Zhang, J., Zheng, X., Yu, T., 2018. Autoclaved yeast enhances the resistance against Penicillium expansum in postharvest pear fruit and its possible mechanisms of action. Biol. Control. 119, 51–58.

Szczęsna-Antczak, M., Kamińska, J., Florczak, T., Turkiewicz, M., 2014. Cold-active yeast lipases: recent issues and future prospects. Cold-Adapted Yeasts. Springer, Berlin, Heidelberg, pp. 353–375.

Takahashi, T., Kakehi, J.I., 2009. Polyamines: ubiquitous polycations with unique roles in growth and stress responses. Ann. Bot. 105 (1), 1–6.
Tian, S.P., Qin, G.Z., Xu, Y., Wang, Y.S., 2004. Application of antagonistic yeasts under field conditions and their biocontrol ability against postharvest diseases of sweet cherry. Acta Bot. Sin. – Eng. Ed. 46 (11), 1324–1330.
Trollmo, C., André, L., Blomberg, A., Adler, L., 1988. Physiological overlap between osmotolerance and thermotolerance in Saccharomyces cerevisiae. FEMS Microbiol. Lett. 56 (3), 321–325.
Tsai, P.W., Yang, C.Y., Chang, H.T., Lan, C.Y., 2011. Human antimicrobial peptide LL-37 inhibits adhesion of Candida albicans by interacting with yeast cell-wall carbohydrates. PLoS One 6 (3), e17755.
Türkel, S., Korukluoğlu, M., Yavuz, M., 2014. Biocontrol activity of the local strain of Metschnikowia pulcherrima on different postharvest pathogens. Biotechnol. Res. Int. 2014.
Urquhart, E.J., Punja, Z.K., 2002. Hydrolytic enzymes and antifungal compounds produced by Tilletiopsis species, phyllosphere yeasts that are antagonists of powdery mildew fungi. Can. J. Microbiol. 48 (3), 219–229.
Vadkertiová, R., Molnárová, J., Vránová, D., Sláviková, E., 2012. Yeasts and yeast-like organisms associated with fruits and blossoms of different fruit trees. Can. J. Microbiol. 58 (12), 1344–1352.
Varela, J.C., Van Beekvelt, C., Planta, R.J., Mager, W.H., 1992. Osmostress-induced changes in yeast gene expression. Mol. Microbiol. 6 (15), 2183–2190.
Verstrepen, K.J., Klis, F.M., 2006. Flocculation, adhesion and biofilm formation in yeasts. Mol. Microbiol. 60 (1), 5–15.
Wang, W., Vinocur, B., Altman, A., 2003. Plant responses to drought, salinity and extreme temperatures: towards genetic engineering for stress tolerance. Planta 218 (1), 1–14.
Wisniewski, M., Droby, S., 2012. Biopreservation of food and feed by postharvest biocontrol with microorganisms. Beneficial Microorganisms in Agriculture, Food and the Environment: Safety Assessment and Regulation. CAB International, Oxfordshire, pp. 57–66.
Wisniewski, M., Wilson, C., Droby, S., Chalutz, E., El Ghaouth, A., Stevens, C., 2007. Postharvest biocontrol: new concepts and applications. Biological Control: A Global Perspective. CABI International, United Kingdom, pp. 262–273.
Wisniewski, M., Biles, C., Droby, S., 1991. The use of yeast Pichia guilliermondii as a biocontrol agent: characterization of attachment to Botrytis cinerea. ARS-US Department of Agriculture, Agricultural Research Service (USA).
Xin, G., Glawe, D., Doty, S.L., 2009. Characterization of three endophytic, indole-3-acetic acid-producing yeasts occurring in Populus trees. Mycol. Res. 113 (9), 973–980.
Xu, H., Nobile, C.J., Dongari-Bagtzoglou, A., 2013. Glucanase induces filamentation of the fungal pathogen Candida albicans. PLoS One 8 (5), e63736.
Yan, Q., Shi, M., Ng, J., Wu, J.Y., 2006. Elicitor-induced rosmarinic acid accumulation and secondary metabolism enzyme activities in Salvia miltiorrhiza hairy roots. Plant Sci. 170 (4), 853–858.
Yehuda, H., Droby, S., Bar-Shimon, M., Wisniewski, M., Goldway, M., 2003. The effect of under-and overexpressed CoEXG1-encoded exoglucanase secreted by Candida oleophila on the biocontrol of Penicillium digitatum. Yeast 20 (9), 771–780.
Zajc, J., Gostinčar, C., Černoša, A., Gunde-Cimerman, N., 2019. Stress-tolerant yeasts: opportunistic pathogenicity versus biocontrol potential. Genes 10 (1), 42.

Zha, D., Xu, L., Zhang, H., Yan, Y., 2014. Molecular identification of lipase LipA from Pseudomonas protegens Pf-5 and characterization of two whole-cell biocatalysts Pf-5 and Top10lipA. J. Microbiol. Biotechnol. 24 (5), 619–628.

Zhang, D., Spadaro, D., Garibaldi, A., Gullino, M.L., 2011. Potential biocontrol activity of a strain of Pichia guilliermondii against grey mold of apples and its possible modes of action. Biol. Control. 57 (3), 193–201.

Zhang, D., Spadaro, D., Valente, S., Garibaldi, A., Gullino, M.L., 2012. Cloning, characterization, expression and antifungal activity of an alkaline serine protease of Aureobasidium pullulans PL5 involved in the biological control of postharvest pathogens. Int. J. Food Microbiol. 153 (3), 453–464.

Zhang, H., Zheng, X., Wang, L., Li, S., Liu, R., 2007. Effect of yeast antagonist in combination with hot water dips on postharvest Rhizopus rot of strawberries. J. Food Eng. 78 (1), 281–287.

Zhang, S., Gan, Y., Xu, B., 2014. Efficacy of Trichoderma longibrachiatum in the control of Heterodera avenae. BioControl 59 (3), 319–331.

Zhu, J.K., 2001. Plant salt tolerance. Trends Plant Sci. 6 (2), 66–71.

Ziedan, E.S.H.E., Farrag, E.S., 2011. Application of yeasts as biocontrol agents for controlling foliar diseases on sugar beet plants. J. Agric. Tech. 7 (6), 1789–1799.

Złotek, U., Świeca, M., 2016. Elicitation effect of Saccharomyces cerevisiae yeast extract on main health-promoting compounds and antioxidant and anti-inflammatory potential of butter lettuce (Lactuca sativa L.). J. Sci. Food Agric. 96 (7), 2565–2572.

Further reading

Wisniewski, M., Droby, S., Norelli, J., Liu, J., Schena, L., 2016. Alternative management technologies for postharvest disease control: the journey from simplicity to complexity. Postharvest Biol. Technol. 122, 3–10.

Chapter 18

Phytostimulation and ISR responses of fungi

Puja Kumari, Arti Singh and Ravindra Nath Kharwar*
Mycopathology and Microbial Technology Laboratory, CAS in Botany, Institute of Science, Banaras Hindu University, Varanasi, India
**Corresponding author*. Email: *rnkharwar@gmail.com*

Introduction

For decades, plant–microbe symbiotic relationship is not only important for the plant, but also very beneficial to our agricultural crops. Arbuscular mycorrhizal fungi get attached with the extra radical mycelium of the roots of the fungi and helps the host plant in various ways. The current attention of the researchers is upon the role of the fungal endophytes which live inside the plant tissues that maintains mutualistic interaction with their host plant. Fungal endophytes help the plant in various types of phytostimulations which ameliorate the plant against different stress conditions and provide the protection from various invading pathogens, pests, they also help in phytostabilization that make them enable to survive under harsh stressed conditions (Chadha et al., 2015). This is one of the reasons behind the plant living and flourishing in nature due to these endo-symbiotic microbes. Some fungi present in rhizospheric region due to the constant secretion of nutrient by plant root mucilage. Plant growth-promoting fungi (PGPFs) mainly live in there vicinity of the plant root and help the plant growth promotions, but there are some fungi which live inside the healthy plant tissues and play a vital role in the plant growth development and their protection against biotic and abiotic stresses as plant growth- promoting endophytic fungi. There is a wide study on the resistivity of plant against the pathogens, insects and other enemies due to the effects of the rhizospheric and endophytic fungi on the host plant (Herre et al., 2007; Miao et al., 2006). It is considered that the role of rhizospheric and endophytic fungal communities is very important for the growth and protection of plants and the fertility of soil. PGPFs can get the economic importance to modern agriculture, and can reduce the use of expensive mineral or chemical fertilizers and they may also improve the water utilization.

Fungi Bio-Prospects in Sustainable Agriculture, Environment and Nano-Technology.
DOI: https://doi.org/10.1016/B978-0-12-821394-0.00018-4

PGPFs may not only improve the plant growth, but by lowering the chemical fertilizers use and improving the water use efficiency, it will lower down the production costs. It may also reduce the harmful pollution from the environment while raising in high yield production. It is reported that an input of 1 kg of fertilizers enhances up to 10 kg of additional cereal at least for the initial fertilizers, but excessive use of chemical fertilizers after a thresh hold, it may cause environmental pollution in addition to damage of soil mycoflora. The damage to the crops by pathogens may be reduced and the quality can be controlled by the inoculation of PGPFs instead of using the harmful synthetic chemical fertilizers. Deficiency of water, nutrition, phosphate, carbon and other minerals may be the major limiting factors in the growth of crops. PGPFs have important contributions in plant immunity by inducing the systemic resistance of the plant. Therefore, plant growth promoting fungi have fascinated the interest as biological fertilizer for not only the quality and quantity improvement of the plants but also the positive relationship between the plant, fungi and our bionomical environment. This chapter will discuss the role of several groups of fungi in phytostimulations and their effect in the ISR responses (Table 18.1).

Phytostimulation

Phytostimulation is the process of enhancement in the microbial activity by some signaling molecules secreted by the plant root part in the rhizospheric sites of the soil for the sequestration of the soil contaminants. This method can be useful in the rhizo-filtering of the harmful organic contaminants like the aromatic and polycyclic aromatic hydrocarbons (PAHs), pesticides from the soil sites. This method can become the cost effective way of phytoremediation of the organic pollutants from soil (Miller, 1996). The fungi are also being used as a biosorption of metals (Dhankhar and Hooda, 2011). The process not only stimulates the microorganisms under the soil like phytostabilization, rhizofiltration and phytodegradation but also activates the process above the ground like phytoextraction and phytovolatilization. The sites that contains the impurities of heavy metal toxicity, poor soil conditions having low nutrient content, organic pollutants such as PAH (polycyclic aromatic hydrocarbon) and the major microbial diversity, don't let the plant established its vegetation or proper growth in heavy metal polluted sites. The process in which the group of fungi help the plant to get established in heavy metals stress condition and immobilize the metals or the other contaminants (migrated via wind, leaching, soil dispersion, etc.) in a nontoxic form by the accumulation, absorption or precipitation within the rhizospheric area of plant root is termed as phytostabilization by fungi. In plant, AMF helps in the alleviation of metal toxicity by reducing metal translocation from the root to shoot (Leyval et al., 1997). Fungi can be used as a complement as it can become the cheapest source of bioremediation of the trace metals, PAH

TABLE 18.1 Various groups of fungi having phytostimulating activity.

S. no.	Plant	Fungi	Phyto-stimulating activity of fungi	References
1.	*Brassica campestris* L.	*Mucor* sp. MHR-7 (endophytic)	Alleviation of heavy metal toxicity in mustard	Zahoor et al. (2017)
2.	*Glycine max* L.	AMF	Enhancement in leaf proline, photosynthetic leaf area index, relative growth rate. Drought tolerant	Pavithra and Yapa (2018)
3.	*Olea europea*	AMF	Alleviation in drought impact and increased turgor potential and uptake of minerals	Sara et al. (2018)
4.	*Poncirus trifoliate*	*Funneliformis mossease, Paraglomus accultum*	Increase in hyphal length, hyphal, water absorption rate, ad leaf water potential	Zhang et al. (2018)
5.	*Zea mays*	*Rhizophagus intraradices, Funneliformis mosseae, F. geosporum*	Increase in leaf lenth, plant height, leaf number, chlorophyll a, photosynthetic rate, stomatal conductance, and transpiration rate	Mathur et al. (2016)
6.	*Glycine max* L.	*Paecilomyces formosus* LHL10, *Paecilomyces formosus* LHL6	Heavy metal (Ni, Cd and Al), high temperature and drought stress resistant	Bilal et al. (2019)
7.	*Triticum* sp.	Fungal endophytes	Heat and drought tolerant	Hubbard et al. (2014)
8.	*Triticum aestivum* L.	Fungal endophytes	Salinity, drought, high temperature and heavy metal stress tolerant, antibiotic sensitive, IAA production	Ripa et al. (2019)

(*Continued*)

TABLE 18.1 (Continued)

S. no.	Plant	Fungi	Phyto-stimulating activity of fungi	References
9.	*S. surattense* Burm. f.	*Pencillium roqueforti* Thom., Fungal endophyte	Enhancement in stress tolerance and nutrient uptake, IAA production, heavy metal tolerance (Ni, Cd, Cu, Zn and Pb)	Ikram et al. (2018)
10.	*Arabidopsis thaliana*	*Piriformospora indica*	Stimulation of drought related genes in leaves, drought tolerance capacity	Sherameti et al. (2008)
11.	*Arabidopsis thaliana*	*Piriformospora indica*	Enhanced production of ascorbate	Vadassery et al. (2009)
12.	Barley	*Piriformospora indica*	Salt stress tolerance, disease resisitance, higher yield production.	Waller et al. (2005)
13.	Tomato	*Fusarium equiseti* GF191	*Fusarium* crown and root rot/ *Fusarium oxysporum* f.sp. *radices-lycopersici*	Horinouchi et al. (2007)

transformation or degradation, its potential of cost savings attribute compared to the costly amendments. The traditional costly amendments using for many decades were limestone, fly-ash, zeolite, steel slag, organic or inorganics, the conventional non-biological process for the utilization of the heavy metal stabilization (Fig. 18.1) (Curaqueo et al., 2014; Zhao and Masaihiko, 2007; Verdugo et al., 2010; Chen et al., 2000; Gu et al., 2011; Lai et al., 2012).

Role of fungi in plant growth promotions and development

Modulating phytohormone level

Plant hormones are the biochemicals or signaling molecules which play a key role in the plant growth and development under various conditions. Auxin, gibberellin, cytokinin, abscisic acid, ethylene, brassinosteroids,

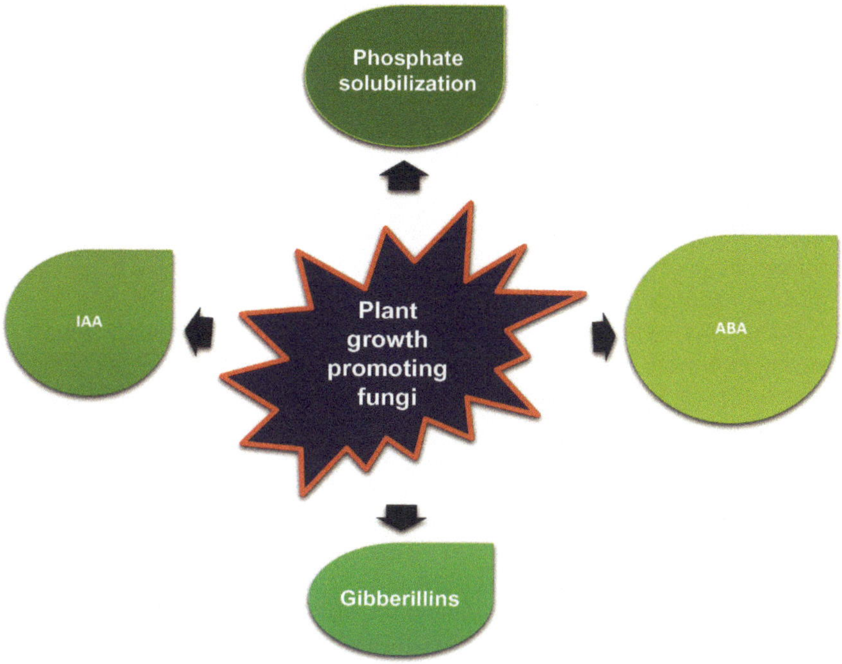

FIGURE 18.1 Plant growth promoting activities of fungi.

jasmonic acid related compound, salicylic acid, strigolactones are some plant hormones that affect the plant growth under stressed conditions and also regulate different functions in the plant at molecular and cellular level (Miransari, 2016).

PGPFs alleviating the immunity of the plant

Plants have very limited potential to grow under biotic and abiotic conditions which increase the chances of wooing damage and disease that are responsible for the reduction in plant growth and yield. In such cases, PGPFs contribute the plant immunity by inducing Induced systemic resistance (ISR) such as *Trichoderma* sp., *Pythium oligandrum*, *Fusarium oxysporum*, *Penicillium* sp., and *Piriformospora indica* that reduce the diseases in upper parts of plant. Some of the PGPFs are being extensively used as a biocontrol agent in agricultural system due to their useful contribution in the regulation of various kind of diseases caused by fungal pathogens such as *Trichoderma* spp. which has the ability to antagonize and parasitize other fungi (mycoparasite), mainly by the secretion of cell wall-degrading enzymes and other compound responsible to kill the pathogens. Along with the mycoparasitism property, it produces antibiotics, peptaibols and many compounds having antibiosis activity. Various kinds of crops are saved and promoted in USA by the use of *Trichoderma* spp.

Colonization of fungi to the plant root and their role in the plant growth promotion

Arbuscular mycorrhizal fungi (AMF) are the soil microbes that colonize majority of the plant root and establish a connection between the plant and the substrate, help in production of plant growth hormones, increase the nutrient availability and also inhibit the root pathogens. AMF and their host plant have the mutualistic symbiotic relationship since time immemorial (Fig. 18.2) (Sangers et al., 1977). AMF help the plant to grow in various abiotic stressful conditions such as heavy metals, drought, salinity, temperature and several kinds of plant diseases (Rodriguez et al., 2008). AMF create connection between plant and fungus via its hyphal growth pattern i.e. extra radical mycelium (ERM) that helps in the enhancement of root growth in the soil with large surface area as well as for water transport, nutrient uptakes (that helps in nutrient improvement), soil quality improvement by decomposing the soil organic matter, and CO_2 fixation by the plants (Bowles et al., 2016; Rouphael et al., 2015; Zou et al., 2016; Paterson et al., 2016). Root colonized AMF may also lead in the alleviation of biotic and abiotic stresses from the concerned host plants which depends only on their functional diversity (Tuinen et al., 1998; Li et al., 2013; Kothari et al., 1991; Thygesen et al., 2004; Lax et al., 2011; Oliveira et al., 2006; Seguel et al., 2013). AMF facilitate with countless benefits to the soil by increasing the soil biota, nutrients translocation and water transport, etc. Therefore, AMF can also be used as bio-fertilizer replacing the inorganic or chemical fertilizers (Ortas, 2012).

FIGURE 18.2 The symbiotic association of Arbuscular mycorrhizal fungi with their host plant and their role in plant growth.

In addition to AMF, endophytic fungi also support the improvement in the agricultural development, by enhancing the plant growth and nutrition uptake with different plant growth promoting attributes like production of growth hormones (IAA, GA, cytokinin), solubilization of phosphate, potassium and zinc (Suman et al., 2016; Verma et al., 2017; Paenibacilius, 2016; Yadav and Yadav, 2018). Endophytic fungi are ubiquitously present in all parts of the plants without causing any detectable disease symptoms. Some studies have also revealed the significant role of fungal endophytes responsible for the production of bioactive substances either responsible to inhibit the plant pathogens or to promote the growth in plants (Krings et al., 2007; Gunatilaka, 2006). This group of fungi get entered in the host plant via two methods i.e. vertical (via seed) and horizontal (via air, stem and leaf) transmission (Fig. 18.3) (Yadav, 2019).

Induced resistance system in plants

Stress and development responses are controlled by a very complex network in the plant system. The defense mechanism of plants includes Systemic Acquired Resistance (SAR) and Induced Systemic Resistance (ISR). SAR is induced by the plant pathogenic microbes and insects whereas the ISR is mediated by the beneficial root colonizing mutualistic microbiota residing in the rhizospheric region of soil like *Pseudomonas simiae, P. fluorescence, Paenibacillius pomymyxa* or *Trichoderma* spp. Induced systemic resistance (ISR) is the defense response of a plant, when the plant get treated with

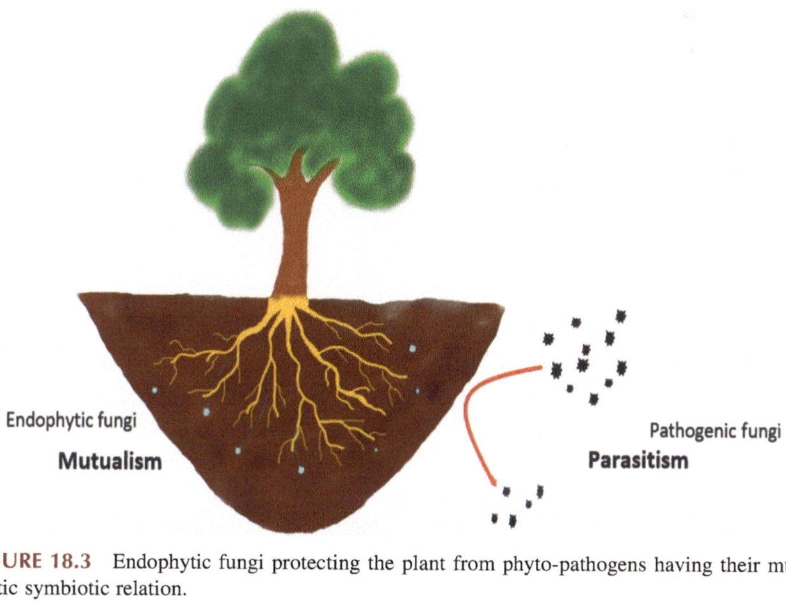

FIGURE 18.3 Endophytic fungi protecting the plant from phyto-pathogens having their mutualistic symbiotic relation.

certain chemicals, inoculated with some pathogenic avirulent strain, a continuous irritation with certain substance or wound from a long period or stressful environment, plant secretes signaling molecules like Jasmonic acid, salicylic acid (by fungi and bacteria) and ethylene (bacteria only).

Plant not only responses to the abiotic stress like drought, salinity, high temperature, heavy metal contamination but also when triggered by the biological agents like pathogenic microbes and insects (Romera et al., 2019; Pieterse et al., 2014; Martínez-Medina et al., 2017; Verbon et al., 2017). ISR is provoked by the mutualistic microbiota of the soil and it is mediated by the SA- independent pathway which relies on the JA (Jasmonic acid) and ET (Erthylene) signaling molecules. Redox- regulated protein NON EXPRESSOR OF PR GENES1 (NPR1), is required for the cytosolic function in JA/ET signaling and thus it typically functions without *PR*- gene activation (Pieterse et al., 1998, 2002, 2012, 2014; Choudhary et al., 2007). Plant root is safe from the pathogenic disease due the induced resistance system by some beneficial microbes in the soil. Induced resistance system is a defense mechanism which includes an induced-resistance system which is very complex. SAR is elucidated in some of the model plants viz., *Arabidopsis.* SAR has some pathways in which, there is direct production of pathogenesis- related (PR) protein by the plant when attacked by the pathogens and this pathway depends on the secretion of a signaling molecule (Salicylic acid) by the plant. In the other pathway, PR protein is produced due to wounding or necrosis inducing plant pathogens that is mediated by the secretion of another signaling molecule JA. This JA induced pathway is referred as induced systemic resistance (Fig. 18.4). Plant have evolved chemicals for the defense strategies to protect themselves from wounding by the herbivorous insects. These strategies includes the formation of the proteinaceous defense compound, proteinase inhibitor (PIRs) that affect the consumption of the nutrient by the herbivorous, secretion of some toxic compounds like nicotine, production of volatile compounds to attract the insect predators and jasmonic acid is being found to be an essential compounds which helps the plant the most in healing upon wounding.

PR-proteins

PR-proteins or pathogenesis-related proteins are the proteinaceous compound that is induced by the plant in the form of defense response system in presence of different harsh chemicals, poor atmospheric conditions, wounding by insect and microbes, etc. But also, in some plants different pathogenesis-related protein has been reported that is induced by the plant itself either in desiccation stressed conditions or in wounding by the insect and microbes. However, there is constitutively expression of these PR-proteins by some of the plants. A total of 18 families of PR- proteins have been reported so far.

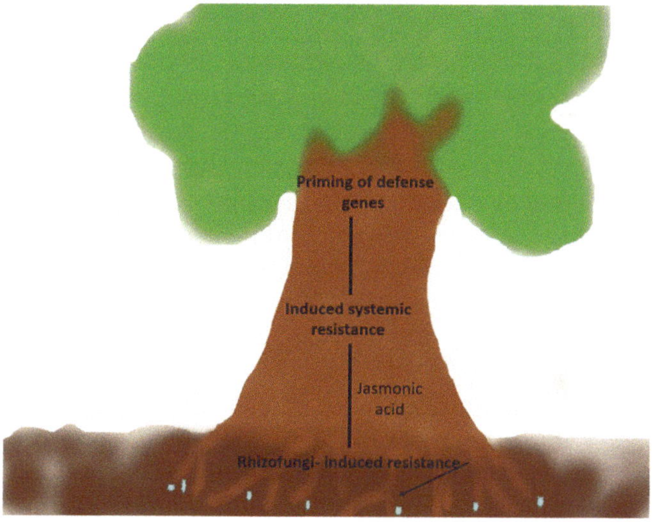

FIGURE 18.4 Induced systemic resistance by rhizofungi.

These proteins are mainly resistant to protease enzyme and show a constant stability in low pH (Sinha et al., 2014).

Induced systemic resistance responses by fungi

The disease resistance capacity of the plant can also be elevated by the non-pathogenic microbes. Protection of above ground plant tissues against disease were first seen in the experiment in which the roots of plants were colonized with plant growth promoting rhizobium (PGPR) (Van Loon et al., 1998). Observing the studies of the last decades, it is clear that fungi are also responsible for inducing a similar broad-spectrum ISR. Some fungi responsible for triggering ISR: non-pathogenic strain *Trichoderma* spp. (Vinale et al., 2008), *F. oxysporum* (Paparu et al., 2007) and mycorrhizal fungi (Pozo and Azcon-Aguilar et al., 2007). Fungal endophyte isolated from root tissue of the plant also triggers ISR in plant. (Vu et al. 2006; Waller et al. 2005; Soichiro et al. 2003).

Signal transduction of ISR-SA

Local and system defense responses are induced by the beneficial and parasitic microbes. Plant hormones like salicylic acid (SA), jasmonic acid (JA) and ethylene (ET), responsible for the signaling molecules that control the beneficial colonization of microbes with plant root and triggering the induced systemic responses in plant. SA, JA and ET pathways cross communicate each other and help the plant to give better responses according to the invading attack. There is an ample of evidences which show that fungi involved in JA-dependent, ET-dependent and SA-independent pathways (Hossain et al., 2008; Lavicoli et al., 2003; Tran et al., 2007; Djonovic et al., 2007; Ahn et al., 2007; Ryu et al., 2004; Shoresh et al., 2005; Van wees et al., 1997), except a few ones which are totally dependent of SA signaling (Tjamos et al., 2005; Audenaert et al., 2002). Although, the plant defense response is regulated by Salicylic acid, jasmonic acid, ethylene, abscisic acid and these signaling molecules are involved in the plant defense mechanisms (Fujita et al., 2006; Beckers and Spoel, 2006; Lorenzo & Solano, 2005).

Fungi: an emerging bioinoculant

There is a global demand for more food, more fiber as well as more bio-resources, and that seems one of the major challenge to the farmers, agricultural system, researcher and the policy makers. In the recent years, use of fungal inoculation is satisfying the plants by the alleviation in the key nutrients like phosphorous (P) and also enhanced the quantity to the crop plants. Fungi helps the plant in the resource allocation, mineral uptake. They have role in improving the host plant reproductive fitness by enhancing the growth in plants. They provide the increase in biomass and grain yield of crops like wheat, rice and barley.

Conclusion

Plant growth promoting fungi (PGPFs) are the heterogeneous group of non-pathogenic fungi that can be obtained in rhizosphere, at the root surfaces or inside the roots of plant. The PGPFs not only provide the plant in the root extension, plant growth development (seed germination and seedling vigor and photosynthetic efficiency) and protection from various kinds of phytopathogens, but also helps in the soil improvements. The exact mechanism of the plant growth promoting fungi are not much understood but yes, they are thought to have the potential to help the high yield production of plants, agricultural crops, herbal and rare medicinal plants and of course they will be beneficial to the human health and our ecological system.

References

Ahn, I.P., Lee, S.W., Suh, S.C., 2007. Rhizobacteria-induced priming in Arabidopsis is dependent on ethylene, jasmonic acid, and NPR1. Mol. Plant-Microbe Interact. 20 (7), 759–768.

Audenaert, K., Pattery, T., Cornelis, P., Höfte, M., 2002. Induction of systemic resistance to Botrytis cinerea in tomato by Pseudomonas aeruginosa 7NSK2: role of salicylic acid, pyochelin, and pyocyanin. Mol. Plant-Micr. Inter. 15 (11), 1147–1156.

Beckers, G.J.M., Spoel, S.H., 2006. Fine-tuning plant defencesignalling: salicylate versus jasmonate. Plant Biol. 8 (01), 1–10.

Bilal, S., Shahzad, R., Imran, M., Jan, R., Kim, K.M., Lee, I.J., 2019. Synergistic association of endophytic fungi enhances Glycine max L. resilience to combined abiotic stresses: heavy metals, high temperature and drought stress. Ind. Crop. Prod. 143, 111931. Available from: https://doi.org/10.1016/j.indcrop.2019.111931.

Bowles, T.M., Barrios-Masias, F.H., Carlisle, E.A., Cavagnaro, T.R., Jackson, L.E., 2016. Effects of arbuscularmycorrhizae on tomato yield, nutrient uptake, water relations, and soil carbon dynamics under deficit irrigation in field conditions. Sci. Total Environ. 566, 1223–1234.

Chadha, N., Mishra, M., Rajpal, K., Bajaj, R., Choudhary, D.K., et al., 2015. An ecological role of fungal endophytes to ameliorate plants under biotic stress. Arch. Microbiol. 197 (7), 869–881.

Chen, H.M., Zheng, C.R., Tu, C., Shen, Z.G., 2000. Chemical methods and phytoremediation of soil contaminated with heavy metals. Chemosphere 41 (1–2), 229–234.

Choudhary, D.K., Prakash, A., Johri, B.N., 2007. Induced systemic resistance (ISR) in plants: mechanism of action. Indian J. Microbiol. 47, 289–297.

Curaqueo, G., Schoebitz, M., Borie, F., Caravaca, F., Roldán, A., 2014. Inoculation with arbuscularmycorrhizal fungi and addition of composted olive-mill waste enhance plant establishment and soil properties in the regeneration of a heavy metal-polluted environment. Environ. Sci. Pollut. Res. 21 (12), 7403–7412.

Dhankhar, R., Hooda, A., 2011. Fungal biosorption – an alternative to meet the challenges of heavy metal pollution in aqueous solutions. Environ. Technol. 32 (5), 467–491.

Djonović, S., Vargas, W.A., Kolomiets, M.V., Horndeski, M., Wiest, A., Kenerley, C.M., 2007. A proteinaceous elicitor Sm1 from the beneficial fungus Trichoderma virens is required for induced systemic resistance in maize. Plant Physiol. 145 (3), 875–889.

Doares, S.H., Syrovets, T., Weiler, E.W., Ryan, C.A., 1995. Oligogalacturonides and chitosan activate plant defensive genes through the octadecanoid pathway. Proc. Natl Acad. Sci. 92 (10), 4095–4098.

Fujita, M., Fujita, Y., Noutoshi, Y., Takahashi, F., Narusaka, Y., Yamaguchi-Shinozaki, K., et al., 2006. Crosstalk between abiotic and biotic stress responses: a current view from the points of convergence in the stress signaling networks. Curr. Opin. Plant Biol. 9 (4), 436–442.

Gu, H.H., Qiu, H., Tian, T., Zhan, S.S., Chaney, R.L., Wang, et al., 2011. Mitigation effects of silicon rich amendments on heavy metal accumulation in rice (Oryza sativa L.) planted on multi-metal contaminated acidic soil. Chemosphere 83 (9), 1234–1240.

Gunatilaka, A.A., 2006. Natural products from plant-associated microorganisms:distribution, structural diversity, bioactivity, and implications of theiroccurrence. J. Nat. Prod. 69, 509–526.

Hase, S., Takahashi, S., Takenaka, S., Nakaho, K., Arie, T., Seo, S., et al., 2008. Involvement of jasmonic acid signalling in bacterial wilt disease resistance induced by biocontrol agent Pythium oligandrum in tomato. Plant. Pathol. 57, 870–876.

Herre, E.A., Mejía, L.C., Kyllo, D., Rojas, E., Maynard, Z., Butler, A., et al., 2007. Implications of observed anti-pathogen effects of tropical fungal endophytes and mycorrhizae. Ecology 88, 550–558.

Horinouchi, H., Muslim, A., Suzuki, T., Hyakumachi, M., 2007. Fusarium equiseti GF191 as an effective biocontrol agent against Fusarium crown and root rot of tomato in rock wool systems. Crop. Prot. 26 (10), 1514–1523.

Hossain, M.M., Sultana, F., Kubota, M., Hyakumachi, M., 2008. Differential inducible defense mechanisms against bacterial speck pathogen in Arabidopsis thaliana by plant-growth-promoting-fungus Penicillium sp. GP16-2 and its cell free filtrate. Plant Soil 304, 227–239.

Hubbard, M., Germida, J.J., Vujanovic, V., 2014. Fungal endophytes enhance wheat heat and drought tolerance in terms of grain yield and second-generation seed viability. J. Appl. Microbiol. 116 (1), 109–122.

Ikram, M., Ali, N., Jan, G., Jan, F.G., Rahman, I.U., Iqbal, A., et al., 2018. IAA producing fungal endophyte Penicillium roqueforti Thom., enhances stress tolerance and nutrients uptake in wheat plants grown on heavy metal contaminated soils. PLoS One 13 (11), e0208150.

Korolev, N., Rav David, D., Elad, Y., 2008. The role of phytohormones in basal resistance and Trichoderma-induced systemic resistance to Botrytis cinerea in Arabidopsis thaliana. BioControl 53, 667–683.

Kothari, S.K., Marschner, H., Römheld, V., 1991. Effect of a vesicular–arbuscularmycorrhizal fungus and rhizosphere micro-organisms on manganese reduction in the rhizosphere and manganese concentrations in maize (Zea mays L.). N. Phytol. 117 (4), 649–655.

Krings, M., Taylor, T.N., Hass, H., Kerp, H., Dotzler, N., Hermsen, E.J., 2007. Fungalendophytes in a 400-million-yr-old land plant: infection pathways, spatialdistribution, and host responses. N. Phytol. 174, 648–657.

Lai, T., Cao, A., Zucca, A., Carucci, A., 2012. Use of natural zeolites charged with ammonium or carbon dioxide in phytoremediation of lead-and zinc-contaminated soils. J. Chem. Tech. Biotech. 87 (9), 1342–1348.

Lavicoli, A., Boutet, E., Buchala, A., Métraux, J.P., 2003. Induced systemic resistance in Arabidopsis thaliana in response to root inoculation with Pseudomonas fluorescens CHA0. Mol. Plant-Micr. Inter. 16 (10), 851–858.

Lax, P., Becerra, A.G., Soteras, F., Cabello, M., Doucet, M.E., 2011. Effect of the arbuscular mycorrhizal fungus Glomus intraradices on the false root-knot nematode Nacobbus aberrans in tomato plants. Biol. Fertil. Soils 47 (5), 591–597.

León, J., Rojo, E., Sánchez-Serrano, J.J., 2001. Wound signalling in plants. J. Exp. Bot. 52 (354), 1–9.

Leyval, C., Turnau, K., Haselwandter, K., 1997. Interactions between heavy metals and mycorrhizal fungi in polluted soils: physiological, ecological and applied aspects. Mycorrhiza 7, 139–153.

Li, Y., Liu, Z., Hou, H., Lei, H., Zhu, X., Li, et al., 2013. Arbuscular mycorrhizal fungi-enhanced resistance against Phytophthorasojae infection on soybean leaves is mediated by a network involving hydrogen peroxide, jasmonic acid, and the metabolism of carbon and nitrogen. Actaphysiologiaeplantarum 35 (12), 3465–3475.

Lorenzo, O., Solano, R., 2005. Molecular players regulating the jasmonatesignalling network. Curr. Opin. Plant Biol. 8 (5), 532–540.

Martínez-Medina, A., Van Wees, S.C.M., Pieterse, C.M.J., 2017. Airborne signals from Trichoderma fungi stimulate iron uptake responses in roots resulting in priming of jasmonic acid dependent defences in shoots of Arabidopsis thaliana and Solanumlycopersicum. Plant Cell Env. 40, 2691–2705.

Mathur, S., Sharma, M.P., Jajoo, A., 2016. Improved photosynthetic efficacy of maize Zea mays plants with arbuscularmycorrhizal fungi (AMF) under high temperature stress. J. Photochem. Photobiol. B 180, 149–154.

Miao, Z.Q., Li, S.D., Liu, X.Z., Chen, Y.J., Li, Y.H., Wang, Y., et al., 2006. The causal microorganisms of Panax notoginseng root rot disease. Sci. Agric. Sin. 39, 1371–1378 (Chinese).

Miller, R., 1996. Phytoremediation, Technology Overview Report, Ground-Water Remediation Technologies Analysis Center, Series O, Vol. 3, October.

Miransari, M., 2016. Soybeans and plant hormones. Environmental Stresses in Soybean Production. Academic Press, pp. 131–156.

Oliveira, R.S., Castro, P.M., Dodd, J.C., Vosátka, M., 2006. Different native arbuscularmycorrhizal fungi influence the coexistence of two plant species in a highly alkaline anthropogenic sediment. Plant. Soil. 287 (1–2), 209.

Ortas, I., 2012. The effect of mycorrhizal fungal inoculation on plant yield, nutrient uptake and inoculation effectiveness under long-term field conditions. Field Crop. Res. 125, 35–48.

Paparu, P., Dubois, T., Coyne, D., Viljoen, A., 2007. Defense-related gene expression insusceptible and tolerant bananas (Musa spp.) following inoculation with non pathogenic Fusarium oxysporum endophytes and challenge with Radopholus similis. Physiol. Mol. Plant. Pathol. 71, 149–157.

Paenibacilius, B., 2016. Microbial diversity of extreme regions: an unseen heritage and wealth. Indian J. Plant Genet. Resour. 29, 246–248.

Parchmann, S., Gundlach, H., Mueller, M.J., 1997. Induction of 12-oxo-phytodienoic acid in wounded plants and elicited plant cell cultures. Plant Physiol. 115 (3), 1057–1064.

Paterson, E., Sim, A., Davidson, J., Daniell, T.J., 2016. Arbuscular mycorrhizal hyphae promote priming of native soil organic matter mineralization. Plant Soil 408, 243–C254. Available from: https://doi.org/10.1007/s11104-016-2928-8.

Pavithra, D., Yapa, N., 2018. Arbuscularmycorrhizal fungi inoculation enhances drought stress tolerance of plants. Groundw. Sustain. Dev. 7, 490–494.

Pieterse, C.M.J., Van der Does, D., Zamioudis, C., Leon-Reyes, A., Van Wees, S.C.M., 2012. Hormonal modulation of plant immunity. Annu. Rev. Cell Dev. Biol. 28, 489–521.

Pieterse, C.M.J., Van Wees, S.C.M., Ton, J., Van Pelt, J.A., Van Loon, L.C., 2002. Signalling in rhizobacteria-lnduced systemic resistance in Arabidopsis thaliana. Plant Biol. 4, 535–544. Available from: https://doi.org/10.1055/s-2002-35441l.

Pieterse, C.M.J., Van Wees, S.C.M., van Pelt, J.A., Knoester, M., Laan, R., Gerrits, H., et al., 1998. A novel signaling pathway controlling induced systemic resistance in Arabidopsis. Plant. Cell 10, 1571–1580.

Pieterse, C.M.J., Zamioudis, C., Berendsen, R.L., Weller, D.M., Van Wees, S.C.M., Bakker, P. A.H.M., 2014. Induced systemic resistance by beneficial microbes. Annu. Rev. Phytopathol. 52, 347–375. Available from: https://doi.org/10.1146/annurev-phyto-082712-102340.

Pozo, M.J., Azcon-Aguilar, C., 2007. Unraveling mycorrhiza-induced resistance. Curr. Opin. Plant. Biol. 10, 393–398.

Ripa, F.A., Cao, W.D., Tong, S., Sun, J.G., 2019. Assessment of plant growth promoting and abiotic stress tolerance properties of wheat endophytic fungi. BioMed. Res. Int.

Rodriguez, R.J., Henson, J., Van Volkenburgh, E., Hoy, M., Wright, L., Beckwith, F., et al., 2008. Stress tolerance in plants via habitat-adapted symbiosis. Int. Soc. Microb. Ecol. 2, 404–416. Available from: https://doi.org/10.1038/ismej.2007.106.

Romera, F.J., García, M.J., Lucena, C., Martinez Medina, A., Aparicio, M.Á., Ramos, et al., 2019. Induced systemic resistance (ISR) and Fe deficiency responses in dicot plants. Front. Plant Sci. 10, 287.

Rouphael, Y., Franken, P., Schneider, C., Schwarz, D., Giovannetti, M., Agnolucci, M., 2015. Arbuscular mycorrhizal fungi act as bio-stimulants in horticultural crops. Sci. Hort. 196, 91–108.

Ryu, C.M., Murphy, J.F., Mysore, K.S., Kloepper, J.W., 2004. Plant growth-promoting rhizobacteria systemically protect Arabidopsis thaliana against Cucumber mosaic virus by a salicylic acid and NPR1-independent and jasmonic acid-dependent signaling pathway. The Plant J. 39 (3), 381–392.

Sangers, F.E., Tinker, Black, R.L.B., Palmerley, S.M., 1977. The development of endopmycorrhizal root system. N. Phytolol. 78, 257–268.

Sara, O., Ennajeh, M., Zrig, A., Gianinazzi, S., Khemira, H., 2018. Estimating the contribution of Arbuscular mycorrhizal fungi to drought tolerance of potted olive trees (Olea europaea). Acta Physiol. Plant 40, 1–81. Available from: https://doi.org/10.1007/s11738-018-2656-1.

Segarra, G., Casanova, E., Bellido, D., Odena, M.A., Oliveira, E., Trillas, I., 2007. Proteome, salicylic acid and jasmonic acid changes in cucumber plants inoculated with Trichoderma asperellum strain T34. Proteomics 7, 3943–3952.

Seguel, A., Cumming, J.R., Klugh-Stewart, K., Cornejo, P., Borie, F., 2013. The role of arbuscular mycorrhizas in decreasing aluminium phytotoxicity in acidic soils: a review. Mycorrhiza 23 (3), 167–183.

Sherameti, I., Tripathi, S., Varma, A., Oelmüller, R., 2008. The root-colonizing endophyte Pirifomospora indica confers drought tolerance in Arabidopsis by stimulating the expression of drought stress-related genes in leaves. Mol. Plant–Microbe Interact. 21 (6), 799–807.

Shoresh, M., Yedidia, I., Chet, I., 2005. Involvement of jasmonic acid/ethylene signaling pathway in the systemic resistance induced in cucumber by Trichoderma asperellum T203. Phytopathol. 95 (1), 76–84.

Sinha, M., Singh, R.P., Kushwaha, G.S., Iqbal, N., Singh, A., Kaushik, et al., 2014. Current overview of allergens of plant pathogenesis related protein families. Sci. World J.

Suman, A., Yadav, A.N., Verma, P., 2016. Endophytic microbes in crops: diversity and beneficial impact for sustainable agriculture. Microbial Inoculants in Sustainable Agricultural Productivity. Springer, New Delhi, pp. 117–143.

Thygesen, K., Larsen, J., Bodker, L., 2004. Arbuscular mycorrhizal fungi reduce development of pea root-rot caused by Aphanomyces euteiches using oospores as pathogen inoculum. Eur. J. Plant. Pathol. 110 (4), 411–419.

Tjamos, S.E., Flemetakis, E., Paplomatas, E.J., Katinakis, P., 2005. Induction of resistance to Verticillium dahliae in Arabidopsis thaliana by the biocontrol agent K-165 and pathogenesis-related proteins gene expression. Mol. Plant-Microbe Interact. 18 (6), 555–561.

Tuinen, V.D., Jacquot, E., Zhao, B., Gollotte, A., Gianinazzi-Pearson, V., 1998. Characterization of root colonization profiles by a microcosm community of Arbuscular mycorrhizal fungi using 25S rDNA-targeted nested PCR. Mol. Ecol. 7 (7), 879–887.

Vadassery, J., Tripathi, S., Prasad, R., Varma, A., Oelmüller, R., 2009. Monodehydroascorbatereductase 2 and dehydroascorbate reductase 5 are crucial for a mutualistic interaction between Piriformospora indica and Arabidopsis. J. Plant Physiol. 166 (12), 1263–1274.

Van Loon, L.C., Bakker, P.A.H.M., Pieterse, C.M.J., 1998. Systemic resistance induced by rhizosphere bacteria. Annu. Rev. Phytopathol. 36, 453–483.

Van Wees, S.C., Pieterse, C.M., Trijssenaar, A., Van't Westende, Y.A., Hartog, F., Van Loon, L.C., 1997. Differential induction of systemic resistance in Arabidopsis by biocontrol bacteria. Mole. Plant-Microbe Interact. 0 (6), 716–724.

Vinale, F., Sivasithamparam, K., Ghisalberti, E.L., Marra, R., Woo, S.L., Lorito, M., 2008. Trichoderma–plant–pathogen interactions. Soil Biol. Biochem. 40, 1–10.

Verbon, E.H., Trapet, P.L., Stringlis, I.A., Kruijs, S., Bakker, P.A.H.M., Pieterse, C.M.J., 2017. Iron and immunity. Annu. Rev. Phytopathol. 55, 355–375. Available from: https://doi.org/10.1146/annurev-phyto-080516-035537.

Verdugo, C., Sanchez, P., Santibanez, C., Urrestarazu, P., Bustamante, E., et al., 2010. Efficacy of lime, biosolids, and mycorrhiza for the phytostabilization of sulfidic copper tailings in Chile: a greenhouse experiment. Int. J. Phytoremediat. 13 (2), 107–125.

Verma, P., Yadav, A.N., Kumar, V., Singh, D.P., et al., 2017. Beneficial plant–microbes interactions: biodiversity of microbes from diverse extreme environments and its impact for crop improvement. Plant–Microbe Interactions in Agro-Ecological Perspectives. Springer, Singapore, pp. 543–580.

Vu, T., Sikora, R., Hauschild, R., 2006. Fusarium oxysporum endophytes induced systemic resistance against Radopholus similis on banana. Nematology 8 (6), 847–852.

Waller, F., Achatz, B., Baltruschat, H., Fodor, J., Becker, K., Fischer, M., et al., 2005. The endophytic fungus Piriformospora indica reprograms barley to salt stress tolerance, disease resistance, and higher yield. Proc. Natl. Acad. Sci. USA 102, 13386–13391.

Yadav, A.N., 2019. Endophytic fungi for plant growth promotion and adaptation under abiotic stress conditions. Acta Sci. Agric. 3, 91–93.

Yadav, A.N., Yadav, N., 2018. Stress-adaptive microbes for plant growth promotion and alleviation of drought stress in plants. Acta Sci. Agric. 2 (6), (ISSN: 2581-365X).

Zahoor, M., Irshad, M., Rahman, H., Qasim, M., Afridi, S.G., Qadir, M., et al., 2017. Alleviation of heavy metal toxicity and phytostimulation of Brassica campestris L. by endophytic Mucor sp. MHR-7. Ecotoxicol. Environ. Saf. 142, 139–149.

Zhang, F., Jia-Dong, H.E., Qiu-Dan, N.I., Qiang-Sheng, W.U., Zou, Y.N., 2018. Enhancement of drought tolerance in trifoliate orange by mycorrhiza: changes in root sucrose and proline metabolisms. Not. Bot. Horti. Agrobot. Cluj-Napoca 46, 270. Available from: https://doi.org/10.15835/nbha46110983.

Zhao, X.L., Masaihiko, S., 2007. Amelioration of cadmium polluted paddy soils by porous hydrated calcium silicate. Water Air Soil Pollut. 183 (1–4), 309–315.

Zou, Y.N., Srivastava, A.K., Wu, Q.S., 2016. Glomalin: a potential soil conditioner for perennial fruits. Int. J. Agric. Biol. 18, 293–297. Available from: https://doi.org/10.17957/IJAB/15.0085.

Chapter 19

Bio-management of *Fusarium* spp. associated with fruit crops

Sampat Nehra[1,*], Raj Kumar Gothwal[1,2], Alok Kumar Varshney[1,2], Pooran Singh Solanki[1,2], Shivani Chandra[3], Poonam Meena[4], P.C. Trivedi[4] and P. Ghosh[1]

[1]*Birla Institute of Scientific Research, Statue Circle, Jaipur, India,* [2]*Department of BioEngineering, Birla Institute of Technology, Mesra, Jaipur, India,* [3]*Amity Institute of Biotechnology, Amity University, Noida, India,* [4]*Department of Botany, University of Rajasthan, Jaipur, India*

**Corresponding author.* Email: nehrasampat@gmail.com

Introduction

Soil is a much complex and diverse ecosystem on earth having diverse microbial population. 'Microbial population' is also known as one of the major component of soil. Soil microbial populations are important because they play important role in the mineralization of nutrients into forms available for plant uptake, and a number of other advantages as well. The associations existing between different soil microbial populations, whether it is symbiotic or antagonistic in nature, influence the activities of microorganisms in the soil. Some possible interactions occurring among the microbial populations in soil can be divided as: (i) Beneficial, (a) Mutualism, (b) Commensalisms and (c) Protocooperation or (ii) Detrimental/Harmful (a) Amensalism, (b) Antagonism, (c) Competition, (d) Parasitism and (e) Predation.

Among soil microbial population, soil fungi are most abundant and diverse taxonomic groups on the Earth (Bar-On et al., 2018; Egidi et al., 2019), are pathogens and mutualistic symbionts of both plants and animals (Frąc et al., 2018), and play essential roles in ecosystems, such as pedogenesis, nutrient cycling, and disease suppression (Bardgett and Van der Putten, 2014). The basic properties of soil microfungi emphasized that they have (i) wide dispersibility, indicating soil microfungi are very much generalized and cosmopolitan in distribution (ii) heterotrophy indicating dependency of fungi on other living host and (iii) they have superior survival capacities to survive under adverse conditions (Singh and Dwivedi, 2004; Rohilla and Salar, 2012).

Fungi Bio-Prospects in Sustainable Agriculture, Environment and Nano-Technology.
DOI: https://doi.org/10.1016/B978-0-12-821394-0.00019-6

It is well known that soil microbial populations have a wide range of tolerance to the physio-chemical properties e.g., temperature, soil moisture, soil texture, organic matter and organic carbon however, the dominance and presence of soil fungi have been reported from India and different parts of the world (Nannipieri et al., 2003; State of world's fungi report, 2018).

Soil fungi depict their importance in terrestrial ecosystem by regulating the carbon flow through their effects on organic matter decomposition and cycling of nutrients (Toal et al., 2000; Dwivedi et al., 2019). When a disease causing agent survives and spread through soil is known as "soil-borne pathogen". Thus the diseases caused by these pathogens is considered as "soil-borne diseases". These soil borne pathogens cause significant economic losses in crop production. *Fusarium oxysporum* is a pathogenic soil borne ascomycete fungi affecting many plants in the world by causing wilt, which is a lethal vascular syndrome in plants (Flood, 2006; Dita et al., 2018).

Over 100 host-specific species (formae speciales: f. spp) of *Fusarium oxysporum* have been reported to be distributed around the world (Gordon Thomas, 2017; Diane et al., 2017). Economic losses caused by *Fusarium* wilt of common stock pathogens are estimated at 50–75% of the attainable yield for most of the crops (Lewis and Papavizas, 1991; Navas Cortes et al., 2000; Mihajlovic et al., 2017). *Fusarium* Head Blight (FHB) is a severe disease caused by different *Fusarium* species, that affects a wide range of cereal crops, including wheat. It determines 10–30% yield loss in Europe (Baffoni et al., 2015). The yield and quality of pulses are also adversely affected due to wilt disease amounting around 100% yield losses (Choudhary et al., 2013). *Fusarium* wilt is one of the most important yield limiting diseases in potato production worldwide (Mejdoub-Trabelsi et al., 2016). About 15–70% of potato fields have been reported to be infected with *Fusarium* wilt causal agents and mainly *F. oxysporum* around the world (Ommati and Sharif, 2008; Saremi and Amiri, 2010; Gachango et al., 2012; Mejdoub-Trabelsi et al., 2016).

During pathogenesis the pathogen produces toxic substances/enzymes and other growth regulators. These substances react biochemically with plant cells and causes disease. Due to these reactions the physical and physiological characteristics of plants get affected adversely that result in to the disease. However, the severity varies with the year and the location, and is greatly influenced by farming practices and environmental conditions. Some fungi are important pathogens, responsible for plant disease and they may cause high yield losses in the crops (Park et al., 2005; Shenoy et al., 2007; Soares and Barreto, 2008; Than et al., 2008; Baffoni et al., 2015; Sinha et al., 2018).

In a survey, Dean et al. (2012) nominated 'Top 10' fungal pathogens based on scientific/economic importance. This generated 495 votes from the international community, and resulted in the generation of a Top 10 fungal plant pathogen list for *Molecular Plant Pathology* (JMPP). The Top 5 fungi amongst 10, list includes, in rank order, (1) *Magnaporthe oryzae*; (2) *Botrytis cinerea*; (3) *Puccinia* spp.; (4) *Fusarium graminearum*; (5) *Fusarium oxysporum*. In fourth and fifth places are two *Fusarium* species, but with contrasting host ranges, with *F. raminearum* causing significant damage predominantly to cereals and a few noncereal species, and *F. oxysporum* having a wide host range, with severe losses in crops as diverse as tomato, cotton and banana.

Mushtaq and Hashmi (1997) reported 16 genera and 28 species of fungi from root, stem, leaves and seeds of infected plants of bell pepper (*Capsicum annuum*). *Fusarium* spp. i.e. *F. anthohilum, F. moniliformae, F. oxysporum, F. solani* and *F. proliferatum* were most dominant and showed symptoms of wilting in pepper plants. Latiffah et al. (2007) reported *Fusarium* species from 12 cultivated soil planted with different crops in Penang. A total of 42 *Fusarium* isolates have been recovered in which four *Fusarium* species were identified as *F. solani, F. semitectum, F. equiseti* and *F. oxysporum*. The most prevalent *Fusarium* species recovered was *F. solani* (84%) followed by *F. semitectum* (7%), *F. equiseti* (7%) and *F. oxysporum* (2%).

Nur Ain Izzati et al. (2011) stated that *Fusarium* species associated with maize are widely distributed in Malaysia. During their work they identified 167 *Fusarium* isolates from seven locations throughout Malaysia on the basis of micro and macro morphological features i.e. growth rates, colony features, mode of production of microconidia, macroconidia, conidiophores and chlamydospores. Out of 167 species eight *Fusarium* spp. were found most dominant e.g. *F. proliferatum, F. semitectum, F. verticillioides, F. subglutinan, F. oxysporum, F. solani, F. equiseti* and *F. pseudograminearum*. *Fusarium* spp. is known to attack a wide variety of plants causing root-rot, stem-rot and wilt diseases. The main drawback of this species is that it can easily survive for a long period without any host due to the formation of specialized resistant structures called chlamydospores (Nelson, 1981; Mihajlovic et al., 2017). The pathogen of *Fusarium* sp. can be dispersed through a number of ways which include; moving of infected planting materials and seeds, farm tools and machinery especially when used in the infected soils and later used in healthy fields. Air and water plays a role in the spread of this fungus (Gamliel-Atinsky et al., 2009; Joshi, 2018). Characteristic features like resistant spore structure, variety in strains, wide dispersibility, heterotrophy and survivorship under adverse condition, make it most pernicious and promotes to cause a variety of diseases in a variety of crops.

Silvestro et al. (2013), examined the occurrence and distribution of soil *Fusarium* spp. at different soil depths in a zero tillage system after the wheat was harvested. One hundred twenty two isolates of *Fusarium* were obtained and identified as *F. equiseti, merismoides, F. oxysporum, F. scirpi* and *F. solani*. The most prevalent species was *F. oxysporum*, which was observed in all sequences and depths.

Some important diseases associated with soil fusaria are as follows: (1) Wilt disease (*F. solani, F. oxysporum, F. oxysporum* f. sp. *pisi, F. oxysporum* f. sp. *dianthi, F. oxysporum* f. sp. *psidii, F. oxysporum* f. sp. *niveum, F. oxysporum* f. sp. *lycopersicii*), root-rot (*F. solani*), cob-rots (*F. graminearum, F. verticillioides*) and collar rot (*F. solani*). The fungal pathogens affect the plants by producing macro and micro conidia that infect the roots of the plants feeding and infecting the vascular tissues leading to wilting and then gradual death of the host plants (Joshi, 2018).

Indiscriminate use of fungicides has resulted in accumulation of toxic compounds potentially hazardous to living beings and environment and also in the build up of resistance of the pathogens. In order to tackle these problems, effective alternatives to chemical control are being employed. In this context, biological control practices are a nature friendly approaches by using specific microorganisms that interfere with plant pathogens to overcome the problems caused by chemical methods of plant protection (Dwivedi and Dwivedi, 1999; Cohen et al., 2000; Marasas et al., 2006; Wu et al., 2009; Gupta and Misra, 2009; Alabouvette et al., 2009; Iqbal et al., 2010; Zakaria et al., 2012, Baffoni et al., 2015; Mejdoub-Trabelsi et al., 2016; Nosir, 2016; Mihajlovic et al., 2017; Joshi, 2018; Bjelic et al., 2018; Sinha et al., 2018).

Most of the *Fusarium* species are known as harmful fungi that cause vascular disease in plants (Armstrong and Armstrong, 1981; Nelson et al., 1983; Gordon and Martyn, 1997; Gordon Thomas, 2017; Bodah, 2017; Srinivas et al., 2019) by penetrating their hyphae into host vascular tissues and by producing deleterious secondary metabolites (Marasas et al., 1984), secretion of hydrolytic enzymes related to pathogenesis, mycotoxin production and cellular apoptosis of host plant cells in the progression of the infection (Abraham, 1986; Abbas et al., 1997; Pavlovkin et al., 2004; Mudili et al., 2014; Srinivas et al., 2019). *Fusarium* toxins are the most abundant natural contaminants of diets containing cereals and other grains (Venkataramana et al., 2014; Divakara et al., 2014; Kalagatur et al., 2017; Kalagatur et al., 2018). So for the management of any pathogen, host–pathogen relationship is very important for the researcher to plan the most appropriate management practice.

Host–pathogen interaction: penetration, establishment, disease development and disease symptomatology

Fusarium spp. are pathogenic soil borne ascomycetes fungi infecting many economically important plants in the world by causing wilt, which is a lethal vascular syndrome in plants (Flood, 2006). There are more than 100 host-specific species of only *Fusarium oxysporum* of which they are widely distributed around the world (Gordon Thomas, 2017).

Penetration and disease establishment

According to Joshi (2018), *Fusarium oxysporum* fungi depend on plants for nutrients and have developed mechanisms that enable them to get into the plants. Some of the penetration mechanisms that have been studied and identified by researchers include structural morphology adjustment for instance hyphae development that helps in attachment and penetration of the epidermal cells of both the wounded and un-wounded roots. Production of cell wall degrading enzymes such as cutinase, pectinase and polygalacturonase by the fungi supports to aid in penetration of the host (Mendgen Kurt et al., 1996).

The fungi produce hyphae to attach to the root surface, produce cell wall degrading enzymes to break the cell wall, fungus colonize the root cells and allows the mycelium to grow through the cortex into the vascular tissue (phloem and xylem) (Mendgen Kurt et al., 1996). The spreading of fungi will accomplish with the help of upstream flow of water also called the transpiration stream pull of water in the vascular tissues specifically the xylem. The fungus grows and produces micro and macro conidia while spreading in the vascular tissues (Bishop and Richard Cooper, 1983).

The mycelia growth in the vascular region of the host leads to blockage of the vessels, resulting in to water stress and development of wilt symptoms. The process is accelerated as the host plants try to respond by producing defence factors such as tyloses, gels and proliferation of the parenchyma cells resulting to further blockage of the vessels (Ortiz et al., 2014).

Resistant plants push to counter the spread of the fungi by producing antifungal compound such as tyloses, gels and gums to inhibit the growth and spread of the fungi. Early productions of the antifungal compounds before the fungi reach the stem vessels may enable the plant to survive but without symptoms. In susceptible plants, production of antifungal compounds is inactive, compared to fungal growth hence, leads to development of symptoms.

Disease development and symptomatology

Fusarium wilt is a warm weather soil borne fungal disease that grow optimally at temperatures of about 28 °C, temperatures above 34 °C or below 17 °C arrest the disease development. *Fusarium oxysporum* is a parasitic pathogen that interacts with the host plants resulting into development of fusarium wilt disease. Commercially *F. oxysporum* strains infect, establish, grow, cause disease symptoms and kill a wide range of plants. *Fusarium* wilt disease affects various crops ranging from the commercial legumes, melons, maize, tomatoes, bananas and oil palms (Joshi, 2018). *Fusarium* wilt becomes evident once the fungal pathogen has colonized and spread to various parts of the plant and causing blockage of the vessels as a result of mycelia growth. The plants also produce antifungal compounds that contribute to vessels blockage in the process of trying to inhibit fungal spread (Ortiz et al., 2014). Vascular vessel blockage and nutrient absorption leads to acute disease development and symptoms.

According to Joshi (2018), *Fusarium* disease occurs in two forms also called syndromes. The first form of the disease manifestation is the "acute wilt" a case where the plants leaves dry out very quickly and die while attached to the original position, while the second form of the syndrome is the "chronic wilt" where the plants lives with the fungi for a longer period which can even proceed to years and remain stunted in chronic wilt syndrome (Flood, 2006). The fungi infect the plants at any stage and some of the symptoms that are exhibited include root rots and damping off in young plants, marginal yellowing and sometimes full chlorosis on leaves of mature plants (Zitter Thomas, 1998). The fungi also affect the plants leading to gradual stunted growth of the plants over time (Flood, 2006).

Internal discoloration of the vascular bundles is the main symptom that is used as a diagnostic measure for fusarium wilt. Production of gummy red ooze from the lesion is another common symptom of fusarium wilt but according to various reviews on the symptoms this should not be mistaken with ooze as a result of stem blight (Zitter Thomas, 1998). In some rare occasions fusarium wilt has been associated with sudden collapse of some plants. The pathogen produces asexuality through spore formation that can remain in the dormant state in the soil for as long as 30 years, an important factor which makes it difficult to control.

Management of fusarium and fusarium wilt disease is a difficult task

According to Joshi (2018), *F. oxysporum* is a soil borne pathogen and is able to stay dormant in the soil for several seasons (Flood, 2006). The fungal pathogen produces spores by Asexual reproduction. The spores can remain in the dormant state in the soil for as long as >30 years, a factor which makes it difficult to control. Secondly, the existence of perennial host plants has contributed to the difficulty for its management.

Epidemiology/disease dispersal of fusarium pathogens

Epidemiology is the mode or a way in which a disease or a pest is spread from one place to another or transmission from one organism to another. *F. oxysporum* is a soil borne fungal pathogen and its spread is mainly facilitated through contact with infected plant parts or seeds (Flood, 2006). Farm cultivation tools and machinery contributes to high percentage of pathogen locomotion from one field to another especially when the implements are used in an infected field and hygiene of the implements is not checked before using them in another clean field (Joshi, 2018). The pathogen can be dispersed from one area to another through a number of ways, for instance infected planting materials and seeds are the most common method of spread and allows long distance spreading of the disease (Wong, 2003). Air and water also plays a role in the spread of this fungus especially floods and irrigation water. Environmental and soil factors are crucial for plant infection and disease manifestation. From different research the effects of fusarium wilt are high in soil infected with nematodes. The disease is also enhanced by the presence of phosphorus and ammonium nitrogen and slightly acidic soils (Wong, 2003).

Fusarium spp. as plant pathogens

Ingle (2017), stated that the members of the genus *Fusarium* are important pathogens associated with different host plants including fruits, crop plants, vegetables, etc. *Fusarium* species are primary cause for wilts, blights, root-rots and cankers in these plants. It produces different types of mycotoxins, that can unfavorably affect human and animal health. Ingle (2017), recovered 48 isolates of *Fusarium* spp. from various fruits, vegetables and food grains. Out of these, 23 isolates were recovered from vegetables, 14 from food grains and 11 from fruits.

F. oxysporum, found in its many pathogenic forms, is the most damaging species of the genus in plants (Larkin and Fravel, 1998; Polizzi et al., 2010; Perveen and Bokhari, 2010; Felgueiras et al., 2010; Khan et al., 2017; Gordon Thomas, 2017; Bodah, 2017, Srinivas et al., 2019). Sibtain et al. (2001) stated that culture filtrate containing *Fusarium* phytotoxins are helpful in studying level of resistance or susceptibility of hosts to *Fusarium*. Fusaric acid, a well known phytotoxin, produced by several pathogenic strains of *F. oxysporum*, the causal organism of wilt in an enormous variety of plants (Martyn, 1996; Ravikumar and Babu, 2007; Wu et al., 2008; Crutcher et al., 2017). Besides fusaric acid, Jedryczka (1996) isolated two other compounds from pathogenic strains of *F. oxysporum* f. sp. *pisi*. These were phenylacetic acid and methyl p-hydroxyl phenylacetate but their phytotoxicity was lower than that of fusaric acid. Liu et al. (2016) worked on chemical investigation of the endophytic fungus *F. oxysporum* isolated from fruits of *Drepanocarpus lunatus* afforded eight new fusaric acid derivatives, fusaricates A-G, 1–7, and 10-hydroxy-11-chlorofusaric acid, 8, along with four known compounds. Fusaricates A-G, 1–7, represent the first examples of fusaric acid linked to a polyalcohol moiety via an ester bond. All isolated fusaric acid derivatives 1–8 showed significant phytotoxicity to leaves of barley.

Though *Fusarium* has wide range of host plants, the formae speciales are host specific. For instance, f. sp. *cubense* is specific to banana, f.sp. *passiflorae* for passion fruits, f. sp. *cucumerinum* for cucumbers, f.sp. *cepae* for onions, f. sp. *phaseoli* for beans, *Fusarium* sp. *lycopersici* for tomatoes and

f. sp. *elaeidis* for palm oil (Joshi, 2018). Therefore to get rid from such soil borne pathogen, there is an urgent need to (a) to identify different *Fusarium* spp. associated with different type of plants as pulses, fruits, cereals, oil yielding crops etc, (b) to study host-parasite relationship deeply, (c) to study mode of action of the biocontrol agents in the host, (d) selection and application of suitable/environment friendly management practices for reduction in disease incidence in several agricultural crops of economic importance.

Fruit crops: *Fusarium* spp. associated with fruit crops

Fusarium spp are serious pest affecting various fruit trees/crops in the world. *Fusarium* spp., the causative agent of fusarium wilt, a major fungal disease devastating fruits, cereals, vegetables and leading to huge loses around the world. Many ways have been formulated to try and manage the spread of the fungi but have proven to be difficult because of its ability to remain in a dormant state for a long period in the soil, existence of perennial host and its wide range of hosts. Some important soil borne fusaria are: *F. solani, F. longipes, F. moniliforme, F. oxysporum* f.sp. *psidii, F. coeruleum* and *F. oxysporum.* Mango (*Mangifera indica* L.) is a prominent commercial fruit of worldwide. It is susceptible to various biotic and abiotic stresses (Shahbaz et al., 2009; Pandey et al., 2017). This disease has been reported from India, Bangladesh, Burma, UAE, Indonesia, Malaysia, USA, Australia, Egypt, South Africa, Sudan, Cuba, Central America, Brazil, Mexico, Israel, Philippines and Thailand (Kumar et al., 1993) in mango. The details of Fusaria associated with some important fruit crops have been presented in table (Table 19.1).

Consumption of *Fusarium* mycotoxins contaminated food (vegetables, fruits and food grains) can cause illness and various disorders in human beings and animals (Ingle, 2017). The most important *Fusarium* mycotoxins, which can frequently occur at biologically significant concentrations in various food products are fumonisins, zearalenone and trichothecenes (deoxrnivalenol and nivalenoland T-2 toxin; Ferrigo et al., 2016; Tima et al., 2016; Tola and Kebede, 2016; Ingle, 2017). Elhassan (2016) reported the association of *F. oxysporum* in *Fusarium*-wilt disease of date palm (*Phoeniex dactylifera* L.) in Northern State, Sudan and identified different isolates of *F. oxysporum* from infected plants. Dragon fruit (*Hylocereus polyrhizus*) plants were also found severely infected with *F. semitectum* leading to the heavy economic loss of many cultivars (Hawa et al., 2010).

Banana (*Musa sapientum*) is very susceptible to the infection by several fungal diseases of which *Fusarium*-wilt caused by *F. oxysporum* is considered to be important and responsible for severe economic losses in banana-producing countries (Butler, 2013). Leong et al. (2009) reported the fungal infection caused by *F. oxysporum* in banana and isolated 13 isolates of *F. oxysporum* from Malaysia and two isolates from Indonesia. Triest et al.

TABLE 19.1 Fusaria associated with some important fruit crops.

S. no.	Fruits	Diseases	Pathogens	References
1.	Watermelon (*Cucumis melo* L.)	*Fusarium* wilt	*F. oxysporum* f. sp. *niveum*	Lee (1994), Traka-Mavrona et al. (2000), Wu et al. (2009), Zhou et al. (2010), Everts et al. (2014)
2.	Guava (*Psidium guajava* L.)	*Fusarium* wilt, Guava decline	*F. oxysporum* f. sp. *psidii*, *F. solani*, *F. oxysporum*, *F. moniliformae*, *F. coeruleum*, *F. longipes*	Edward (1960), Dwivedi and Dwivedi (1999), Srivastava et al. (2011), Bokhari et al. (2008), Dolly et al. (2006), Gupta and Misra (2009), Zakaria et al. (2012), Hussain et al. (2018)
3.	Banana (*Musa acuminata*)	*Fusarium* wilt/ Panama disease, Fruit rot	*F. oxysporum* f. sp. *cubense*, *F. semitectum*, *F.verticillioides*, *F. solani*	Navajothy et al. (2011), Sathiamoorthy et al. (2001), Zakaria et al. (2012), Ploetz (2015)
4.	Cucumber (*Cucumis sativus*)	*Fusarium* wilt	*F. oxysporum* f. sp. *conglutinans*, *F. oxysporum* f. sp. *cucumerinum*	Cohen et al. (2000), Kwon et al. (2004), Yao and Wu (2010), Dong et al. (2019)
5.	Mango (*Mangifera indica*)	Mango malformation disease	*F. sterilihyphosum*, *F. proliferatum*, *F. mangiferae*, *F. pseudocircinatum*, *F. mexicanum*, *F. subglutinans*	Marasas et al. (2006), Freeman et al. (2004), Otero-Colina et al. (2010), Iqbal et al. (2010), Katoch et al. (2019)
6.	Strawberry (*Fragaria*)	*Fusarium* wilt	*F. oxysporum* f. sp. *conglutinans*, *F. oxysporum* f. sp. *fragariae*	Hartz et al. (1993), Kasperbauer (2000), Smith et al. (2001), White et al. (1993), Koike et al. (2009), Koike and Gordon (2015)
7.	Papaya (*Carica papaya*)	Fruit rot, stem rot	*F. oxysporum*, *F. semitectum*, *F. solani*	Zakaria et al. (2012), Kumar and Singh (2016)

(2016) reported that *F. musae* was infective for banana fruits and such infected fruits may cause fusariosis in human being.

India is the world's largest producer of tropical and subtropical fruits. A large variety of fruits are grown in India, of which mango, banana, citrus, guava, grape, pineapple and apple are the major ones. Beyond commercial estimates, fruits play a vital role in maintaining subsistence for by providing vitamins and other vital nutrients. *Fusarium* is responsible for causing several plant diseases such as head blights, vascular wilts, scabs, crown-rots, cankers and root-rots in fruits, vegetables and cereals (Leslie and Summerell, 2006).

Banana (*Musa acuminata*) has occupied a top position in India's booming fruit industry with an annual production of 13.5 mt from an area of 4.0 lakh ha. Panama disease, also known as *Fusarium* wilt of banana (*Musa* spp), is one of the most serious of all plant diseases. It is caused by *F. oxysporum* f. sp. *cubense* continues to pose threat among commercial Indian cultivars. Similarly, guava (*Psidium guajava* L) is one of the most common and popular fruit crop due to its high nutritional value, comparatively low price and good in taste, ranked fourth position in India as a growing crop. Bihar falls under largest producer of guava followed by Maharashtra, Uttar Pradesh, Karnataka and Andhra Pradesh. But its successful production is hampered by a group of pathogens that cause variety of diseases. Gupta et al. (2009) reported the Guava wilting caused by *F. solani* in different parts of Northern India and isolated 6 isolates of *F. solani*, and found that it is the most prevalent fungus than other pathogen responsible for wilting.

During the last two decades the problem of mango malformation disease (MMD) has reached to an alarming situation. Most of the attempts to control malformation have met with little success. The incidence of controls is little known, but these are limited, and largely confined to cover sprays of insecticide, which provide poor control and incur environmental and health risks and prolonged economic losses.

Management practices

Chemical control

Chemical practices include the use of pesticides, heavy metals, phenolic compounds and antibiotics, etc. Dwivedi (1990) studied the effect of several antibiotics against *F. oxysporum* f. sp. *psidii* causing guava wilt and showed that Actidion was found to be effective at all concentrations, while 800 ppm concentration of Norfloxacin, streptomycin, griseofulvin, mycostatin and aureofungin checked the growth of the test pathogen by 52.3%, 52.3%, 60.3%, 63.4% and 69.8% respectively. Chloramphenicol, garamycin, cephalexin and kannamycin reduced the growth by 20%, 21%, 30.7% and 46.1% at 800 ppm concentration respectively. Upadhyaya (1978) and Singh (1975) also reported the efficacy of mycostatin and griseofulvin against pathogenic fungi *in vitro*. Dwivedi (1990) reported that some phenolics i.e. phenol and salicylic acid exerted complete inhibition of *Fusarium* f. sp. *psidii* causing wilt disease in guava at 0.1% and 0.15% concentration. Dolly et al. (2006) reported that tested fungicides (carbendazim and carboxin in 0.1% and 0.2%), reduced the colony growth of *Fusarium.* f. sp. *psidii* and *Verticillium (*causing wilt in guava) by 73%, 83% and 70%, 60%, respectively.

Iqbal et al. (2010) evaluated eight fungicides for their *in vitro* effects on colony growth of *Fusarium mangiferae* after 3, 8 and 16 d of inoculation. The fungicides showed variable response in inhibiting the colony growth of the pathogen according to their nature and specificity at different minimum inhibitory concentrations (MICs). Benlate 50 WP and Carbendazim proved to be the best fungicides giving 100% suppression of the colony growth. Benlate and Carbendazim showed 100% decrease over control after 3, 8 and 16 d of inoculation as compared to the control. The fungicides, Score 250EC, Daconil W75 and Captan 50WP were found comparatively less effective. Within the three classes of fungicides, *F. mangiferae* has been found highly sensitive to type-I fungicides (Benlate 50WP, Carbendazim, Topsin-M 70WP and Copper oxychloride 50WP) with 100% suppression. The studies may be helpful as suitable control strategy to curb malformation in mango orchards (Bilgrami and Dube, 1976; Kumar, 1983; Pujol et al., 1997; Wada et al., 1990).

Pandey and Chakrabarti (2004) tested Carbendazim against *F. moniliforme*. It inhibited the growth of the germ tubes produced by conidia and also reduced the infection rate. Gaur and Chakrabarti (2009) found that Captan and Carbendazim was most effective in reducing the mycelial growth of *F. mangiferae*. In a similar study, Mamza et al. (2010) who tested six fungicides *in vitro* against *F. pallidoroseum* and found that benomyl completely inhibited mycelial growth of the fungus at 1.5 ×, 1.0 × and 0.5 × mg a.i/mL concentration.

Wu et al. (2007a, 2007b, 2009) studied the efficiency of gallic acid against *F. oxysporum* f. sp. *niveum* causing wilt in watermelons *in vitro* and found that it effectively inhibited the colony growth of the test pathogen up to 22.9%, as compared to the control. Two other phenolic compound showed similar results i.e. coumaric and cinnamic acid inhibited the hyphal growth of *F. oyxsporum* f. sp. *niveum* effectively (Wu et al., 2007a, 2007b).

Biological control

In an *in vitro* study, Dwivedi (1992) reported that *Aspergillus flavus* was least effective in case of *F. oxysporum* f. sp. *psidii* reducing the growth only by 16.6% whereas, *Trichoderma viride* inhibited the growth of *F. oxysporum* f. sp. *psidii* by 71.4%, while *T. harzianum* reduced the growth of *F. oxysporum* f. sp. *psidii* and *F. moniliformae* by 60% and 60.4%, respectively.

The isolates of *Aspergillus* spp, *Trichoderma* spp and *Penicillium citrinum* were found superior in inhibiting the growth of *Fusariun solani in vitro* (Singh et al., 2002; Misra and Prasad, 2003). Misra et al. (2004) reported that some antagonistic fungi have shown their effectiveness against wilt pathogen of guava i.e. *F. oxysporum* f. sp. *psidii*. Sultana and Ghaffar (2010) reported that microbial antagonists have potential to inhibit the seed infection in cucumber. *T. viride* was found effective against wilt pathogen of bitter gourd (*Fusarium solani*).

Dwivedi and Dwivedi (2012a) reported that plant extracts of *Eugenia caryophyllata* (Clove), *Moringa oleifera* (Sehjan), *Trachyspermum captivum* (Ajwain) and *Zingiber officinale* (Ginger) were found effective in reducing the growth of the *F. solani*. Among all four plant extracts tested, *E. caryophyllata* was found to be the most effective against *F. solani* on 3rd, 5th and 7th d at 50% concentration and suppress the colony growth by 100%, as compared to the control followed by *T. captivum* and *M. oleifera* whereas, *Z. officinale* showed least efficiency. Similarly, Srivastava et al. (2011) checked the efficacy of three plant leaf extracts viz. neem (*Azadirachta indica*), datura (*Datura stramonium*) and lantana (*Lantana camara*) against guava wilt pathogen (*Fusarium oxysporum* f. sp. *psidii*). Among them, neem leaf extract significantly reduced the mean radial growth of *Fusarium oxysporum* f. sp. *psidii* followed by lantana leaf extract.

Madhanraj et al. (2010) evaluated the efficacy of antagonists against *F. solani* causing wilt disease in banana. Among antagonists tested *T. viride* showed maximum inhibition of mycelial growth of the test pathogen followed by *A. niger* (75%), *A. terreus* (65%), *A. fumigatus* (60%), *A. flavus* (55%), *P. chrysogenum* and *P. javanicum* (44%). They also reported that in case of plant extract tested, the leaves of medicinal plant extract such as *Adathoda vasica, Azadirachta indica* and *Vitex negundo* were more effective at 20% concentration against the pathogen. The leaf extracts of these three plants along with culture filtrate of *T. viride* in 1:1 ratio was more effective at 20% level against *F. solani*. However, the combined treatment of *A. vasica* and *T. viride* was more effective in controlling *F. solani* that cause wilt disease in banana. The soil application of talc based formulations of *Pseudomonas fluorencens* at 15 g/plant in banana found to suppress *Fusarium* wilt significantly as compared to *F. oxysporum* f. sp. *cubense* (FOC) alone inoculated plants (Saravanan et al., 2003).

Dwivedi and Dwivedi (2012b) reported the efficacy of five fungal antagonists viz., *Aspergillus flavus, A. luchuensis, Penicilliun citrinum P. chrysogenum* and *T. viride* against guava wilt pathogen (*F. solani*) at 10%, 25%, 50%, 75% and 100% concentration following "Poisoned food technique". All the antagonists were found effective in controlling the growth of the test pathogen at different levels. Among the antagonists tested, *A. luchuensis* was most effective followed by *A. flavus* and *P. citrinum*. However, *P. chrysogenum* and *T. viride* gave almost similar results and inhibited the mycelial growth of the pathogen by 79.50% and 79.95% on 7th d after inoculation at 100% concentration, respectively.

Kumar and Singh (2016), reported integrated management of root rot disease of Papaya (*Carica papaya* L.) caused by *F. solani* with the integration of plant extract, organic cake and bio-agent in different combinations under field conditions. They were able to manage the disease incidence up to 29.60% as compared to the control where disease incidence was recorded up to 81.5%. MMD is an ambiguous disease of mango with the tremendous economic importance throughout the mango growing regions. Strong antifungal concoction derived from different plant extracts were reported to be beneficial to manage MMD. As the malformed mango lacks some plant growth regulators (PGRs) compared to healthy ones, researchers found the spray of the PGRs to be useful to increase the yield by reducing disease incidence (Katoch et al., 2019).

Mode of action of biological control agents

Plant diseases caused by soil-borne pathogens are among the most difficult to control. In the absence of more effective chemical control methods, there is renewed attention in biological control based on use of populations of antagonistic micro-organisms. *Pseudomonas* spp. and *Trichoderma* spp. are the two most widely studied groups of biological control agents now a days. Searching the mechanisms involved in the protective capability of these biocontrol agents is not only necessary for their development as commercial biocontrol agents but raises many basic questions related to the determinism of pathogenicity versus biocontrol capacity in the *F. oxysporum* species complex. The successful outcome of biological control depends not only on plant–microbial interactions but also on the ecological fitness of the biological control agents (Alabouvette et al., 2009).

It is very important to define what is meant by 'biological control', as there is a tendency to consider all methods using natural products to promote growth, by stimulating plant defense responses to stresses of abiotic or biotic origin, as part of the biological control approach. Eilenberg (2006) said, 'biological control is the use of living organisms to suppress the population density or impact of a specific pest organism, making it less abundant or less damaging than it would otherwise be'. The major difference between biological control and other control methods is the use of living populations of beneficial organisms, which have several modes of action and thus avoid the risk of rapid appearance of resistance in the target population (Alabouvette et al., 2009).

In this review we will mainly discuss on mode of action of 'microbiological control' of plant diseases, which is based on the use of avirulent, nonpathogenic or antagonistic microorganisms. Before going into adopting biological control one should analyze the progress and failures in biological control research and to identify the bottlenecks that prevent more rapid success in the application of microbial control.

Many modes of action are involved in the protective potential of BCAs, and generally a given strain acts through several mechanisms expressed successively, simultaneously or synergistically.

Microbial antagonism

Microbial antagonism results from direct interactions between two microorganisms sharing the same ecological niche. Three main types of direct interaction may be characterized: parasitism, competition for nutrients competition for plant tissues, and antibiosis (Alabouvette et al., 2009).

Parasitism

Parasitism of a plant pathogen by other microorganisms, including viruses, is a widely distributed phenomenon. The parasitic activity of strains of *Trichoderma* spp. towards pathogens such as *Rhizoctonia solani* has been extensively studied by Chet and Baker (1981). It involves specific recognition between the antagonist and its target pathogen and several types of cell wall-degrading enzymes (CWDEs) that enable the parasite to penetrate the hyphae of the pathogen in to the host. This type of antagonism, which causes death of the target organism, mainly results in a decrease in the inoculum density. Parasitism of fungal pathogens by viruses or virus-like particles can also induce hypovirulence. For example, hypovirulent strains of *Cryphonectria parasitica* are used to control chestnut blight. Hypovirulence is contagious; dsRNAs can be transmitted from a hypovirulent strain to a virulent compatible strain and, under favorable conditions, hypovirulence can spread naturally in diseased forests (Milgroom and Cortesi, 2004).

Competition for nutrients

Regulating the population dynamics of microorganisms sharing the same ecological niche and having the same physiological requirements when resources are limited, is defined as competition for nutrients. Competition for nutrients in soil is certainly one of the modes of action of many BCAs such as *Trichoderma* spp. For example, the reduction in the germination rate of chlamydospores of *F. oxysporum* in the rhizosphere of cotton *Gossypium* sp. and melon *C. melo* in the presence of *T. harzianum* T35 was attributed to competition for nutrients (Sivan and Chet, 1989).

Competition for nutrients, especially for carbon, is common in an oligotrophic milieu such as soil, and is considered to be responsible for fungistasis (Lockwood, 1977; De Boer et al., 2003), which is one of the reason of inhibition of fungal spore germination in soil. Competition for carbon between pathogenic and nonpathogenic *F. oxysporum* is one of the main modes of action of biocontrol strains of *F. oxysporum*. By studying the quantitative interactions in soil between a pathogenic strain and several nonpathogenic strains of *F. oxysporum*, Çouteaudier and Alabouvette (1990) established that some nonpathogenic strains were more competitive for a carbon source than others and should be selected for biological control. However, the intrinsic capacity of a strain to use the carbon source efficiently corresponds to a competitive advantage only when the two populations are of the same size.

Larkin and Fravel (1999) proposed the dose–response relationships between pathogenic and biocontrol strains of *F. oxysporum* governing biological control and defined an effective biocontrol dose for each protective strain. These researchers attributed to differences in mode of action of the strains, the differences in dose–response relationships and concluded that the fungal strain Fo47 expressed its biological control effect mainly through competition for nutrients, while the strain CS-20 acted primarily by inducing resistance. In practice the BCA Fo47 needs to be treated at an inoculum density much higher than that of the pathogen to be controlled.

Competition for minor elements also frequently occurs in soil. For example, competition for iron is one of the important modes of action by which siderophore-producing *Pseudomonas* spp. limit the growth of plant pathogenic fungi and reduce disease incidence or severity (Schippers et al., 1987; Bakker et al., 1991; Loper and Henkels, 1997). Moreover, iron competition from the siderophore-producing *Pseudomonas* spp. was shown to enhance the antagonistic effect of the protective strain Fo47 by making the pathogen more susceptible to competition for carbon (Lemanceau et al., 1992; Lemanceau and Alabouvette, 1993).

Competition for colonization of the plant tissues

Competition may also occurs for colonization of the root surface and plant tissues. It has been observed and proposed that intense colonization of the root surface by an antagonist would prevent access of the pathogen to the infection sites. Using several experimental approaches to quantify root colonization by a nonpathogenic and a pathogenic strain of *F. oxysporum*, Eparvier and Alabouvette (1994) concluded that the glucuronidase activity of the GUS-transformed pathogen was reduced in the presence of the protective strain and observed that these strains were competing for root colonization. However, another interpretation of these results have proposed: the presence of the protective strain might directly or indirectly inhibit the metabolic activity of the pathogen without reducing its biomass.

Antibiosis

Antibiosis can be defined as, 'The antagonism resulting from the toxicity of secondary metabolites produced by one microorganism for other microorganisms'. Antibiosis is a very common and known phenomenon responsible for the biocontrol activity of many BCAs such as fluorescent *Pseudomonas* spp., *Bacillus* spp., *Streptomyces* spp. and *Trichoderma* spp. Very diverse molecules have been documented and their role in the suppression of several plant pathogens have been described (Weller and Thomashow, 1993). They included antibiotics like bacteriocines, enzymes such as CWDEs, and volatile compounds with an antifungal activity. Strain of BCAs may produce several types of secondary metabolites, having different functions and also effective to control different species of fungal pathogens.

For example, the strain CHAO of *Pseudomonas fluorescens* produces siderophores, cyanide, phenazines and 2,4-diacetylphloroglucinol, different combinations of these metabolites being responsible for the antagonism expressed against *Gaeumannomyces graminis* var. *tritici* and *Chalara elegans* (Defago and Haas, 1990). Strains of *Trichoderma* spp. produce many types of secondary metabolite (Sivasithamparam and Ghisalberti, 1998) including antibiotics (Howel, 1998) and CWDEs (Kubicek and Penttila, 1998; Lorito, 1998), the role of which has been clearly established in biocontrol activity (Vinale et al., 2008). It is important to state here that a single antifungal metabolite does not account for all the antagonistic activity of BCA. Different secondary metabolites produced by a given strain of BCA might be responsible for antagonistic activities toward different types of pathogens. Therefore, one cannot become generalize from one patho-system to another patho-system, and from one BCA to another BCA. The best example comes from the work of Woo and Lorito (2007), who demonstrated that a strain of *T. harzianum* produced different secondary metabolites depending not only on the host plant to which it was applied but also on the target pathogen infecting that particular plant.

Induced resistance

At present, most of the researchers are focused on the plant–microbial interactions leading to enhanced disease resistance for biocontrol. Plant reacts to physical stresses such as drought, salt, heat, frost, inoculation with pathogenic or nonpathogenic microorganisms and chemical molecules of natural or synthetic origin by expressing defense reactions. The first experimental evidence of systemic protection induced by a microorganism was reported by Kuc (1987), who observed that cucumber (*Cucumis sativus*) protection against *Colletotricum orbiculare* after pre-inoculation of the cotyledons with this same pathogen.

Most of the researches dealing with induced resistance (IR) in plants have focused on signaling pathways at the biochemical and molecular levels. On basis of the data obtained, a distinction between induced systemic resistance (ISR) and systemic acquired resistance (SAR) has been proposed. SAR is activated by necrotizing pathogens, resulting in enhanced resistance to a broad spectrum of pathogens in organs distant from the site of infection (Ryals et al., 1996; Hammerschmidt, 1999).

The expression of SAR depends on the accumulation of salicylic acid (SA) and is associated with the induction of pathogenesis-related (PR) proteins. ISR corresponds to the resistance induced by plant growth-promoting rhizobacteria (PGPR) inoculated onto the roots and involving the jasmonic acid and ethylene pathways (Pieterse et al., 1998). These two main pathways are not independent and there are some common nodes; for example, both SAR and ISR are controlled by the regulatory protein non-expressor of pathogenesis-related genes1 (NPR1). Cross-communication between defense pathways provides a regulatory potential that allows the plant to fine-tune its defense responses (Pieterse and Van Loon, 2004).

It was assumed that protection conferred by IR was based on direct activation of defense reactions. However, based on the most recent research on ISR, the phenomenon of 'priming' appears to be a common feature of the plant's immune system that offers protection (Van Hulten et al., 2006). According to Goellner and Conrath (2008), priming is a component of IR in plants and corresponds to a physiological state in which an induced plant shows a faster or greater activation of defense responses after infection with a challenging pathogen.

Most of our knowledge of plant–microbial interactions has been obtained from studies on plant–pathogen interactions, which might not be representative of the interactions between plants and BCAs. Many studies of the signaling pathways induced in the plant by BCAs have been carried out, *Trichoderma* spp. and fluorescent *Pseudomonas* spp. being the two groups of BCAs on which most attention has been focused in this field. Research on ISR-inducing *Pseudomonas* showed that some of the microbe-associated molecular patterns (MAMPs), such as flagellin and lipopolysaccharides (LPSs), that are known to play a role in the interactions between pathogenic *Pseudomonas* spp. and the plant, are involved in the protection conferred by beneficial *Pseudomonas* spp. (Van Wees et al., 2008). Purified flagellin and LPS of the nonpathogenic resistance-inducing strains of *P. fluorescens* or *Pseudomonas putida* showed the capability to induce resistance in the plant. Some of the antibiotics produced by the beneficial bacteria could also function as inducers of the immune response. For example, the 2,4-diacetylphloroglucinol produced by many fluorescent *Pseudomonas* spp. was demonstrated to induce resistance (Weller et al., 2002). Similarly, many of the secondary metabolites involved in hyperparasitism or antibiosis of *Trichoderma* spp. are also able to trigger plant resistance. Inoculation of

Trichoderma spp. on to the roots resulted in the up-regulation of different endogenous defense-related proteins and enzyme activities in the plant, a situation typically occurring when the plant faces pathogen attack (Woo and Lorito, 2007).

The protective strains of *F. oxysporum*

The idea of using nonpathogenic strains of *F. oxysporum* to control *Fusarium* diseases came from studies of soils naturally suppressive to *Fusarium* wilts, such as the soil from Châteaurenard (Louvet et al., 1976). These soils harbor high populations of nonpathogenic *F. oxysporum* and *Fusarium solani* whose involvement in the mechanism of soil suppressiveness was confirmed experimentally using a variant of 'Koch's postulates' (Rouxel et al., 1979). Strains of *F. oxysporum* were much more efficient in establishing suppressiveness in soil than other species of *Fusarium* (Tamietti and Alabouvette, 1986). There is a great diversity among soil-borne nonpathogenic strains of *F. oxysporum* in their capacity to protect plants against their specific pathogens (Alabouvette et al., 1987; Forsyth et al., 2006), and some effective strains have not been isolated from soil but from the stems of healthy plants (Ogawa and Komada, 1984; Postma and Rattink, 1992). Moreover, it is well established that a pathogenic strain applied to a nonhost plant is able to protect it against further infection by its specific *forma specialis* . This phenomenon was first described by Matta (1971) and defined as 'cross-protection' or 'premunition'. Recently, Minerdi et al. (2008) showed that a protective strain of *F. oxysporum* was in fact a pathogenic strain belonging to the *forma specialis lactucae* , associated with several bacteria. The cured strain was pathogenic on the plant, suggesting that the protective effect was not a fungal trait but was dependent on the interaction with the ectosymbiotic bacteria.

The protective strains of *F. oxysporum* have been shown to control *Fusarium* wilt in many crops, including asparagus (*Asparagus officinalis*; Elmer, 2004), banana (*Musa* sp; Forsyth et al., 2006), basil (*Ocimum basilicum*; Larkin and Fravel, 1999), carnation (*Dianthus caryophyllus*; Garibaldi et al., 1986), chickpea (*Cicer arietinum*; Hervas et al., 1997), cucumber (*Cucumis sativus*; Mandeel and Baker, 1991), cyclamen (*Cyclamen persicum*; Minuto et al., 1995), gladiolus (*Gladiolus* sp.; Magie, 1980), melon (*Cucumis melo*; Rouxel et al., 1979), tomato (*Solanum lycopersicum*; Olivain et al., 1995; Fuchs et al., 1997) spinach (*Spinacia oleracea*; Katzube et al., 1994) and watermelon (*Citrullus lanatus*; Larkin et al., 1996). In contrast to other BCAs such as *Pseudomonas* spp. or *Trichoderma* spp., the protective strains of *F. oxysporum* are effective mostly against pathogenic *F. oxysporum*.

The protective strains are usually more effective when they are applied a few days before inoculation of the pathogen; and the protection is often improved when the strains are associated with rhizobacteria, especially fluorescent pseudomonads (Lemanceau and Alabouvette, 1993; Saman, 2009).

How to make biological control successful in the field application?

The success of microbiological control of the plant diseases needs a sufficient understanding of the modes of action of the antagonist and also of its interactions with the plant, the pathogen and the rest of the microbiota. How to improve the performance of BCAs in the fields is a major question that scientists should address. The use of registered BCAs is limited, mostly because of the lack of efficacy and consistency of biological control (Alabouvette et al., 2009).

The first approach to accomplish the target successfully, would be to combine several modes of action to either improve efficacy or increase the activity spectrum of the BCAs. This can be achieved by improving the expression of different mechanisms involved in the biocontrol capacities of a BCA. For example, strains of *P. fluorescens* have been manipulated to produce overproduction of pyrrolnitrin, an antifungal compound (Ligon et al., 2000), or to produce simultaneously phenazine and phloroglucinol, resulting in a greater capacity to suppress *Rhizoctonia* root rot (Huang et al., 2004). It is also possible to use associations of BCAs having complementary modes of action or targeting different pathogens. The efficiency of Fo47 is usually improved by association with strains of fluorescent *Pseudomonas*, which can be produced in the same substrate as that used to grow Fo47 (Olivain et al., 2004; Alabouvette et al., 2009).

The success of biological control also depends on the ecological fitness of the BCAs, especially if they target soil-borne plant pathogens. To make it more effective, a BCA must establish in soil and express its protective potential. Failure of biological control has often been observed with lack of fitness of the BCA, the total population of which does not reach the threshold level needed to be effective to suppress the pathogen.

After introduction into the soil, the BCA must be compatible with the physico-chemical properties and with the biological characteristics of the soil. In a study by Edel-Hermann et al. (2009), the population dynamics of Fo47 was applied in two soils of contrasting physico-chemical characteristics. In the sterilized soils, where the native biota was absent, Fo47 established at similar high levels (close to 10^6 CFU/gram of soil) in an acidic sand (pH 4.3) and a clay-loam (pH 8.2), whatever the initial concentration at which it was introduced. Under such conditions, the water potential and temperature of the soils did not influence the population dynamics. In the presence of the native microflora, Fo47 established in both soils at lower densities than those at which it was introduced. These results demonstrate that Fo47 is more susceptible to the available biota than to the physico-chemical properties of the soils.

According to Cook and Baker (1983), the concept of soil suppressiveness of diseases that describes the potential of the soil to control establishment of the pathogen and expression of its virulence, also applies to biological control. In other words, some soils are comparatively more favorable than others to the establishment and expression of the protective capacity of a given BCA. Also the ecological fitness of the biocontrol agents depends on the production and formulation processes. The BCA propagules, industrially produced on rich media and under aseptic conditions, are probably less well adapted to the soil's natural environment than the native propagules of the pathogen. Therefore, one can hypothesize that, in soil, competitive interactions among the native microflora, the pathogen and the BCA will favor the pathogen to survive. Several approaches have been proposed to produce and formulate BCAs at low cost, and solid-state fermentation is favored for producing propagules with higher fitness than propagules produced in submerged fermentation. To give a competitive advantage to the BCA, it has been proposed that a food base be added to promote its growth and activity after release.

From the industries, some biological control products are made of fungal conidia 'coated' with glucose or saccharose. However, it would be preferable, to select a food base required more specifically by the BCA than sugars, which can be consumed by most soil microorganisms. Also, the success of biological control in the field depends on the dose and mode of application. In soilless cultures, the application of BCA is quite easy, either by mixing it with the growth substrate before planting or by using the drip irrigation system. This method is not economically viable in open fields to apply the BCA evenly at the dose needed for efficacy. Bennett and Whipps (2008) concluded that seed coating is the best approach, as it enables the BCA to be positioned close to the seedlings and good colonization of the rhizosphere of the plant can be expected.

Whatever, the modes of action of the BCA, the success of biological control totally depends on the activity of the BCA at the root level, in the rhizosphere. Rhizosphere competence is very important criterion that must be deeply observed early during the screening process. There are great differences in the ability to colonize the rhizosphere of the plant even among strains belonging to the same species. Besides, this capacity is host plant specific, and therefore a given BCA might be effective in controlling a given pathogen in one crop but not in another.

The last very important issue to be discussed is that of integration of biological control agents with other cultural practices that requires a holistic approach and taking into account all the components of agro-ecosystem, at farm scale or even at the landscape level. To conclude, in our opinion research work done on biological control of pathogens, if biological control has to make successful more research is required, not only to elucide modes of action at the molecular level, but also to characterize the ecological fitness of the BCAs, and to improve the production, formulation and application processes.

References

Abbas, H.K., Mirocha, C.J., Kommedahl, T., Vesonder, R.F., Golinski, P., 1997. Comparison of ceramide synthase inhibitor with other phytotoxins produced by *Fusarium* species. J. Nat. Toxin 6, 163–181.

Abraham, J.Z., 1986. *Fusarium* Species: Their Biology and Toxicology. John Wiley & Sons, Inc, New York, p. 588.

Alabouvette, C., De La Broise, D., Lemanceau, P., Couteaudier, Y., Louvet, J., 1987. Utilisation de souches non pathogènes de *Fusarium* pour luttrer contre les fusarioses: situation actuelle dans la pratique. Bull. OEPP-EPPO 17, 665–674.

Alabouvette, C., Olivain, C., Migheli, Q., Steinberg, C., 2009. Microbiological control of soil-borne phytopathogenic fungi with special emphasis on wilt-inducing *Fusarium oxysporum*. N. Phytologist 184, 529–544.

Armstrong, G.M., Armstrong, J.K., 1981. *Formae speciales* and races of *Fusarium oxysporum* causing wilt diseases. In: Nelson, P.E., Toussoun, T.A., Cook, R.J. (Eds.), Fusarium: Diseases, Biology and Taxonomy. Pennsylvania State University Press, University Park and London, pp. 391–399.

Baffoni, L., Gaggia, F., Dalanaj, N., Prodi, A., Paola Nipoti, P., Pisi, A., et al., 2015. Microbial inoculants for the biocontrol of *Fusarium* spp. in durum wheat. BMC Microbiol. 15, 242.

Bakker, P.A.H.M., Van Peer, R., Schippers, B., 1991. Suppression of soil-borne plant pathogens by fluorescent pseudomonads: mechanisms and prospects. In: Beemster, A.B.R., Bollen, G.H., Gerlach, M., Ruissen, M.A., Schippers, B., Tempel, A. (Eds.), Development in Agriculturally Managed-Forest Ecology. Elsevier, Amsterdam, The Netherlands, pp. 217–230.

Bardgett, R.D., Van der Putten, W.H., 2014. Below ground biodiversity and ecosystem functioning. Nature 515, 505–511.

Bar-On, Y.M., Phillips, R., Milo, R., 2018. The biomass distribution on Earth. Proc. Natl. Acad. Sci. USA 115, 6506–6511.

Bennett, A.J., Whipps, J.M., 2008. Beneficial microorganism survival on seed, roots and in rhizosphere soil following application to seed during drum priming. Biol. Control. 44, 349–361.

Bilgrami, K.S., Dube, H.C., 1976. A Text Book of Modern Plant Pathology. Vikas Publishing House, New Delhi, India, p. 344.

Bishop, C.D., Richard Cooper, M., 1983. An ultrastructural study of vascular colonization in three vascular wilt diseases 1. Colonization of susceptible cultivars. Physiol. Plant Pathol. 23, 323–343.

Bjelic, D., Ignjatov, M., Marinkovic, J., Karaman, M., 2018. *Bacillus* isolates as potential biocontrol agents of *Fusarium* clove rot of garlic. Zemdirbyste-Agriculture 105, 369–376.

Bodah, E.T., 2017. Root rot disease in plants: a review of common causal agents and management strategies. Agric. Res. Technol. 5 (3). Available from: https://doi.org/10.19080/ARTOAJ.2017.04.555661.

Bokhari, A.A., Sahi, T.S., Khan, A.M., Ahmad, R., 2008. *In vivo* studies on the biological and chemical control of guava decline caused by different soil borne pathogens. Pak. J. Agric. Sci. 45, 54–56.

Butler, D., 2013. Fungus threatens top banana. Nature 504, 195–196.

Chet, I., Baker, R., 1981. Isolation and biocontrol potential of *Trichoderma hamatum* from soil naturally suppressive to *Rhizoctonia solani*. Phytopathology 71, 286–290.

Choudhary, A.K., Kumar, S., Patil, B.S., Bhat, J.S., Sharma, M., Kamal, S., et al., 2013. Narrowing yield gaps through genetic improvement for *Fusarium* wilt resistance in three pulse crops of the semi arid tropics. SABRAO J. Breed. Genet. 45, 341–370.

Cohen, R., Pivonia, S., Burger, Y., Edelstein, M., Gamliel, A., Katan, J., 2000. Toward integrated management of *Monosporascus* wilt of melons in Israel. Plant Dis. 84, 496–505.

Cook, R.J., Baker, K.F., 1983. The Nature and Practice of Biological Control of Plant Pathogens. American Phytopathological Society, St. Paul, MN.

Couteaudier, Y., Alabouvette, C., 1990. Quantitative comparison of *Fusarium oxsporum* competitiveness in relation with carbon utilization. FEMS Microbiol. Ecol. 74, 261–268.

Crutcher, F.K., Puckhaber, L.S., Bell, A.A., Liu, J., Duke, S.E., Stipanovic, R.D., et al., 2017. Detoxification of fusaric acid by the soil microbe, *Mucor rouxii*. J. Agric. Food Chem. 65 (24), 4989–4992.

De Boer, W., Verheggen, P., Klein Gunnewiek, P.J.A., Kowalchuk, G.A., Van Veen, J.A., 2003. Microbial community composition affects soil fungistasis. Appl. Environ. Microbiol. 69, 835–844.

Dean, R., Van Kan, J.A., Pretorius, Z.A., Hammond-Kosack, K.E., Di Pietro, A., Spanu, P.D., et al., 2012. The Top 10 fungal pathogens in molecular plant pathology. Mol. Plant Pathol. 13, 414–430. Available from: https://doi.org/10.1111/j.1364-3703.2011.00783.x.

Defago, G., Haas, D., 1990. Pseudomonads as antagonists of soil-borne plant pathogens: modes of action and genetic analysis. In: Bollag, J.M., Stotsky, G. (Eds.), Soil Biochemistry. Marcel Dekker Inc, New York, USA, pp. 249–291.

Diane, M., Molina, A.B., Daniells, J., et al., 2017. The distribution and host range of the banana *Fusarium* wilt fungus, *Fusarium oxysporum* f. sp. *cubense*. Asia. Public Lib. Sci. PLoS One 12 (7), e0181630.

Dita, M., Barquero, M., Heck, D., Mizubuti, E.S.G., Staver, C.P., 2018. *Fusarium* wilt of banana: current knowledge on epidemiology and research needs toward sustainable disease management. Front. Plant Sci. Available from: https://doi.org/10.3389/fpls.2018.01468.

Divakara, S.T., Santosh, P., Aiyaz, M., Ramana, M.V., Hariprasad, P., Nayaka, S.C., et al., 2014. Molecular identification and characterization of *Fusarium* spp. associated with sorghum seeds. J. Sci. Food Agriculture 94, 1132–1139.

Dolly, J., Gupta, P., Gupta, V., 2006. Etiology and management of guava wilt disease in Allahabad and Kaushambi Districts of Uttar Pradesh, India. J. Mycol. Plant Pathol. 36, 365–372.

Dong, J., Xu, J., Xu, X., Xu, Q., Chen, X., 2019. Inheritance and quantitative trait locus mapping of *Fusarium* wilt resistance in cucumber. Front. Plant. Sci. Available from: https://doi.org/10.3389/fpls.2019.01425.

Dwivedi, D., Tang, J., Bouskill, N., Georgiou, K., Chacon, S.S., Riley, W.J., 2019. Abiotic and biotic controls on soil organo-mineral interactions: developing model structures to analyze why soil organic matter persists. Rev. Mineral. Geochem. 85, 329–348.

Dwivedi, S.K., Dwivedi, N., 2012a. Antifungal activity of some plant extracts against guava wilt pathogen. Int. J. Environ. Sci. 3, 412–420.

Dwivedi, S.K., Dwivedi, N., 2012b. *In vitro* bioefficacy of some selected fungal antagonists against guava wilt pathogen. IOSR J. Eng. 2, 1217–1223.

Dwivedi, S.K., Dwivedi, P., 1999. Wilt disease of guava: a national problem. J. Appl. Hortic. 1, 151–154.

Dwivedi, S.K., 1990. Antifungal activity of some phenolic compounds on *Fusarium oxysporum* f.sp. *psidii* causing wilt disease of guava. Hindustan Antibiotics Bull. 32, 33–35.

Dwivedi, S.K., 1992. Effect of culture filtrate of soil microbes of pathogens inciting wilt disease of guava (*Psidium guajava* L.) under *in vitro* conditions. Natl Acad. Sci. Lett. 15, 33–35.

Edel-Hermann, V., Brenot, S., Gautheron, N., Aime, S., Alabouvette, C., Steinberg, C., 2009. Ecological fitness of the biocontrol agent *Fusarium oxysporum* Fo47 in soil and its impact on the soil microbial communities. FEMS Microbiol. Ecol. 68, 37–45.

Edward, J.C., 1960. Penetration and establishment of *Fusarium oxysporum* f. sp. *psidii* in Guava root. Indian.Phytopathology 13, 168–170.

Egidi, E., Delgado-Baquerizo, M., Plett, J.M., Wang, J., Eldridge, J.D., Bardgett, Richard D., et al., 2019. A few *Ascomycota* taxa dominate soil fungal communities worldwide. Nat. Commun. 10, 2369. Available from: https://doi.org/10.1038/s41467-019-10373-z.

Eilenberg, J., 2006. Concepts and visions of biological control. In: Eilenberg, J., Hokkanen, H. M.T. (Eds.), An Ecological and Societal Approach to Biological Control. Springer, Dordrecht, The Netherlands, pp. 1–11.

Elhassan, A.M., 2016. Isolation, identification and characterization of *Fusarium oxysporum*, the causal agent of *Fusarium* wilt disease of date palm *Phoeniex dactylifera* L. in Northern State. Int. J. Curr. Microbiol. Appl. Sci. 5, 381–386.

Elmer, W.H., 2004. Combining nonpathogenic strains of *Fusarium oxysporum* with sodium chloride to suppress *Fusarium* crown rot of asparagus in replanted fields. Plant Pathol. 53, 751–758.

Eparvier, A., Alabouvette, C., 1994. Use of ELISA and GUS-transformed strains to study competition between pathogenic and non-pathogenic *Fusarium oxysporum* for root colonization. Biocontrol Sci. Technol. 4, 35–47.

Everts, K.L., Egel, D.S., Langston, D., Zhou, X.-G., 2014. Chemical management of *Fusarium* wilt of watermelon. Crop. Prot. 66, 114–119.

Felgueiras, M., Dias, A., Chicau, G., Berbega, M., León, M., Armengo, J., 2010. First report of *Fusarium* Wilt caused by *Fusarium oxysporum* f. sp. *basilici* on *Ocimum minimum* in Portugal. Plant Dis. 94 (9), 1170.1. Available from: https://doi.org/10.1094/PDIS-01-19-0134-PDN.

Ferrigo, D., Raiola, A., Causin, R., 2016. *Fusarium* toxins in cereals: occurrence, legislation, factors promoting the appearance and their management. Molecules 21, 627. Available from: https://doi.org/10.3390/molecules21050627.

Flood, J., 2006. A review of *Fusarium* wilt of oil palm caused by *Fusarium oxysporum* f sp. Elaeidis. Phytopathology 96, 660–662.

Forsyth, L.M., Smith, L.J., Aitken, E.A.B., 2006. Identification and characterization of non-pathogenic *Fusarium oxysporum* capable of increasing and decreasing *Fusarium* wilt severity. Mycol. Res. 110, 929–935.

Frąc, M., Hannula, S.E., Bełka, M., Jędryczka, M., 2018. Fungal biodiversity and their role in soil health. Front. Microbiol. 9 (707), (2018).

Freeman, S., Klein-Gueta, N.K.D., Sztejnberg, A., 2004. Epidemiology and survival of *Fusarium mangiferae*, the causal agent of mango malformation disease. Acta Hortic. 645, 487–491.

Fuchs, J.G., Moenne-Loccoz, Y., Defago, G., 1997. Non-pathogenic *Fusarium oxysporum* strain Fo47 induces resistance to *Fusarium* wilt in tomato. Plant Dis. 81, 492–496.

Gachango, E., Kirik, W., Schafer, R., Wharton, P., 2012. Evaluation and comparison of biocontrol and conventional fungicides for control of post harvest potato tuber diseases. Biol. Control. 63, 115–120.

Gamliel-Atinsky, E., Sztejnberg, A., Maymon, M., Shtienberg, D., Freeman, S., 2009. Inoculum availability and conidial dispersal patterns of *Fusarium mangiferae*, the causal agent of mango malformation disease. Phytopathology 99, 160–166.

Garibaldi, A., Brunatti, F., Gullino, M.L., 1986. Suppression of *Fusarium* wilt of carnation by competitive non-pathogenic strains of *Fusaria*. Mededelingen van. de. Faculteit Landbouwwetenschappen, Rijksuniv. Gent 51, 633–638.

Gaur, V.P., Chakrabarti, D.K., 2009. Incidence of malformation in mango (*Mangifera indica*) nursreries in eastern Uttar Pradesh. Indian J. Agric. Sci. 79, 160–162.

Goellner, K., Conrath, U., 2008. Priming: it's all the world to induced disease resistance. Eur. J. Plant Pathol. 121, 233–242.

Gordon Thomas, R., 2017. *Fusarium oxysporum* and the *Fusarium* wilt syndrome. Annu. Rev. Phytopathol. 55, 23–39.

Gordon, T.R., Martyn, R.D., 1997. The evolutionary biology of *Fusarium oxysporum*. Annu. Rev. Phytopathol. 35, 111–128.

Gupta, V.K., Misra, A.K., 2009. Efficacy of bioagents against *Fusarium* wilt of guava. J. Mycol. Plant. Pathol. 39, 101–106.

Gupta, V.K., Misra, A.K., Gaur, R., Pandey, R., Chauhan, U.K., 2009. Studies on genetic polymorphism in isolates of *Fusarium solani*. Austral. J. Crop. Sci. 3, 101–106.

Hammerschmidt, R., 1999. Induced disease resistance: how do induced plants stop pathogens? Physiol. Mol. Plant Pathol. 55, 77–84.

Hartz, T.K., Devay, J.E., Elmore, C.L., 1993. Solarization is an effective soil disinfestation technique for strawberry production. HortScience 28, 104–106.

Hawa, M.M., Salleh, B., Latiffah, Z., 2010. Characterization and intraspecific variation of *Fusarium semitectum* (Berkeley and Ravenel) associated with red-fleshed dragon fruit (*Hylocereus polyrhizus* [Weber] Britton and Rose) in Malaysia. Afr. J. Biotechnol. 9, 273–284.

Hervas, A., Landa, B., Jimenez-Diaz, R., 1997. Influence of chickpea genotype and *Bacillus* sp. on protection from *Fusarium* wilt by seed treatment with nonpathogenic *Fusarium oxysporum*. Eur. J. Plant. Pathol. 103, 631–642.

Howel, C.R., 1998. The role of antibiosis in biocontrol. In: Harman, G.E., Kubicek, C.P. (Eds.), Trichoderma & Gliocladium. vol. 2, Enzymes, Biological Control and Commercial Application. Taylor and Francis Ltd, London, UK, pp. 173–180.

Huang, Z.Y., Bonsall, R.F., Mavrodi, D.V., Weller, D.M., Thomashow, L.S., 2004. Transformation of *Pseudomonas fluorescens* with genes for biosynthesis of phenazine-1-carboxylic acid improves biocontrol of rhizoctonia root rot. FEMS Microbiol. Ecol. 49, 243–251.

Hussain, M., Usman, M., Fatima, B., Nawaz-ul-Rehman, M.S., 2018. Assessment of guava wilt disease (GWD) and varietal susceptibility in Punjab-Pakistan. Pak. J. Agric. Sci. 56, 393–399.

Ingle, A., 2017. Diversity and identity of *Fusarium* species occurring on fruits, vegetables and food grains. *Nusantara*. Bioscience 9, 44–51.

Iqbal, Z., Pervez, M.A., Ahmad, S., Iftikhar, Y., Yasin, M., Nawaz, A., et al., 2010. Determination of minimum inhibitory concentrations of fungicides against fungus *fusarium mangiferae*. Pak. J. Bot. 42, 3525–3532.

Jedryczka, M. 1996. Methods of resistance evaluation and obtaining pea (*Pisum sativum* L.) genotypes resistant to Fusarium Wilt (Fusarium oxysporum Schlecht. f. sp. pisi [Van Hall] Snyd. Et Hans.) (Ph.D. thesis), 1995. Institute of Plant Genetics, Polish Academy of Sciences, Poznan, Poland. Grain Legumes, 13, 9.

Joshi, R., 2018. A review of *Fusarium oxysporum* on its plants interaction and industrial use. J. Med. Plants Stud. 6, 112–115.

Kalagatur, N.K., Kamasani, J.R., Mudili, V., 2018. Assessment of detoxification efficacy of irradiation on zearalenonemycotoxin in various fruit juices by response surface methodology and elucidation of its *in-vitro* toxicity. Front. Microbiol. 9, 2937.

Kalagatur, N.K., Karthick, K., Allen, J.A., Ghosh, N., Sivaraman, O., Chandranayaka, S., et al., 2017. Application of activated carbon derived from seed shells of Jatrophacurcasfor decontamination of zearalenonemycotoxin. Front. Pharmacol. 8, 760.

Kasperbauer, M.J., 2000. Strawberry yield over red versus black plastic mulch. Crop. Sci. 40, 171–174.

Katoch, P., Katoch, A., Dangi, B., 2019. An overview on mango malformation and the potential approaches to their management. J. Pharmacognosy Phytochem. 8, 621–626.

Katzube, K., Akasaka, Y., Nakatani, F., 1994. Biocontrol of *Fusarium* wilt of spinach by using nonpathogenic *Fusarium oxysporum.* 2. Investigation of inoculation methods. Ann. Rep. Plant Prot. N. Jpn. 445, 72–75.

Khan, N., Maymon, M., Hirsch, A.M., 2017. Combating *Fusarium* infection using *Bacillus*-based antimirobials. Microorganisms 5 (75). Available from: https://doi.org/10.3390/microorganisms 5040075.

Koike, S.T., Gordon, T.R., 2015. Management of *Fusarium* wilt of strawberry. Crop. Prot. 73, 67–72.

Koike, S.T., Kirkpatrick, Gordon, T.R., 2009. *Fusarium w*ilt of strawberry caused by *Fusarium oxysporum* in California. Am. Phytopathol. Soc. (APS) . Available from: https://doi.org/ 10.1094/PDIS-93-10-1077A.

Kubicek, C.P., Penttila, M.E., 1998. Regulation of production of plant polysaccharide degrading enzymes by *Trichoderma.* In: Harman, G.E., Kubicek, C.P. (Eds.), Trichoderma and Gliocladium. Enzymes, Biological Control and Commercial Applications. Taylor & Francis Ltd, London, UK, pp. 49–71.

Kuc, J., 1987. Plant immunization and its applicability for disease control. In: Chet, I. (Ed.), Innovative Approaches to Plant Disease Control. John Wiley and Sons, New York, USA:, pp. 255–274.

Kumar, J. 1983. Studies on Symptomatology, Etiology and Control of Mango Malformation (Ph. D. thesis), G. B. Plant University of Agriculture and Technology, Pantnager, India, 199.

Kumar, J., Singh, U.S., Beniwal, S.P.S., 1993. Mango malformation: one hundred years of research. Annu. Rev. Phytopathol. 31, 217–232.

Kumar, R., Singh, S.K., 2016. Integrated management of root rot disease of papaya (*Carica papaya* L.) caused by *Fusarium solani.* J. AgriSearch 3 (2), 106–109.

Kwon, D.H., Choi, J.H., Jung, H.K., Lim, J.H., Joo, G.J., Kim, S.D., 2004. Selection and identification of Auxin-producing plant growth promoting rhizobacteria having phytophathogen antagonistic activity. J. Korean Soc. Appl. Biol. Chem. 47, 17–21.

Larkin, R.P., Fravel, D.R., 1998. Efficacy of various fungal and bacterial biocontrol organisms for control of *Fusarium* wilt of tomato. Plant Dis. 82, 1022–1028.

Larkin, R.P., Fravel, D.R., 1999. Mechanisms of action and dose-response relationships governing biological control of *Fusarium* wilt of tomato by non-pathogenic *Fusarium* spp. Phytopathology 89, 1152–1161.

Larkin, R.P., Hopkins, D.L., Martin, F.N., 1996. Suppression of *Fusarium* wilt of watermelon by non-pathogenic *Fusarium oxysporum* and other microorganims recovered from a disease suppressive soil. Phytopathology 86, 812–819.

Latiffah, Z., Mohd, Z.M., Baharuddin, S., 2007. Malaysian Diversity of *Fusarium* species in cultivated soils in Penang. J. Microbiol. 3, 27–30.

Lee, J.M., 1994. Cultivation of grafted vegetables. 1. Current status, grafting methods and benefits. HortScience 29 (4), 235–239.

Lemanceau, P., Alabouvette, C., 1993. Suppression of *Fusarium* wilts by fluorescent *Pseudomonads*: mechanisms and application. Biocontrol Sci. Technol. 3, 219–224.

Lemanceau, P., Bakker, P.A.H.M., De Kogel, W.J., Alabouvette, C., Schippers, B., 1992. Effect of pseudobactin 358 production by *Pseudomonas putida* WCS358 on suppression of *Fusarium* wilt of carnations by nonpathogenic *Fusarium oxysporum* Fo47. Appl. Environ. Microbiol. 58, 2978–2982.

Leong, S.K., Latiffah, Z., Baharuddin, S., 2009. Molecular characterization of *Fusarium oxysporum* f. sp. cubense of banana. Am. J. Appl. Sci. 6, 1301–1307.

Leslie, J.F., Summerell, B.A., 2006. The *Fusarium* Laboratory Manual. Blackwell Publishing Ltd, Oxford, UK, p. 387.

Lewis, J.A., Papavizas, G.C., 1991. Biocontrol of cotton damning off caused by *Rhizoctonia solani* in the field with formulations of *Trichoderma* spp and *Gliocladium virens*. Crop. Prot. 10, 396–402.

Ligon, J.M., Hill, D.S., Hammer, P.E., Torkewitz, N.R., Hofmann, D., Kempf, H.J., et al., 2000. Natural products with antifungal activity from *Pseudomonas* biocontrol bacteria. Pest Manag. Sci. 56, 688–695.

Liu, S., Dai, H., Orfali, R.S., Lin, W., Liu, Z., Proksch, P., 2016. New fusaric acid derivatives from the endophytic fungus *Fusarium oxysporum* and their phytotoxicity to barley leaves. J. Agric. Food Chem. 64, 3127–3132.

Lockwood, J.L., 1977. Fungistasis in soils. Biol. Rev. 52, 1–43.

Loper, J.E., Henkels, M.D., 1997. Availability of iron to *Pseudomonas fluorescens* in rhizosphere and bulk soil evaluated with an ice nucleation reporter gene. Appl. Environ. Microbiolol. 63, 99–105.

Lorito, M., 1998. Chitinolytic enzymes and their genes. In: Harman, G.E., Kubicek, C.P. (Eds.), Trichoderma & Gliocladium. Volume 2, Enzymes, Biological Control and Commercial Application. Taylor and Francis Ltd, London, UK:, pp. 73–92.

Louvet, J., Rouxel, F., Alabouvette, C., 1976. Recherches sur la résistance des sols aux maladies. I - Mise en évidence de la nature microbiologique de la résistance d'un sol au développement de la fusariose vasculaire du melon. Ann. Phytopathol. 8, 425–436.

Madhanraj, P., Ambikapathy, V., Panneerselvam, 2010. A biological control of banana wilt caused by Fusarium solani (mart.) sacc. Int. J. Appl. Biol. Pharm. Technol. 1, 1032–1039.

Magie, R.O., 1980. Fusarium disease of gladioli controlled by inoculation of corms with nonpathogenic Fusaria. Proc. Florida State Hortic. Soc. 93, 172–175.

Mamza, W.S., Zarafi, A.B., Alabi, O., 2010. *In vitro* evaluation of six fungicides on radial mycelial growth and regrowth of *Fusarium pallidoroseum* isolated from castor (*Ricinus communis*) in Samaru, Nigeria. Arch. Phytopathol. Plant Prot. 43, 116–122.

Mandeel, Q., Baker, R., 1991. Mechanisms involved in biological control of Fusarium wilt of cucumber with strains of nonpathogenic *Fusarium oxysporum*. Phytopathology 81, 462–469.

Marasas, W.F.O., Nelson, P.E., Tousson, T.A., 1984. Toxigenic *Fusarium* Species: Identity and Mycotoxicology. Pennsylvania State University Press., University Park, Pennsylvania.

Marasas, W.F.O., Ploetz, R.C., Wingfield, M.J., Wingfield, B.D., Steenkamp, E.T., 2006. Mango malformation disease and the associated *Fusarium* species. Phytopathology 96, 667–672.

Martyn, R.D., 1996. *Fusarium* wilt of watermelon. In: Zitter, T.A., Hopkins, D.L., Thomas, C.E. (Eds.), Compendium of Cucurbit Diseases. The American Phytopathology Society, St. Paul, MN, pp. 13–14.

Matta, A., 1971. Microbial penetration and immunization of uncongenial host plants. Annu. Rev. Phytopathol. 9, 387–410.

Mejdoub-Trabelsi, B., Abdallah, R.A.B., Ammar, N., Kthiri, Z., Hamada, W., Daami-Remadi, M., 2016. Bio-suppression of *Fusarium* wilt disease in potato using nonpathogenic potato-associated fungi. J. Plant Pathol. Microbiol. 7, 4. Available from: https://doi.org/10.4172/2157-7471.1000347.

Mendgen Kurt, M., Hahn, M., Holger, D., 1996. Morphogenesis and mechanisms of penetration by plant pathogenic fungi. Annu. Rev. Phytopathol. 34, 367–386.

Mihajlovic, M., Rekanovic, E., Hrustic, J., Grahovac, M., Tanovic, B., 2017. Methods for management of soilborne plant pathogens. Pesticides Phytomed. (Belgrade) 32, 9–24.

Milgroom, M.G., Cortesi, P., 2004. Biological control of chestnut blight with hypovirilence: a critical analysis. Annu. Rev. Phytopathol. 42, 331–638.

Minerdi, D., Moretti, M., Gilardi, G., Barberio, C., Gullino, M.L., Garibaldi, A., 2008. Bacterial ectosymbionts and virulence silencing in a *Fusarium oxysporum* strain, Environ. Microbiol., 10. pp. 1725–1741.

Minuto, A., Migheli, Q., Garibaldi, A., 1995. Evaluation of antagonistic strains of *Fusarium* spp. in the biological and integrated control of *Fusarium* wilt of cyclamen. Crop. Prot. 14, 221–226.

Misra, A.K., Prasad, B., 2003. Relative efficacy of different bio-agents for the control of guava wilt. J. Mycol. Plant Pathol. 33, 494.

Misra, A.K., Prasad, D., Prasad, B., Shukla, S.K., 2004. Effective management of wilt disease of guava. Proceedings of the National Symposium on Crop Surveillance: Disease Forecasting and Management, Feb. 19–21, Division of Plant Pathology, IARI New Delhi, pp. 92–93.

Mudili, V., Siddaih, C.N., Nagesh, M., Garapati, P., Naveen Kumar, K., Murali, H.S., et al., 2014. Mould incidence and mycotoxin contamination in freshly harvested maize kernels originated from India. J. Sci. Food Agric. 94, 2674–2683.

Mushtaq, M., Hashmi, M.H., 1997. Fungi associated with wilt disease of capsicum in Sindh Pakistan. Pak. J. Bot. 29, 217–222.

Nannipieri, P., Ascher, J., Ceccherini, M.T., Landi, L., Pietramellara, G., Renella, G., 2003. Microbial diversity and soil functions. Eur. J. Soil Sci. 54, 655–670.

Navajothy, F., Narayanaswamy, A., Ponniah, R.D., Irudayaraj, V., 2011. Spectroscopical differences among soil samples of selected susceptible and resistant varieties of banana in relation to wilt disease. Int. J. Multidiscip. Res. 1 (3), 20–28.

Navas Cortes, J.A.N., Hau, B., Jimenez-Diaz, R.M., 2000. Yield loss in chickpeas in relation to development of *Fusarium* wilt epidemics. Phytopathology 90, 1269–1278.

Nelson, P.E., 1981. Life cycle and epidemiology of *Fusarium oxysporum*. In: Mace, M.E., Bell, A.A., Beckman, C.H. (Eds.), Fungal Wilt Diseases of Plants. Academic Press, New York, USA, pp. 51–80.

Nelson, P.F., Toussoun, T.A., Morasas, 1983. *Fusarium* Species: An llustrated Manual for Identification. Pennsylvania State University Press, University Park, PA, p. 203.

Nosir, W.S., 2016. *Trichoderma harzianum* as a growth promoter and bio-control agent against *Fusarium oxysporum* f. sp. *tuberosi*. Adv. Crop. Sci. Technol. 4 (2). Available from: https://doi.org/10.4172/2329-8863.1000217.

Nur Ain Izzati, M.Z., Azmi, A.R., Siti Nordahliawate, M.S., Norazlina, J., 2011. Contribution to the knowledge of diversity of *Fusarium* associated with maize in Malaysia. Plant Prot. Sci. 47, 20–24.

Ogawa, K., Komada, H., 1984. Biological control of *Fusarium* wilt of sweet potato by non-pathogenic *Fusarium oxysporum*. Ann. Phytopathol. Soc. Jpn. 50, 1–9.

Olivain, C., Alabouvette, C., Steinberg, C., 2004. Production of a mixed inoculum of *Fusarium oxysporum* Fo47 and *Pseudomonas fluorescens* C7 to control *Fusarium* diseases. Biocontrol Sci. Technol. 14, 227–238.

Olivain, C., Steinberg, C., Alabouvette, C., 1995. Evidence of induced resistance in tomato inoculated by non-pathogenic strains of *Fusarium oxysporum*. In: Manka, M. (Ed.), Environmental Biotic Factors in Integrated Plant Disease Control. The Polish Phytopathological Society, Poznam, Poland, pp. 427–430.

Ommati, F., Sharif, K., 2008. Determination of species and dispersal of potato Fusarium wilt in Semman province. In: Proceedings of the 18th Iranian Plant Protection Congress. August 24–27, 2008. Hamadon, Iran, p. 70.

Ortiz, E., Cruz, M., Melgarejo, L.M., Marquinez, X., Hoyos-Carvajal, L., 2014. Histopathological features of infections caused by *Fusarium oxysporum* and *F.solani* in purple passion fruit plants (*Passiflora edulis* Sims). Summa Phytopathol. 40, 134–140.

Otero-Colina, G., Rodríguez-Alvarado, G., Fernández-Pavía, S., Maymon, M., Ploetz, R.C., Aoki, T., et al., 2010. Identification and characterization of a novel etiological agent of mango malformation disease in Mexico, *Fusarium mexicanum* sp. Phytopathology 100, 1176–1184.

Pandey, M.K., Chakrabarti, D.K., 2004. Management of malformation of mango (*Mangifera indica*). J. Mycol. Plant Pathol. 34 (3), 881–888.

Pandey, P., Irulappan, V., Bagavathiannan, M.V., Senthil-Kumar, M., 2017. Impact of combined abiotic and biotic stresses on plant growth and avenues for crop improvement by exploiting physio-morphological traits. Front. Plant Sci. 8, 537. PMID: 28458674.

Park, J.H., Choi, G.J., Jang, S.K., Lim, K.H., Kim, T.H., Cho, Y.K., et al., 2005. Antifungal activity against plant pathogenic fungi of chaetoviridins isolated from *Chaetomium globosum*. FEMS Microbiol. Lett. 252, 309–313.

Pavlovkin, J., Mistrik, I., Prokop, M., 2004. Some aspects of the phytotoxic action of fusaric acid on primary Ricinus roots. Plant Soil Environ. 50 (9), 397–401.

Perveen, K., Bokhari, N., 2010. First report of *Fusarium* wilt of *Lavandula pubescens* caused by *Fusarium oxysporum* in Saudi Arabia. Plant Dis. 94 (9), 1,163.2.

Pieterse, C.M.J., Van Loon, L.C., 2004. NPR1: the spider in the web of induced resistance signalling pathways. Curr. Opin. Plant Biol. 7, 456–464.

Pieterse, C.M.J., Van Wees, S.C.M., Van Pelt, J.A., Knoester, M., Laan, R., Gerrits, H., et al., 1998. A novel signaling pathway controlling induced systemic resistance in Arabidopsis. Plant Cell 10, 1571–1580.

Ploetz, R.C., 2015. Review on *Fusarium* wilt of Banana. APS Publications. Available from: https://doi.org/10.1094/PHYTO-04-15-0101-RVW.

Polizzi, G., Aiello, D., Guarnaccia, V., Vitale, A., Perrone, G., Stea, G., 2010. First report of *Fusarium* wilt of paper flower (*Bougainvillea glabra*) caused by *Fusarium oxysporum* in Italy. Plant Dis. 94 (4), 483.1.

Postma, J., Rattink, H., 1992. Biological control of *Fusarium* wilt of carnation with a non-pathogenic isolate of *Fusarium oxysporum*. Can. J. Bot. 70, 1199–1205.

Pujol, I., Guarro, J., Gene, J., Sala, J., 1997. *In vitro* antifungal susceptibility of clinical and environmental *Fusarium* spp., strains. J. Antimicrob. Chemotherapy 39 (2), 163–167.

Ravikumar, R.L., Babu, D.R., 2007. *In vitro* screening of chickpea genotypes for *Fusarium* wilt resistance through root feeding of pathotoxin. Curr. Sci. 93 (1), 20–22.

Rohilla, S.K., Salar, R.K., 2012. Isolation and characterization of various fungal strains from agricultural soil contaminated with pesticides. Res. J. Recent Sci. 1, 297–303.

Rouxel, F., Alabouvette, C., Louvet, J., 1979. Recherches sur la sistance des sols aux maladies. IV-Mise en vidence du rle des Fusarium autochtones dans le sistance d'un sol la fusariose vasculaire du melon. Annu. Rev. Phytopathol. 11, 199–207.

Ryals, J.A., Neuenschwander, U.H., Willits, M.G., Molina, A., Steiner, H.Y., Hunt, M.D., 1996. Systemic acquired resistance. Plant Cell 8, 1808–1819.

Saman, A., 2009. Use of nonpathogenic *Fusarium oxysporum* and rhizobacteria for suppression of *Fusarium* root and stem rot of *Cucumis sativus* caused by *Fusarium oxysporum* f. sp. *radicis-cucumerinum*. Arch. Phytopathol. Plant Prot. 42, 73–82.

Saravanan, T., Muthusamy, M., Marimuthu, T., 2003. Development of integrated approach to manage the fusarial wilt of banana. Crop Prot. 22, 1117–1123.

Saremi, H., Amiri, M.E., 2010. Exploration of potato cultivar resistant to the major fungal pathogen wilting disease in Iran. J. Food Agric. Environ. 8, 821–826.

Sate of World Fungi Report, 2018. Royal Botanic Garden kews State of World Fungi. http://stateoftheworldsfungi.org/reports/SOTWFungi_2018_FullReport.

Sathiamoorthy, S., Uma, S., Selvarajan, R., 2001. Banana research and development programme in India and highlights of NRCB-INIBAP collaborative projects. In: Monila, A.B., Roa, V. N., Maghuyop, M.A.G. (Eds.), Advancing Banana and Plantain R&D In Asia and the Pacific Proceedings of the 10th INIBAP-APNET Regional Advisory Committee Meeting Held at Bangkok. INIBAP-APNET, Los Banos, pp. 67–76. , 2000/11/10-11.

Schippers, B., Bakker, A.W., Bakker, P.A.H.M., 1987. Interactions of deleterious and beneficial rhizosphere microorganims and the effect of cropping practices. Annu. Rev. Phytopathol. 25, 339–358.

Shahbaz, M., Iqbal, Z., Saleem, A., Anjum, M.A., 2009. Association of *Lasiodiplodia theobromae* with different decline disorders in mango (*Mangifera indica* L.). Pak. J. Bot. 41 (1), 359–368.

Shenoy, B.D., Jeewon, R., Hyde, K.D., 2007. Impact of DNA sequence-data on the taxonomy of anamorphic fungi. Fungal Divers. 26, 1–54.

Sibtain, M., Ilyas, M.B., Khan, I.A., Alam, S.S., 2001. Screening of chickpea germplasm against *Fusarium* wilt in pot soil water culture and cultural filtrate. J. Biol. Sci. 1 (4), 229–231.

Silvestro, L.B., Stenglein, S.A., Forjan, H., Dinolfo, M.I., Arambarri, A.M., Manso, L., et al., 2013. Occurrence and distribution of soil *Fusarium* species under wheat crop in zero tillage. Span. J. Agric. Res. 11 (1), 72–79.

Singh, D.P., Dwivedi, S.K., 2004. Environmental Microbiology and Biotechnology. New Age International Pvt. Ltd. Publisher, New Delhi.

Singh, I.D., 1975. Taxonomic and Cultural Studies on Some Parasitic Fungi of Medicinal Plants of Varanasi (Ph.D. thesis), Banaras Hindu University, Varanasi, India.

Singh, R., Singh, B.K., Upadhyay, R.S., Rai, B., Lee, Y.S., 2002. Biological control of *Fusarium* wilt disease of pigeonpea. Plant Pathol. J. 18, 279–283.

Sinha, P., Rizvi, G., Parashar, R., 2018. Management of wilt disease of pulses: a review. Int. J. Pure Appl. Biosci. 6 (4), 696–708.

Sivan, A., Chet, I., 1989. The possible role of competition *Trichoderma harzianum* and *Fusarium oxysporum* on rhizosphere colonization. Phytopathology 79, 198–203.

Sivasithamparam, K., Ghisalberti, E.L., 1998. Secondary metabolism in *Trichoderma* and *Gliocladium*. In: Harman, G.E., Kubicek, C.P. (Eds.), Trichoderma & Gliocladium. Volume 1, Basic Biology, Taxonomy and Genetics. Taylor and Francis Ltd, London, UK:, pp. 139–192.

Smith, R.F., Bendixen, W.E., Fennimore, S.A., 2001. UC Pest Management Guidelines: Strawberry Herbicide Treatment Table. University of California at Davis., UC ANR Publication 3468.

Soares, D.J., Barreto, R.W., 2008. Fungal pathogens of the invasive riparian weed *Hedychium coronarium* from Brazil and their potential for biological control. Fungal Divers. 28, 85–96.

Srinivas, C., Nirmala Devi, D., Narasimha Murthy, K., Dhananjaya Mohan, C., Lakshmeesha, T. R., Kalagatur, B.P., et al., 2019. *Fusarium oxysporum* f. sp. *lycopersici* causal agent of vascular wilt disease of tomato: Biology to diversity- a review. Saudi J. Biol. Sci. Available from: https://doi.org/10.1016/j.sjbs.06.0022019.

Srivastava, S., Singh, V.P., Kumar, R., Srivastava, R., Sinha, A., Simon, S., 2011. *In vitro* evaluation of carbendazim 50%, WP, Antagonists and Botanicals against *Fusarium oxysporum* f. sp. *psidii* associated with rhizosphere soil of guava. Asian J. Plant Pathol. 5 (1), 46–53. Sultana, N., Ghaffar, A., 2010. Effect of fungicides, microbial antagonists and oil cakes in control of *Fusarium solani*, the cause of seed rot, seedling and root infection of bottle gourd and cucumber. Pak. J. Bot. 42 (4), 2921–2934.

Tamietti, G., Alabouvette, C., 1986. Resistance des sols aux maladies: XIII - Role des *Fusarium oxysporum* non pathogenes dans les mecanismes de resistance d'un sol de Noirmoutiers aux fusarioses vasculaires. Agronomie 6, 541–548.

Than, P.P., Prihastuti, H., Phoulivong, S., Taylor, P.W.J., Hyde, K.D., 2008. Chilli anthracnose disease caused by *Colletotrichum* species. J. Zhejiang Univ. Sci. B 9, 764–778.

Tima, H., Bruckner, A., Mohacsi-Farkas, C., Kisko, G., 2016. *Fusarium* mycotoxins in cereals harvested from Hungarian fields. Food Addit. Contaminants Part. B: Surveill. 9 (2), 127–131.

Toal, M.E., Yoemans, C., Killham, K., Meharg, A.A., 2000. A review of rhizosphere carbon flow modelling. Plant Soil 222, 263–281.

Tola, M., Kebede, B., 2016. Occurrence, importance and control of mycotoxins: a review. Cogent. Food Agric. 2. Available from: https://doi.org/10.1080/23311932.2016.1191103.

Traka-Mavrona, E., Koutsika-Sotiriou, M., Pintsa, T., 2000. Response of squash (Cucw-bita spp.) as rootstock for melon (Cucurn/s nw/a L.). Sci. Hurt. 83, 353–362.

Triest, D., Pierard, D., De-Cremer, K., Hendrickx, M., 2016. *Fusarium musae* infected banana fruits as potential source of human fusariosis: may occur more frequently than we might think and hypotheses about infection. Commun. Integr. Biol. 9 (2), e1162934. Available from: https://doi.org/10.1080/19420889.2016.1162934.

Upadhyaya, R.K., 1978. Biology of Microfungi Associated With a Medicinal Plant (Ph.D. thesis), Banaras Hindu University, Varanasi, India.

Van Hulten, M., Pelser, M., Van Loon, L.C., Pieterse, C.M.J., Ton, J., 2006. Costs and benefits of priming for defense in Arabidopsis. Proc. Natl. Acad. Sci. USA 103, 5602–5607.

Van Wees, S.C.M., Van Der Ent, S., Pieterse, C.M.J., 2008. Plant immune responses triggered by beneficial microbes. Curr. Opin. Plant Biol. 11, 443–448.

Venkataramana, M., Navya, K., Chandranayaka, S., Priyanka, S.R., Murali, H.S., Batra, H.V., 2014. Development and validation of an immunochromatographic assay for rapid detection of fumonisin B1 from cereal samples. J. Food Sci. Technol. 51 (9), 1920–1928.

Vinale, F., Sivasithamparam, K., Ghisalberti, E.L., Marra, R., Woo, S.L., Lorito, M., 2008. *Trichoderma*-plant-pathogen interactions. Soil Biol. Biochem. 40, 1–10.

Wada, T., Kuzuma, S., Takenaka, M., 1990. Sensitivity of *Fusarium moniliforme* isolates to Pefurazoate. Annu. Phytopathol. Soc. 56, 449–456.

Weller, D.M., Thomashow, L.S., 1993. Microbial metabolites with biological activity. In: Lumsden, R.D., Vaughn, J.L. (Eds.), Pest Management: Biologically Based Technologies. American Chemical Society, Washington, DC, USA, pp. 173–180.

Weller, D.M., Raaijmakers, J.M., Gardener, B.B.M., Thomashow, L.S., 2002. Microbial populations responsible for specific soil suppressiveness to plant pathogens. Annu. Rev. Phytopathology 40, 309–348.

White, T.J., Lee, S.B., Taylor, J.W., 1993. PCR Protocols: A Guide to Methods and Application. Academic Press, NY, p. 315.

Wong, M.-Y., 2003. *Fusarium oxysporum* f. sp. *lycopersici* (Sacc.) WC Snyder and HN Hans. Soilborne Plant Pathogen Class Project, [cit. 201009–1O] 2003, 728.

Woo, S.L., Lorito, M., 2007. Exploiting the interactions between fungal antagonists, pathogens and the plant for biocontrol. In: Vurro, M., Gressel, J. (Eds.), Novel Biotechnologies for Biocontrol Agent Enhancement and Management. Springer, Dordrecht, The Netherlands, pp. 107–130.

Wu, H.S., Waseem, R., Liu, D.Y., Wu, C.L., Mao, Z.S., Shen, Q.R., 2007a. Allelopathic impact of artificially applied coumarin on *Fusarium oxysporum* f. sp. Niveum. World J. Microbiol. Biotechnol. 24, 1297–1304.

Wu, H.-S., Bao, W., Liu, D.-Y., Ling, N., Ying, R.-R., Raza, W., et al., 2008. Effect of fusaric acid on biomass and photosynthesis of watermelon seedlings leaves. Caryologia-Firenze 61 (3), 258–268.

Wu, H.S., Wang, Y.C., Zhang, Y., Gu, M., Liu, Y.X., Chen, G., et al., 2009. Physiological and biochemical responses of *in vitro Fusarium oxysporum* f.sp. *niveum* to benzoic acid. Folia Microbiol. 54 (2), 115–122.

Wu, H.S., Waseem, R., Fan, J.Q., Sun, Y.G., Bao, W., Shen, Q.R., 2007b. Cinnamic acid inhibits growth but stimulates production of pathogenesis factors by *in vitro* cultures of *Fusarium oxysporum* f. sp. *niveum*. J. Agric. Food Chem. 56 (4), 316–1321.

Yao, H., Wu, F., 2010. Soil microbial community structure in cucumber rhizosphere of different resistance cultivars to *Fusarium* wilt. FEMS Microbiol. Ecol. 72 (3), 456–463.

Zakaria, L., Chik, M.W., Heng, K.W., Salleh, B., 2012. *Fusarium* species associated with fruit rot of banana (*Musa* spp.), papaya (*Carica papaya*) and guava (*Psidium guajava*). Malaysian J. Microbiol. 8 (2), 127–130.

Zhou, X.G., Everts, K.L., Bruton, B.D., 2010. Race 3, a new and highly virulent race of *Fusarium oxysporum* f. sp. niveum causing *Fusarium* wilt in watermelon. Plant Dis. 94, 92–98.

Zitter Thomas, A., 1998. Fusarium Dseases of Cucurbits. Published by Cornell Cooperative Extension, New York, pp. 1–4.

Index

Note: Page numbers followed by "*f*" and "*t*" refer to figures and tables, respectively.

B

D

E

G

H

N

O

P

S

U

V

W

X

Y

Z

FSC
www.fsc.org
MIX
Papier | Fördert gute Waldnutzung
FSC® C083411

Kolibri 360 GmbH
Ferdinand-Jühlke-Straße 7
99095 Erfurt, Deutschland
produktsicherheit@kolibri360.de